Spectroscopy of Polymers

Spectroscopy of Polymers

Jack L. Koenig
Case Western Reserve University

American Chemical Society, Washington, DC 1992

Library of Congress Cataloging-in-Publication Data
Koenig, Jack L.
Spectroscopy of Polymers/Jack L. Koenig

p. cm.

Includes bibliographical references and index.

ISBN 0-8412-1904-4.-ISBN 0-8412-1924-9 (pbk.)
1. Polymers-Analysis. 2. Spectrum analysis. I. Title

QD139.P6K64 1991
547.7'046-dc20

91-13352
CIP

The paper used in this publication meets the minimum requirements of American National Standard for Information Sciences—Permanence of Paper for Printed Library Materials, ANSI Z39.48-1984. ∞

PRINTED IN THE UNITED STATES OF AMERICA

1992 ACS Books Advisory Board

My life has been enriched by the joy of being a grandfather.
Therefore, I would like to dedicate this book
to my grandchildren

Robert
Brian
Chelsea
Christopher
Dustin
Alexander
and, of course, the grandcat,
Shaun

About the Author

Dr. Jack Koenig is the J. Donnell Institute Professor of the Departments of Macromolecular Science and Chemistry at Case Western Reserve University in Cleveland, Ohio. Dr. Koenig received his B.A. in Chemistry and Mathematics from Yankton College, Yankton, South Dakota, in 1955. His Ph.D. in theoretical spectroscopy was granted in 1959 by the University of Nebraska.

After receiving his Ph.D., Dr. Koenig was first employed as a research chemist in the Plastics Department at the E. I. du Pont de Nemours and Company in Wilmington, Delaware, from 1959 to 1963. In 1963, Dr. Koenig became an assistant professor of polymer science at Case Institute of Technology, and three years later he became an associate professor. In 1979, Dr. Koenig was promoted to professor in the Department of Macromolecular Science at Case Western Reserve University. He was awarded the J. Donnell Institute Endowed Chair in 1990.

During 1972 and 1973, Dr. Koenig was the first program manager of the Polymer Science Program of the Materials Research Division of the National Science Foundation in Washington, D.C. Upon his return to Case Western Reserve University, Dr. Koenig became the director of the Molecular Spectroscopy Laboratory and of the Materials Research Laboratory. More recently, he became associate director of the National Science Foundation Science and Technology Center, which is a consortium including Case Western Reserve University, the University of Akron, and Kent State University.

Dr. Koenig has received many awards and honors during his career. In 1966, he received the Sigma Xi Research Award to Young Faculty. He became a fellow in the American Physical Society in 1970. Dr. Koenig received the Pittsburgh Society Spectroscopy Award in 1984. In 1986, he won the Alexander Von Humboldt Award for Senior U.S. Scientists and spent the year in Germany as a visiting professor at the Institute for Macromolecular Chemistry at the University of Freiburg. Dr. Koenig also won the ACS Morley Medal in 1986. More recently, he received the Society of Plastics Engineers Research Award and the Cleveland Technological Society Distinguished Service Award. In 1992, he will receive the ACS Doolittle Award and the Gold Medal of the Eastern Analytical Society.

Dr. Koenig is well known for his theoretical and experimental work in infrared and Raman spectroscopy of polymers. His monograph (with P. Painter and M. Coleman) is considered the standard reference in the field. More recently, he has become involved in solid-state NMR spectroscopy and has made a number of contributions to this new area. He has published over 400 papers and has been the editor or coauthor of five monographs.

Dr. Koenig is married and has four children and six grandchildren. He resides in Chagrin Falls, Ohio.

Contents

Preface

John Steinbeck has written:

> *A man writes a book—why?—Because he wants to. His impulse is probably the same one that makes him sing or try to go to bed with all beautiful ladies. It's a kind of outpouring. And there's his book—somehow naked and cold standing up while the well-clothed mock at it. Usually they are angry with it before they start. Now the purpose of a book I suppose is to amuse, interest, instruct but its warmer purpose is just to associate with the reader. You use symbols that he can understand so that the two of you can be together. The circle is not closed until the trinity is present, —the writer, the book, and the reader. And this works with everyone but critics. Critics dare you to be "great". But this is all after the fact. You didn't want to be great. You just wanted to write a book and have people read it.**

This textbook was written with the objective in mind of answering the question "What is the chemical structure of a polymer system?" in a way that would meet the needs of the student and the practitioner of polymer science alike.

The material in this book is directed to the individual who has a basic knowledge of polymer chemistry and some familiarity with the basic vibrational and resonance spectroscopic methods, but who does not have the particular specialized knowledge that is required to apply the modern techniques of FTIR, Raman, and NMR spectroscopy to the characterization of polymers.

Although not an elementary manual, Spectroscopy of Polymers is not intended to make you an expert polymer spectroscopist with complete theoretical and experimental skills. Rather, it is intended to give you a sufficient knowledge of the spectroscopic techniques to allow you to decide which spectroscopic method to use, the sampling techniques to use, and the type of results that can be expected. Examples are given in the text to demonstrate the strengths and weaknesses of the various vibrational and NMR techniques. An understanding of the material in the text will allow intelligent communication with the "expert" spectroscopists who will ultimately do the work.

My continuing experience teaching a course for incoming graduate students and an ACS short course were the basis for this book. I have previously written a textbook on this subject, but after the lapse of ten years, even some of the most

**From* The New York Times, *6 August, 1990.*

basic material needed updating. This completely new manuscript was prepared with the realization that only a few short years will pass before this book, too, becomes dated.

However, in order to reap the benefits of modern spectroscopic instrumentation and data analysis, the polymer scientist must have a knowledge of the current methods of determining polymer structure. My hope is that this book will partially fulfill this need.

JACK L. KOENIG
Case Western Reserve University
Cleveland, OH 44022

October 17, 1991

Acknowledgments and Apologia

I will take full credit for those portions of the textbook that are readable and helpful. I will also take full blame for the errors and omissions that inevitably occur in an undertaking of this kind.

A comment about the material presented in the textbook is necessary. With the thousands of papers published each year in the area of spectroscopic characterization, it would be impossible to cover all of the results reported in the literature. The material in this book was selected on the basis of two considerations: the utility of the results for demonstrating a concept and my knowledge of the subject. This latter aspect is reflected in the large amount of work cited from my own laboratory. Other workers will surely feel that their work should have been cited. I cannot argue, and I extend apologies to those researchers who feel neglected. The references were also selected on the basis of their utility for extending the material presented here. The older references are "classics" that any interested student should read. Recent literature citations are based on the requirement that the current results reflect the progress of the field.

This textbook reflects 25 years of teaching experience. Any teacher will admit that the teacher learns from the educational process, and the hundreds of students who have been in my classes have taught me a great deal with their inquiring minds and questions. It is impossible to acknowledge their contributions except to say thank you.

One group of students has contributed specifically to this book by acting as critics of the manuscript during the various writing stages. I am referring to my current graduate students, who have not always liked the editing job, but who did it anyway—with some differences in the level of commitment. I would especially like to recognize Maureen Sargent and Shari Tidrick for their outstanding efforts. Ron Grinsted, Mike Krejsa, Steve Smith, Regan Silvestri, Fernando Fondeur, and Jin-who Hong all contributed in their own special way.

My office staff deserves some kind of award for their support. Barbara Leach, who has lived through two previous books, again made major contributions to getting this book out the door. Christine Meyer produced (and reproduced) figures and tables. Margaret Amer did most of the copying of the manuscript.

The ACS Books Department staff has been particularly helpful. I would like to recognize Barbara C. Tansill, who convinced me that the ACS was the only way to go; the copy editor, Julie Poudrier Skinner; and Peggy D. Smith, the Senior Production Specialist who figuratively held my hand during the difficult process of changing a manuscript into a book.

Finally, my family, and particularly my wife, deserves recognition for allowing me to spend mornings and evenings with my computer. Jeanus hopes that the days of being a "computer widow" are over. She has asked that I write an epilogue to the book promising not to write another. I think that my staff and present graduate students would agree. I hope that after you have read the book you will *not* agree.

J.L.K.

1

Theory of Polymer Characterization

The primary motivation for determining the structure of a polymer chain is to relate the structure to the performance properties of the polymer in end use. If a polymer chain is completely characterized and the structural basis of its properties is known, the polymerization can be optimized and controlled to produce the best possible properties from the chemical system.

When the problem of characterizing a polymer chain is considered, the difficulty of the task seems insurmountable. Indeed, one cannot expect to determine the atomic positions or spatial coordinates of every atom in a synthetic polymer chain in the same manner as for simple molecules. Even if only a single molecular structure is involved, and even if the appropriate techniques were available (they are not), it would take centuries to work out the results.

For a single polymer chain with only two different structural elements, 0 and 1, the number of different chain structures possible is 2^N, where N is a very large number, that is, 10,000 or greater. So there is no point in trying to completely define the spatial coordinates of the atoms on a single polymer chain. To make matters even worse, most real synthetic polymer chains have a large number of different structural elements. Some of the possible structural variables found in synthetic polymers are as follows:

- molecular weight
- chemical defects: impurities in feed, monomer isomerization, and side reactions
- enchainment defects: positional, stereospecific, branches, and cyclic isomers
- chain conformations: stiff ordered chains and flexible amorphous chains
- morphological effects: crystal phases, interfacial regions, and entanglements

The list is long, and some polymers exhibit a number of these structural variables simultaneously. Hence the number of possible structures for a single chain is very large, indeed.

The problem is further complicated by the fact that the nature of the distribution of the structural variables along the chain influences the properties. The types of distributions of structural elements that can occur are as follows:

Type	*Nature of Process*
Random	Stochastic
Block $(A)_N - (B)_M$	Designed
Alternating $[(A)_1 - (B)_1]_n$	Controlled

These structural distributions are determined by the nature of the polymerization process; that is, the disordered distributions arise from the type of stochastic processes involved in the polymerization. The alternating ordered distribution can only be obtained by systematic chemical design and the block structure by chemical control of the polymerization process.

1904—4/92/0001$06.00/1

The distribution influences the characterization of the polymer in two ways. First, the chain structure is highly variable because the polymerization process is a statistical process determined by probability considerations. Thus, the polymer sample is always a multicomponent complex structural mixture. Second, detailed pictures of the structure are not possible because our measurements are going to provide only some weighted average structure.

Elements of Polymer Structure

The following basic terms are used for defining a polymer structure.

- The *composition* of a molecule defines the nature of the atoms and their type of bonding irrespective of their spatial arrangement.
- The *configuration* of chemical groups characterizes a chemical state of a molecule. Different configurations constitute different chemical individuals and cannot be converted into one another without rupture of chemical bonds.
- The *conformation* of chemical groups characterizes the geometrical state of a molecule. Different conformations of a molecule can be produced by rotation about single bonds without rupture of chemical bonds. Changes in conformation arise from physical considerations such as temperature, pressure, or stress and strain.

Polymer chains are made up of sequences of chemical repeating units that may be arranged regularly or irregularly on the backbone.

- The chemical *microstructure* is defined as the internal arrangement of the different sequences on the polymer chain.
- The polymer *morphology* defines the intermolecular packing of the polymer molecules as crystals or spherulites.

From a structural point of view polymers are chainlike molecules. From the simplest perspective, the structural elements of a polymer molecule with a single structural repeating unit can be represented by the molecular formula, $X(A)_nY$, where X and Y are the end-group units, and A represents the repeating unit of the polymer molecule. The repeating unit can be very simple in chemical structure (e.g., CH_2 for polyethylene) or very complicated. The number of connected repeating units, n, can range from 2 to 100,000.

The end-group units X and Y can be substantially different in structure from A or very similar depending on the nature of the polymerization process (i.e., condensation or addition). Determination of the nature and number of end groups is a common spectroscopic measurement when the molecular-weight range is sufficiently low to produce detectable signals.

Structural variations within the structure of the chain can be represented by the letters B, C, etc., to indicate the differences in the chemical, configurational, or conformational structure.

The polymerization reaction converts the initial bifunctional monomers into a chain of chemically connected repeating units. The two general classes of polymerization reactions are: (1) condensation polymerization, in which any monomer can react or connect with any other monomer in the system; and (2) addition polymerization, in which the monomers react or connect only at a growing active site.

The process of condensation polymerization can be written in the following form:

$$nM \rightarrow (-A-A-A-A-A \cdots)_n \tag{1.1}$$

where M is the monomer. We have neglected for the moment the fact that the end groups X and Y are the residual monofunctional structures of A rather than A itself. However, this representation of the reaction is incomplete, as the polymerization reaction is statistical in nature and does not generate a single molecule of a specified length n. Rather millions of condensation reactions are occurring simultaneously, generating millions of molecules of various lengths ranging from 1 to a very large number (e.g., 100,000) depending on how many condensation reactions have occurred between the individual molecules during the polymerization. So more precisely, the polymerization reaction must be written

$$\sum nM \rightarrow \sum_{n=1}^{\infty}(-A-A-A-A-A \cdots)_n \tag{1.2}$$

Thus, the condensation polymerization batch,

contains a mixture of chain molecules ranging in length from very short to very long. For a simple polymer system, the only structural variables are the lengths n of the chains and the number of molecules of these various lengths N_A. In other words, we know the fractions of molecules having different specific lengths.

Approach to Structure Determination Using Probability Considerations

A simple example is the degree of polymerization. With the polymerization model just described, and assuming an equal likelihood for the selection of any polymer molecule from the mixture, it is possible to calculate the probability of finding a molecule with a given n and ultimately the fraction of all molecules that possess the stated chain length, n.

> You can visualize the probability approach as one of reaching into the reaction mixture and pulling out a single polymer molecule. You must then calculate the probability that the molecule selected has a specified length.

Let p be the probability that a condensation reaction has occurred and $(1 - p)$ be the probability of termination; that is, the molecule has not undergone condensation. Let $P(n)$ be the probability that a molecule of length n has been formed, and let $n = 2$. If the condensation polymerization probabilities are independent (i.e., occur randomly and do not depend on chain length) the probability of forming a molecule of length $n = 2$ is the probability P that one condensation reaction coupling two monomer units has occurred times the probability $(1 - P)$ that termination or no further reaction has occurred. Hence,

$$P(2) = P(1 - P) \tag{1.3}$$

A chain molecule of $n = 3$ is formed by two condensations and a termination; the probability of the first condensation, P, times the probability of the second condensation, P, forming the trimer times the probability of termination, $(1 - P)$, or that no reaction has occurred. Hence,

$$P(3) = P^2(1 - P) \tag{1.4}$$

A chain molecule of $n = 4$ is formed by three condensations, P^3, and a termination. So

$$P(4) = P^3(1 - P) \tag{1.5}$$

A general trend is emerging. The number of condensations is always one less than the length of the chain, and the termination step is always required, so the general formulation for any chain length can be written

$$P(n) = P^{n-1}(1 - P) \tag{1.6}$$

Equation 1.6 is the probability distribution function for the structural parameter of interest, which in this case is the chain length n of the polymer. From the probability distribution function for n, any desired information or property arising from the chain-length distribution of the system can be obtained by appropriate mathematical calculations. Inversely and equally important from our perspective, experimental determinations of the molecular weights and molecular-weight distributions allow a determination of the probability of condensation, P, which is the characteristic quantity controlling the ideal condensation polymerization process we are using as a model.

Self Test: Calculate the probability distribution function for the ideal addition polymerization with termination by disproportionation. The chemistry involved is as follows:

$$A_n^* + A \rightarrow A_{n+1} \tag{1.7}$$

$$A_n^* + A_m^* \rightarrow A_n + A_m \tag{1.8}$$

(*Hint:* Start with A_n^* and calculate the probability of addition compared to termination.)

These chain-length probabilities must sum to 1 (i.e., the complete condition, meaning that something must happen):

$$\sum_{n=1}^{\infty} P(n) = 1 \tag{1.9}$$

because n must have some value between 1 and infinity. This relationship indicates that the individual

$P(n)$ are simply the number fraction of individual molecules in the mixture:

$$P(n) = \frac{N_n}{N} \quad (1.10)$$

where N_n is the number of molecules of length n, and N is the total number of polymer molecules in the polymerization batch.

Structure Calculations Using the Probability Distribution Function

A Simple Example: Degree of Polymerization (1). The structural variable in this case is the length of the polymer chain or the *degree of polymerization*, DP, defined as the number of similar structural units linked together to form the polymer molecule. This number is converted to molecular weight by multiplying by the molecular weight of a single structural or repeating unit. Measurement of the molecular weight of the polymer system by any physical means yields numbers representing the weighted averages of the DPs of all the molecules present.

A colligative property measures the number of molecules in solution. A colligative property measurement, for example, osmotic pressure, freezing-point depression, boiling-point elevation, or vapor-pressure lowering, of a polymer solution yields a "number-average" DP or molecular weight (simply by multiplying by the molecular weight of a single repeating unit). The effective DP measured is the sum of the DPs of all the molecules divided by the number of molecules present. This can be written as

$$\overline{DP}_n = \frac{\sum nN_n}{\sum N_n}$$
$$= \frac{1N_1 + 2N_2 + 3N_3 + 4N_4 + \cdots}{N_1 + N_2 + N_3 + N_4 + \cdots} \quad (1.11)$$

where N_n is the number of molecules present whose DP is n. The (*$\sum N_n = N$ is the total number of molecules in the system. The number of molecules having a specific length N_n corresponds to the probability of finding molecules of this length multiplied by the total number of molecules

$$N_n = NP(n) = NP^{n-1}(1 - P) \quad (1.12)$$

Substituting eq 1.12 in eq 1.11 yields

$$\overline{DP}_n = \frac{\sum nNP(n)}{N}$$
$$\sum nP^{n-1}(1 - P) = \frac{1}{1 - P} \quad (1.13)$$

You are probably wondering about the "magic" of the summation made at the end of equation (1.13); the method is given in detail in my previous book (2).

Equation 1.13 demonstrates that a measurement of the average degree of polymerization allows a determination of the probability of propagation for this model of the condensation polymerization. The DP_n can also be written

$$DP_n = \frac{N_o}{N} = \frac{1}{1 - P} \quad (1.14)$$

where N_o is the number of monomer units at the start of the polymerization, so

$$N = N_o(1 - P) \quad (1.15)$$

Therefore, the number distribution function can be written

$$N_n = N_o(1 - P)^2 P^{n-1} \quad (1.16)$$

With a knowledge of P, the derived probability distribution function can be used to calculate the other parameters of the polymerization, including the various types of average molecular weights such as the weight and z average and the moments of the molecular-weight distribution. For this simple condensation model, a determination of P is all that is required for a complete structural evaluation because the only structural variable is the length of the chains.

Number-Average Molecular Weights

For a monodisperse system, the molecular weight is given by

$$M_n = \frac{W}{N} \quad (1.17)$$

where W is the weight of the sample in grams and N is the number of molecules. For a polydispersed system,

$$W = \sum_n N_n M_n \tag{1.18}$$

and

$$N = \sum_n N_n \tag{1.19}$$

so the number-average molecular weight is given by

$$\overline{M_n} = \frac{\sum N_n M_n}{\sum_n N_n} \tag{1.20}$$

Now,

$$M_n = nM_o \tag{1.21}$$

where M_o is the molecular weight of a repeating unit. The number distribution function can be used in these equations to calculate $\overline{M_n}$ in terms of the probability of condensation, P,

$$\overline{M_n} = \frac{M_o}{1 - P} \tag{1.22}$$

Weight-Average Molecular Weights

For light-scattering measurements of polymer solutions, the effect is proportional to the molecular weight of the molecules in solution, so a weight-average molecular weight is measured. The weight-average molecular weight, $\overline{M_w}$, is given by

$$\overline{M_w} = \frac{\sum W_n M_n}{W} \tag{1.23}$$

and

$$W_n = n M_o N_n \tag{1.24}$$

$$W = M_o N_o \tag{1.25}$$

Substitution and evaluation reveal

$$M_w = \frac{M_o(1 + P)}{1 - P} \tag{1.26}$$

With this equation, from a measurement of the weight-average molecular weight, the probability of condensation, P, can be determined.

Distributions of Molecular Weight

The distributions of molecular weight that can be measured with gel permeation chromatography and sedimentation can also be calculated for the simple model given, and in fact correspond to the number distribution function previously described. Moment analysis is commonly used to characterize the distributions, where in our case the first moment is the number-average molecular weight, the second moment is the weight-average molecular weight, and the third moment is the *z*-average molecular weight. The ratio of the weight-average to number-average molecular weights is termed the *dispersity*.

Characterization of Polymer Microstructure (3)

Structural Model of the Polymer Chain

For a microstructural model of the polymer chain, we use a model made up of connected repeating units of similar or different structures. Letters A and B designate the different structural types of the repeating unit.

The chains are structurally presented as sequences of similar or different units, that is A, AA, AAA, B, BB, BBB, AB, AAB, BBB, etc. The different structural components can be indicated by the use of additional letters, C, D, E, and so forth. The application of this model to copolymers A and B and terpolymers A, B, and C is obvious. Positional isomerism, conformation isomerism, configurational isomerism (stereoregularity), and branching and cross-linking are considered as copolymer analogs although they are not generated by copolymerization. In polymer characterization, the goal is to generate the structural sequence distribution function in order to calculate the highest possible weighted-average sequence structure that represents the chain. Our aims are (1) to relate these average sequence structures to the performance properties of the polymer under consideration, and (2) to seek the polymerization mechanism and parameters that generate the sequence structure in order to optimize and control the polymer structure.

For simplicity, we begin with the experimental measurements that are possible on the microstructure of a polymer chain made up of only two structures, A and B. The following is a portion of such a chain:

Chain A – B – B – A – B – B – B – A – B – A –

which is a chain of 10 repeating units. We assume that this portion of the chain is representative of the complete chain.

> In mathematical terms, this assumption is called the stationary condition, that is, the distribution of the two structural elements of the chain does not change as the polymerization proceeds. We could have selected any length of segment that we desired, but the counting process becomes tedious if the segment is too long, and we cannot demonstrate the various points if the chain is too short. So we selected 10. Why not? If you feel the need to select 20 units after we are finished with our 10 units, please do so and send the results to me when you have finished!

Again for simplicity, we also assume that the molecular weight is sufficiently large so that the end groups need not be taken into consideration.

Measurement of Polymer Structure Composition

If we have infinite structural resolution or the ability to detect the differences between the two structural elements A and B, we can measure the individual number of A and B structural elements in the polymer. That is, we can count the relative number of A and B elements taken one at a time, which we will term mono-ads or 1-ads and write $N_1(\mathrm{A})$ as the number of mono-ads of A. For the segment shown previously two types of mono-ads are

$$N_1(\mathrm{A}) = 4 \tag{1.27}$$

and

$$N_1(\mathrm{B}) = 6 \tag{1.28}$$

Spectroscopically, the number of mono-ads cannot actually be counted, but the fraction of units in the chain can be measured. This result can be expressed in terms of the number-fraction probabilities $P_1(\mathrm{A})$ or $P_1(\mathrm{B})$ for mono-ads by dividing by the total number of segments ($N = 10$). So

$$P_1(\mathrm{A}) = \frac{N_\mathrm{A}}{N} \quad P_1(\mathrm{B}) = \frac{N_\mathrm{B}}{N} \tag{1.29}$$

and for our system

$$P_1(\mathrm{A}) = \frac{N_\mathrm{A}}{N} = \frac{4}{10} = 0.4$$

$$P_1(\mathrm{B}) = \frac{N_\mathrm{B}}{N} = \frac{6}{10} = 0.6 \tag{1.30}$$

This result, $P_1(\mathrm{A})$ or $P_1(\mathrm{B})$, is the fractional composition in terms of A and B units of the polymer when we assume that the 10-segment portion of the polymer is representative of the total chain. Also,

$$P_1(\mathrm{A}) + P_1(\mathrm{B}) = 1 \tag{1.31}$$

as is required because for our model polymer only A and B are allowed.

Measurement of Polymer Structure Using Dyad Units

On a slightly more sophisticated level, we can analyze this 10-unit segment by counting sequences two at a time, that is, counting the number of dyads (2-ads) in the chain:

Chain A – B – B – A – B – B – B – A – B – A –
| – A – next sequence

Dyads AB – BB – BA – AB – BB – BB – BA –
AB – BA – AA –

The final AA dyad arises because the adjacent segment that is identical to this segment starts with an A. There are four possible types of dyads, AA, BB, AB, and BA, and for this chain

$$N_2(\mathrm{AA}) = 1$$
$$N_2(\mathrm{BB}) = 3$$
$$N_2(\mathrm{AB}) = 3$$
$$N_2(\mathrm{BA}) = 3$$

Can we truly experimentally distinguish between the heterodyads AB and BA? In this case they are differentiated by the direction of the counting (right to left), but what if the chain segment were reversed? The AB dyads would become BA dyads and vice versa. Which direction of the chain is the proper one? Clearly, both directions are equally likely, and we will never know from examination of the final polymer chain in which direction it grew during the polymerization. Therefore, AB and BA are equally likely outcomes and so (*4*)

$$N_2\,(\mathrm{AB}) = N_2\,(\mathrm{BA}) \qquad (1.32)$$

The equality saves the day for spectroscopic measurements, because no known method of spectroscopic measurement can distinguish between AB and BA, and so from a spectroscopic point of view they are indistinguishable. Consequently any measurement of dyad units measures the sum $N_2\,(\mathrm{AB}) + N_2\,(\mathrm{BA})$, which we will designate as $N_2\,\overline{\mathrm{AB}})$:

$$N_2\,(\overline{\mathrm{AB}}) = N_2\,(\mathrm{AB}) + N_2\,(\mathrm{BA}) \quad (1.33)$$

The number fraction of each dyad $P_2(\mathrm{AA})$, $P_2(\mathrm{BB})$ and $P_2(\overline{\mathrm{AB}})$ can be determined by using $N_2(\text{total}) = 10$.

$$P_2(\mathrm{AA}) = \frac{N_2\,(\mathrm{AA})}{N} \qquad (1.34)$$

$$P_2(\mathrm{BB}) = \frac{N_2\,(\mathrm{BB})}{N} \qquad (1.35)$$

$$P_2(\overline{\mathrm{AB}}) = P_2(\mathrm{AB}) + P_2(\mathrm{BA}) = \frac{N_2\,(\overline{\mathrm{AB}})}{N} \qquad (1.36)$$

For the 10-segment polymer chain

$$P_2(\mathrm{AA}) = \frac{N_2\,(\mathrm{AA})}{N} = 0.1 \qquad (1.37)$$

$$P_2(\mathrm{BB}) = \frac{N_2\,(\mathrm{BB})}{N} = 0.3 \qquad (1.38)$$

$$P_2(\overline{\mathrm{AB}}) = P_2(\mathrm{AB}) + P_2(\mathrm{BA}) = \frac{N_2\,(\overline{\mathrm{AB}})}{N} = 0.6 \qquad (1.39)$$

The normalization condition which is

$$P_2(\mathrm{AA}) + P_2(\mathrm{AB}) + P_2(\mathrm{BA}) + P_2(\mathrm{BB}) = 1 \qquad (1.40)$$

is satisfied. Or, in terms that can be measured spectroscopically

$$P_2(\mathrm{AA}) + P_2(\overline{\mathrm{AB}}) + P_2(\mathrm{BB}) = 1 \qquad (1.41)$$

Measurement of Polymer Structure Using Triad Segments

We can now proceed to dissect the polymer structure in terms of higher *n*-mers in spite of the fact that we might introduce a certain amount of boredom with the process. Let us determine the structure of the polymer chain in terms of triads or 3-ads, that is, counting the units as threes. For triads, there are a total of eight possible structures.

Chain
A − B − B − A − B − B − B − A − B − A| − A − B −
next segment

Triads
− ABB − BBA − BAB − ABB − BBB − BBA −
BAB − ABA − BAA − AAB −

Once again, the final two triads are recognized by the adjoining segment that corresponds to the first two units in the beginning of this segment. Well, how many do we have?

$$N_3(\mathrm{AAA}) = 0$$
$$N_3(\mathrm{AAB}) = 1$$
$$N_3(\mathrm{ABA}) = 1$$
$$N_3(\mathrm{BAA}) = 1$$
$$N_3(\mathrm{ABB}) = 2$$
$$N_3(\mathrm{BBA}) = 2$$
$$N_3(\mathrm{BAB}) = 2$$
$$N_3(\mathrm{BBB}) = 1$$

with $N_3(\text{total}) = 10$.

The two "reversibility relations" for the triads are

$$N_3\,(\mathrm{AAB}) = \mathrm{N}_3\,(\mathrm{BAA}) \qquad (1.42)$$

and

$$N_3\,(\mathrm{BBA}) = N_3\,(\mathrm{ABB}) \qquad (1.43)$$

and a normalization condition representing the sum of all of the triads. We can now calculate the number fraction of each triad

$$P_3(\text{AAA}) = 0$$

$$P_3(\overline{\text{AAB}}) = 0.2$$

$$P_3(\text{ABA}) = 0.1$$

$$P_3(\overline{\text{ABB}}) = 0.4$$

$$P_3(\text{BAB}) = 0.2$$

$$P_3(\text{BBB}) = 0.1$$

Measurement of Polymer Structure Using Higher *N*-ad Segments

We could continue with tetrads, pentads, etc., up to the 10-ads for the case at hand, but you have now gotten the idea of the process (and at this stage you are completely bored). We could derive the necessary relations observing that the total number of different combinations for the higher *n*-ads goes up as 2^n where n is the *n*-ad length. Therefore, the number of combinations increases very rapidly, and the task gets tedious although not difficult. Heptamers have been observed in high-resolution NMR spectroscopy, so it may be necessary to plow through the derivation process (or look it up in the literature).

The ability to count the fraction of the different *n*-ads is a function of the sensitivity of the spectroscopic method involved. In some cases, the spectroscopic method allows the observation of only isolated units, in others dyads, and in some triads, etc. We will discuss the nature of the spectroscopic *n*-ad sensitivity and how it can be determined for the polymer system being studied at the appropriate time.

Relationships Between the Various Orders (Lengths) of Sequences

As might be expected, all of the *n*-ad sequences are related to all the others in some fashion through the structural distribution function. But at the moment, we haven't the slightest idea what this structural distribution function is. Rather there are some required relationships between the different levels of *n*-ads, and these relationships are easily derived.

Take the simplest case first: the relationship between the 1-ads (isolated units) and the 2-ads (dyads). We know that the dyads are formed from the 1-ads by the addition of another unit and that there are only two possible ways of making the addition—before and after the unit. So we start with an A unit:

$$P_1(\text{A}) = P_2(\text{AA}) + P_2(\text{AB}) \tag{1.44}$$

units added after the A unit

$$P_1(\text{A}) = P_2(\text{AA}) + P_2(\text{BA}) \tag{1.45}$$

units added before the A unit

Experimentally, we have

$$P_1(\text{A}) = P_2(\text{AA}) + \frac{1}{2}P_2(\overline{\text{AB}}) \tag{1.46}$$

In a similar fashion,

$$P_1(\text{B}) = P_2(\text{BB}) + \frac{1}{2}P_2(\overline{\text{AB}}) \tag{1.47}$$

We have derived the reversibility relationship (eq 1.32) to which we had previously alluded

$$P_2(\text{AB}) = P_2(\text{BA}) \tag{1.48}$$

The relationships between the various *n*-ads are shown in Table 1.1 (*3*). These relationships between the various orders of the *n*-ads are useful because they represent a quantitative requirement to be met by the experimental measurements and a test of the structural assignments made by the spectroscopists (who are not always infallible).

Calculation of Polymer Structural Parameters from Sequence Measurements

The measured information on the *n*-ads can now be used to calculate structural information about the polymer, which will be useful in comparing one polymer system to another. Calculations of various average quantities as those derived for the degree of polymerization will be useful in understanding the type of polymer we are studying in terms of the distribution of its microstructure.

Structural Composition

The structural composition of the polymer can be defined as the ratio of the various types of structure.

Table 1.1. Necessary Relations Between Relative Concentrations of Comonomer Sequences of Different Lengths

Sequence	*Probability for Individual Units*	*Total Number-Fraction Probability*
Dyad-monad	$P(\mathrm{AA}) + (1/2)P(\overline{\mathrm{AB}})$	$P(\mathrm{A})$
	$P(\mathrm{BB}) + (1/2)P(\mathrm{AB})$	$P(\mathrm{B})$
Triad-dyad	$P(\mathrm{AAA}) + (1/2)P(\overline{\mathrm{AAB}})$	$P(\mathrm{AA})$
	$P(\mathrm{BAB}) + (1/2)P(\mathrm{AAB}) + P(\mathrm{ABA}) + (1/2)P(\overline{\mathrm{ABB}})$	$P(\mathrm{AB})$
	$P(\mathrm{BBB}) + (1/2)P(\mathrm{ABB})$	$P(\mathrm{BB})$
Tetrad-triad	$P(\mathrm{AAAA}) + (1/2)P(\overline{\mathrm{AAAB}})$	$P(\mathrm{AAA})$
	$P(\mathrm{BAAB}) + (1/2)P(\mathrm{AAAB}) + (1/2)P(\overline{\mathrm{AABA}}) + (1/2)P(\overline{\mathrm{AABB}})$	$P(\mathrm{AAB})$
	$(1/2)P(\mathrm{ABAB}) + (1/2)P(\mathrm{BABB})$	$P(\mathrm{BAB})$
	$(1/2)P(\mathrm{ABAB}) + (1/2)P(\mathrm{AABA})$	$P(\mathrm{ABA})$
	$P(\mathrm{ABBA}) + (1/2)P(\mathrm{ABBB}) + (1/2)P(\overline{\mathrm{AABB}}) + (1/2)P(\overline{\mathrm{BABB}})$	$P(\mathrm{ABB})$
	$P(\mathrm{BBBB}) + (1/2)P(\mathrm{ABBB})$	$P(\mathrm{BBB})$

If the number fraction of single units A or B can be measured, the composition can be calculated from the ratio of these measurements:

$$\text{structural composition} = \frac{P_1(\mathrm{A})}{P_1(\mathrm{B})} \quad (1.49)$$

For our example,

$$\frac{P_1(\mathrm{A})}{P_1(\mathrm{B})} = \frac{0.4}{0.6} = 0.67 \quad (1.50)$$

Sequence Order Parameter

If both mono-ads and dyads are measured, a statistical parameter that can then be determined yields information about whether the distribution of the structures is random, and whether it tends towards an alternating or block distribution. This order parameter, χ, is defined as

$$\chi = \frac{1}{2}\frac{P_2(\overline{\mathrm{AB}})}{P_1(\mathrm{A})P_1(\mathrm{B})} \quad (1.51)$$

If

$$\frac{1}{2}P_2(\overline{\mathrm{AB}}) = P_1(\mathrm{A})P_1(\mathrm{B}) \quad (1.52)$$

independent probabilities exist, and the result is termed random.

When

- $\chi = 1$ random distribution of A and B
- $\chi > 1$ more alternating tendency
- $\chi < 1$ more block tendency
- $\chi = 2$ completely alternating A and B
- $\chi = 0$ complete blocks of A and B

Number-Average Sequence Lengths

The number of $N_A(n)$ fractions of the sequences of A units are defined as

$$N_A(n) = \frac{P_{n+2}(\mathrm{BA}_n\mathrm{B})}{\sum_{n=1}^{\infty} P_{n+2}(\mathrm{BA}_n\mathrm{B})} \quad (1.53)$$

Also,

$$P_2(\mathrm{AB}) = \sum_{n=1}^{\infty} P_{n+2}(\mathrm{BA}_n\mathrm{B}) \quad (1.54)$$

and

$$P_2(\mathrm{BA}) = \sum_{n=1}^{\infty} P_{n+2}(\mathrm{AB}_n\mathrm{A}) \quad (1.55)$$

so

$$N_A(n) = \frac{P_{n+2}(\mathrm{BA}_n\mathrm{B})}{P_2(\mathrm{BA})} \quad (1.56)$$

The number-average sequence length, $\bar{l}_A$, has the definition

$$\bar{l}_A = \frac{\sum_{n=1}^{\infty} nN_A(n)}{\sum_{n=1}^{\infty} N_A(n)} = \sum_{n=1}^{\infty} nN_A(n) \quad (1.57)$$

Substitution yields

$$\bar{l}_A = \frac{\sum_n nP_{n+2}(BA_nB)}{P_2(BA)} \quad (1.58)$$

and, because

$$\sum_n nP_{n+2}(BA_nB) = P_1(A) \quad (1.59)$$

the number-average sequence length is obtained in terms of simple measurable sequences

$$\bar{l}_A = \frac{P_1(A)}{\frac{1}{2}P_2(\overline{AB})} \quad (1.60)$$

and likewise

$$\bar{l}_B = \frac{P_1(B)}{\frac{1}{2}P_2(\overline{AB})} \quad (1.61)$$

With the required n-ad relationships, this result can be expressed in terms of the measured dyads only:

$$\bar{l}_A = \frac{\left[P_2(AA) + \frac{1}{2}P_2(\overline{AB})\right]}{\frac{1}{2}P_2(BA)} \quad (1.62)$$

and likewise

$$\bar{l}_B = \frac{\left[P_2(BB) + \frac{1}{2}P_2(\overline{AB})\right]}{\frac{1}{2}P_2(BA)} \quad (1.63)$$

If triads are being measured, the number-average sequence length can be calculated by using

$$\bar{l}_A = \frac{[P_3(BAB) + P_3(\overline{AAB})\, P_3(AAA)]}{P_3(BAB) + \frac{1}{2}P_3(\overline{AAB})} \quad (1.64)$$

and

$$\bar{l}_B = \frac{[P_3(ABA) + P_3(\overline{BBA})\, P_3(BBB)]}{P_3(ABA) + \frac{1}{2}P_3(\overline{BBA})} \quad (1.65)$$

Of course, if tetrads or higher n-ads are measured, corresponding relationships can be derived for calculating the number-average sequence lengths.

Relating the Polymer Structure to the Polymerization Parameters

What factors in the polymerization reaction are responsible for the sequence structure of the polymer? If the sequence distribution of the polymer chain can be completely characterized, and if proper control has been exercised, it should be possible to relate the structure to the basic polymerization process. In the ideal circumstance, discovering the relationships between the chemistry and the structure is quite easy. However, in the real case, it is quite difficult but possible, particularly with some of the modern computational techniques. In every case, it is possible to gain some insight into the polymerization process. Here, we will demonstrate the ideal case and later point the direction for the cases observed in the real world.

Microstructure of Terminal Copolymerization Model

For a simple copolymerization model, the terminal copolymerization model, the rate of addition of comonomer depends on the nature of the terminal group. This model is represented by the following reactions and their corresponding rates in terms of kinetic rate constants and concentrations of reacting species, where A* and B* are the propagating terminal species.

Terminal Group	*Added Group*	*Rate*	*Final*
~A*	[A]	K_{AA}[A*][A]	~AA*
~B*	[A]	K_{BA}[B*][A]	~BA*
~A*	[B]	K_{AB}[A*][B]	~AB*
~B*	[B]	K_{BB}[B*][B]	~BB*

To know the nature of the microstructure resulting from this polymerization, we need to know the probability that an A unit is next to another A, or perhaps next to a B. That is, what is the probability that an A unit has added an A or a B? These kinds of probabilities are termed *conditional probabilities* and will be designated by the notation P(B/A) which represents the conditional probability that a B unit has been added to an A unit.

Notice the reversal of the notation: An AB unit has been formed, but the notation is P(B/A). Why such a reversed notation? I don't know.

These conditional probabilities start with the assumption that one has an A unit and then asks the question what is the probability that a B unit is added, or alternatively that an A unit is added. These conditional probabilities are determined by the relative rates of addition of the A or B units to the terminal A unit. The terminal conditional probabilities can be written as the ratio of the rate of formation of the desired product (AA or AB) to the rates of the reaction of the starting unit (A*):

$$P(\mathrm{A/A}) = \frac{K_{AA}[\mathrm{A}][\mathrm{A^*}]}{K_{AA}[\mathrm{A}][\mathrm{A^*}] + K_{AB}[\mathrm{B}][\mathrm{A^*}]} \quad (1.66a)$$

$$P(\mathrm{A/A}) = \frac{\dfrac{K_{AA}[\mathrm{A}]}{K_{AB}[\mathrm{B}]}}{\left\{\dfrac{K_{AA}[\mathrm{A}]}{K_{AB}[\mathrm{B}]}\right\} + 1} = \frac{r_A\, x}{1 + r_A\, x} \quad (1.66b)$$

and

$$P(\mathrm{B/A}) = \frac{1}{1 + r_A\, x} \quad (1.67)$$

where r_A is the reactivity ratio for [A*]:

$$r_A = \frac{k_{AA}}{k_{AB}} \quad (1.68)$$

and χ is the monomer feed ratio

$$x = \frac{[\mathrm{A}]}{[\mathrm{B}]} \quad (1.69)$$

Likewise, a treatment of the reactions of B* yields

$$P(\mathrm{B/B}) = \frac{\dfrac{r_B}{x}}{1 + \dfrac{r_B}{x}} \qquad P(\mathrm{A/B}) = \frac{1}{1 + \dfrac{r_B}{x}} \quad (1.70)$$

where r_B is the reactivity ratio for [B*]

$$r_B = \frac{k_{BB}}{k_{BA}} \quad (1.71)$$

There are only two independent terminal conditional probabilities, that is

$$P(\mathrm{A/A}) + P(\mathrm{B/A}) = 1 \quad (1.72)$$

and

$$P(\mathrm{B/B}) + P(\mathrm{A/B}) = 1 \quad (1.73)$$

We will use P(A/B) and P(B/A) in further calculations.

We need to relate these conditional probabilities to the measured n-ads. For example, Bayers' theorem yields the relative probability of the dyad sequence as

$$P_2(\mathrm{BA}) = P_1(\mathrm{B})P(\mathrm{A/B}) \quad (1.74)$$

concluding that the dyad fraction is given by the probability of finding a B, which is P_1(B), times the conditional probability of adding an A given a B unit, which is P(A/B). With this in mind, there are a number of routes we could take, but let us begin by using the reversibility relationship eq 1.48

$$P_2(\mathrm{AB}) = P_2(\mathrm{BA}) \quad (1.75)$$

Writing this equation in terms of the terminal conditional probabilities,

$$P_1(\mathrm{A})P(\mathrm{B/A}) = P_1(\mathrm{B})P(\mathrm{A/B}) \quad (1.76)$$

and rearranging, we obtain our first useful result (that is, the composition in terms of the conditional probabilities)

$$\frac{P_1(\mathrm{A})}{P_1(\mathrm{B})} = \frac{P(\mathrm{A/B})}{P(\mathrm{B/A})} \quad (1.77)$$

This result is known as the Mayo–Lewis equation when expressed in terms of the polymerization parameters

$$\frac{P_1(\mathrm{A})}{P_1(\mathrm{B})} = \frac{1 + r_\mathrm{A}x}{1 + \frac{r_\mathrm{B}}{x}} \quad (1.78)$$

The dyad concentrations can be derived by substituting in the following

$$P_2(\mathrm{AA}) = P_1(\mathrm{A})P(\mathrm{A/A}) = \frac{P(\mathrm{A/B})[1 - P(\mathrm{B/A})]}{P(\mathrm{B/A}) + P(\mathrm{A/B})} \quad (1.79)$$

$$P_2(\mathrm{AB}) = P_2(\mathrm{BA}) = P_1(\mathrm{A})\,P(\mathrm{B/A}) \quad (1.80a)$$

$$P_2(\mathrm{AB}) = \frac{P(\mathrm{A/B})P(\mathrm{B/A})}{P(\mathrm{B/A}) + P(\mathrm{A/B})} \quad (1.80b)$$

$$P_2(\mathrm{BB}) = 1 - 2P(\mathrm{AB}) - P_2(\mathrm{AA}) \quad (1.81)$$

The triad fraction can also be written

$$F_{\mathrm{AAA}} = \frac{P_3(\mathrm{AAA})}{P_1(\mathrm{A})} = [1 - P(\mathrm{B/A})]^2 \quad (1.82)$$

$$F_{\mathrm{BAB}} = [\mathrm{P(B/A)}]^2 \quad (1.83)$$

$$F_{\mathrm{BAA}} = F_{\mathrm{AAB}} = P(\mathrm{B/A})[1 - P(\mathrm{B/A})] \quad (1.84)$$

In fact, all of the experimental parameters can be calculated in terms of the two conditional probabilities $P(\mathrm{B/A})$ and $P(\mathrm{A/B})$, or the corresponding r_A, r_B, and x, which are the basic parameters of the terminal model copolymerization mechanism.

The order parameter χ is given by

$$\chi = P(\mathrm{A/B}) + P(\mathrm{B/A}) \quad (1.85)$$

and the $r_\mathrm{A}r_\mathrm{B}$ product determines whether the copolymer is alternating, random, or block. When $r_\mathrm{A}r_\mathrm{B} < 1$, then $\chi > 1$ and the polymer has an alternating tendency. If $r_\mathrm{A}r_\mathrm{B} = 1$, the polymer is random. If $r_\mathrm{A}r_\mathrm{B} > 1$, then $\chi < 1$ and the polymer has a block character.

The number-average sequence in terms of the terminal conditional probabilities becomes

$$\bar{l}_\mathrm{A} = \frac{1}{P(\mathrm{B/A})} = 1 + r_\mathrm{A}x \quad (1.86)$$

$$\bar{l}_\mathrm{B} = \frac{1}{P(\mathrm{A/B})} = 1 + \frac{r_\mathrm{B}}{x} \quad (1.87)$$

As shown previously, $\bar{l}_\mathrm{A}$ or $\bar{l}_\mathrm{B}$ can be determined from the measurement of the various n-ads. Consequently, the terminal model can be tested by the relationships in eqs 1.86 and 1.87; that is, plotting $\bar{l}_\mathrm{A}$ versus x. If a linear relationship is found, then r_A is the slope. Similarly, plotting $\bar{l}_\mathrm{B}$ versus $1/x$ yields r_B as the slope.

Likewise, the number fraction of A (or B) sequences that allows the calculation of the amount of As (or Bs) of varying sequence length in the copolymer chain can be calculated for the terminal model from

$$N_\mathrm{A}(n) = \left(\frac{[r_\mathrm{A}x]^{n-1}}{1 + r_\mathrm{A}x}\right)\left(1 - \frac{r_\mathrm{A}x}{1 + r_\mathrm{A}x}\right) \quad (1.88)$$

$$N_\mathrm{B}(n) = \left(\frac{\left(\frac{r_\mathrm{B}}{x}\right)^{n-1}}{1 + \frac{r_\mathrm{B}}{x}}\right)\left(1 - \frac{\frac{r_\mathrm{B}}{x}}{1 + \frac{r_\mathrm{B}}{x}}\right) \quad (1.89)$$

By jove, I think we've done it! We have related the observed structure to the type of polymerization mechanism and devised a method to determine the fundamental parameters, at least for the terminal model. Admittedly, a number of assumptions have been made including isothermal conditions, reactivities independent of chain length, and instantaneous conditions (no conversion) but it has demonstrated the approach we are taking. However, before we become enraptured with our success, let us examine a polymerization model that is slightly more complicated, that is, the penultimate polymerization model.

Microstructure of the Penultimate Polymerization Model

When the nature of the penultimate (next to last) unit has a significant effect on the absolute rate constant in the copolymerization, eight reactions yielding the penultimate conditional probabilities can be derived in terms of the absolute rate constants and the monomer feed. The penultimate conditional probabilities can be written in terms of the relative number of chemical reactions occurring for each penultimate unit. The reactions and their corresponding rates are as follows:

Penultimate Group	*Added Group*	*Rate*	*Final*
~AA*	[A]	k_{AAA}[AA*][A]	~AAA*
~AA*	[B]	k_{AAB}[AA*][B]	~AAB*
~BA*	[A]	k_{BAA}[BA*][A]	~BAA*
~BA*	[B]	k_{BAB}[BA*][B]	~BAB*
~AB*	[A]	k_{ABA}[AB*][A]	~ABA*
~AB*	[B]	k_{ABB}[AB*][B]	~ABB*
~BB*	[A]	k_{BBA}[BB*][A]	~BBA*
~BB*	[B]	k_{BBB}[BB*][B]	~BBB*

The conditional probabilities are calculated in the same fashion as previously, that is, by determining the relative rates of addition to the penultimate group. In other words, we must calculate the probability of AA rather than A, as was used for the terminal model

$$P(\mathrm{A/AA}) = \frac{k_{AAA}[\mathrm{A}][\mathrm{AA}^*]}{k_{AAA}[\mathrm{A}][\mathrm{AA}^*] + k_{AAB}[\mathrm{B}][\mathrm{AA}^*]} \quad (1.90a)$$

$$P(\mathrm{A/AA}) = \frac{\dfrac{k_{AAA}[\mathrm{A}]}{k_{AAB}[\mathrm{B}]}}{\left\{\dfrac{k_{AAA}[\mathrm{A}]}{k_{AAB}[\mathrm{B}]}\right\} + 1} \quad (1.90b)$$

Four independent conditional probabilities can be written, with four monomer reactivity ratios, in a manner similar to the terminal model:

$$P(\mathrm{B/AA}) = \frac{1}{1 + r_A x}$$
$$P(\mathrm{A/AA}) = 1 - P(\mathrm{B/AA}) \quad (1.91)$$

$$P(\mathrm{A/BA}) = \frac{r'_A x}{1 + r'_A x}$$
$$P(\mathrm{B/BA}) = 1 - P(\mathrm{A/BA}) \quad (1.92)$$

$$P(\mathrm{B/AB}) = \frac{\dfrac{r'_B}{x}}{1 + \dfrac{r'_B}{x}}$$
$$P(\mathrm{A/AB}) = 1 - P(\mathrm{B/AB}) \quad (1.93)$$

$$P(\mathrm{A/BB}) = \frac{1}{1 + \dfrac{r_B}{x}}$$
$$P(\mathrm{B/BB}) = 1 - P(\mathrm{A/BB}) \quad (1.94)$$

where

$$r_A = \frac{k_{AAA}}{k_{AAB}} \qquad r'_A = \frac{k_{BAA}}{k_{BAB}} \quad (1.95)$$

$$r_B = \frac{k_{BBB}}{k_{BBA}} \qquad r'_B = \frac{k_{ABB}}{k_{ABA}} \quad (1.96)$$

The composition of the copolymer in terms of the conditional penultimate probabilities is found by substitution:

$$\frac{P_1(\mathrm{A})}{P_1(\mathrm{B})} = \frac{P(\mathrm{A/B})}{P(\mathrm{B/A})} = \frac{1 + \dfrac{P(\mathrm{A/BA})}{P(\mathrm{B/AA})}}{1 + \dfrac{P(\mathrm{B/AB})}{P(\mathrm{A/BB})}} \quad (1.97)$$

These results can be expressed in terms of the reactivity coefficients and monomer feed:

$$\frac{P_1(\mathrm{A})}{P_1(\mathrm{B})} = \frac{1 + \dfrac{r'_A x(1 + r_A x)}{(1 + r'_A x)}}{1 + \dfrac{\dfrac{r'_B}{x}\left(1 + \dfrac{r_B}{x}\right)}{\left(1 + \dfrac{r'_B}{x}\right)}} \quad (1.98)$$

All of the other parameters can be calculated for the penultimate model and are shown in Table 1.2. The relationships are not particularly difficult, but they are cumbersome, partly because four different

Table 1.2. Relations Between Experimental Parameters, Conditional Probabilities, and Reactivity Ratios for the Penultimate Model of Copolymerization

Experimental	General	Conditional Probabilities	Reactivity Ratios and Monomer Feed
$P_2(\mathrm{AB})$	$2P_2(\mathrm{BA})$	$\dfrac{2}{P(\mathrm{A/BA})/P(\mathrm{B/AA})} + 2 + \dfrac{P(\mathrm{B/AB})}{P(\mathrm{A/BB})}$	$\dfrac{2}{\dfrac{r'_A\chi(1 + r_A\chi)}{1 + r'_A\chi} + 2 + \dfrac{(r'_B/\chi)(r_B + \chi)}{r'_B + \chi}}$
χ	$\dfrac{P_2(\mathrm{BA})}{P_1(\mathrm{B})P_1(\mathrm{A})}$	$\dfrac{P(\mathrm{A/BA})/P(\mathrm{B/AA}) + 2 + P(\mathrm{B/AB})/P(\mathrm{A/BB})}{P(\mathrm{A/BA})/P(\mathrm{B/AA}) + 1 + P(\mathrm{A/BA})/P(\mathrm{B/AA})P(\mathrm{B/AB})P(\mathrm{A/BB})/P(\mathrm{B/AB})P(\mathrm{A/BB})}$	$\dfrac{r'_A\chi\left(\dfrac{1 + r_A\chi}{1 + r_A\chi}\right) + 2 + \dfrac{r'_B}{\chi}\left(\dfrac{r_B + \chi}{r'_B + \chi}\right)}{r'_A\left(\dfrac{1 + r_A\chi}{1 + r'_A\chi}\right) + 1 + r'_A r'_B\left(\dfrac{1 + r_A\chi}{1 + r_A\chi}\right)\left(\dfrac{r_B + \chi}{r'_B + \chi}\right) + \dfrac{r'_B}{\chi}\left(\dfrac{r_B + \chi}{r'_B + \chi}\right)}$
$\bar{l}_A$	$\dfrac{P_1(\mathrm{A})}{P_2(\mathrm{AB})}$	$1 + \dfrac{P(\mathrm{A/BA})}{P(\mathrm{B/AA})}$	$1 + \left[r'_A\chi\left(\dfrac{1 + r_A\chi}{1 + r'_A\chi}\right)\right]\chi$
$\bar{l}_B$	$\dfrac{P_1(\mathrm{B})}{P_2(\mathrm{BA})}$	$1 + \dfrac{P(\mathrm{B/AB})}{P(\mathrm{A/BB})}$	$1 + \dfrac{r'_B\left(\dfrac{r_B + \chi}{r'_B + 1\chi}\right)}{\chi}$
$1/\alpha_A$	$\dfrac{P_1(\mathrm{A})}{P_2(\mathrm{BA})}$	$1 + \dfrac{P(\mathrm{B/AA})}{P(\mathrm{A/BA})}$	$1 + \dfrac{1}{\left[r'_A\left(\dfrac{1 + r_A\chi}{1 + r_A\chi}\right)\right]\chi}$
$1/\alpha_B$	$\dfrac{P_1(\mathrm{B})}{P_2(\mathrm{BB})}$	$1 + \dfrac{P(\mathrm{A/BB})}{P(\mathrm{B/AB})}$	$1 + \dfrac{\chi}{r'_B\left(\dfrac{1r_B + \chi}{r'_B + \chi}\right)}$

parameters are involved, as opposed to two parameters for the terminal model.

Higher Order Models of Copolymerization and Other Complications

Of course, the actual polymerization mechanism can be more complicated in a number of ways. Any intramolecular action or intermolecular interaction will change the results. The effects of longer terminal units influencing the polymerization are possible. In some cases, one or both of the monomers may associate with the active growing end or the solvent. It is also possible for one of the monomers to show a tendency for depolymerization to occur simultaneously with polymerization. The individual radicals can react with two different energy states, for example, when the monomer isomerizes, it will have different reactivities for the different isomers. Mathematical methods have been devised to work with these types of complications (*5, 6*). The major modification is the necessity to use matrix calculations, if for no other reason than to make the bookkeeping tractable.

The probabilistic method described in the preceding sections does not account for time-dependent or conversion processes, but these processes can be introduced through integration procedures (*2*).

Differentiation Between Model Mechanisms of Copolymerization

Most investigations of the mechanism of copolymerization involve an analysis of the composition of the initial copolymers formed, but it is often difficult to distinguish between the various models for copolymerization on the basis of their fit to the composition data alone. The distribution of monomer sequences in the copolymer contains more information about the polymerization system than does the copolymer composition.

The general approach is to prepare copolymers over a range of comonomer compositions and measure the sequence distributions obtained (*2*). The sequence distributions are mathematically fitted recognizing the nonlinear nature of the problem to the various calculated structural parameters, and a "best-fit" model is determined. Care must be taken to account for the effect of conversion on the polymer sequence distribution in performing the calculations.

Objective tests have been described to assess the accuracy of the copolymerization models, including the statistical F test (*7*). This test is based on the ratio F of the residual sums of the squares of the models A and B to be differentiated, where B is a special case of the model A. The equation involved is

$$F = \frac{\dfrac{[SS_B - SS_A]}{[p_A - p_B]}}{\dfrac{SS_A}{[n - p_A]}} \qquad (1.99)$$

where SS is the sum of the squares, p_A and p_B are the number of parameters for each model, and n is the number of experimental observations. The ratio F is compared with critical values of $F(\alpha)$ that are available in tables (*8*). This test is highly recommended.

The problem with determining the mechanism of the polymerization is due primarily to the difficulty in making accurate analytical measurements of the composition and structure. There are only relatively small differences in composition expected theoretically between the two models, as is shown in Figure 1.1 (*9*). This figure shows the molar ratio in the

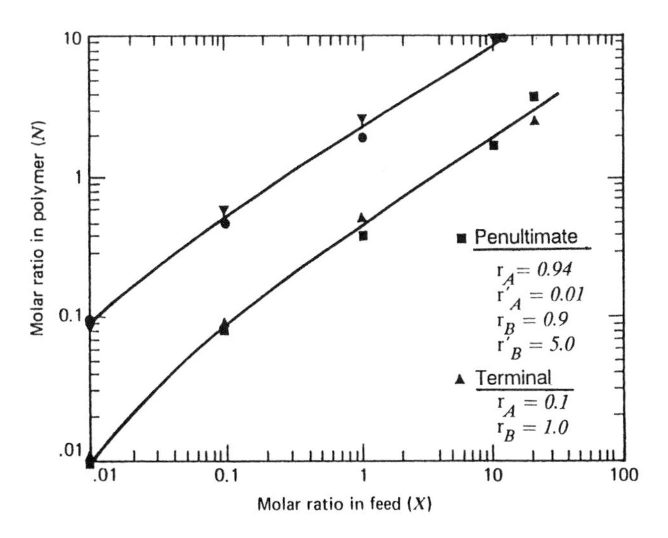

Figure 1.1. Comparison of the molar ratio in the polymer as a function of the molar ratio in the feed for the terminal and penultimate models with two sets of reactivity ratios. The reactivity ratios for the terminal model were $r_A = 0.1$ and $r_B = 1.0$. For the penultimate model, the reactivity ratios were $r_A = 0.94$, $r_A' = 0.01$, $r_B = 0.9$, and $r_B' = 5.0$. (Reproduced with permission from reference 9. Copyright 1964 John Wiley & Sons, Inc.)

polymer versus the molar ratio in the feed, x, calculated for reactivity ratios in the intermediate range for the terminal and penultimate models. This figure clearly shows the small differences (~2%) in composition between the two mechanisms of polymerization. This result, might seem unimportant. Let us examine the sequence distribution using the same reactivity ratios for the two models. The number distribution of sequences is shown in Figure 1.2, and the weight-average distribution is shown in Figure 1.3 (*9*). Now the differences are quite substantial, particularly for the shorter sequence lengths. Furthermore, these sequences influence the physical and mechanical properties of a polymer. Therefore, when the sequence distribution is altered, differences in the performance properties of the polymer will be observed.

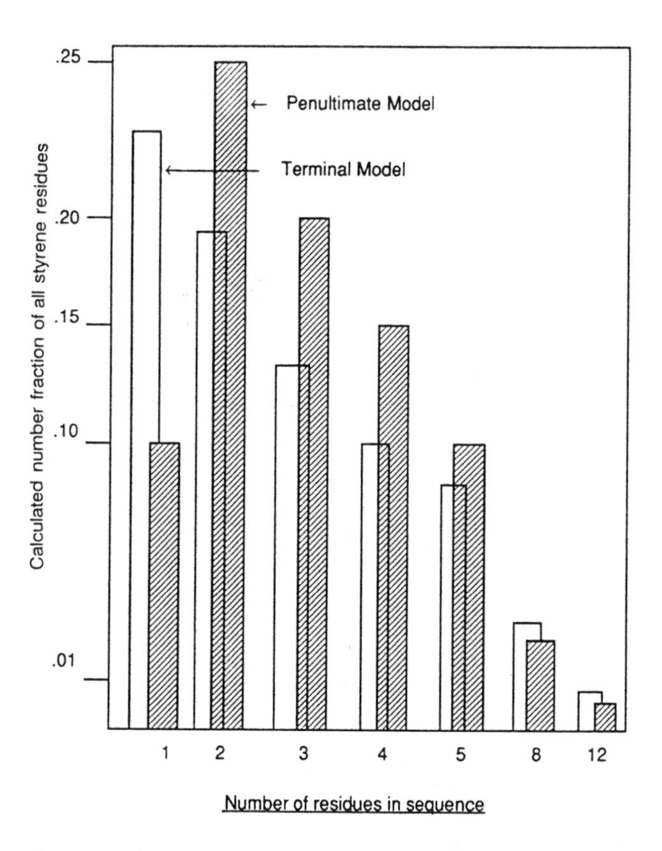

Figure 1.2. Number distribution of monomer sequences calculated from terminal and penultimate models. The reactivity ratios for both models were the same as in Figure 1.1. (Reproduced with permission from reference 9. Copyright 1964 John Wiley & Sons, Inc.)

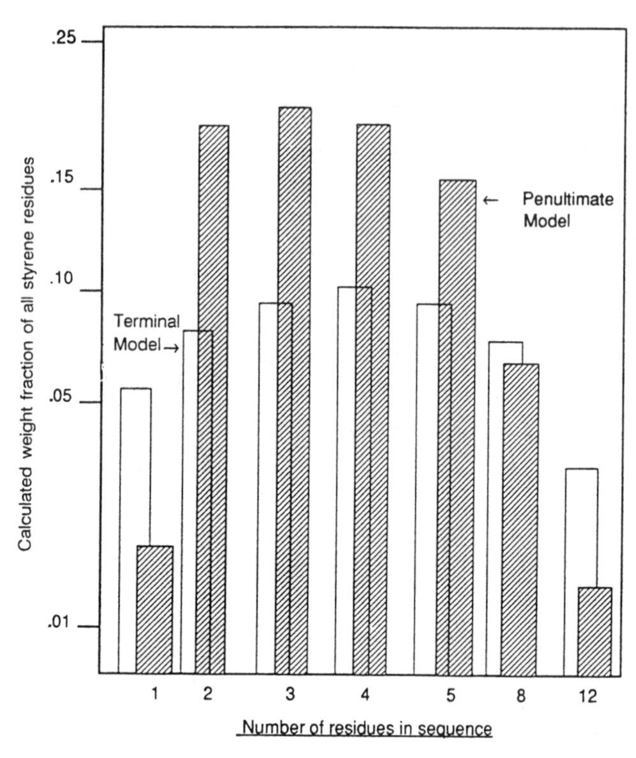

Figure 1.3. Weight distribution of monomer sequences calculated from terminal and penultimate models. The reactivity ratios for both models were the same as in Figure 1.1. (Reproduced with permission from reference 9. Copyright 1964 John Wiley & Sons, Inc.)

From another point of view, the results depicted in Figures 1.2 and 1.3 point to the possibility of measuring these short sequences for determining the nature of the polymerization model. Although the compositional differences are small, the number and kind of comonomer sequences are substantially different. It is also clear that the short sequences dominate the microstructure and therefore are amenable to measurement using sequence-sensitive analytical techniques.

In the early days of polymer chemistry, researchers were justifiably concerned with the effect of composition on properties. There was a time when copolymers were made with the same composition but exhibited substantial differences in properties. Then the researchers began to become interested in the nature of the sequence distribution. Techniques were sought to determine the sequence distributions and relate the results to the differences in end-use properties.

Determination of the Polymerization Parameters

Once the mechanism of the copolymerization process is determined, the next step is to determine the polymerization parameters, or more precisely, the reactivity ratios of the comonomer system. Generally, the ratios are determined from a plot of the data obtained from samples with different feed ratios. In most cases, the equations are reorganized to yield linear relationships whose slopes and intercept values reflect the reactivity ratios. These linearized relations, although apparently useful, can yield misleading results because of the different weighting of the data and the corresponding experimental error (*10*). Nonlinear methods have also been used, and some of their difficulties and advantages have been discussed (*11*).

Whatever the method you use, remember that the calculated reactivity ratios are not universal constants. That is, their value depends on the environment of the polymerization, including the temperature, pressure, ionic strength, and so forth. However, if your numbers are substantially different from those recognized values in the literature for polymerizations similar to your own, I think you might reexamine your data.

Summary

This chapter has demonstrated, in the simplest possible terms, the approach that is necessary to go from spectroscopic measurements of the microstructure of the polymer chain to a determination of the fundamental parameters of the polymerization process. After all, that is what we really want to know! Although it would be interesting to know the position and geometric relationship between every atom in a polymer chain, what would we do with the information? Spectroscopically, we can measure only the average structure of the polymer chains in the sample. Therefore, our only choice is to try to relate these structural parameters to the performance properties of the polymer and to the polymerization route that gave the polymer structure. Believe it or not, this step is possible using the spectroscopic techniques to be described in the later chapters of this book.

The Shape of Things To Come

Spectroscopic techniques for the study of polymers must yield high-resolution, narrow linewidth spectra that provide selectivity and structural information. Because polymer systems are always complex mixtures of structure and molecules, the spectroscopic probe must permit selective monitoring of more than one structural type at a time. It must possess sufficient sensitivity to detect and monitor very low levels of structure in the polymer, as small structural changes produce much larger effects on the physical and mechanical properties. The spectroscopic probe must be very specific in its informational content, as we will need to determine not only the structure of the single repeating units but also how they are connected together and to what extent the units are ordered. The technique should be nondestructive and noninvasive, because the polymer samples will need to be evaluated by other methods besides spectroscopic characterization. The probe should be capable of studying the polymer in its useful form, be it fiber, film, composite, coating, or adhesive.

The majority of the spectroscopic techniques, such as UV and visible or mass spectroscopy, do not meet the specifications of the spectroscopic probe. However, three spectroscopic techniques have evolved for polymer analysis that do fit these criteria: Fourier transform infrared (FTIR), Raman, and high-resolution NMR spectroscopy of the solid state. These techniques used individually or in combination can provide detailed structural information on polymers for research, analysis, and quality control. High-resolution NMR spectroscopy of solutions does not meet the specification of being capable of studying the polymer in its final engineering form, but the contributions that NMR spectroscopy has made to our understanding of the structure of polymers are so great that it must be considered in any textbook on spectroscopy of polymers.

References

1. Lowry, G. G. In *Markov Chains and Monte Carlo Calculations in Polymer Science;* Lowry, G. G., Ed.; Dekker: New York, 1969.

2. Koenig, J. L. *Chemical Microstructure of Polymer Chains;* Wiley: New York, 1982. Reprinted in 1990 by Kreiger, Melbourne, FL.

3. Randall, J. C. *Polymer Sequence Determination;* Academic: New York, 1977.

4. Flory, P. J. *Principles of Polymer Chemistry;* Cornell University: Ithaca, NY, 1953; Chapter 3.

5. Price, F. P. In *Markov Chains and Monte Carlo Calculations in Polymer Science;* Lowry, G. G., Ed.; Dekker: New York, 1969.

6. Lopez-Serrano, F.; Castro, J. M.; Macosko, C. W.; Tirrell M. *Polymer* **1980,** *24,* 263.

7. McFarlane, R. C.; Reilly, P. M.; O'Driscoll, K. F. *J. Polym. Sci. Polym. Lett. Ed.* **1980,** *18,* 81.

8. Wilson, E. B. *An Introduction to Scientific Research;* McGraw-Hill: New York, 1952; p 204.

9. Burger, E.; Kuntz, I. *J. Polym. Sci.* **1964,** *2A,* 1687.

10. Kennedy, J. P.; Kelen, T.; Tudos, F. *J. Polym. Sci., Polym. Chem. Ed.* **1975,** *13,* 2277.

11. Joshi, R. M. *J. Macromol. Sci., Chem.* **1973,** *A7,* 1231.

2

Vibrational Spectroscopy of Polymers

Infrared (IR) spectroscopy is one of the most often used spectroscopic tools for the study of polymers. There are a number of reasons for the success of IR. The IR method is rapid and sensitive with sampling techniques that are easy to use. Also, the instrumentation is inexpensive, the operation of the equipment is simple, and service and maintenance of the equipment are not difficult. Finally, interpreting the spectra is not particularly difficult and can be learned easily, although better results may be obtained by an infrared professional.

The primary limitation of IR spectroscopy is in quantitative measurements. Although IR measurements are precise about the relative ranking of the amount of specific structures in a set of samples, making accurate absolute quantitative infrared measurements is a demanding process.

In this chapter, we will briefly review the basis of IR spectroscopy and the type of results to be expected from polymer samples. Much of the theory and practice of Raman spectroscopy is the same as IR, but for simplicity of presentation we will deal directly with IR. More specific comparisons of the two forms of vibrational spectroscopy are in Chapter 5.

Infrared Spectroscopy as an Identification Tool

Molecules consist of atoms held together by valence forces. These atoms vibrate by thermal energy, giving every molecule a set of resonance vibrations analogous to the resonance modes of mechanical structures. Accordingly, when impinging radiation passes through the material it is absorbed only at frequencies corresponding to molecular modes of vibration, and a plot of transmitted radiation intensity versus frequency shows absorption bands (absorption spectrum). IR spectroscopy measures the vibrational energy levels of molecules. The characteristic band parameters measured in IR spectroscopy are frequency (energy), intensity (polar character), band shape (environment of bonds), and the polarization of the various modes, that is, transition-moment directions in the molecular framework. Because the vibrational energy levels are distinctive for each molecule (and its isomers), the IR spectrum has often been called the *fingerprint* of a molecule. In the sense that the IR spectrum registers the most specific information concerning the molecule, this description is useful (*1* – *3*).

> *The IR absorption spectrum of a compound is probably its most unique physical property. Except for optical isomers, no two compounds having different structures have the same IR spectrum. In some cases, such as with polymers differing only slightly in molecular weight, the differences may be virtually indistinguishable but, nevertheless, they are there. In most instances the IR spectrum is a unique molecular fingerprint that is easily distinguished from the absorption patterns of other molecules.*
>
> –A. Lee Smith (*4*)

IR spectroscopy should be used whenever chemical specificity and selectivity are needed. As an identification tool, IR has no close spectroscopic competitor. Commercial laboratories have thousands of

1904—4/92/0019$07.00/1

spectra filed on computers, and "seek and identify" programs have been written to aid in the identification process (*5, 6*).

All of the identification procedures are based on the assumption that the compound is pure, but routine samples are seldom pure. Therefore, it is advisable to ascertain the purity of the sample before spectral analysis is undertaken. For polymer samples, additives such as fillers, antioxidants, lubricants, and mold release agents can generate spectral interferences (*7*). However, except for fillers and plasticizers, the process and stabilizer additives generally amount to 1% by weight of total polymer and contribute very little to the total spectrum.

Basis of Infrared Spectroscopy as a Structural Tool

Because our primary interest is in determining the structure of polymers, not polymer identification, except as a secondary responsibility, by IR spectroscopy we need to understand the molecular basis of IR spectroscopy.

Structural Dependence of Infrared Frequencies. An IR spectrum is ordinarily recorded in wavenumbers, ν, which is the number of waves per centimeter. The relationship between ν and the wavelength, λ, is

$$\nu\,(\text{cm}^{-1}) = \frac{10^4}{\lambda}\,(\mu\text{m}) \qquad (2.1)$$

which can also be written

$$\nu\,(\text{cm}^{-1}) = 3 \times 10^{10}\ \text{Hz} \qquad (2.2)$$

The wavenumber scale is directly proportional to the energy and vibrational frequency of the absorbing unit.

In wavenumbers

$$\Delta E_{\text{vib}} = hc_s\nu\,(\text{cm}^{-1}) \qquad (2.3)$$

where ΔE_{vib} is the vibrational energy level separation, h is Planck's constant (6.62×10^{-27} erg s), and c_s is the speed of light (3×10^{10} cm/s). The *fundamental* IR region arbitrarily extends from 4000 cm^{-1} to approximately 300 cm^{-1}. The *far-IR* region extends from 300 to 10 cm^{-1}, but the low IR source energy available makes this region generally inaccessible except with special instrumentation. These definitions of the different regions, which were originally based on instrumental requirements of dispersion and detectors, are less important now with Fourier transform IR (FTIR) spectroscopy, but the terms are still in common usage.

The vibrational energy levels can be calculated from first principles by using a technique called *normal coordinate analysis* (*8*), and as a result some of the factors influencing spectra have been discovered. These factors include bond stiffness and atomic mass, as well as the geometry and interaction between neighboring chemical bonds.

There is no obvious relationship between the observed IR frequencies and the chemical reactivity of a molecule. The bond stiffness is the important parameter, and this parameter has little to do with the chemical reactivity of the bond. However, every year someone discovers a new correlation between IR frequencies and reactivity. Don't let it fool you!

In Figure 2.1, the IR stretching frequencies are given for a series of common bonds between the hydrogen atom and other elements (*9*). This figure

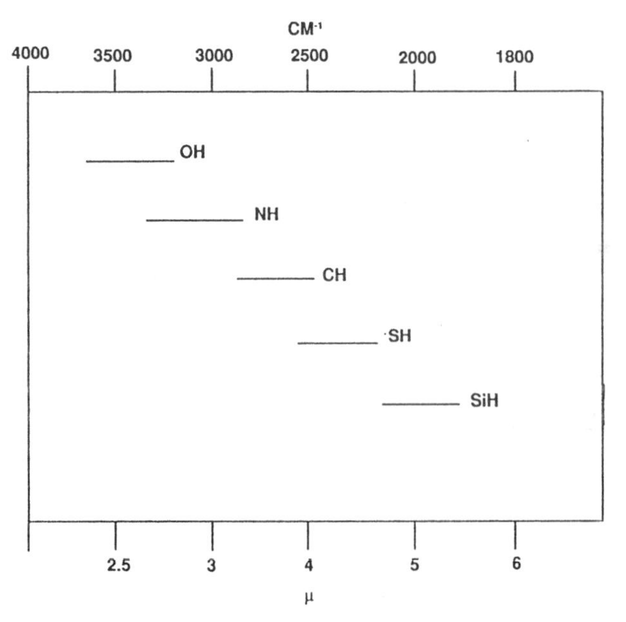

Figure 2.1. The IR stretching frequencies for a series of common bonds between the hydrogen atom and other elements. (Reproduced with permission from reference 9. Copyright 1948 The Royal Society of Chemistry.)

Figure 2.2. In IR spectroscopy, the model of a chemical bond is one of point masses connected by a harmonic spring. (Reproduced with permission from reference 2, Chapter 1. Copyright 1980 John Wiley Sons, Inc.)

shows the range of frequencies observed for OH, NH, CH, SH, and SiH bonds. What is the basis for these different frequencies?

The model of a chemical bond in IR spectroscopy is one of point masses connected by a harmonic spring, as shown in Figure 2.2 (*8*). The vibrational frequency of the stretching mode of a diatomic molecule A – B can be easily calculated by using eq 2.4

$$\nu\,(\mathrm{cm}^{-1}) = \left(\frac{1}{2\pi c}\right)\left[k_f \frac{[M_A + M_B]}{M_A M_B}\right]^{\frac{1}{2}} \quad (2.4)$$

where k_f is the force constant, or stiffness of the bond, and M_A and M_B are the masses of the two atoms. The force constant can be expressed in the form

$$k_f \propto N_b \left[\frac{X_A X_B}{d^2}\right]^{\frac{3}{4}} \quad (2.5)$$

where N_b is the bond order; d is the bond length; and X_A and X_B are the electronegativities of atoms A and B, respectively. An increase in the stiffness, k_f, of the bond increases the frequency. One obvious way to increase bond stiffness is to increase the bond order from sp^3 to sp^2 to sp hybridization. When this occurs, the frequency of the stretching vibration will inevitably increase.

If all molecules were diatomic, then IR spectra would not be very interesting, because there would be only a single IR mode when the atoms are different and no observable mode when the atoms are the same (see the heading *Infrared Selection Rules*). A molecule made up of N atoms can be visualized as a set of N masses connected by springs (chemical bonds). The vibrations can be resolved into $3N - 6$ normal modes. *Normal modes* are defined as modes of vibration where the respective atomic motions of the atoms are in "harmony"; that is, they all reach their maximum or minimum displacement at the same time. These normal modes can be expressed in terms of bond stretches and angle deformations (termed the internal coordinates) and can be calculated by using a procedure called normal coordinate analysis (*8*).

I am compelled to relate a story about one of the more famous IR spectroscopists—the late Professor Bellamy, who wrote several books on characteristic group frequencies. He said that when he looked at a new molecule and its complex IR spectrum, he would often have the urge to perform a normal coordinate analysis to help him with the spectral interpretation. He also said that when this happened he would lie down until the feeling passed! With modern computers, the normal coordinate analysis problem is much simpler but still is nontrivial to the inexperienced and untrained, so we will take the approach of Bellamy whenever we feel the need for normal coordinate analysis.

Consider molecules that are only slightly more complicated than a diatomic molecule, for example a methylene group. In fact, consider a special type of methylene group: one that is attached rigidly to something, perhaps a brick. This special molecule will be called "methylene brick". It has three atoms: a carbon and two hydrogens. Because it is attached to a brick, the special methylene unit cannot translate freely (this will simulate the situation in a polyethylene chain).

As you have probably guessed, I want to examine the vibrational motions of the methylene units of a polyethylene chain, but for now we will ignore the adjoining carbons.

Because normal coordinate analysis will not be used to solve the problem (Painter et al. have written a very useful book on this subject [*10*]) the vibrational analysis must begin with intuitive thinking. First, the C – H bonds stretch or compress like springs in relation to each other when vibrating.

We can simulate the motion by using our bodies and letting the hands be the hydrogen atoms and the body be the carbon. Don't laugh until you have tried it!

Because the carbon is attached firmly to the brick, the atoms cannot fly off into space but must reach their maximum extension and return to their minimum point in harmony, and continue this oscillation or vibrational motion. This motion is called vibrational C – H stretching. But, there are two C – H bonds, and each undergoes the aforementioned vibrational motion in phase with the other while they are connected to the same carbon atom. Because there are two C – H bonds, there must be two different types of stretching vibrations resulting from two identical C – H groups connected together. Consequently, there is some restriction on their individual motions, and the result is two normal modes. In one case, the maximum displacements of the C – H groups occur at the same time (like moving our hands out together at the same time), and the motions are in-phase, as shown in Figure 2.3. This vibrational mode is the *symmetric* C – H stretching mode. The second C – H stretching mode occurs when the two C – H bonds reach their maximum displacement completely out of phase; that is, one is extended by stretching, and the other is compressed.

> One of the C – H bonds is compressed to its minimum, and the other is extended to its maximum. One hand moves away from the body, while at the same time the other hand is pulled toward the chest.

This mode is the *asymmetric* C – H stretching mode.

The allowed motions are restricted by the fact that the vibrational motion cannot displace the center of

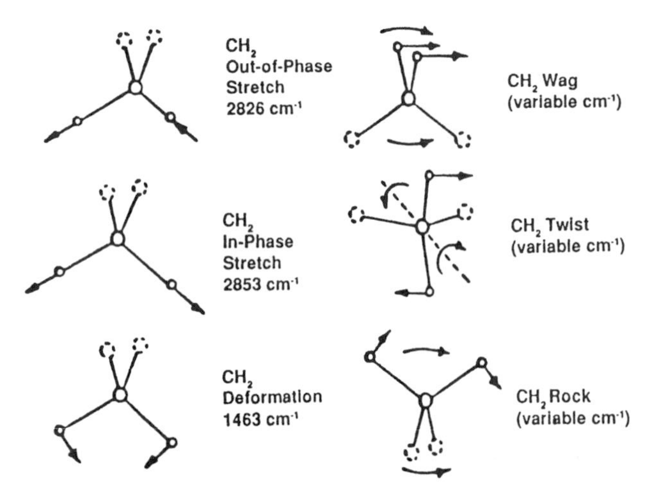

Figure 2.3. Vibrational motions of the methylene group with the IR frequencies. (Reproduced with permission from reference 2, Chapter 1. Copyright 1980 John Wiley & Sons, Inc.)

mass of the molecule. Considering this restriction, the displacement of the carbon atom attached to the two moving hydrogen atoms must compensate for the mass displacement of the hydrogens. Therefore, the displacement of the carbon is quite different in the in-phase and out-of-phase stretching modes described, even though the hydrogens are moving in an identical fashion, except that they have different phases relative to each other (Figure 2.3). Because the carbon is moving in a different manner to compensate for the hydrogen motion, the resulting vibrational energies of these two C – H stretching modes are different, and the two stretching modes are not degenerate; that is, they absorb at different IR frequencies. This intuitive result is confirmed by experiment. The symmetric stretching mode of the methylene in polyethylene occurs at 2853 cm^{-1}, but the asymmetric mode appears at 2926 cm^{-1}.

The motion of the C – H groups away from the carbon atom is motion in the *z* direction. Now consider the motion of the hydrogens toward each other–a bending of the H – C – H angle of the methylene group. (In the plane of the methylene group, this would be the *y* direction, to be more specific.) From a molecular point of view, this involves the bending or deformation of the angle between the two C – H bonds. This motion may be visualized by extending both your arms and moving the hands toward each other in a clapping motion–but the hands cannot touch because of natural restrictions of atomic size and packing. Spectroscopists call this vibrational motion the *methylene bending motion* because the angle between the two hydrogens is being bent. This motion involves the *asymmetric* movement of the hydrogens toward each other.

There also is a motion where the hydrogens move in the *y* direction together. If you use your hands at the end of your extended arms to represent the hydrogens for this motion, you will notice that it is necessary for you (the carbon atom) to wag or rotate your body to accomplish the motion. Hence, the spectroscopists call this mode the *methylene wagging mode*, and it has a different vibrational frequency than the bending mode previously described. These two modes are the *symmetric* and *asymmetric bending modes*.

Additional vibrational modes, with their observed IR and Raman frequencies, are shown in Figure 2.3.

The point here is that the same chemical group can have a number of vibrational modes with different energies arising from the different internal modes. Thus, to identify a given chemical group in a polymer, the different IR modes can be used to confirm the existence of the group without having to rely on a single observed mode (*3*).

For the study of polymers, the nature of the coupling, not only between two C−H groups on the same carbon, but also between methylene groups on the same carbon chain, must be understood (*10*). See the heading *Coupled Infrared Vibrations as a Polymer Structure Probe*.

Infrared Selection Rules. The first requirement for IR absorption is that a frequency in the impinging source of IR radiation must correspond to the frequency of the vibration as expressed in eq 2.3. Frequency matching is a necessary condition for absorption but not the only one. An additional requirement is some mode of interaction between the impinging IR radiation and the molecule (there must be something to catch the radiation). Even if the IR radiation has the same frequency as the normal vibration of the molecule, it will be absorbed only under certain conditions. The rules determining optical absorption are known as *selection rules*.

For simplicity, let us take a classical approach to selection rules. An oscillating dipole can emit or absorb radiation. Consequently, the periodic variation of the dipole moment of a vibrating molecule results in the absorption or emission of the same frequency as that of the oscillation of the dipole moment (Figure 2.4). The intensity of the absorption or emission is proportional to the square of the change in the dipole moment. So The IR intensity, I, of the kth mode is

$$I_k = C\left[\frac{\partial \mu}{\partial Q}\right]^2 \quad (2.6)$$

where C is a proportionality constant, μ is the dipole moment, and Q is the displacement coordinate of the motion. If the dipole moment is large, (i.e., a highly polar group of atoms with different electronegativities), then the change in the dipole moment with bond stretching will be large and the IR absorption intensity will be high. If the dipole moment is small, (i.e., atoms of similar electronegativity), the IR absorption intensity will be small. If there is no dipole moment, then no IR absorption can occur. In other words, it is possible for molecular vibrations having the proper IR frequency to occur, but be totally invisible to the IR beam. For example, consider the two in-plane stretching vibrations of CO_2 (Figure 2.5). The molecular dipole moment of the symmetric unperturbed molecule is zero when the dipole displacements of both bonds are added. In the totally symmetric vibration (Figure 2.5a), the two oxygen atoms move in phase successively from and toward the carbon atom. The change in dipole moment of one C−O bond is balanced by the change in the opposite C−O bond.

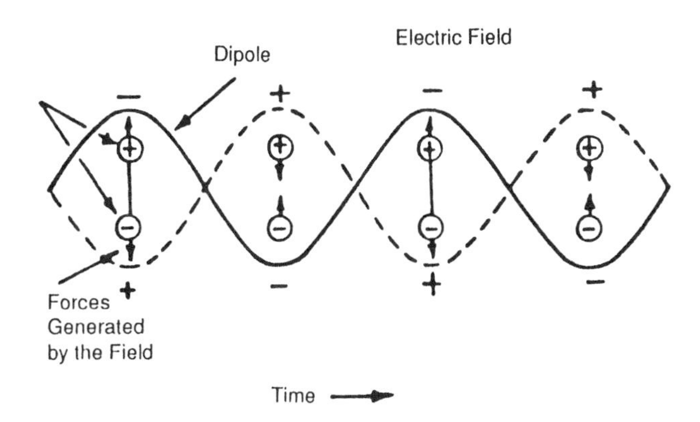

Figure 2.4. The periodic variation of the dipole moment of a vibrating molecule results in the absorption or emission of the same frequency as that of the oscillation of the dipole moment. (Reproduced with permission from reference 2, Chapter 1. Copyright 1980 John Wiley & Sons, Inc.)

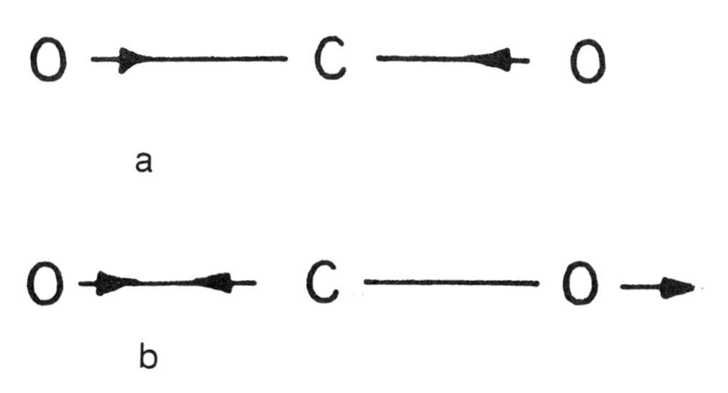

Figure 2.5. The two in-plane stretching vibrations, symmetric (a) and asymmetric (b), of CO_2. (Reproduced with permission from reference 10. Copyright 1982 John Wiley & Sons, Inc.)

The symmetry of the molecule is maintained in this vibration, and there is no total change in dipole moment. For this mode of vibration, there is no IR absorption (but as will be seen later, strong Raman scattering). However, in the asymmetric stretching vibration (Figure 2.5b), the symmetry of the molecule is perturbed, there is a change in the net dipole moment, and IR absorption occurs.

For the CO_2 molecule, the IR selection rules can be obtained by mere inspection of the vibrations occurring. For more complicated molecules, the selection rules can be determined from symmetry considerations (*8*), because symmetry is helpful in determining the type of motion that is allowed. In fact, the normal modes fall into different symmetry species. For symmetrical molecules, the normal modes and the symmetry modes are nearly the same. The symmetry analysis is quite simple but highly tedious and generally uninteresting. For simple types of low-molecular-weight molecules, tables are available with the results for the various geometries (*11*). However, if you wish to learn the process, a number of excellent books are available (*11, 12*). For polymers, the symmetry relations for most simple vinyl systems also are available (*10*).

A simple rule is that polar bonds yield strong IR absorption, and nonpolar bonds do not. If this rule is applied to polymer molecules, it can be determined that for hydrocarbon polymers with carbon backbones, the IR absorption of the backbone will be weak, and this result has been observed under experimental conditions. However, substituted groups like C–H, C–F, and C=O will have polar bonds because of differences in electronegativity, and the IR absorption will be strong. Again this speculation has been observed under experimental conditions.

To further complicate the interpretation of the observed IR spectra, unwanted bands appear in the spectra. These bands are called *overtone* and *combination bands*, and they result from the effect of anharmonic terms in the potential energy. Overtone bands appear at approximately $2\nu_i$, where i is one of the fundamental modes. Combination bands appear at approximately $\nu_i + \nu_j$ where i and j are fundamental modes. Some methods of symmetry analysis can help determine which overtones might be allowed, but these methods will not be investigated here (*11*). Before making structure assignments to the observed IR bands, it is usually a good idea to consider the possibility that the band under consideration may be an overtone. To determine if this is the case, multiply the frequencies of some of the stronger lower frequency bands by 2 and see if a match is found. The overtone and combination bands are always weaker than the corresponding fundamental bands.

Infrared Intensities. The IR intensity usually refers to the integrated absorbance

$$A_s = \left[\frac{1}{cb}\right] \int_{\text{band}} \ln\left(\frac{I_0}{I}\right) d\nu \qquad (2.7)$$

If c is the concentration in moles per liter (millimoles per cubic centimeter), b is the path length in centimeters, and $d\nu$ is in reciprocal centimeters, A_s is given in centimeters per millimole (sometimes called *darks*). In these units, the value of A_s is approximately 1000 cm/mmol for a very intense absorption and approximately 0.1 cm/mmol for a very weak absorption.

When IR radiation is absorbed by a molecule, the intensity of the absorption depends on the movement of the electronic charges during the molecular vibrations. Therefore, the IR intensities should provide information about the electronic charge distributions in molecules and about how the electrons redistribute themselves during molecular vibrations.

The electric dipole moment of a molecule is given by

$$\mu_0 = \sum_i e_i X_i \qquad (2.8)$$

where e_i is the charge of the ith particle and X_i is the position of the ith particle given in the space-fixed coordinate system. The electric dipole moment, μ, for a molecule undergoing vibration can be expanded into a power series in the normal coordinates, Q_s,

$$\mu = \mu_0 + \sum \left[\frac{\partial \mu}{\partial Q_s}\right]_0 Q_s + \ldots \qquad (2.9)$$

where the subscript 0 denotes the value of the equilibrium position, and μ_0 is the static molecular dipole moment. In practice, only the first two terms are necessary, so μ may be evaluated by knowing the values of μ_0 and the parameters $\partial\mu/\partial Q_s$.

The IR intensity A_s is related to the molecular parameters as follows (*13*):

$$A_s = \left[\frac{8\pi^2 N_A}{3hc_s} \right] \nu_s \left| \frac{\partial \mu}{\partial Q_s} \right|^2 \qquad (2.10)$$

where N_A is Avogadro's number, h is Planck's constant, c_s is the speed of light, and ν_s is the frequency of the band center (cm^{-1}). This relationship can be expressed inversely as

$$\left| \frac{\partial \mu}{\partial Q_s} \right| = 0.03200\ (A_s)^{\frac{1}{2}} \qquad (2.11)$$

Hence, each IR intensity can be related to the dipole moment derivative with respect to the Q_s normal coordinate. The sign of the derivative of the bond moment is always ambiguous, because it is a squared function of the IR intensities. The determination of the sign must come from evidence outside the intensity measurement itself.

Each valence bond is characterized by a bond dipole moment, μ_n, and the derivative of this bond moment with respect to bond length, $\partial\mu_n/\partial r_n$. For the C – H bond, μ_0 (CH) is positive in the sense $^-C-H^+$, and $\partial\mu/\partial r$ (CH) is negative, indicating that μ (CH) decreases when the bond stretches.

The IR intensities can be calculated for some molecules, but the calculations are difficult and involve a number of assumptions about the signs and orientations of the dipole moments (*13*).

> As we have taken the Bellamy approach to calculation of the frequencies using normal coordinate analysis, it will not surprise you that we are taking a similar approach to the calculation of the IR intensities. This is not to suggest that such calculations are not useful or rewarding; such detail is not required for the types of structural investigations that are possible with polymers.

For quantitative analysis, the IR intensities for each vibrational mode are given by Beer's law, $A = abc$, where A is the absorbance, a is the absorptivity, b is the path length, and c is the concentration of the characteristic group of the vibration. It is necessary to determine a before a quantitative analysis can be made. This determination can be done by calibration using an independent chemical or physical technique or by proper deconvolution of the spectra of mixtures containing the desired component. We will discuss the quantitative aspects of IR spectroscopy later (see Chapter 3).

As a tool for the discrimination of structural elements, the IR intensities are significantly more sensitive to small changes in structure or bond environment than are the IR frequencies. Intensities would be expected to be a most valuable structural tool in IR spectroscopy. The problem with using IR intensities to determine structural elements in solids and polymers is the difficulty in obtaining accurate measurements and in transferring such measurements from one IR instrument to another. With the advent of FTIR and the analog-to-digital converter, the linearity of the absorbances is much greater than with the optical comb used in dispersion instruments.

> My belief is that we are approaching the time in IR spectroscopy where measurements of IR intensities should be made with the objective of using the results for the refinement of structure and as an additional tool to complement the IR frequency correlations. Work must be done in this area, but the rewards will be high!

However, a reasonable IR spectrum for a polymer may be constructed empirically by using the measured intensity values, A_j, of the characteristic vibrational modes of similar systems. The integrated intensities measured by transmission spectroscopy(*14*) are indicative only of the possibility of combining characteristic group frequencies with characteristic group intensities for further specificity in identification of polymer structural components.

Infrared Dichroism in Solids. For each vibrational mode of a molecule, there is a transition-moment vector, **M**. Because the transition moment is a vector quantity, it has both magnitude and direction. Each of the IR vibrational modes will have a transition-moment vector at some angle, α_ν, to the major axis of the molecule. In Figure 2.6, the transition-moment directions of some vibrational modes are illustrated (the symbols + and – denote the movement of atoms perpendicular to the plane of the paper) (*15*). The intensity of the IR absorption band depends on the angle that the electric vector of the incident radiation makes with the transition moment. For gases, liquids, and solutions, the movement of

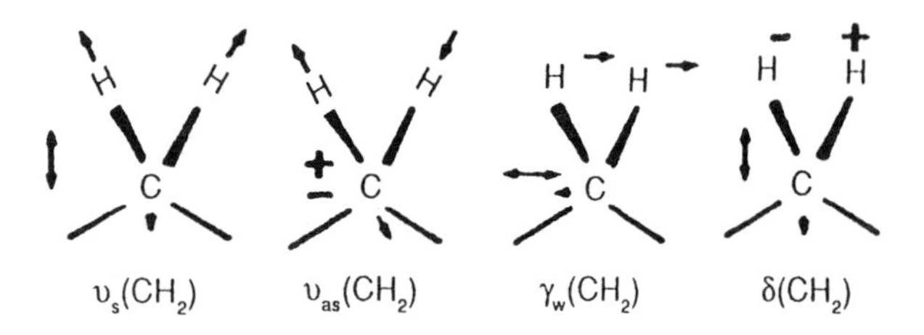

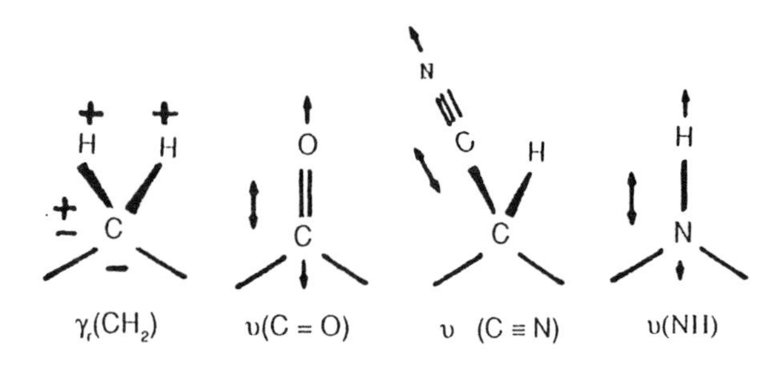

Figure 2.6. The transition-moment directions of some vibrational modes. (Reproduced with permission from reference 15. Copyright 1980 Marcel Dekker, Inc.)

the molecules is sufficient to yield a random orientation of these transition-moment vectors. For solids, on the other hand, there is often a preferred orientation of the molecules.

When IR radiation is plane-polarized such that when the electric vector, **E**, is parallel to the transition-moment vector, **M**, strong absorption occurs and when **E** is perpendicular to **M**, no absorption occurs. For solids, there can be an orientation selection rule. If the matching frequency of the incident light is known, and if the proper change in the dipole moment for IR absorption occurs, it is possible to observe no absorption with linearly polarized light if the sample is perfectly oriented.

The absorption of each mode (I) is proportional to the square of the dot product of the electric (**E**) and transition-moment (**M**) vectors

$$I = C(\mathbf{E} \cdot \mathbf{M})^2 = (EM \cos \Theta)^2 \quad (2.12)$$

where C is a proportionality constant and E and M are the magnitudes of the electric field of the incident beam and the transition moment, respectively. The angle, Θ, is the angle between the two vectors. For an oriented solid sample, in order for the light to be absorbed, a component of the oscillating electric field vector of the incident light must be oriented in a plane parallel to the electric dipole transition moment (Figure 2.7). Light polarized perpendicular to the dipole transition moment will not be absorbed.

When the absorbing groups are oriented as in solids, they exhibit IR absorptions that depend not only on how many groups are present in the sample but also on how the groups are oriented with respect to the beam.

> A simple test of whether a solid sample is oriented can be carried out in the IR instrument by simply tilting the sample out of the plane of the beam, recording the spectrum, and comparing the observed IR bands. If the sample has random orientation, the ratios will not change when the film is tilted, but if orientation is present, the ratios will be a function of the angle of tilt (*16*). This simple test should be performed on any solid sample which is being considered for quantitative analysis. If orientation is present, throw the sample out and prepare a new one. If you cannot prepare an unoriented sample, carry out IR dichroic measurements to correct for the inherent orientation.

By using linearly polarized IR radiation, the orientation of the functional groups in a polymer system can be measured.

Measurement of IR linear dichroism requires light polarized both parallel and perpendicular to a fixed reference direction of the sample. For parallel polarized light, the absorbance is termed $A_{\parallel}$, and the absorbance with perpendicular polarized light is termed $A_{\perp}$. The dichroic ratio, R, is defined as

$$\mathrm{R} = (A_{\parallel}) / (A_{\perp}) \quad (2.13)$$

For random orientation, $R = 1$. The measured absorption bands are generally classified as parallel (π bands), or as perpendicular (σ bands) depending on whether R is greater or less than 1. This classification of the dichroic behavior is helpful in assigning the various modes to the symmetry types of the normal modes in solids (*17, 18*).

For unidirectional molecular orientations, such as for uniaxially drawn polymers, these dichroic parameters can be related to the Herman orientation function, F. This quantity is equivalent to the second moment of the orientation distribution function for the molecular axis and is given by (*19*)

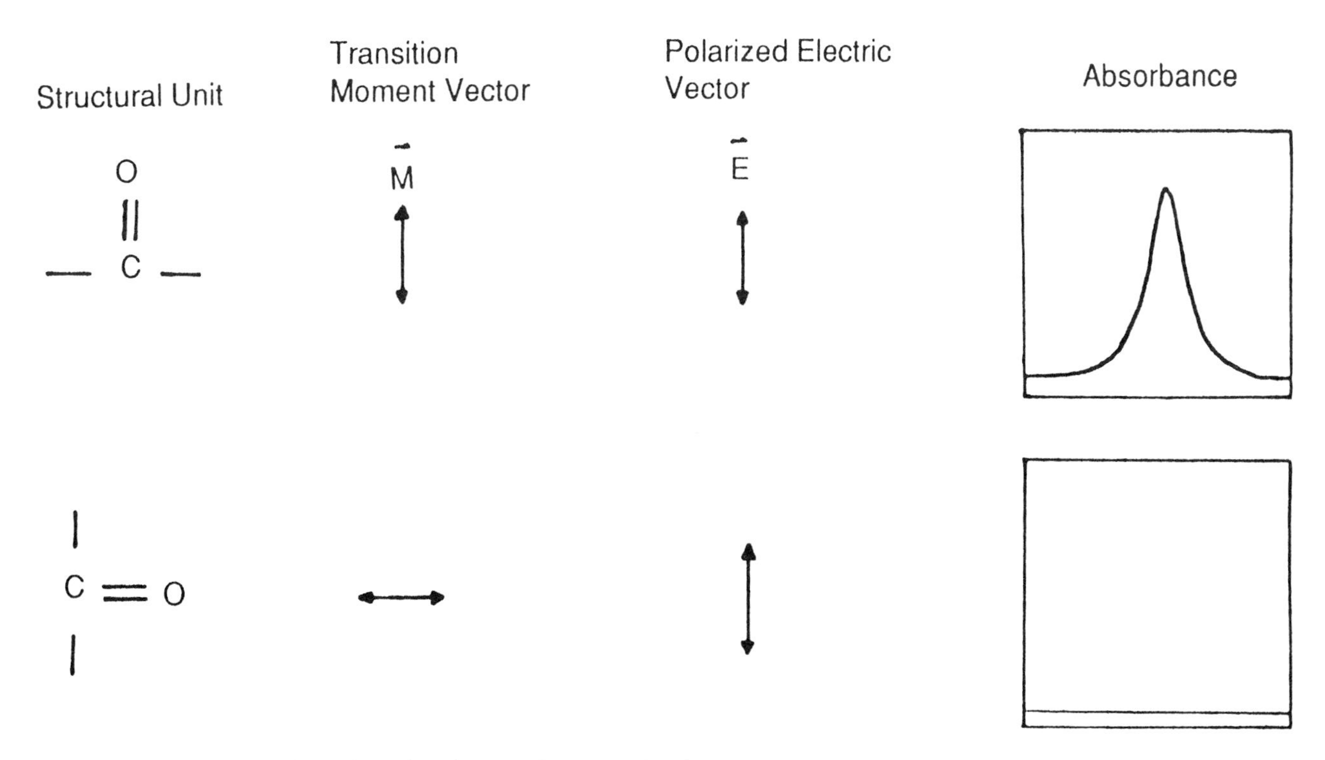

Figure 2.7. Linearly polarized IR absorbance of structural units.

$$F = \frac{3 < \cos^2 \theta > - 1}{2} \tag{2.14}$$

The orientation angle θ is the angle between the draw direction and the local molecular chain axis (Figure 2.8). The Herman function is equal to 1 when the chain axis is parallel to the film orientation, 0 when the system is randomly oriented, and ½ when the chain axis is perpendicular to the film orientation direction. This function can be calculated from measurements of the dichroic ratio by using (*20*)

$$F = \frac{(R - 1)(R_0 + 2)}{(R + 2)(R_0 - 1)} \tag{2.15}$$

where R_0 is the dichroic ratio for perfect uniaxial order. The value of the constant R_0 is unknown and can be different for every IR band studied (*20*).

For perfect uniaxial order, it is assumed that the polymer chains are all oriented parallel to the draw direction, and the transition moments associated with the vibrations lie on a cone with a semiangle ψ to the chain axis direction (Figure 2.8) (*21*). The dichroic ratio is then expressed by

$$R_0 = 2 \cot^2 \psi \tag{2.16}$$

As ψ varies from 0 to $\pi/2$, R_0 varies from ∞ to 0. No dichroism ($R_0 = 1$) will be observed for $\psi = 54°44'$ (the *magic angle*).

In practice, the orientation of the polymer chains is imperfect and is often described in terms of f_p, the fraction of polymer that is perfectly oriented, while the remaining fraction $(1 - f_p)$ is randomly distributed. The dichroic ratio is

$$R = \frac{[f_p \cos^2 \psi + (1/3)(1 - f_p)]}{[(1/2 f_p) \sin^2 \psi + (1/3)(1 - f_p)]} \tag{2.17}$$

To solve this problem an independent, quantitative experimental method for determining the unknown constant, f_p, is needed.

An alternate method of defining the imperfect orientation is to suppose all the molecular chains are displaced from parallelism with the draw direction by the same angle θ (*15*). The expression for R then becomes

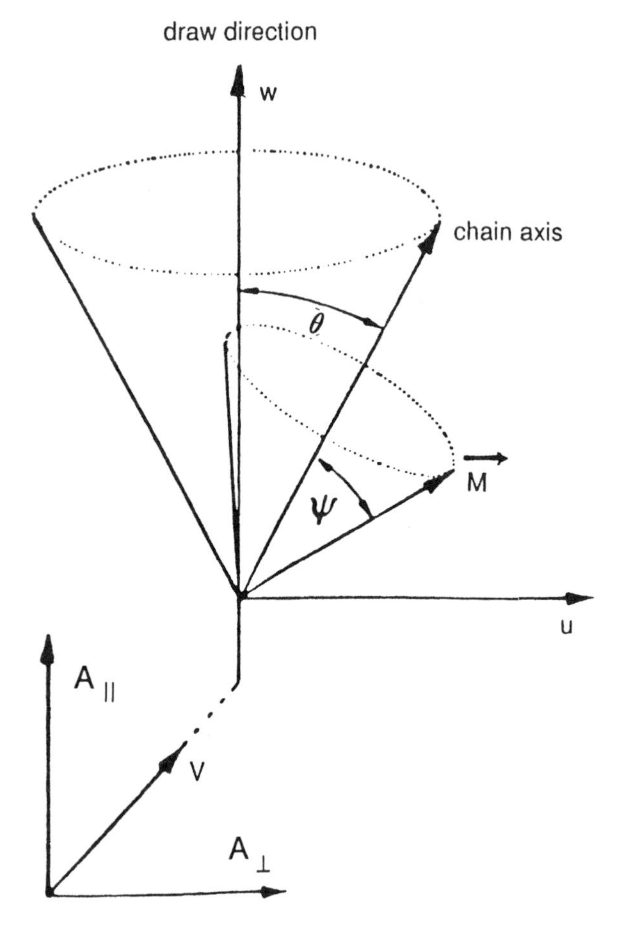

Figure 2.8. The orientation angle, θ, is the angle between the draw direction and the local molecular-chain axis. (Reproduced with permission from reference 21. Copyright 1984 Springer – Verlag.)

$$R = \frac{[\,2\cot^2\psi\cos^2\theta + \sin^2\theta\,]}{\left[\cot^2\psi\sin^2\theta + \dfrac{(1+\cos^2\theta)}{2}\right]} \quad (2.18)$$

If the direction of the transition moment with respect to the chain axis is known, the average orientation of the chain segments can be determined from the measured dichroic ratio.

Fortunately, the Herman orientation function, F, can be determined by a number of independent techniques, including X-ray diffraction, birefringence, sonic modulus, and refractive index measurements. These methods, coupled with IR dichroic measurements of absorption, can be used to calculate quantitative values for the transition-moment angles by using eq 2.15. A plot of F measured by X-ray diffraction versus $(R - 1)/(R + 2)$ will be linear with a zero intercept. A least-squares evaluation of the data from this line will yield the slope, $(R_0 + 2)/(R_0 - 1)$. When α_ν is 0° (parallel to the molecular chain axis) the slope equals 1.0 and when α_ν equals 90°, the slope equals −2.0. A quantitative value of the transition-moment angle can then be calculated. Transition-moment angle measurements have been published for isotactic polypropylene and some of the results are as follows (*22*):

Frequency (ν)	α_ν	*Frequency* (ν)	α_ν
928	90°	1220	90°
973	18°	1256	0°
998	18°	1307	0°
1045	0°	1363	90°
1103	90°	1378	70°
1168	0°		

SOURCE: Reproduced with permission from reference 22. Copyright 1981.

If it is assumed that the transition moments are parallel to the chemical bonds, the results can be used to determine the structural angles of the molecule or polymer.

Once the transition-moment angles have been identified for a particular chemical group of the polymer, it is possible to measure the orientation of the polymer molecules in a sample by measuring the IR dichroism at the appropriate absorption frequencies. By using the data for polypropylene just presented, the Herman orientation function for samples of polypropylene that are oriented can be calculated from the dichroic measurements. The results are shown in Table 2.1 (*23*).

The advantage of using IR spectroscopy for measuring orientation in polymers is that it can detect the orientation of the geometric isomers in the disordered or amorphous regions, as well as the orientation of the preferred isomer found in the ordered or crystalline phase, because they generally appear as separate IR bands. The other experimental techniques such as x-ray diffraction (crystalline phase only) and refractive index (crystalline and amorphous phases) cannot easily separate the contributions of the separate phases. This attribute of IR dichroism is particularly

Table 2.1. Dichroic Ratios, Orientation Function, and Average Angles of Orientation for Drawn Polypropylene Films

Draw Ratio[a]	R	F	θ (°)
1	1.015	−0.010	55.1
1.5	0.745	0.186	
4	0.536	0.366	40.6
7	0.056	0.918	13.5
8	0.51	0.926	12.9
9	0.036	0.947	10.8

NOTE: The frequency was 1220 cm^{-1}, and the band angle was 90°.

[a]The number shown is the ratio of the length of the drawn film to the initial length of the film.

SOURCE: Adapted from reference 23. Copyright 1987.

important in multiphase systems (*24*) and blends (*25*).

The classical dichroic ratio is a two-dimensional measurement, but the actual sample is usually oriented in three dimensions. To completely characterize the orientation, three-dimensional IR measurements are required (*26–29*). The macroscopic coordinate directions of a uniaxially drawn film are defined by y in the stretching direction, x in the transverse direction, and z in the direction that is normal to the film (Figure 2.9). The orientation in three dimensions is completely characterized if the A_z component of the absorption is determined together with the A_x and A_y components. The classical dichroic ratio in this reference system is

$$R_{yx} = (A_y)/(A_x) \tag{2.19}$$

If either R_{zy} or R_{zx} is known in addition to R_{yx}, the remaining ratio can be found as a function of the other two: $R_{zy} = R_{zx}/R_{yx}$.

The important parameter to be determined is the structure factor A_0 (*26*)

$$A_0 = \frac{A_x + A_y + A_z}{3} \tag{2.20}$$

which represents the absorbance of the band exclusive of contributions due to orientation of the polymer. This factor is proportional to the amount of chemical structure in the oriented solid giving rise to the IR absorption and is the same quantity measured

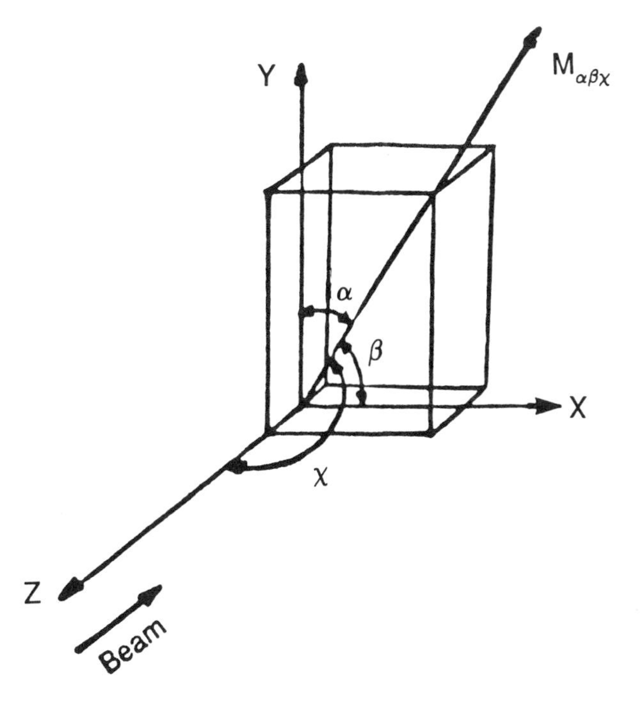

Figure 2.9. The macroscopic coordinate directions of a uniaxially drawn film are defined by y *in the stretching direction,* x *in the transverse direction, and* z *in the direction normal to the film.*

in gases and liquids, which have no preferred orientation. The difficulty in measuring A_0 in solids is that it is necessary to know the absorbance, A_z, in the thickness direction, which for polymer films is very small. This thickness direction absorbance may be obtained by tilting the sample at some angle, α, (Figure 2.10) and determining the component in the z direction through the relationship (*26*)

$$A_z = \frac{A_\alpha \cos \alpha_t - A_y}{\sin^2 \alpha_t} + A_y \tag{2.21}$$

A_α and A_y are calculated from the spectra. The true angle, α_t, that the beam makes with the sample is found by correcting the measured angle for the refraction resulting from the refractive index (*26*).

For three-dimensional measurements, the orientation parameters $A_x/3A_0$, $A_y/3A_0$, and $A_z/3A_0$ can be used to determine the fraction of molecules oriented in the three mutually perpendicular directions. In uniaxially oriented samples, which are often encountered by drawing in one direction, the structure factor becomes

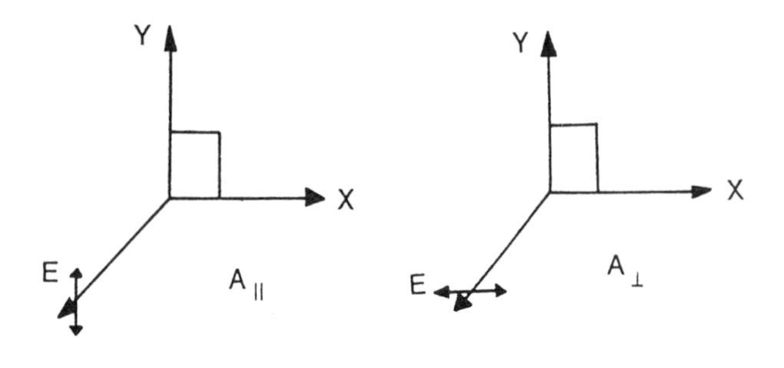

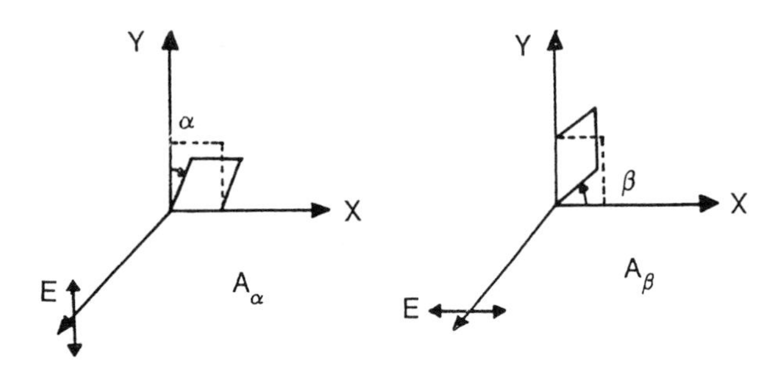

Figure 2.10. The thickness-direction absorbance may be obtained by tilting the sample at some angle, α, and determining the component in the z direction. (Reproduced with permission from reference 26. Copyright 1967 John Wiley Sons, Inc.)

$$A_0 = (1/3)(A_{\parallel} + 2A_{\perp}) \qquad (2.22)$$

When a low-molecular-weight liquid substance is adsorbed or mixed into a polymer sample, and the polymer is oriented by stretching or other methods, the entrapped substances become aligned with the polymer chains and exhibit dichroic behavior (*30, 31*), even though these substances may be liquid. The stretched polymers behave as anisotropic solvents, and the polarization properties of the molecules can be determined. Such measurements are useful for determining the structures of organic materials, and should be useful for the study of polymer–solute interactions because increased interaction results in increased orientation in the solute (*25*).

Characteristic Group Frequencies in Infrared Spectroscopy and Interpretation of Polymer Spectra Using Group Frequencies

Each polyatomic molecule is expected to have 3*N* vibrations (actually 3*N* − 6 when the three nonabsorbing or zero energy translations and rotations of the molecule are counted) where *N* is the number of atoms in the molecule. As indicated previously, these vibrational modes result in a unique spectrum or fingerprint of the molecule. Careful cataloging of the IR spectra reveals an interesting correlation between the presence of certain chemical groups in the molecules and the appearance of specific IR absorption frequencies.

> In spite of the apparent complexity of the IR spectrum, whenever some specific chemical groups are present in a molecule, the resulting spectrum has an absorption in a narrow and predictable region. It appears that whenever a specific chemical group is present, it identifies itself. IR spectroscopists started using these "characteristic group frequencies" as structural probes for chemical groups in newly prepared molecules with unknown chemical structure. How is this possible? The spectrum is highly specific to the molecular framework of the molecule and at the same time is highly reproducible for specific functional groups within the molecular framework. This paradox of IR needs further investigation.

The vibrational modes of complex molecules fall quite naturally into two distinct categories: *internal modes* and *external modes*. As the names imply, the internal modes involve predominantly only a few selected atoms, and the external modes generally involve the motions of all of the atoms in the entire molecule. The internal motions, such as stretching of chemical bonds or bending of bond angles in triatomic groups, are relatively unaffected by the nature of the rest of the molecule and have nearly the same vibrational energy regardless of the attached molecular skeleton. When the internal modes are widely separated from each other and from the external modes, they have a characteristic group frequency that is relatively unaffected by the chemical nature or architecture of the rest of the molecule. In other words, these characteristic group modes are *uncoupled* from the rest of the molecule and vibrate as if no other atoms were around. When a specific group is found in a number of different molecules, its uncoupled internal vibrations appear in the same narrow frequency region regardless of the type of molecular attachment. For example, a mode appears between 3300 and 2700 cm^{-1} for a C–H bond regardless of the nature of the rest of the molecule. The particular

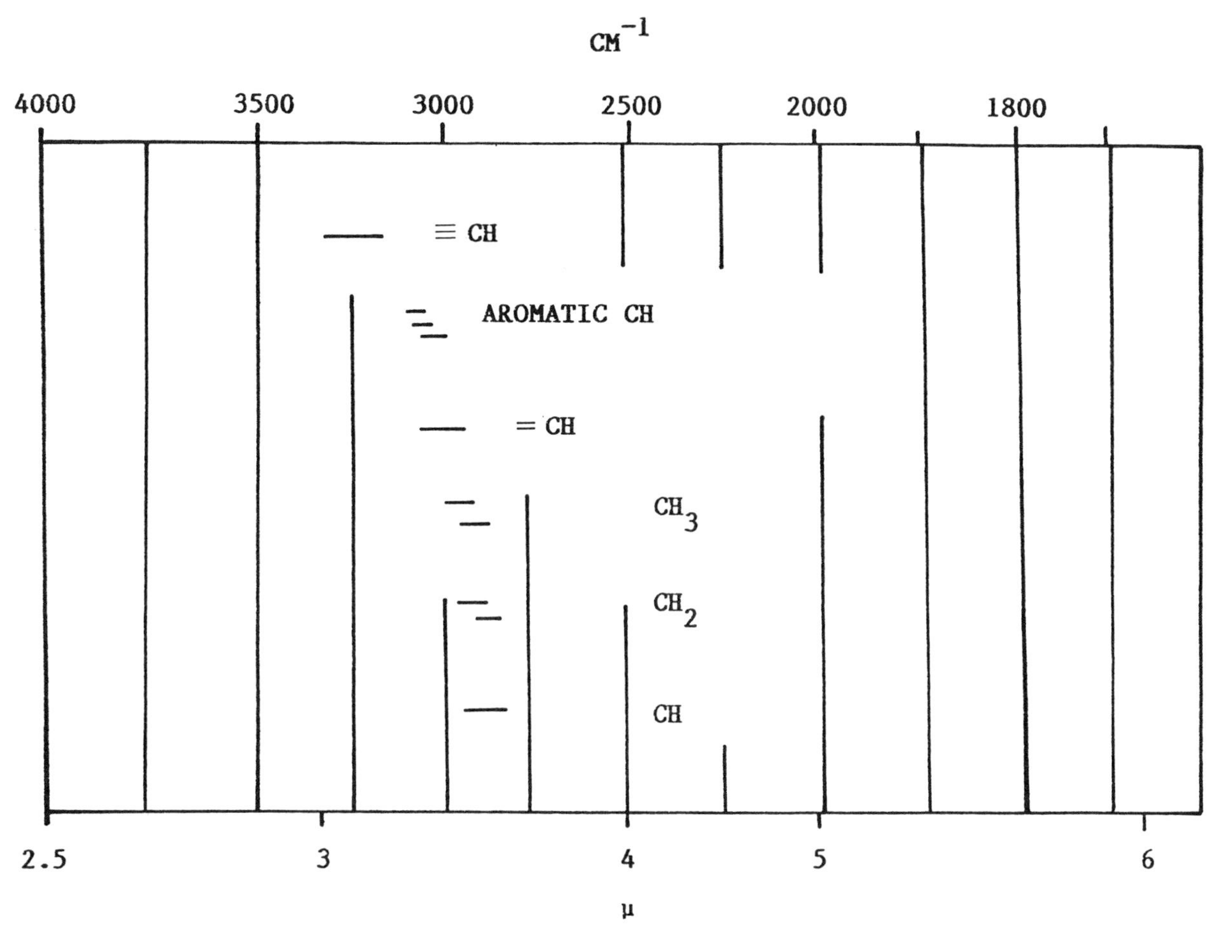

Figure 2.11. The particular type of C – H bond results in a specific absorption frequency within the overall C – H absorption range. (Reproduced with permission from reference 9. Copyright 1948 The Royal Society of Chemistry.)

type of C – H bond, that is, aryl or alkane, results in a more specific absorption frequency within this range, as shown in Figure 2.11.

This group frequency concept popularized IR spectroscopy. Catalogs of characteristic group frequencies were reported in the literature, and spectroscopists tried to outdo each other with new and exciting structural assignments of group frequencies. Entire books were written on this subject (*32 – 34*), and IR spectroscopy became a valuable tool for determining the chemical structure of molecules. The polymer chemists benefited as well because the characteristic group frequencies generally appear in the same region for polymers as for other molecules (*1, 7*). Sometimes, specific couplings can occur with regularly ordered chemical functional groups in polymers, and this condition shifts the group frequencies and leads to valuable microstructural information about polymers. This will be covered in detail in the next section, *Coupled Infrared Vibrations as a Polymer Structure Probe*.

> *Someone once said that there are three approaches to interpreting the IR spectrum of an unknown material. The "idiot" approach is to simply plug the spectral data into the computer and let it search the file for a matching spectrum. This works if the unknown is in the file. The "genius" method is for the spectroscopist to derive the structure from his knowledge of group frequencies. It works well if the spectroscopist is a genius (or at least has an excellent memory). The "autistic" approach is to send the sample to mass spectroscopy. In practice, a combination of all three works best.*
>
> – A. Lee Smith (*4*)

The first approach to interpretation of an IR spectrum is to consider the spectrum as the superposition of a number of group frequencies (*35*). The second is to confirm the existence of specific chemical groups by looking for other modes of this group, that is, bending, twisting, and wagging modes.

It is unlikely that you will be able to identify all of the IR bands in any spectrum of a polymer—there are simply too many. Let me return to our interpretive expert, Professor Bellamy. When he was asked for the assignments of all the bands in an IR spectrum, he responded that the questioner needed a psychiatrist instead of a spectroscopist. Nobody in his right mind expects to know the source of all of the bands.

Coupled Infrared Vibrations as a Polymer Structure Probe

To determine the microstructure of polymers, IR modes that reflect the "connectivity" between the repeating units are necessary, as was indicated in Chapter 1. When chemical groups with the same structure and energy are repeated on a chain in a regular fashion, they are potentially capable of resonating or coupling their vibrational motions. This intramolecular vibrational coupling can lead to the development of a series of resolvable vibrational modes characteristic of the length of the *N*-ads (*36*). This phenomenon has been confirmed by analysis of the spectra of ordered polymers.

Vibrations of the Infinite Linear Monatomic Chain

Consider an infinite linear array of atoms of mass m separated by a distance d, as shown in Figure 2.12. This model has no IR spectrum because it has no dipole moment, but it will serve to demonstrate an important point. Consider the longitudinal displacement of the nth particle from the equilibrium position, and let the potential energy of the displacements of the atoms be determined by the force f acting between adjacent particles. Analysis yields the frequency of the motion as (*36*)

$$\nu\,(\mathrm{cm}^{-1}) = (1/c_s\pi)\,(f/m)^{\frac{1}{2}}\,|\,(\sin\phi)/2\,| \quad (2.23)$$

where ϕ is the phase angle with the range $-\pi < \phi < \pi$. The phase angle defines the difference in phase between the displacements of particle n and those of particle $n + 1$. The phase angle is also given by

$$\phi = 2\pi\mathbf{k}d \quad (2.24)$$

where $\mathbf{k}$ is the wave vector. The wave vector ranges from $-\frac{1}{2}\phi < \mathbf{k} < \frac{1}{2}\phi$. A plot of ν versus $\mathbf{k}$ or ϕ is called a *dispersion curve* and is shown in Figure 2.13. The maximum on the dispersion curve is given by

$$\nu\,(\mathrm{cm}^{-1}) = (1/c_s\pi)\,(f/m)^{\frac{1}{2}} \quad (2.25)$$

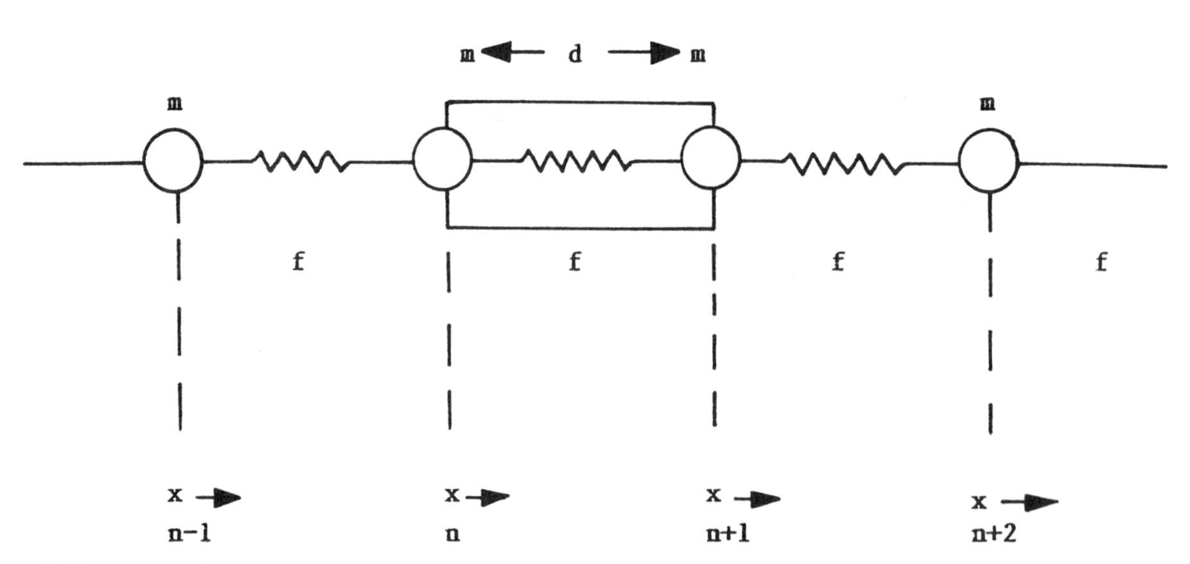

Figure 2.12. A molecular model of an infinite linear array of atoms of mass m *separated by a distance,* d. *(Reproduced with permission from reference 10. Copyright 1982 John Wiley Sons, Inc.)*

Vibrations of the Infinite Diatomic Chain

When a lattice consists of two or more types of atoms, new features in the dispersion relationship that appear throw additional light onto the vibrations of polymer chains. Consider a chain consisting of two different atoms arranged alternately as shown in Figure 2.14. Assuming only nearest neighbor interactions, one obtains the following equation:

$$\nu^2 = \frac{f}{4\pi^2}\left[\left(\frac{1}{m_1}+\frac{1}{m_2}\right) \pm \left(\frac{1}{m_1}+\frac{1}{m_2}\right)^2 - \frac{4}{m_1 m_2}\sin^2 \pi \mathbf{k} d\right]^{\frac{1}{2}} \quad (2.26)$$

where m_1 and m_2 are the masses of the two atoms.

The two possible solutions result in two frequency branches of the dispersion relationship, as shown in Figure 2.15. A positive sign results in the upper curve, called the *optical branch* because it gives rise to fundamentals in the IR and Raman spectra. Solving the equation by using the negative sign gives the lower curve, called the *acoustic branch* because the frequencies fall in the region of sonic or ultrasonic waves. The acoustic branches are not optically active in the IR or Raman spectra for the infinite chain, but these modes do appear for finite chain lengths.

The displacements observed at various points on the dispersion curve are illustrated (rotated 90°) in Figure 2.16. At $\phi = 0$, the optical branch has a normal vibration consisting of a simple stretching of the bond between the two atoms. Thus, the center of mass of the two atoms remains fixed and has a frequency equal to $(1/2c_s\pi)(2f/m_r)^{1/2}$, where m_r is the reduced mass $[(1/m_1) + (1/m_2)]$. There is a frequency range between the optical branch and the acoustic branch that contains no frequencies, corresponding to the region between $(1/2c_s\pi)(2f/m_1)^{1/2}$ and $(1/2c_s\pi)(2f/m_2)^{1/2}$. This region is referred to as the *frequency gap*. A vibrational frequency located in this forbidden region or in the frequency gap must correspond to a normal mode of a defect.

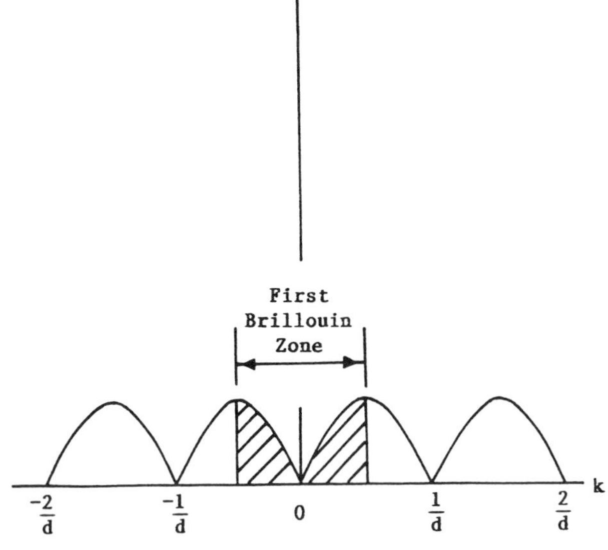

Figure 2.13. A dispersion curve showing a plot of ν versus k or ϕ. (Reproduced with permission from reference 10. Copyright 1980 John Wiley & Sons, Inc.)

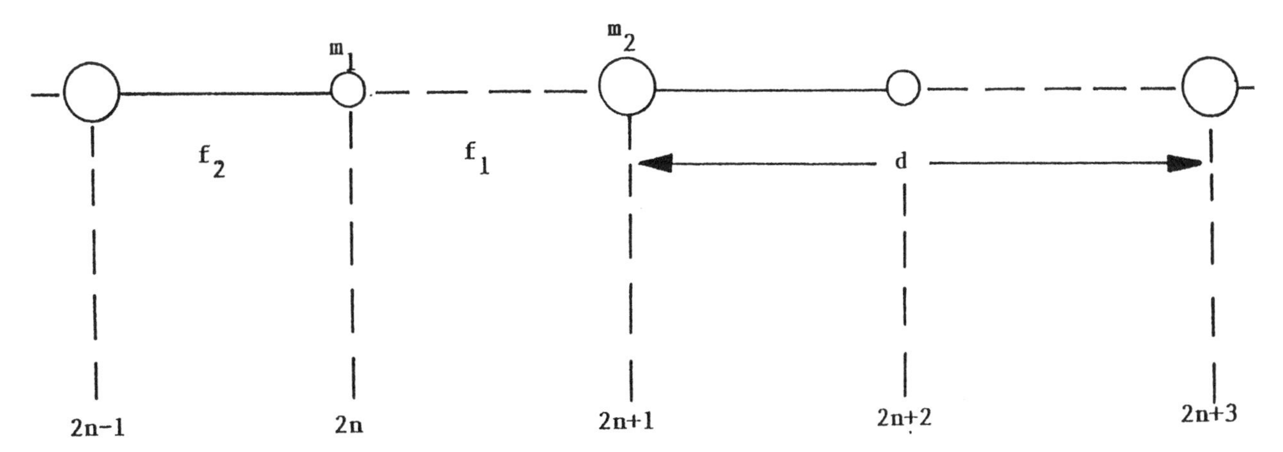

Figure 2.14. A molecular model of a chain consisting of two different atoms arranged alternately. (Reproduced with permission from reference 10. Copyright 1980 John Wiley & Sons, Inc.)

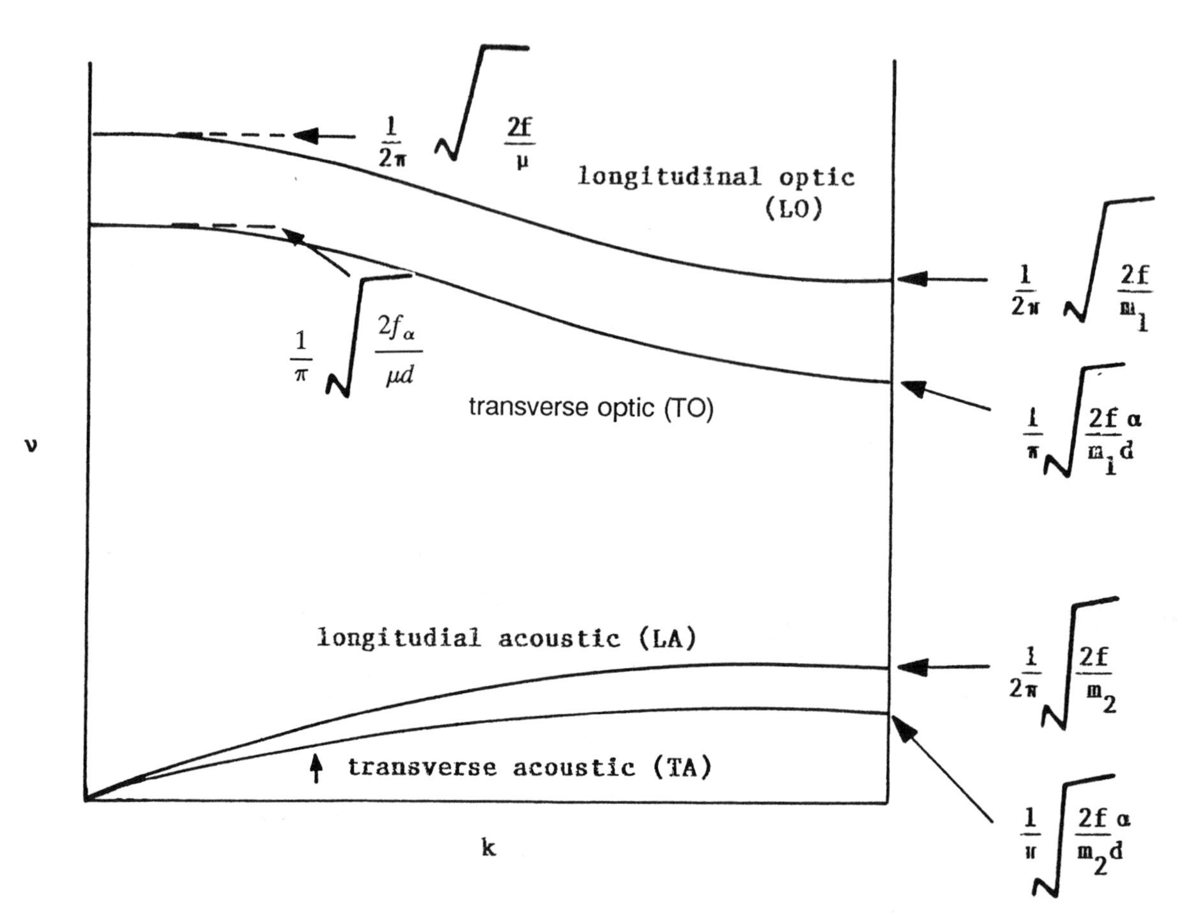

Figure 2.15. The two frequency branches of the dispersion relationship of a chain consisting of two different atoms arranged alternately. (Reproduced with permission from reference 10. Copyright 1982 John Wiley & Sons, Inc.)

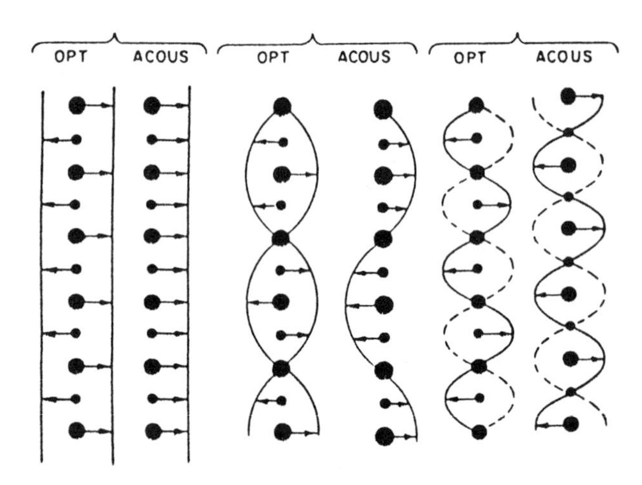

Figure 2.16. The displacements observed at various points on the dispersion curve (rotated 90°). (Reproduced with permission from reference 10. Copyright 1982 John Wiley & Sons, Inc.)

What happens if only weak coupling occurs between adjacent molecules? This condition can be accomplished by simply changing the values of the force constants such that $f_1 > f_2$. The force constant f_1 corresponds to the "stiffness" of the covalent bond linking pairs of atoms, and f_2 represents the much weaker bonding of the molecules together. Now the frequencies become

$$\nu_{\text{int}}\,(\text{cm}^{-1}) = \frac{1}{2c_s\pi}\left(\frac{f_1}{m_r}\right)^{\frac{1}{2}} \tag{2.27}$$

$$\nu_{\text{ext}}\,(\text{cm}^{-1}) = \frac{1}{c_s\pi}\left[\frac{f_2}{(m_1 + m_2)}\right]^{\frac{1}{2}} |\sin \pi \mathbf{k} d| \tag{2.28}$$

For the *internal* vibrations, ν_{int} is approximately equal to the vibrational frequency of an isolated uncoupled diatomic molecule, and for the *external* vibrations, ν_{ext} corresponds to a vibration of the molecules as a whole, each molecule moving as a rigid body against the other. Motions of this type are also referred to as *lattice vibrations* and can be separated into two types: translational-like modes and rotational or vibrational modes.

Vibrations of the Infinite Chain in Three Dimensions

This simple result in the one-dimensional case can be extended to three dimensions (*10*) (but we aren't going to do that here). The transverse motions are independent of the longitudinal motions, and in addition, transverse motions at right angles to one another are also mutually independent (*10*). The resulting dispersion curve is shown in Figure 2.17. In the three-dimensional motions of a diatomic chain, the transverse directions are equivalent. Consequently, only one transverse optic and acoustic dispersion curve is displayed, as they are degenerate.

Considering the vibrations of a three-dimensional lattice, complications arise because the transverse vibrations of molecular crystals are not degenerate. There are three branches for each particle in the unit cell. If there are m_c molecules in the cell, and each

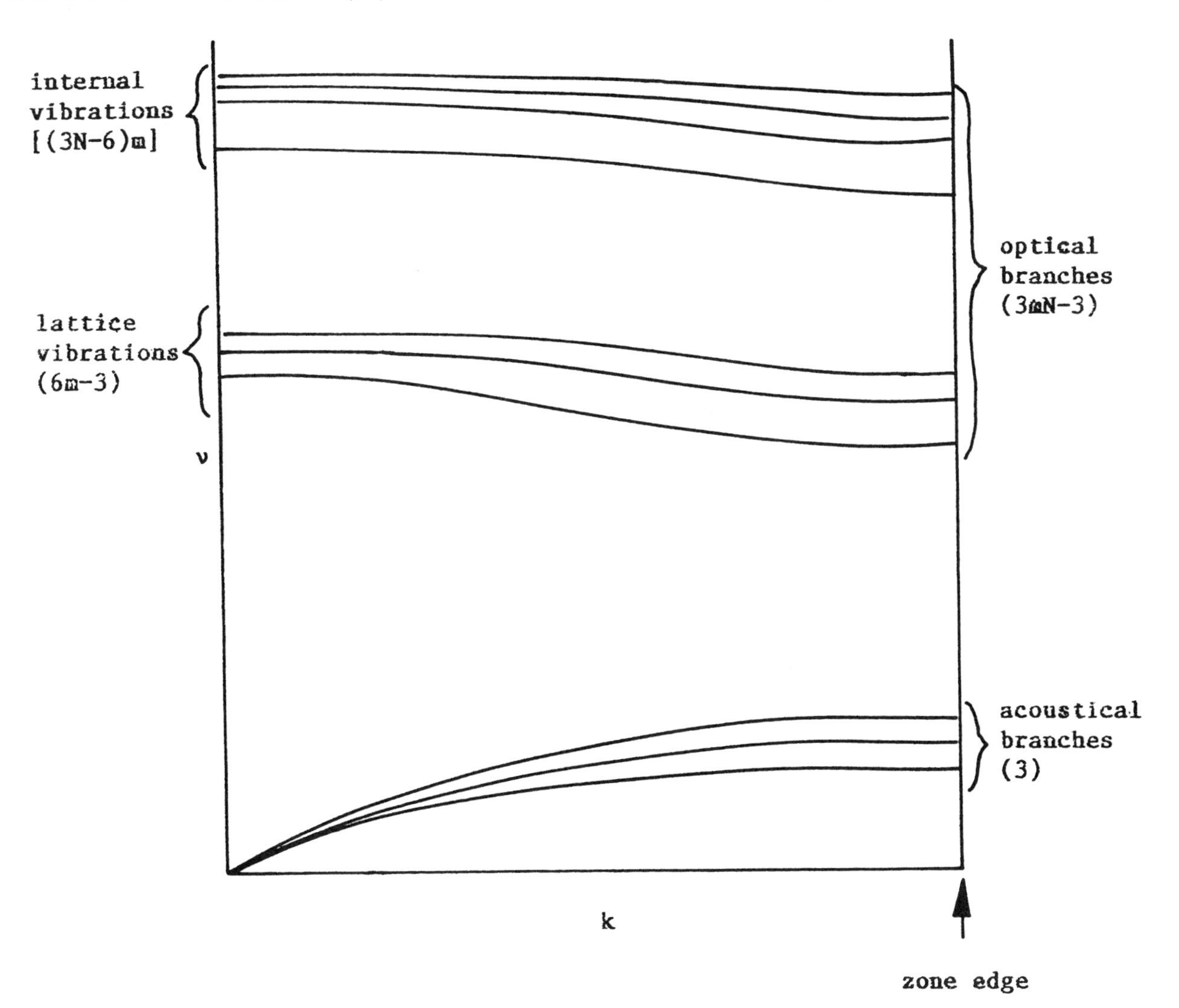

Figure 2.17. The dispersion curve for a three-dimensional crystal. The variable m *is equal to the number of atoms in the repeat unit, and* N *is the number of oscillators. (Reproduced with permission from reference 10. Copyright 1982 John Wiley & Sons, Inc.)*

molecule is composed of N atoms, then there are $3m_c N$ branches. Three of these branches are acoustic and are inactive in the IR region. The general nature of the results is shown in Figure 2.18. In many cases, the vibrations of complex three-dimensional crystals cannot be described as simply longitudinal or transverse because the nature of the waves depends on the details of the intra- and intermolecular forces.

The dispersion curve for an infinite isolated chain of methylene units has been calculated (*36*) and is shown in Figure 2.19. Because the planar zigzag conformation of polyethylene is a 2_1 helix, the observed IR and Raman lines for the various modes are active at values of $\phi = 0$ and π, respectively, depending on the nature of the motion involved. For the infinite chain, no other modes are observable because for all $\phi > 0$, there is a corresponding $\phi < 0$ somewhere on the chain, and the total change, $\Sigma\phi$, will always be zero. The total sum of the dipole changes is zero. This result indicates that no total change in the dipole moment will occur for the infinite chain except for vibrational motions with values of $\phi = 0$ and π, respectively. The IR and Raman activity of the various modes depends on the nature of the motion for $\phi = 0$ and π.

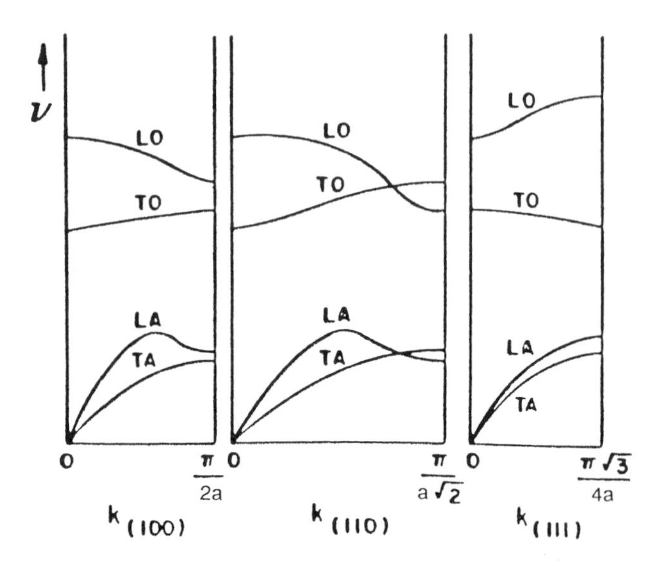

Figure 2.18. The vibrations of a three-dimensional lattice. The distance is defined by a. *(Reproduced with permission from reference 10. Copyright 1982 John Wiley Sons, Inc.)*

This simple "coupled oscillator" approach appears to be quite successful. At a deeper level, it is interesting that the calculated dispersion curve using normal coordinate analysis for an isolated chain of methylene units can be used to predict the frequencies of the spectra of all of the oligomers. It almost makes one want to do normal coordinate analysis! But remember the Bellamy approach.

Vibrations of Finite Chains

Theoretically, the simplest finite polymer is a uniform one-dimensional coupled chain of N point masses. The N frequencies for a linear chain of N atoms acting as parallel dipoles with fixed ends (including only nearest neighbor interactions) are given by the following equations (*36*):

$$\nu_s^2 = \nu_0^2 + \nu_1^2(1 + \cos\theta) \quad (2.29)$$

where

$$\theta = s\pi/(N + 1) \quad s = 1, 2, \cdots N \quad (2.30)$$

with ν_0 being the frequency of the uncoupled or isolated mode, ν_1 the interaction parameter and s an integer from 1 to N. The observed frequency-phase relationship is shown in Figure 2.20 for the methylene wagging mode of paraffins C_3 through C_{30} (*36*). The individual normal vibrational frequencies for the methylene wagging mode of straight-chain paraffins are shown in Figure 2.21.

The dispersion curve (Figure 2.19) for an isolated polyethylene chain shows that some of the vibrational motions are uncoupled from their neighbors as indicated by the independence of the frequencies on the phase angle of the C–H stretching and H–C–H bending modes. Other modes such as C–C–C bending and methylene rocking modes are highly coupled to their neighbors and are very sensitive to the phase angle (which can be reduced to a dependence on sequence length). Although the results for only polyethylene are presented here, experience has shown that the results can be generalized. The frequency dependence as a function of the number of units in the sequence often has the general shape shown in Figure 2.22. However, it is necessary to verify the appropriate shape dependence on sequence length for the mode under investigation.

Distribution of Intensities for Chain Molecules

In addition to the frequency dependence, there is also a dependence of the intensities for the coupled oscillators. Fortunately, only the limiting mode ($k = 1$) is observed, and the remaining oscillator modes are too weak to be observed. The $k = 1$ mode is always the strongest band in the series, and the second strongest

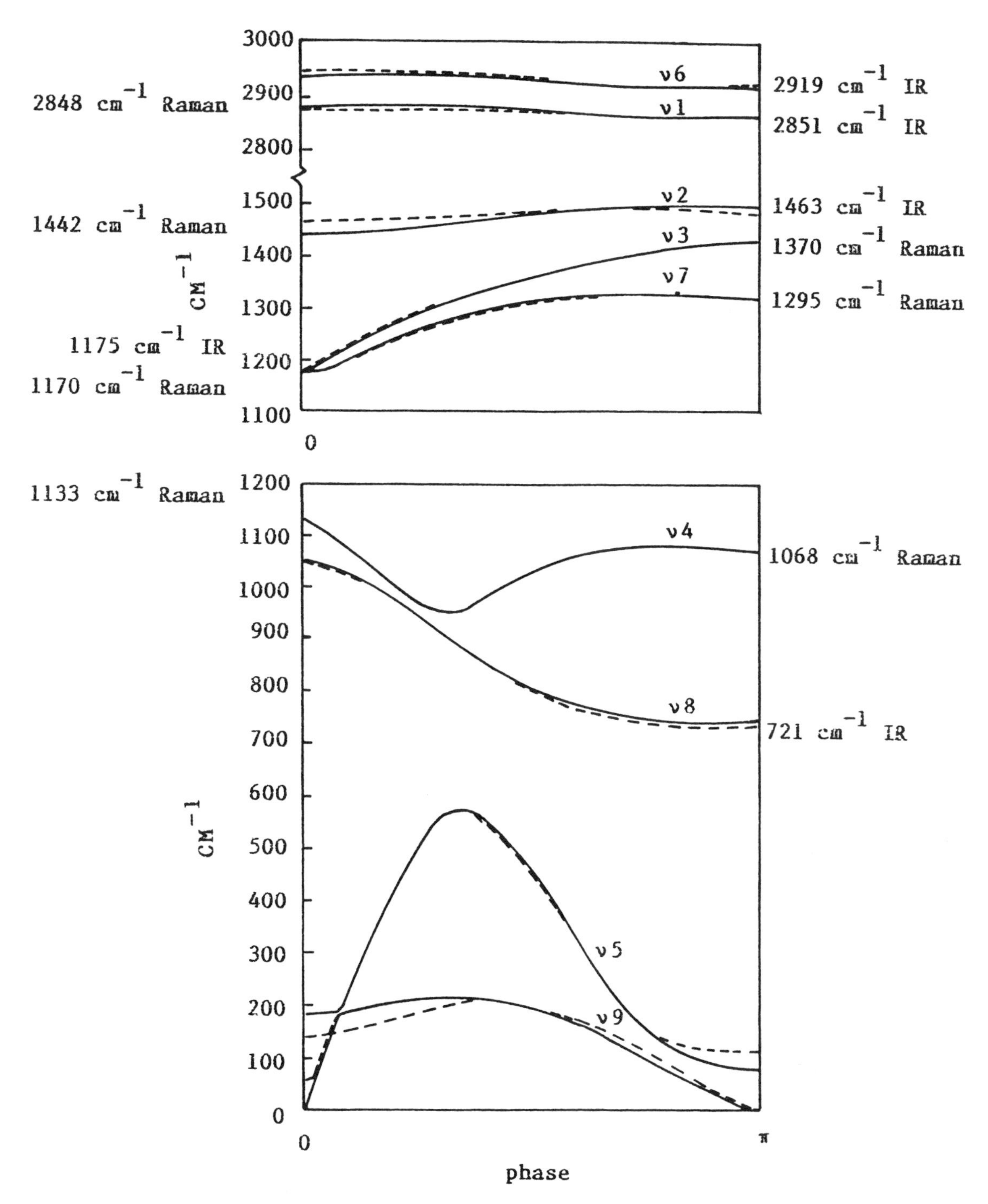

Figure 2.19. The dispersion curve for an infinite isolated chain of methylene units (crystalline polyethylene). (Reproduced with permission from reference 10. Copyright 1982 John Wiley Sons, Inc.)

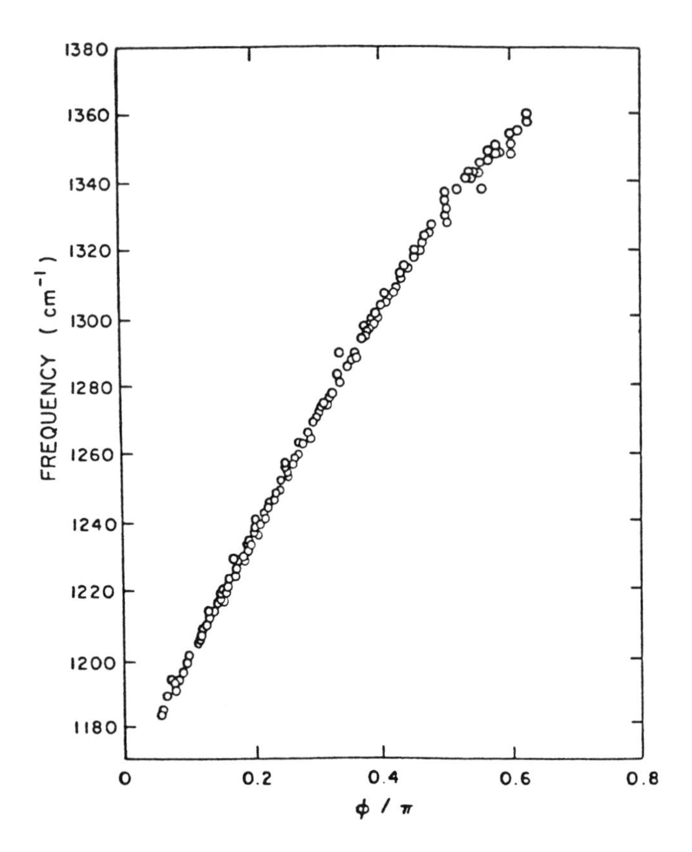

Figure 2.20. The observed frequency-phase relationship for the methylene wagging mode of paraffins C_3 through C_{30}. (Reproduced with permission from reference 10. Copyright 1982 John Wiley & Sons, Inc.)

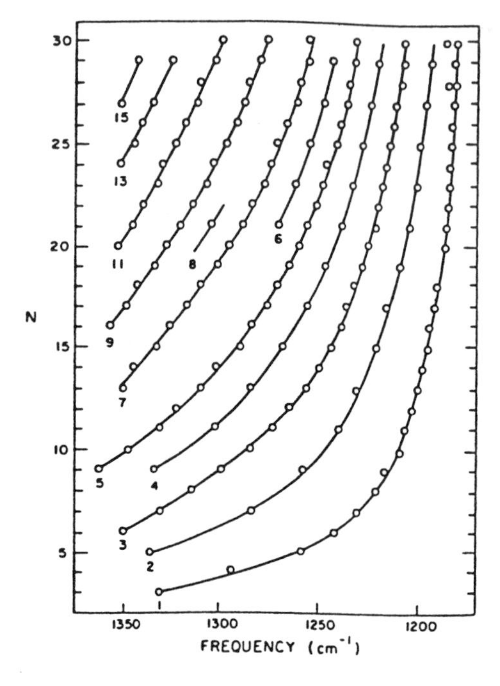

Figure 2.21. The individual normal vibrational frequencies for the methylene wagging modes of straight-chain paraffins. (Reproduced with permission from reference 10. Copyright 1982 John Wiley & Sons, Inc.)

is calculated to be from 3% to 25% of the strongest. This mode is dependent on the ordered chain length up to a limiting value of N that is different for each mode.

The intensities of the modes are more sensitive to structural differences in molecules than are the frequencies. For example, the methyl deformation mode at 1378 cm^{-1} has a constant frequency for the methyl, ethyl, propyl, and butyl groups, but the extinction coefficient decreases in going from methyl to butyl groups.

> Imagine the difficulty this makes in trying to measure the short-chain branch content of polyethylene when the types of branches present are not known. Think about this the next time you see a report of the branches per 1000 carbons measured by IR spectroscopy.

The sensitivity of the vibrational intensities carries over to chain molecules. The IR and Raman intensities are not linearly related to the degree of disorder (*37*). This nonlinearity is a consequence of the fact that the band intensities are associated with modes that are largely delocalized.

Snyder demonstrated the theoretical basis of the nonlinearity by using a particularly simple but ele-

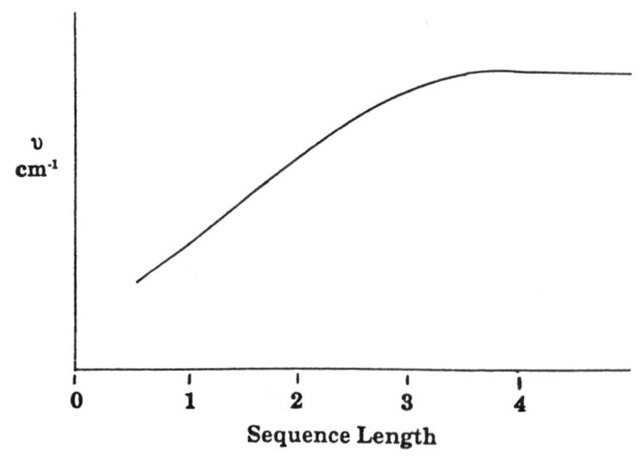

Figure 2.22. The general shape of the plot of the frequency dependence as a function of the number of units in the sequence.

gant model (*37*). The model consists of N identical harmonic oscillators representing the repeat units of a chain molecule (but with only one degree of freedom). Each oscillator has associated with it a vector that represents a local dipole moment derivative. Adjacent oscillators are connected by bonds. Each bond of the model chain can have one of two possible conformations. The conformation of the chain is altered by the introduction of a 180° rotation about one of the bonds as shown in Figure 2.23b. The conformation of the bond between adjoining oscillators is given by the value of the dihedral angle, τ. When $\tau = 0°$, there is parallel alignment of adjacent oscillators, and when $\tau = 180°$, there is antiparallel alignment. As shown in Figure 2.23, the oscillators can be aligned either parallel or antiparallel to the y axis, and their relative alignment determines not only the conformation but also the intensities of the IR modes.

The frequency, ν_k, of the kth vibrational mode of a chain of N oscillators is given by

$$\nu_k = a + 2b_k \cos \phi_k \tag{2.31}$$

where a is the absorptivity and

$$b_k = b \cos \tau_k \tag{2.32}$$

Here b is a constant, and for our model chain $b_k = b$ or $-b$.

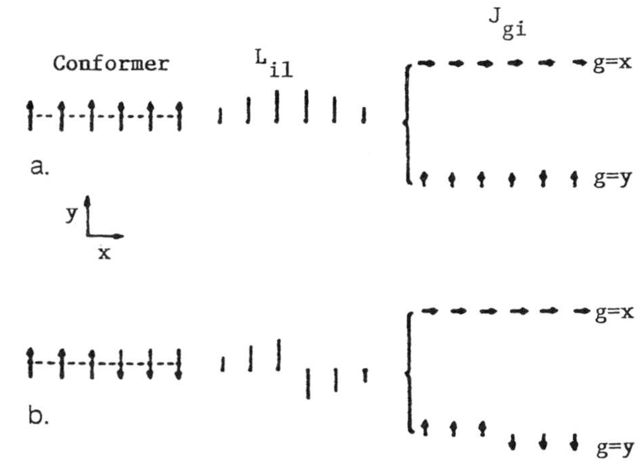

Figure 2.23. Conformations, normal coordinates of the k *= 1 mode, and local dipole moment derivatives for the ordered (τ_i = 0°, 0°, 0°, 0°) conformer (a) and a disordered (τ_i = 0°, 0°, 180°, 0°) conformer (b) of a model chain of six oscillators. (Reproduced from reference 37. Copyright 1990 American Chemical Society.)*

Additionally,

$$\phi = (k\pi)/(n + 1) \tag{2.33}$$

The intensities are given by

$$I^{(k)} = K \Sigma \left(\frac{\partial \mu}{\partial Q_k} \right)^2 \tag{2.34}$$

where μ has x, y, and z components of the dipole moment of the chain, Q_k is the displacement coordinate of the motion, and K is a proportionality constant. Expressing $I^{(k)}$ in terms of the contributions from the individual oscillators and their oscillator coordinates, S_i,

$$I_x^{(k)} = K \Sigma_i \left[\frac{\partial \mu}{\partial S_i} \frac{\partial S_i}{\partial Q_k} \right]^2 \tag{2.35}$$

There is a similar expression for the y component of the intensities.

It is desirable to write this equation in the form

$$I_x^{(k)} = K(\mu_{x'})^2 (\bar{J}_x \bar{L}_k)^2 \tag{2.36}$$

where $\mu_{x'}$ represents the magnitude of the $\partial\mu/\partial S_i$ term and is the same for each oscillator. The term $\bar{L}_k$ is the kth normal coordinate, and the ith element is given by $\partial\mu/\partial S_i/\partial Q_k$. The elements of $\bar{J}_x$ or $\bar{J}_y$ are $+1$ or -1 depending on the sign of $\partial\mu/\partial S_i$. For both conformers,

$$\bar{J}_x(\text{A}) = \bar{J}_x(\text{B}) = (1\ 1\ 1\ 1\ 1\ 1) \tag{2.37}$$

The y component for conformer A is

$$\bar{J}_y(\text{A}) = (1\ 1\ 1\ 1\ 1\ 1) \tag{2.38}$$

but for B

$$\bar{J}_y(\text{B}) = (1\ 1\ 1\ -1\ -1\ -1) \tag{2.39}$$

For conformer A, substitution results in the x- and y-component intensities of the $k = 1$ mode,

$$I_x^{(1)} = K(\mu_{x'})^2(\Sigma \bar{L}_{i1})^2 \tag{2.40}$$

and for the y component

$$I_x^{(1)} = K(\mu_{y'})^2(\Sigma\, \bar{L}_{i1})^2 \quad (2.41)$$

The ith element of the normal-coordinate vector is equal to $\partial S_i/\partial Q_k$. In Figure 2.24, the forms of $\bar{J}_x$ and $\bar{J}_y$ are shown as well as the $\bar{L}_{i1}$ vectors. The magnitude of the ith element of $\bar{L}1$ is $[2/(\nu + 1)]^{1/2} \sin i\phi_1$. The magnitudes of the elements of the normal coordinates are independent of the conformation, but their signs are not; that is, they are inverted at the $i = 4$ oscillator. Adding the x- and y-component intensities for the $k = 1$ mode gives the following results for the two conformers:

$$I_y^{(1)}(\mathrm{B}) = I_y^{(1)}(\mathrm{A}) = K(\mu_{y'})^2 \quad (2.42)$$

However, the x-component intensity of conformer B is zero:

$$I_x^{(1)}(\mathrm{B}) = 0 \quad (2.43)$$

$$I_x^{(1)}(\mathrm{A}) = K(\mu_{x'})^2 \quad (2.44)$$

Hence, depending on the orientation of the dipole moments, the IR intensities reflect differences in conformation.

This type of model has been extended to include an ensemble of disordered model chains based on a parameter, p, which is the probability that a pair of adjacent oscillators are aligned parallel to one another. It is desirable to relate the dependence of the x-component intensity of the $k = 1$ mode to the probability p. Theoretically, the intensity of the $k = 1$ band diminishes with increasing disorder, but the $k = 1$ intensity is not linearly related to p (Figure 2.24). The intensity of the $k = 1$ band is reduced to less than half its value in going from $p = 1$ to 0.9. Theoretically, the intensity loss is distributed among the other k modes so that at maximum disorder ($p = 0.5$) all modes have the same intensity.

The chain-length effect can be added to the relation between the intensity of the $k = 1$ band and the degree of chain disorder (Figure 2.25), where the x component of the $k = 1$ mode is plotted against p for ensembles of chains of different chain lengths. The degree of linearity between intensity and disorder increases as chains become shorter. In the limiting case of $n = 2$, the intensity – disorder relation is linear. This represents the case in which the vibrational mode is highly localized.

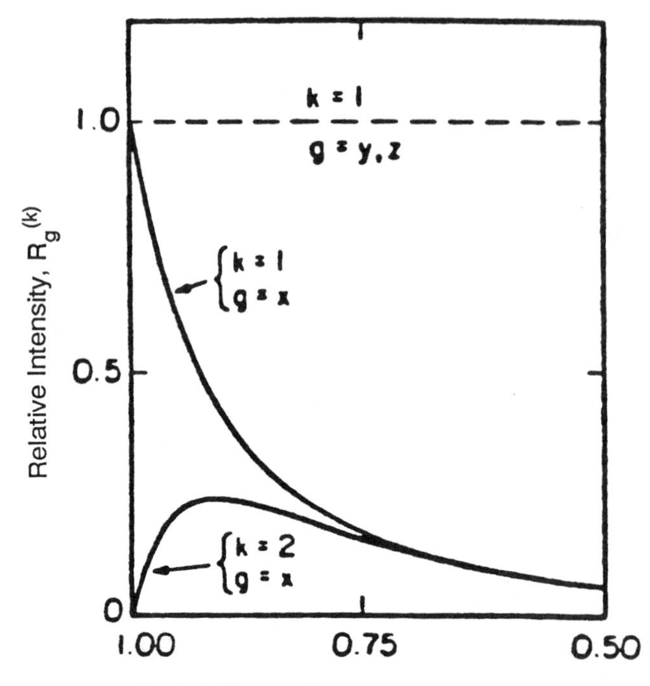

Figure 2.24. Relative IR intensity of the k = *1 mode* (x, y, *and* z *components) and the* k = *2 mode* (x *component) as a function of* p *for an ensemble of 20-oscillator chains. The intensities are scaled by the intensity of the* k = *1 mode in the spectrum of the ordered* n = *20 chain. (Reproduced from reference 37. Copyright 1990 American Chemical Society.)*

To summarize:

> *In practical terms, these results point to the necessity of using localized modes for the quantitative determination of conformational disorder in chain-molecule assemblies. If bands associated with delocalized modes are used and linearity is assumed, the degree of disorder relative to the completely ordered system can be greatly overestimated if the system to be determined is fairly ordered.*
>
> – R. G. Snyder (*37*)

An Example: Vibrational Spectra of Ethylene–Propylene Copolymers

At this point, an example may be useful. Let us look at the spectra of ethylene – propylene (E – P) copolymers (*38*). The structure of the various sequences of this copolymer is shown in the following scheme illustrating the expected origin of the different segments of methylene units.

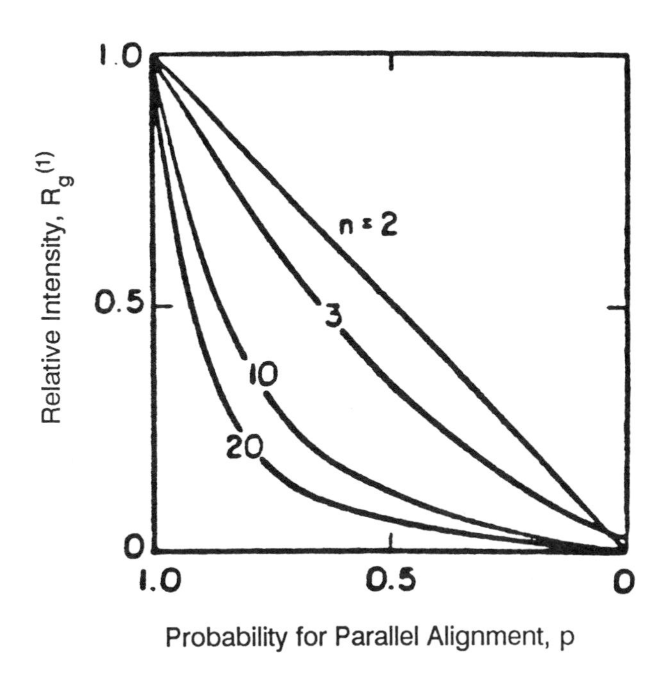

Figure 2.25. IR intensity of the x *component of the* k = *1 mode plotted as a function of* p *for chains with 2, 3, 10, and 20 oscillators. (Reproduced from reference 37. Copyright 1990 American Chemical Society.)*

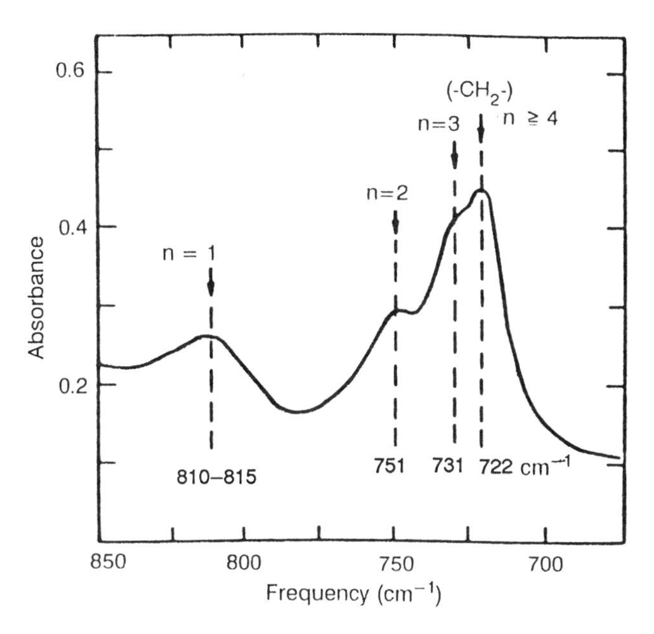

Figure 2.26. An IR spectrum of the methylene rocking region of an ethylene–propylene copolymer (54.3 wt% C_2). (Reproduced with permission from reference 2, Chapter 1. Copyright 1980 John Wiley & Sons, Inc.)

An IR spectrum of the methylene rocking region of an E – P copolymer (54.3 wt% C_2) is shown in Figure 2.26. The assignments of the peaks to the various methylene sequences are also shown. Quantitatively, head-to-tail P – P sequences can be identified by using the 810 – 815-cm^{-1} absorption band. The tail-to-tail P – P sequence is identified by the presence of the band at 751 cm^{-1}, which is assigned to the methylene sequences with $n = 2$. The band at 731 cm^{-1} is assigned to the E – P sequences, and the line at 722 cm^{-1} is associated with the $P(E-E)_{n \geq 2}$ sequences. Quantitative measurements have been made on this system, and the results are shown in Figure 2.27 as a function of the mole percent of propylene.

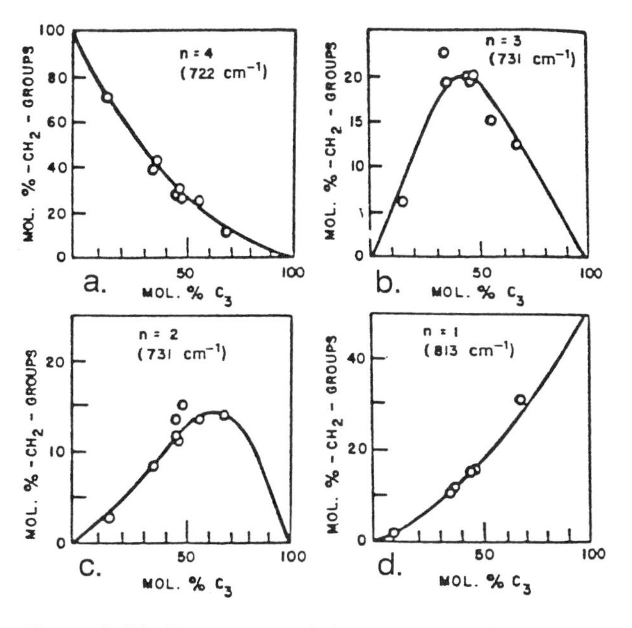

Figure 2.27. Comparison of the experimental results with the theoretical curves as a function of the mole percent propylene: (a) 4 or more contiguous groups, (b) 3 contiguous groups, (c) 2 contiguous groups, and (d) isolated methylene groups. (Reproduced with permission from reference 2, Chapter 1. Copyright 1980 John Wiley & Sons, Inc.)

Summary

This chapter introduced the basics of vibrational spectroscopy as applied to simple molecules and further extended to polymer molecules. An example demonstrated the method used and the nature of the results that can be obtained. The following chapters will discuss the experimental aspects of IR spectroscopy, its applications to polymers, and finally, Raman spectroscopy.

References

1. Henniker, J. C. *Infrared Spectrometry of Industrial Polymers;* Academic: New York, 1967.
2. Hummel, D. O.; Scholl, F. *Infrared Analysis of Polymers, Resins and Additives: An Atlas;* Verlag Chemie: Weinheim, 1978.
3. Colthup, N. B.; Daley, L. H.; Wiberley, S. E. *Introduction to Infrared and Raman Spectroscopy;* 2nd ed.; Academic: New York, 1975.
4. Smith, A. Lee *Appl. Spectrosc.* **1987,** *41,* 1101.
5. Erley, D. S. *Anal. Chem.* **1968,** *40,* 894.
6. Erley, D. S. *Appl. Apec.* **1971,** *25,* 200.
7. Haslam, J.; Willis, H. A. *Identification and Analysis of Plastics;* 2nd ed.; Van Nostrand: New Jersey, 1972.
8. Wilson, E. R., Jr.; Decius, J. C.; Cross, P. C. *Molecular Vibrations;* McGraw Hill: New York, 1955.
9. Thompson, H. W. *J. Chem. Soc.* **1948,** *328.*
10. Painter, P. C.; Coleman, M. M.; Koenig, J. L. *The Theory of Vibrational Spectroscopy and its Application to Polymeric Materials;* Wiley: New York, 1982.
11. Woodward, L. A. *Introduction to the Theory of Molecular Vibrations and Vibrational Spectroscopy;* Oxford (Clarendon Press): New York, 1972.
12. Steele, D. *Theory of Vibrational Spectroscopy;* Saunders: Philadelphia, 1971.
13. Gribov, L. A. *Intensity Theory for Infrared Spectra of Polyatomic Molecules;* Consultants Bureau: New York, 1964.
14. Pacansky, J.; England, C.; Waltman, R. J. *J. Polym. Sci., Polym. Phys. Ed.* **1987,** *25,* 901.
15. Siesler, H. W.; Holland-Moritz, K. *Infrared and Raman Spectroscopy of Polymers;* Dekker: New York, 1980.
16. Koenig, J. L.; Itoga, M. *Appl. Spectrosc.* **1971,** *25,* 355.
17. Krimm, S. *Fortschr. Hochpolym. Forschg.* **1960,** *2,* 51.
18. Zbinden, R. *Infrared Spectrometry of High Polymers;* Academic: New York, 1964.
19. Hermans, J. J.; Hermans, P. H.; Vermaas, D.; Weidinger, A. *Rec. Trav. Chim.* **1946,** *65,* 427.
20. Fraser, R. D. B. *J. Chem. Phys.* **1956,** *24,* 89.
21. Siesler, H. W. *Adv. Polym. Sci.* **1984,** *65,* 2.
22. Samuels, R. J. *Makromol-Chem. Suppl.* **1981,** *4,* 241.
23. Mirabella, F. M., Jr. *J. Polym. Sci., Polym. Phys. Ed.* **1987,** *25,* 591.
24. Jasse, B.; Koenig, J. L. *J. Macromol. Sci. —Rev. Macromol. Chem.* **1979,** *C17(1),* 61-135.
25. Lefebve, D.; Jasse, B.; Monnerie, L. *Polymer* **1981,** *22,* 1616.
26. Koenig, J. L.; Cornell, S. W.; Witenhafer, D. E. *J. Polym. Sci.* **1967,** *A2,* 301.
27. Fina, L. J.; Koenig, J. L. *J. Polym. Sci., Polym. Phys. Ed.* **1986,** *24,* 2509.
28. Fina, L. J.; Koenig, J. L. *J. Polym. Sci., Polym. Phys. Ed.* **1986,** *24,* 2525.
29. Fina, L. J.; Koenig, J. L. *J. Polym. Sci., Polym. Phys. Ed.* **1986,** *24,* 2541.
30. Radziszewski, J. G.; Michl, J. *J. Am. Chem. Soc.* **1986,** *108,* 3289.
31. Michl, J.; Thulstrup, E. W. *Spectroscopy with Polarized Light. Solute Alignment by Photoselection in Liquid Crystals, Polymers and Membranes;* VCH Publishers: Deerfield Beach, FL, 1986.
32. Bellamy, L. J. *The Infrared Spectra of Complex Molecules;* Wiley: New York, 1975.
33. Bellamy, L. J. *Advances in Group Frequencies;* Barnes and Noble: New York, 1968.
34. Dolphin, D.; Wick, A. *Tabulation of Infrared Spectral Data;* Wiley-Interscience: New York, 1977.
35. Smith, A. *Lee Applied Infrared Spectroscopy, Fundamentals, Techniques and Analytical Problem-Solving;* Wiley-Interscience: New York, 1979.
36. Snyder, R. G.; Schachtschneider, J. H. *Spectrochim. Acta.* **1963,** *19,* 85.
37. Snyder, R. G. *Macromolecules* **1990,** *23,* 2081.
38. Drushel, H. V.; Ellerbe, J. J.; Cos, R. C.; Love, L. H. *Anal. Chem.* **1968,** *40,* 370.

3

Experimental IR Spectroscopy of Polymers

The Fourier Transform IR Spectroscopic Method

Originally, IR spectra were measured with a dispersive instrument equipped with an optical element of prisms or gratings to geometrically disperse the IR radiation (*1*). A scanning mechanism passes the dispersed radiation over a slit system that isolates the frequency range falling on the detector. In this manner, the spectrum, that is, the energy transmitted through a sample as a function of frequency, is obtained. This dispersive IR method is highly limited in sensitivity because most of the available energy does not fall on the open slits and hence does not reach the detector. The sensitivity of IR spectroscopy can be improved with a multiplex optical device that allows the continuous detection of all of the transmitted energy simultaneously. The Michelson interferometer is such an optical device, and the IR instrumentation that resulted is termed a Fourier transform infrared (FTIR) spectrometer (*2*).

The Michelson interferometer has two mutually perpendicular arms, as shown in Figure 3.1. One arm of the interferometer contains a stationary plane mirror; the other arm contains a movable mirror. Bisecting the two arms is a beam splitter, constructed from an IR-transparent material, that separates the source beam into two equal beams traversing the two arms of the interferometer. These two light beams travel down their respective arms of the interferometer and are reflected back to the beam splitter where they recombine. This recombined beam is then reflected to the detector. When the two mirrors are at equal distances from the beam splitter ($x = 0$, where x is the optical path difference), the paths of the light beams are identical. Under these conditions, all wavelengths of radiation striking the beam splitter after reflection combine to produce a maximum flux at the detector, and generate a *center burst*. As the movable mirror is displaced from this equidistant point, the optical path length in the arm of the interferometer is changed. This difference in optical path length causes each wavelength of source radiation to destructively interfere with itself at the beam splitter. The resulting flux at the detector, which is the sum of the fluxes for each of the individual wavelengths, rapidly decreases with mirror displacement. Sam-

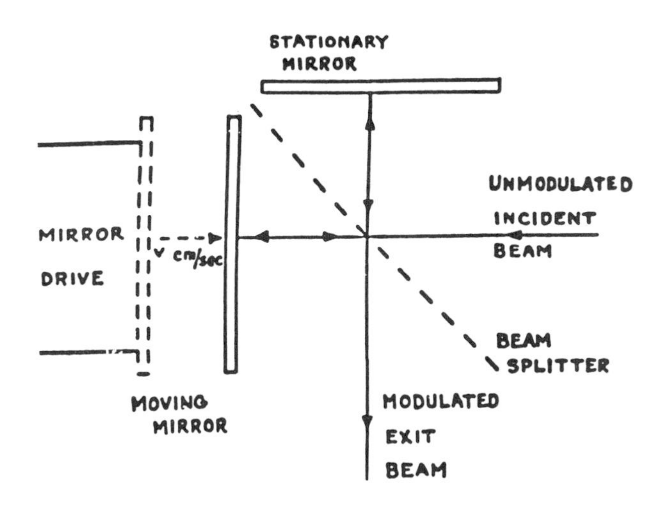

Figure 3.1. Optical diagram of the Michelson interferometer. (Reproduced with permission from reference 1. Copyright 1963 John Wiley & Sons, Inc.)

1904—4/92/0043$09.50/1

pling the flux at the detector provides an interferogram (Figure 3.2a). For a monochromatic source of frequency, ν, the interferogram is a cosine function of the frequency, ν, and x is the path difference. By extension, the interferogram of a polychromatic source appears as the cumulative sum of many individual cosine interference patterns.

The interferogram consists of two parts: a constant (DC) component and a modulated (AC) component. The $I(x)$ or AC component is called the interferogram and is given by

$$I(x) = 2_{\Sigma} I \cos 2\pi\nu x \, d\nu \tag{3.1}$$

where I is the total light flux. An IR detector and an AC amplifier convert this flux into an electrical signal

$$V(x) = \mathrm{re} I(x) \tag{3.2}$$

in volts, where re is the response of the detector and amplifier.

The moving mirror is generally driven on an air-bearing system. For FTIR measurements to be made,

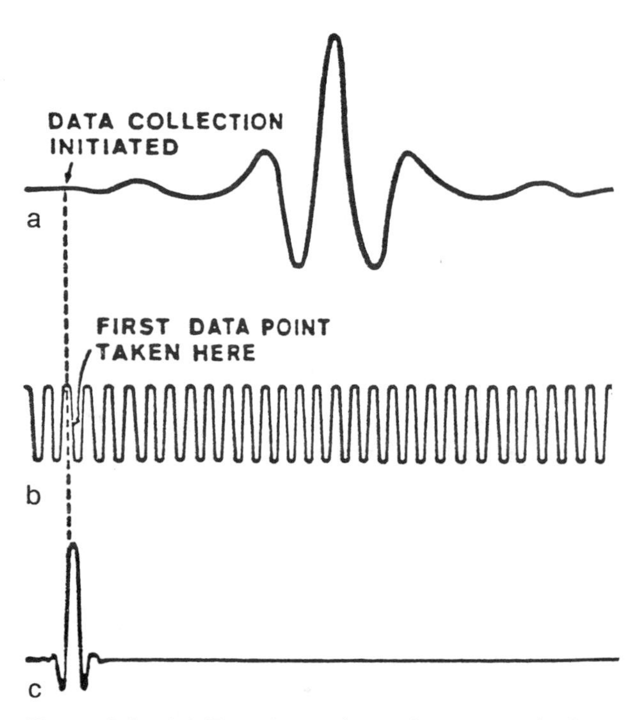

Figure 3.2. (a) Experimental interferogram, (b) laser output, and (c) white light interferogram. (Reproduced with permission from reference 2. Copyright 1978 Plenum Publishing Corp.)

the mirror must be kept in the same plane to within 10 μm-rad for mirror drives up to 10 cm long. It is necessary to have some type of marker to initiate data acquisition at precisely the same mirror displacement every time. The uncertainty in mirror position cannot be greater than 0.1 μm from scan to scan. This precision is accomplished by having a smaller reference interferometer on the moving mirror. A visible white light source is passed through this reference interferometer to produce a sharp *center burst* or *spike*. Data acquisition is initiated when this center burst reaches a predetermined value.

Digitization of the interferogram at precisely spaced intervals is accomplished by using the interference signal of the auxiliary interferometer. The auxiliary interferometer is equipped with a He–Ne laser that yields a monochromatic signal to produce a cosine function from the interferometer (Figure 3.2). The interferogram is sampled at each zero crossing of the laser cosine function. Therefore, the path difference between two successive data points in the digitized interferogram is always a multiple of half wavelengths of the laser, or 0.316 μm. The laser also provides internal calibration of the frequency. To obtain the greatest accuracy in the digitized signal, the maximum intensity in an interferogram should closely match the maximum input voltage of the analog–digital converter (ADC).

The interferogram for a polychromatic source $A(\nu)$ is given by

$$I(x) = \Sigma_x A(\nu) (1 + \cos 2\pi\nu x) \, d\nu \tag{3.3}$$

Evaluating these integrals involves the determination of the values at zero path length and at very long, or infinite, path length. At zero path length difference

$$I(x = 0) = 2\pi A(\nu) \, d\nu \tag{3.4}$$

and for a large path length difference

$$I(x = \infty) = A(\nu) = \frac{I(0)}{2} \tag{3.5}$$

Therefore, the actual interferogram, $F(x)$, is

$$F(x) = I(0) - I(\infty) = \Sigma A(\nu) \cos(2\pi\nu x) \, dx \tag{3.6}$$

From Fourier transform theory

$$A(\nu) = 2\Sigma_x F(x) \cos(2\pi\nu x)\, dx \qquad (3.7)$$

Computation of Fourier Transform Spectra

The Fourier transform process was well known to Michelson and his peers, but the computational difficulty of making the transformation prevented the application of this powerful technique to spectroscopy. An important advance was made with the discovery by Cooley and Tukey (*3*) of the fast Fourier transform (FFT) algorithm. The discovery of the FFT algorithm breathed new life into the field of spectroscopic interferometry by allowing rapid calculation of the Fourier transform.

To Fourier transform a 4096-point array would require $(4096)^2$, or 16.7 million, multiplications. The FFT reduces this to $(4096) \times \log_2 (4096)$, or 4096(12), for a total of 49,152 multiplications. The FFT reduces the number of multiplications by a factor of 341. The advantage of the FFT increases with the number of data points. As computers have improved, the time required for a Fourier transform has been reduced to such an extent that with fast array processors the transformation can now be carried out in less than a second. Thus the spectra can be calculated during the time needed for the moving mirror to return to its starting position.

Sampling Methods for FTIR

IR spectra are dependent on sample preparation procedures and on optical spectroscopic techniques. The quality of the IR spectra can often depend on the skill of the spectroscopist as well. FTIR instrumentation has been improved with high-energy throughput and the multiplex advantage coupled with computerized signal enhancement by signal averaging. This improved instrumentation allows the use of a number of additional sampling techniques for polymers. The availability of these techniques allows the study of polymeric systems in their final-use state, whether as adhesives, fibers, coatings, or injection-molded articles. Figure 3.3 shows diagrams of the sampling techniques that are currently being used (*4*).

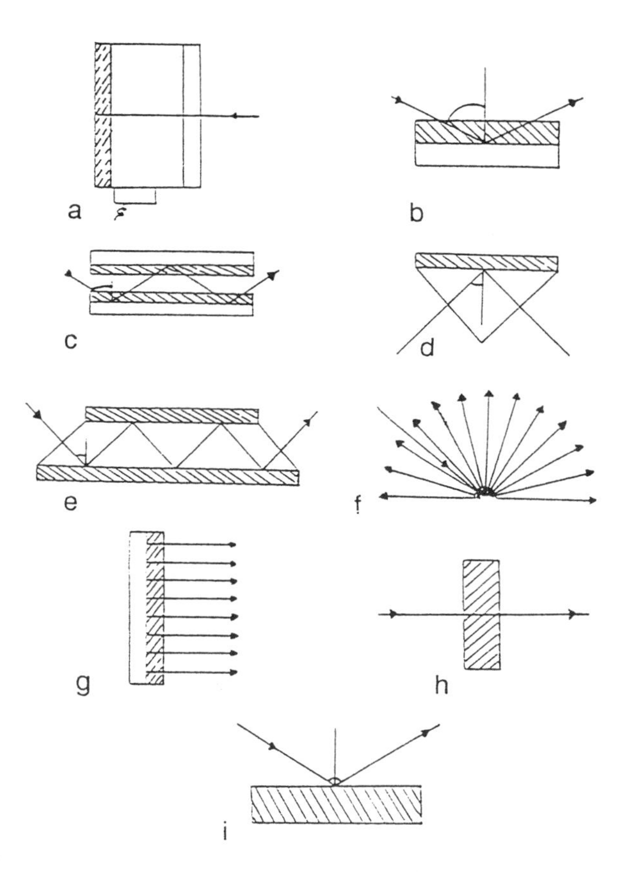

Figure 3.3. Diagrams of the sampling techniques that are currently being used in the FTIR spectroscopy of polymers. In the PAS cell (a), the incident light produces pressure fluctuations that are detected by a sensitive microphone. In single reflection RA spectroscopy (b), light penetrates the sample first and is reflected by the metal mirrors (θ should be 70 to 89.5°). The multiple reflection RA setup is shown in (c). In single reflection IRS (d), light passes through the IRE first and is totally reflected at $\theta > \theta_c$; $\sin \theta_c = n_2/n_1$. The multiple reflection IRS setup is shown in (e). In diffuse reflectance spectroscopy (f), the scattered light is collected by mirrors and directed to the detector. In emission spectroscopy (g), the sample is heated, and the emitted radiation is analyzed. Transmission spectroscopy is shown in (h). In spectral reflection (mirror-like) (i), the angle of incidence equals the angle of reflection. (Reproduced with permission from reference 4. Copyright 1983 Annual Review of Materials Science.)

IR rays impinging on a sample are either reflected, absorbed, transmitted, or scattered. Mathematically,

$$I_o = I_r + I_a + I_t + I_s \qquad (3.8)$$

where I_o is the intensity of the incident ray, I_r is the intensity of the reflected ray, I_s is the intensity of the scattered ray, I_t is the intensity of the transmitted ray, and I_a is the intensity of the absorbed ray. Experimentally, any of these rays can be used to determine the spectrum of the sample. The differences in the sampling methods depend on the angle of incidence and the magnitude of the change in the refractive index, as shown in Figure 3.4. Consequently, transmission spectroscopy is observed at a 90° angle of incidence, and reflectance spectroscopy, including specular, external (both single and multiple), internal (both single and multiple), and diffuse, is observed at smaller angles. Additionally, photoacoustic and emission spectroscopy are used for polymers.

Transmission Spectroscopy

Transmission sampling for IR spectroscopy has several advantages:

- the highest signal-to-noise ratio when the sample has the proper thickness
- no sampling technique convolution of the spectra
- easy quantification

Transmission spectroscopy is primarily useful for *uniform* (no holes, no variation in thickness) thin films of polymeric samples. The polymeric films should be sufficiently thin to allow observation of spectral absorbance values in the linear domain, which is generally an absorbance value of less than 1 for the major portions of the spectra, and particularly

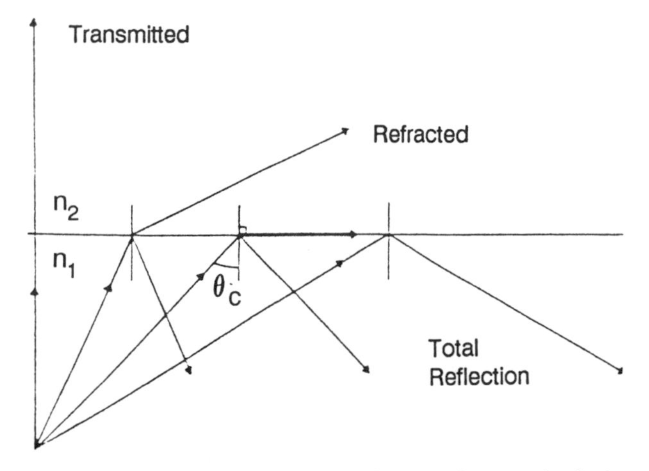

Figure 3.4. The differences in the sampling methods for determining the spectrum of a sample depends on the angle of incidence and the magnitude of the change in the refractive index. (Reproduced with permission from reference 4. Copyright 1983 Annual Review of Materials Science.)

for those frequencies for which spectral subtractions are anticipated.

> For FTIR instrumentation, this absorbance limitation arises from the apodization process. For the common triangular ($sinc^2$) apodization function, a linear relationship is found between the true absorbance and the apparent (observed) absorbance up to about 0.7 absorbance units. Above 0.7 absorbance units, the apparent absorbance becomes nonlinear, with the amount of decrease depending on the true absorbance and the resolution parameter, *r*, where *r* is the ratio of the instrument nominal resolution to the full width at half height. For absorbance values greater than 1, resolution parameters of 0.1 or greater produce nonlinear behavior (*5*).

This need for thin films usually means that polymer film samples should be prepared in the 10 – 100-μm range of thickness. Solvent casting or compression molding of the samples is usually required to prepare such thin films and, unfortunately, these sample preparation techniques transform the sample through melting or dissolution, with the loss of the thermal and process history of the polymer sample. The film samples must be randomly oriented, because nonrandom chain orientation influences the value of the measured absorbances. The basic disadvantage of transmission is just that–transmission; that is, the samples cannot be opaque and must be thin enough to transmit light.

Powders or fluff of polymers can be studied with the transmission technique by forming a sample pellet under high pressure and by compacting the pellet with noninfrared absorbing powders such as KBr, NaCl, or CsI. KBr is used most often. The use of an alkali halide matrix to support and surround a solid sample for IR analysis was first presented by Stimson in 1951 (*6*). This technique has been used widely because of its general applicability. Previously, the only method for recording IR spectra of insoluble solid samples was to suspend or mull the finely divided sample in an inert oil such as mineral oil (Nujol), trifluorovinyl chloride polymer (Fluorolube), or hexachlorobutadiene. Compared to mulling, the pellet technique requires less grinding to achieve good results. The pelleting matrix has no IR spectrum, so only one preparation is required. When the refractive index of the matrix is closely matched to that of the sample, reflective and refractive losses are minimized. Mechanical grinders and vibrating-ball mills reduce the time and tedium of sample grinding. For semicrystalline polymers, however, excessive grinding in a ball-type mill causes changes in the crystalline order. Band narrowing has been observed in studies of poly(ethylene terephthalate) (PET) fibers (*7*). Moisture in the KBr matrix can also be a problem because the moisture results in cloudy pellets that have high scattering.

The KBr pellet method of polymer sampling should be used with caution because the technique involves the use of tremendous pressures that could seriously affect the sample. In addition, intimate contact with the hygroscopic KBr could possibly produce chemical interferences. Under experimental conditions, preparation of clear neat powder pellets is difficult. Imperfect pellets result in light scattering and base line problems. Improper sample dispersion can lead to improper absorbance measurements arising from the wedge shape of the sample.

Internal Reflection Spectroscopy (IRS)

Internal reflection spectroscopy (IRS) (often called *attenuated total reflection* or ATR) is a widely used technique for the analysis of polymer samples with low transmission (*8*). IRS is a contact sampling method involving a crystal with a high refractive index and low IR absorption in the IR region of interest. The depth of penetration depends on a number of factors, including the angle of incidence and refractive index, but is approximately equal to the wavelength of the IR light. The main advantage of IRS is that spectra of opaque samples can be obtained. The length of the IRS crystal determines the sensitivity of the technique because the signal-to-noise ratio is improved with an increase in the number of multiple reflections, which are a function of the length of the crystal.

One problem of IRS is the inability to obtain a reproducible pressure and contact area between the sample and the crystal. The IRS method does not result in a linear plot of signal versus concentration, so careful calibration is required if quantitative information is sought.

A device that has become quite popular is a commercial cell, the ATR CIRCLE Cell (Spectra-Tech), that has a large effective surface area that is approximately 7 – 8 times greater than that of the ordinary rectangular IRS (ATR) crystal. It is very useful for studies of aqueous systems because it is possible to subtract the water absorption in a reproducible fashion (*9*). The depth of penetration, d_p, of the beam can be calculated by using the following equation (*8*):

$$d_p = \frac{\lambda}{2\pi n_1 (\sin^2 \theta - n_{21}^2)^{1/2}} \tag{3.9}$$

where λ is the wavelength of the radiation in air, θ is the angle of incidence, n_1 is the refractive index of the IRS (ATR) crystal, and n_{21} is the ratio of the refractive index of the sample to that of the IRS (ATR) crystal. The *penetration* of the evanescent wave is defined as the distance required for the electric field amplitude to fall to e^{-1} of its value at the surface. For a Ge IRS (ATR) crystal, assuming $\theta = 45°$, $\lambda = 6.1$ μm (the wavelength corresponding to the 1630 cm^{-1} water absorbance band), $n_2 = 4.0$ and $n_{21} = 0.38$ ($n_2 = 1.5$ for water), the effective penetration depth of the electric field is 0.41 μm.

Based on 10 reflections in the IRS (ATR) crystal, the effective transmission-cell path length is 12 μm. The penetration can be further reduced by the addition of a thin metal film on the surface of the IRS (ATR) crystal. This reduction is possible because of the relatively high refractive index and absorptivity of metals with regard to IR radiation (*10*).

Another problem with IRS is the fact that the observed frequencies are shifted from those observed in IR transmission spectroscopy, and the relative intensities can be quite different. This problem can be overcome if the optical constants of the polymer are known.

This apparently little-known fact about ATR has led to some interesting reports in the literature. Because ATR is a contact method, it is generally believed that the technique is a surface analysis technique, and if you believe that a penetration of 1 μm is probing the surface, then perhaps ATR is such a method. However, because frequency shifts are observed as a result of the optical physics of the system, it has often been suggested that the IR spectra of some samples have frequencies that are different for the surface species than for the bulk sample. Well, I certainly believe that this is possible, but generally the observations are a result of IRS-induced frequency shifts rather than differences of a chemical nature.

One of the limitations of IR measurements is the necessity of placing the sample inside the spectrometer. In many cases, this is not possible or desirable. Fiber optics can be used to transport a beam of light from a spectrometer to a remote sampling point. A mid-IR-transmitting optical fiber with a 120-μm core of As – Ge – Se glass has been made, and it transmits light down to about 1000 cm^{-1} (*11, 12*). The fiber has a 90-μm coating of silicon, which limits the useful spectroscopic range to 3250 – 1250 cm^{-1}. These fibers have a high light loss (10 – 15 dB/m), and so the maximum transmission length is 3 m.

The fiber is used as its own internal reflection element to obtain a spectrum. The coating is removed over a small section of the fiber, and this section is immersed into the sample of interest, a liquid or viscous mixture such as an uncured resin (*12*). An accessory that allows these fibers to be used for remote sensing has been developed by Bio-Rad.

External Reflection Spectroscopy

When the sample is deposited on the surface of a smooth mirrorlike substrate, it is possible to use the *specular reflection technique*, or *external reflection spectroscopy* (RA), which is carried out with the beam at near normal incidence (*13*). In RA, the beam makes a high-angle reflection of approximately 88° from the sample. Single or multiple reflections can be used. For the analysis of coatings, multiple reflections are generally required. The RA method has been particularly valuable for examining coatings and adhesives on the surfaces of metals (*14*). The problem with the specular reflection technique is that because of the mirrorlike optical substrate few practical samples can be studied.

When electromagnetic radiation is specularly reflected from a bright surface such as a coated metal, the phase of the wave is shifted as a function of the angle of incidence. For a normal angle of incidence, the phase is shifted by 180°, and the result is a node in the electric field at the surface. Because of this zero field strength, it is impossible to couple the radiation to a thin film on the surface. However, at high angles of incidence and for light polarized in the plane of incidence (*p*-polarization), the phase shift is such that the optical field strength normal to the surface is high. Light polarized normal to the plane of incidence (*s*-polarization) remains phase-shifted by 180°. The number of reflections changes with the angle of incidence, and the optimal angle for obtaining IR spectra with this technique is 88°. Under these conditions, the coupling of the field to the film is between 10 and 50 times more efficient than in a normal incidence transmission measurement.

An approximate expression for the fractional change in reflectivity of the parallel component ($\Delta R_{||}$) is

$$\Delta R_{||} = 1 - \frac{R_1}{R_0} = \frac{abck \tan\theta \sin\theta}{n^3} \quad (3.10)$$

where $k = 4/^{10}\log e$, R_1 and R_0 are the reflectivity with and without the film present, respectively; θ is the angle of incidence; and n is the refractive index. This expression holds for highly reflecting metals such as gold, copper, and silver and for angles of incidence between 0° and 80°.

The polarization sensitivity of RA adds to its usefulness. The incident beam will interact only with those transition moments that are normal to the surface. Therefore, it is possible to determine the orientation of the polymer molecules in the coating relative to the metal. In the same fashion as dichroic measurements are made, the ratio between a spectrum with parallel-polarized radiation and a spectrum with perpendicular-polarized radiation is calculated in order to determine the molecular orientation. In some cases, this orientation sensitivity is also helpful in making band assignments to different functional groups that exhibit a preferred orientation.

The major difficulties with the RA technique are the appearance of asymmetric absorption modes and some shifting of the mode frequencies. Corrections for these spectral differences can be made using the Kramers – Kronig transformation, as will be discussed later in this chapter.

Diffuse Reflectance FTIR Spectroscopy (DRIFT)

The optical phenomenon responsible for diffuse reflectance is multiple reflection (*15*). An IR diffuse reflectance spectrum can be obtained easily if the sample is strongly scattering and weakly absorbing (*15*). Spectra of strongly absorbing samples are obtained by diluting the sample with IR-transparent alkali halide powders. The scattering coefficients depend on the particle size distribution of the alkali halide powders and on the nature of the packing of the sample. For quantitative analyses, the dilution requires accurate weighing and perfect mixing. The main drawback of this method is that the signal is not linearly related to the concentration of the sample, but has to be calculated according to the Kubelka–Munk theory (*16*).

The Kubelka–Munk reflectance, $F(R_\infty)$, is defined as

$$F\,(R_\infty) = \frac{(1 - R_\infty)^2}{2R_\infty} = 2.303\frac{ac}{s} \quad (3.11)$$

where R_∞ is the reflectance of an infinitely thick sample relative to that of a nonabsorbing standard, a is the absorptivity, c is the concentration, and s is a scattering coefficient. Infinite thickness is defined as the sample thickness beyond which any further increase in thickness causes no change in the measured IR spectrum (*17*). In practice, the Kubelka–Munk function is valid only when the amount of *Fresnel reflectance* (defined as radiation that undergoes reflection from the surface of particles but never passes through the particles) is small and the scattering coefficient is independent of the concentration (*18*). Because both Fresnel reflectance and scattering are affected by particle size and packing, reproducible diffuse reflectance spectra require good mixing and sieving of the samples to limit the particle size.

Bulk polymer samples can be studied by using DRIFT (*19, 20*). For the study of the surfaces of bulk samples, a DRIFT accessory should perform the following functions:

- reject specularly reflected radiation
- yield sufficient optical throughput
- accommodate bulky samples
- analyze comparatively small areas

Recently, a DRIFT accessory was developed to examine local areas of bulky samples (*21*).

The basic problem associated with the design of DRIFT cells is the competition between irradiation and detection for the solid angle. With proper considerations, a diffuse reflectance accessory can be designed with efficiencies larger than 0.3 (*21*).

A recent development in DRIFT is the discovery of a simple spectroscopic method of *depth profiling* the surfaces and interfaces of organic–inorganic materials (*22–24*). The new procedure involves the use of a powdered KBr layer on top of the bulk sample to be examined. As the thickness of the overlayer is increased, the ratio of the surface-to-bulk contribution to the recorded spectrum increases. The reflected beam observed at the detector contains contributions from various regions of the sample. By using suitable data processing techniques it is possible to determine the relative contributions of the surface and bulk spectra. This information allows quantitative determinations of differences in structure as a function of depth from the reflecting surface.

My research group applied this technique to the identification and evaluation of 3-aminopropyltriethoxysilane (γ-APS) coupling agent on E-glass fibers (*25*). A circular mat of glass was covered with a uniform layer of KBr, and the DRIFT spectra were recorded. It was then possible to isolate the spectrum of the absorbed γ-APS coupling agent.

Photoacoustic Spectroscopy

Photoacoustic spectroscopy (PAS) is based on the principle that modulated IR radiation striking the surface of a sample will cause the surface to alternately heat and cool with IR absorption (*26*). This cyclic heating is conducted to a coupling gas in the photoacoustic cell. A standing sound wave that can be detected by a microphone develops.

PAS is now widely used for the detection and analysis of various polymeric materials in liquid and solid phases (*27*). The main advantages of PAS are

- minimal sample preparation
- the capability of studying opaque samples
- the capability of depth profiling the sample

Compared to DRIFT, PAS can analyze the neat sample without dilution with KBr powder, but PAS has a lower inherent sensitivity. The limitations of PAS include

- low signal-to-noise ratio per scan
- photoacoustic saturation related to high absorption (the use of thin samples minimizes this effect)
- an artificial increase in the signal due to the thermal dilation of the sample (this dilation is sensitive to the absorption of the whole sample; that is, there is an acoustic background between absorption bands)
- an increase in the heat sources (texture or matrix effects) in heterogeneous samples due to light scattering.

The PAS cell must be sealable so that the coupling gas does not leak out of the chamber during the experiment. One of the utilities of PAS is the ability to obtain a depth profile from the surface by varying the light modulation frequency. The sublayer structure can be studied down to a depth corresponding to half of the thermal diffusion length (*28*).

PAS is a rapidly growing technique for obtaining IR spectra of samples that are hard to prepare as transparent films, that have high internal light scattering, or that are coated onto opaque or strongly light-scattering substrates. Thus the PAS technique is complementary to the IRS (ATR) and diffuse reflectance techniques and has the advantage of no sample preparation. Typically, only 0.25 mL of sample is required, and the specimen usually can be examined "as received". The signal-to-noise ratio for PAS is low, so a longer scanning time (1000 – 6000 scans) relative to transmission FTIR spectroscopy is required.

In the study of polymer surfaces, PAS can enhance the surface modes by using a highly polarizable inert gas (*29, 30*). When a highly polarizable coupling gas is used in the photoacoustic cell, useful information can be obtained regarding the species present and their orientation with respect to the surface of the adsorbate. A distinction between the two types of adsorbed molecules on the surface (parallel and perpendicular to the surface) can be made. This technique is simple and nondestructive, and requires a routine photoacoustic setup. It can be applied to virtually any sample, and it represents a new frontier that should expand upon our traditional ideas of the bonding and structure of surface-adsorbed species.

This surface-specific PAS has been applied to the study of polyamide fibers (Kevlar) before and after oxidative treatment (*31*). The results are shown in Figures 3.5 and 3.6. Kevlar fibers have the unique morphology of a core with an outer layer that can be removed by oxidation. Comparison of these two figures demonstrates that the molecular chains are oriented perpendicularly to each other in the outer layer and in the core. This information can be obtained by comparing the intensities of the C – H modes with the N – H modes before and after the removal of the outer layer by oxidation.

A novel application of PAS is the use of rheophotoacoustic measurements to perform stress – strain studies on polymeric materials (*32*). Propagating acoustic waves are detected as a result of the deformation.

Emission Spectroscopy

Kirchoff's theorem states that the ratio of a body's spectral luminance to its absorbance, L/a, is a func-

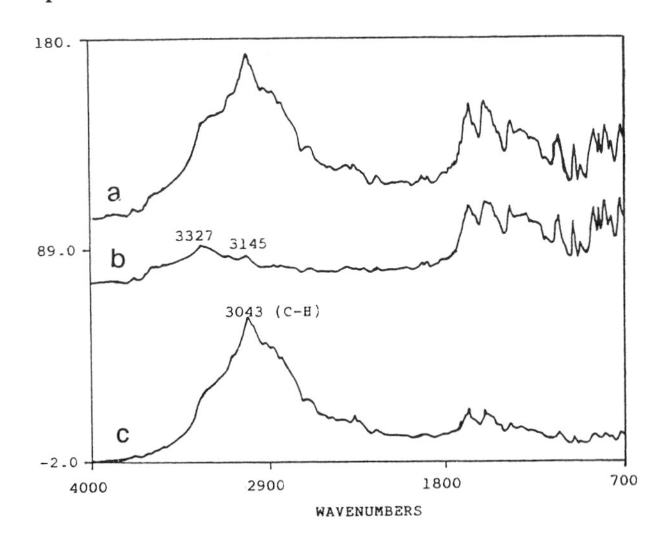

Figure 3.5. PAS spectrum of Kevlar-49 fibers before oxidative treatment. Spectrum a is the PAS spectrum obtained with xenon, spectrum b is the PAS spectrum obtained with helium, and spectrum c is the difference spectrum. (Reproduced with permission from reference 31. Copyright 1986 Hüthig & Wepf Verlag.)

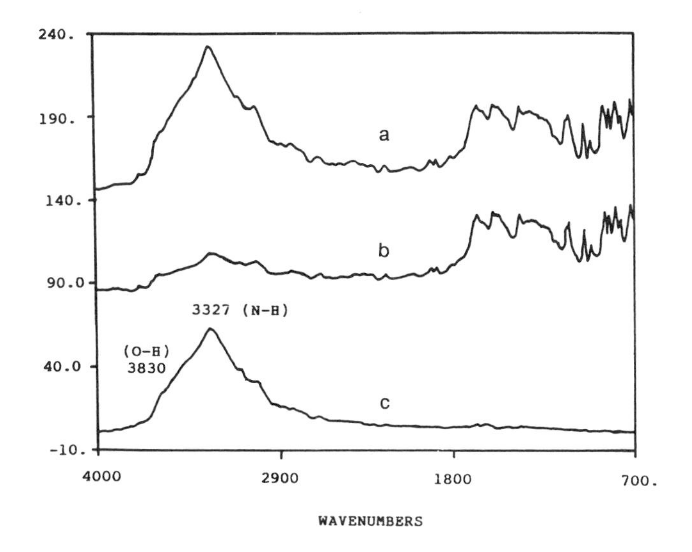

Figure 3.6. PAS spectrum of oxidized Kevlar-49 fibers. Spectrum a is the PAS spectrum obtained with xenon, spectrum b is the PAS spectrum obtained with helium, and spectrum c is the difference spectrum. (Reproduced with permission from reference 31. Copyright 1986 Hüthig & Wepf Verlag.)

tion of the frequency and temperature of the body but not a function of the nature of the body or its geometrical dimensions. This ratio is equal to the spectral luminance of a perfect blackbody

$$L / a = L_{BB} \tag{3.12}$$

By definition, a_{BB} is equal to 1. The emittance, ϵ, is defined as

$$\epsilon = L / L_{BB} \tag{3.13}$$

and therefore,

$$\epsilon = a \tag{3.14}$$

In theory, the emittance and absorbance of a sample are equal.

Under proper conditions, the emission spectra are the mirror images of the absorbance spectra. Therefore, *emission spectroscopy* (EMS) can provide the same information as transmission spectroscopy, but emission spectroscopy can be used on samples such as metals and opaque substrates that cannot be studied by transmission. The signal of the emission spectrum increases with the temperature difference between the sample and the detector according to Stefan's law.

In IR emission spectroscopy, however, several forms of distortion can occur, and of these, background emission is the most important one and requires elimination. Removal of background emission involves measuring the spectra of both the sample and a blackbody reference material at two temperatures (*33*). The measured intensity, $S(\nu, T)$ has several components:

$$S(\nu,T) = R(\nu)[\epsilon(\nu,T)H(\nu,T) + B(\nu) + I(\nu)\rho(\nu) \tag{3.15}$$

where $R(\nu)$ is the instrument response function, $\epsilon(\nu, T)$ is the emittance of the material, $H(\nu, T)$ is the Planck function, $B(\nu)$ is the background radiation, $I(\nu)$ is background radiation reflected off the sample, and $\rho(\nu)$ is the reflectance of the sample. For the blackbody reference material, $\epsilon = 1$ and $\rho = 0$. So the intensities measured on the reference material are

$$S_1(\nu,T_1) = R(\nu)[H(\nu,T_1) + B(\nu)] \tag{3.16}$$

$$S_2(\nu,T_2) = R(\nu)[H(\nu,T_2) + B(\nu)] \tag{3.17}$$

For the sample, the measured intensities are

$$S_3(\nu,T_1) = R(\nu)[\epsilon(\nu,T_1)H(\nu,T_1) + B(\nu) + I(\nu)\rho(\nu)] \tag{3.18}$$

$$S_4(\nu,T_2) = R(\nu)[\epsilon(\nu,T_2)H(\nu,T_1) + B(\nu) + I(\nu)\rho(\nu)] \tag{3.19}$$

Assuming that the sample emittance and reflectance are independent of temperature over the operating range, the emittance is calculated as

$$\epsilon = \frac{S_4 - S_2}{S_3 - S_1} \tag{3.20}$$

Subtracting the interferograms rather than the spectra has been suggested to avoid phase corrections (*33*).

The IR emittance of a sample relative to a blackbody is dependent on the reflectivity at the surface of the sample. With a blackbody as a reference, band distortions such as frequency shift and false splittings resulting from reflectivity variations in the vicinity of strong bands can still be encountered (*34*). Under these circumstances, an opaque sample rather than a blackbody should be used as a reference (*34*).

A pure emission spectrum can originate only from a material that does not have thermal gradients. Readsorption of thermal emission can occur when the radiation emitted by a hot portion of the sample passes through a cooler portion of the sample. The IR emission line shapes for various temperature gradients within the sample are shown in Figure 3.7 (*35*). Self-absorption in optically thick samples can result in emission spectra that resemble those of a blackbody, containing no spectral bands characteristic of the material under analysis.

The usual method of reducing self-absorption is to use a thinner sample so that the thermal gradients are smaller. A novel approach to this problem was made by recognizing that it is necessary to reduce the thickness of only the material actually emitting radiation. By using a laser as the heating source, it is possible to generate a thin, transiently heated surface layer of the material and to collect the radiation emitted from only this layer (*36*). For the experiments reported, the transiently heated layer was created by using a fixed continuous laser beam that was focused on a rotating sample.

> In concert with our love of or preoccupation with acronyms, this technique is called TIRES, meaning *transient IR emission spectroscopy*. Clever? This TIRES technique bears watching as its goes down the road. Oh my!

The emission technique is limited to the range of frequencies greater than 2000 cm^{-1} because of the low flux of energy in the lower frequency region. Measurements down to 400 cm^{-1} are possible if the entire spectrometer and its surroundings are cooled. However, this option is not usually practical.

The sample also may be heated to elevated temperatures, but sample decomposition then becomes a problem.

Because emission occurs primarily from the surface of the sample, the EMS technique has potential for surface analysis. The problem with extending the emission technique to surface analysis is the strong background emission that is superimposed on the weak emission of the surface coating. One method of increasing the emission signal is to collect the emitted radiation over a large solid angle by using an ellipsoidal mirror (*37*). The maximum emission from a surface occurs when the viewing angle is 70 – 80°

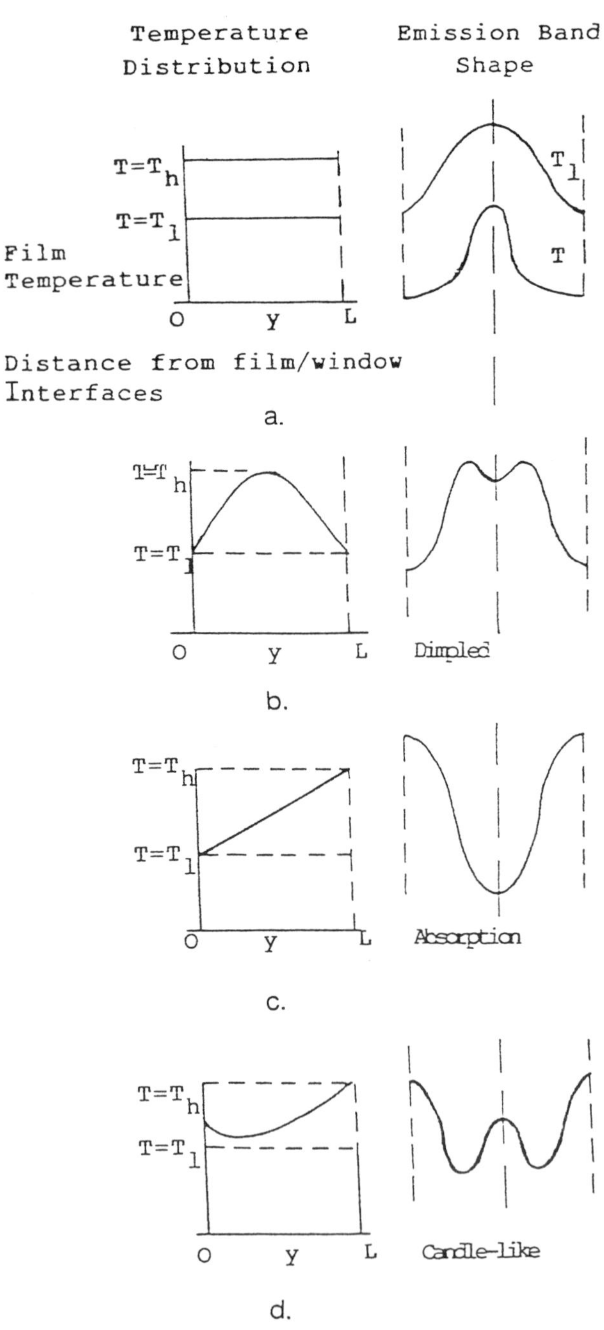

Figure 3.7. IR emission line shapes for various temperature gradients within a sample. (a) no temperature gradient, (b) convex temperature gradient, (c) positive temperature gradient, and (d) concave temperature gradient. (Reproduced with permission from reference 35. Copyright 1981 American Society of Mechanical Engineers.)

from the normal, and the emission is approximately zero in the direction normal to the surface. This zero emission arises because of destructive interference between emitted and reflected light from the interface to the metal surface.

For surface emission studies, the emittance is defined as

$$\epsilon = \frac{S_s - S_b}{S_R - S_b} \tag{3.21}$$

where S is the measured intensity, and the subscripts s, R, and b denote the samples, reference substrate, and background emission from the beam splitter, respectively (*38*). The emission spectra of different thicknesses of poly(acrylonitrile-*co*-styrene) on aluminum at 150°C are shown in Figure 3.8.

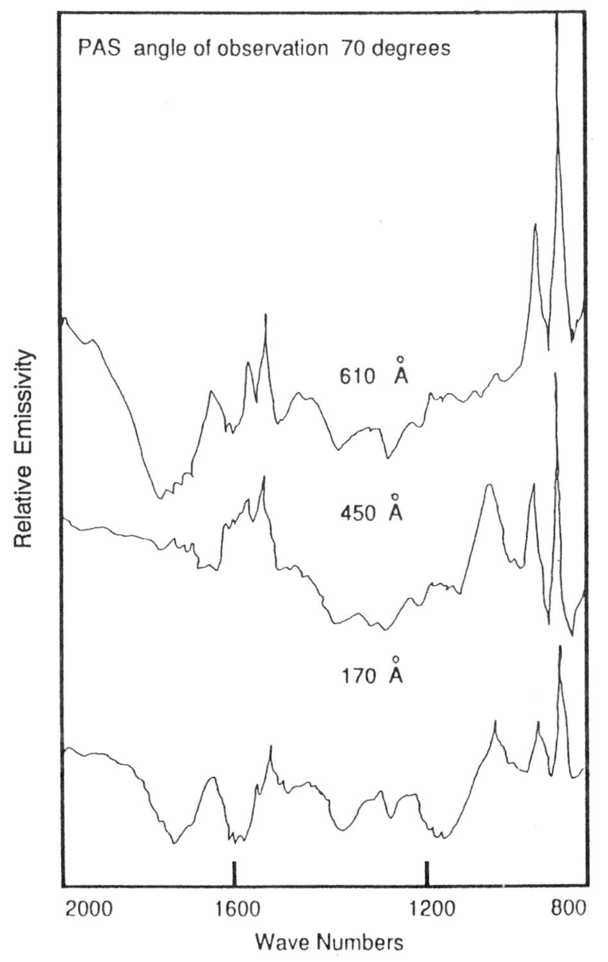

*Figure 3.8. Emission spectra of thin films of poly(acrylonitrile-*co*-styrene) at 150 °C. (Reproduced with permission from reference 38. Copyright 1984 Society for Applied Spectroscopy.)*

The intensity is a linear function of thickness up to 1000 Å, as illustrated in Figure 3.9.

Microsampling Techniques

The applications of IR microspectroscopy have increased as quality IR microscopes with spatial resolution of approximately 50 μm have been developed (*39*). IR microscopes can be used for both transmission and reflection measurements. The characteristics of FTIR spectroscopy that have made these microscopes possible are the energy throughput and multiplex advantages.

A typical IR microscope is all-reflecting, with visible illumination for visual examination of the sample and with a dedicated on-axis small area mercury–cadmium telluride (MCT) detector. The sampling size can vary from less than 50 × 50 μm in linear dimensions to the dimensions of the detector element. The sample is placed on a standard microscope X–Y stage and can be visually examined under a variety of magnifications. The sample area of interest is isolated by placing a variable aperture on an inter-

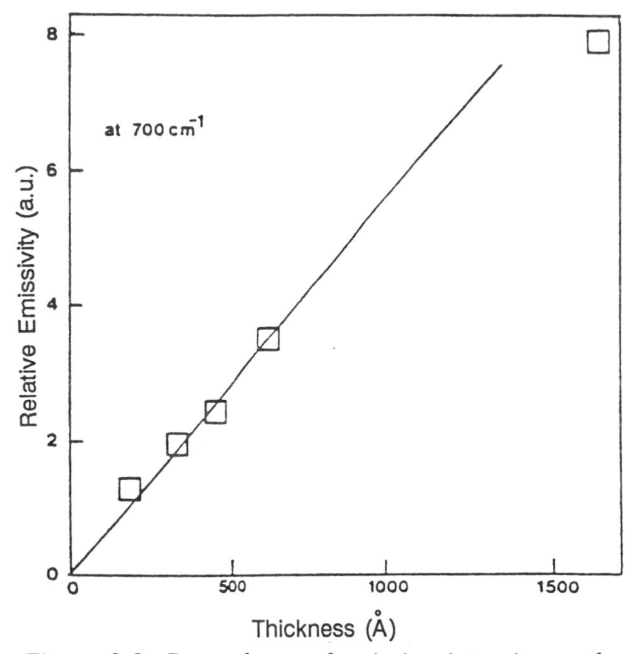

Figure 3.9. Dependence of emission intensity on the thickness of the thin polymer films on aluminum. (Reproduced with permission from reference 38. Copyright 1984 Society for Applied Spectroscopy.)

mediate image of the sample within the barrel of the microscope.

The FTIR microscope is particularly useful for polymer studies because it is relatively easy to obtain a minute polymer specimen, by using a razor blade or a knife, that is thin enough to analyze by transmission. The specimen can be placed in a KBr pellet die or a diamond anvil and pressed to a thickness suitable for analysis.

The IR microscope is a valuable tool for the analysis of fibers, particulates, and inclusions. A dedicated microscope – FTIR instrument with appropriate hardware and software to perform a wide range of microscope experiments has been developed recently.

Recently, IR microspectroscopy has been used to develop two-dimensional functional group images of polymer samples (*40, 41*). A two-dimensional computer-restored step scan was performed on the sample in order to obtain a compositional map. A 250- × 250-μm aperture was used in 250-μm steps in the *x* and *y* directions with the aid of a computer-controlled moving stage. Although the technique is powerful, it has two major disadvantages. The resolution is 3 to 10 times less than with Raman and fluorescence microbeam methods, and the time required for data collection and reduction can be as long as 24 h.

Data Processing of Digitized IR Spectra

Because the FTIR spectral data are recorded in digital form, a variety of digital data processing techniques to eliminate spectral distortions are available. These distortions can arise from sample scattering and reflection or from ATR, photoacoustic, and diffuse reflection sampling devices. Digital data processing techniques also are used to isolate spectral features for study and quantification.

Elimination of Spectral Scattering and Reflection

The main source of the background noise in IR spectroscopy of polymers is the partial nonabsorbing scattering of the IR radiation from inhomogeneities in the polymer sample. Other possible sources of background noise include tilting or improper orientation of samples, interference effects, different size sample holders, and imperfect instrumental optics. Generally, an extraneous background is present in all types of spectral data, and its elimination or correction is necessary before proceeding to other data manipulation procedures.

There is no perfect base line correction, and all corrections are approximates. The simplest base line is a flat line that is parallel to the abscissal axis. Linear base lines with negative or positive slopes have been applied, as well as more complex forms. Several methods are available for base line correction (*42*). As a simple approach, a wedge defined by two points can be subtracted from the absorbance spectrum. Also, the entire spectrum or selected frequency ranges of the spectrum can be fit to a linear least-squares line, repeating the fitting until some predetermined deviation is satisfied. It is important to be able to correct a base line by using a method that avoids as much as possible subjective interaction.

My research group proposed a base line-correction algorithm that is based on the use of a successive least-squares routine and that includes objective criteria on the convergence of a base line that is either linear or quadratic (*42*). The results obtained by using this algorithm are shown in Figure 3.10. The dotted lines are the least-squares fitted lines, and they converge progressively to the line at the bottom. The result of the base line correction of Figure 3.10 is illustrated in Figure 3.11 (*42*).

When a sample is highly uniform in thickness and has smooth surfaces, another irritating spectral ef-

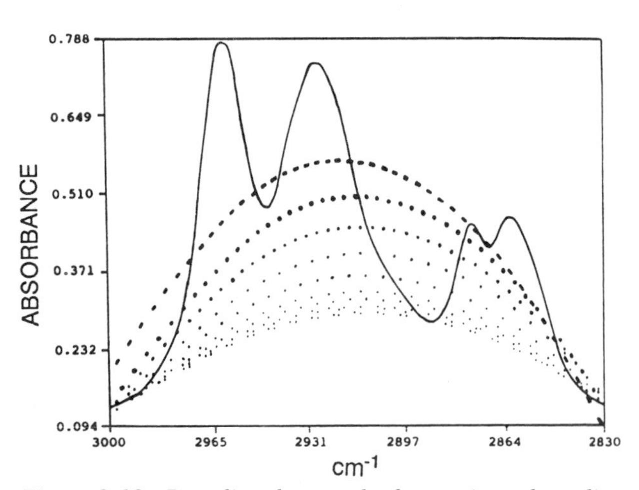

Figure 3.10. Base line that results from using a base-line correction algorithm. (Reproduced with permission from reference 42. Copyright 1987 Society for Applied Spectroscopy.)

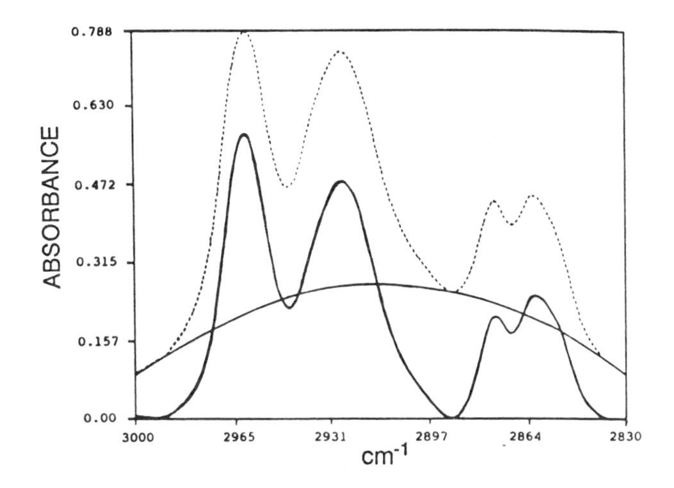

Figure 3.11. Spectrum resulting from the base-line correction of Figure 3.10. (Reproduced with permission from reference 42. Copyright 1987 Society for Applied Spectroscopy.)

fect, *interference fringes*, can be observed. FTIR spectroscopy offers some possibilities for eliminating fringes if the "signature" can be detected in the interferogram. This fringe signature appears as a single spike in the interferogram, and sometimes it can be removed. Consequently, the fringes in the transformed spectra do not appear (*43*). Another approach to removing the fringes is to use direct digital subtraction (*44*). Reflection interference produces a sine wave in the spectrum. A computer file containing a sine wave of the same amplitude and phase as the experimentally observed curve can be generated, and this computer-generated curve can be subtracted from the observed fringes. However, it often easier to simply make the surface of the films rough by rubbing with KBr powder or steel wool.

Elimination of Spectral Distortion Resulting from Sampling Technique

For the analysis of polymers by FTIR spectroscopy, various types of optical sampling experiments are used, depending on the nature of the sample.

Different methods, such as transmission, specular reflection, or attenuated total reflectance, yield spectra that differ in the position, shape, and intensity of the bands, even when all of the experiments are performed on the same sample. In all cases, it would be preferable to obtain an absorbance spectrum that is characteristic of only the chemical structure of the sample without any effects from the optical geometry of the experiment. Furthermore, it is desirable to be able to make direct comparisons between the spectra obtained with different sampling techniques.

The refractive index, n, and absorption index, k, determine the response of a material to incident electromagnetic radiation. Collectively, n and k are referred to as the *optical constants*; however, they are not strictly constants because they vary with frequency. At the interface between two phases, a propagating electromagnetic wave will be both reflected and transmitted. The amount of the wave that is either reflected or transmitted is determined by the relative optical constants of the two phases and by the angle of incidence. By using a Kramers–Kronig transformation, the nature of the response can be determined (*45*). My research group developed an algorithm to simulate ATR and specular reflectance of IR experiments. The results obtained by using these two methods, as well as those obtained by transmission spectroscopy, for a thin film of poly(methyl methacrylate) (PMMA) on a Ge substrate are compared in Figure 3.12. The three spectra show large changes in the peak position and shape. The carbonyl band is shifted to higher frequencies by 12 cm^{-1} with specular reflectance and to lower frequencies by 3 cm^{-1} with ATR. The specular reflec-

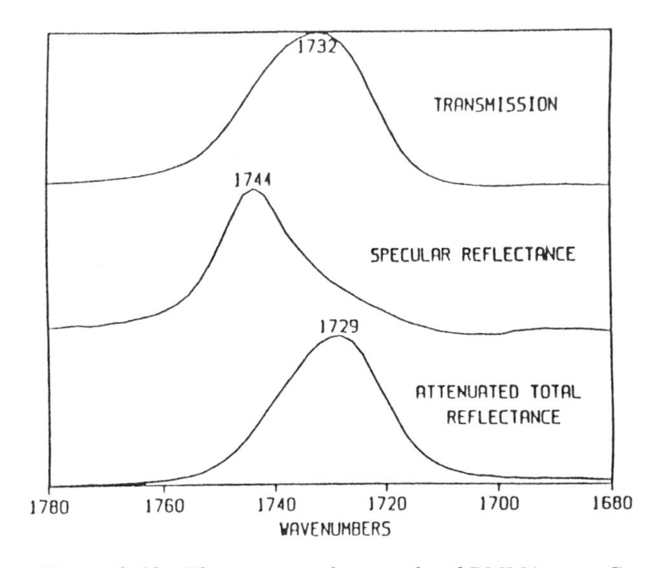

Figure 3.12. The spectra of a sample of PMMA on a Ge substrate obtained by transmission, specular reflection, and ATR spectroscopy. (Reproduced with permission from reference 45. Copyright 1987 Plenum Press.)

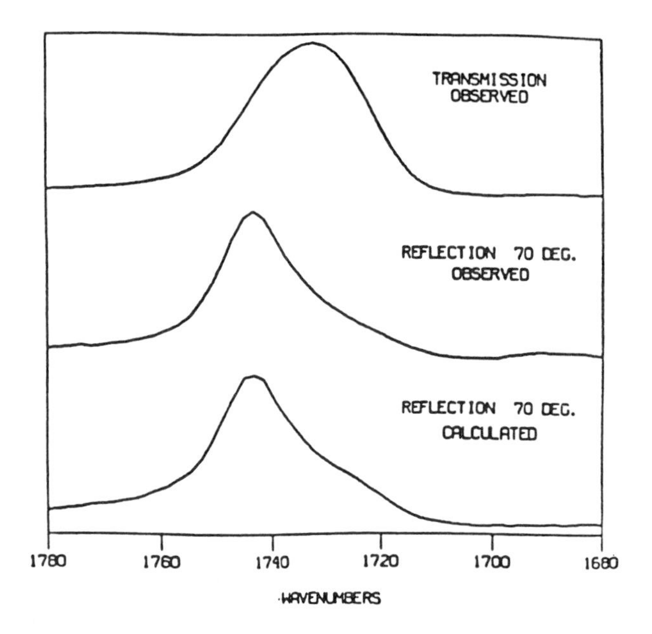

Figure 3.13. Comparison of the transformed spectrum of the specular reflectance experiment with the transmission and reflection spectra of PMMA on a Ge substrate. (Reproduced with permission from reference 45. Copyright 1985 Society for Applied Spectroscopy.)

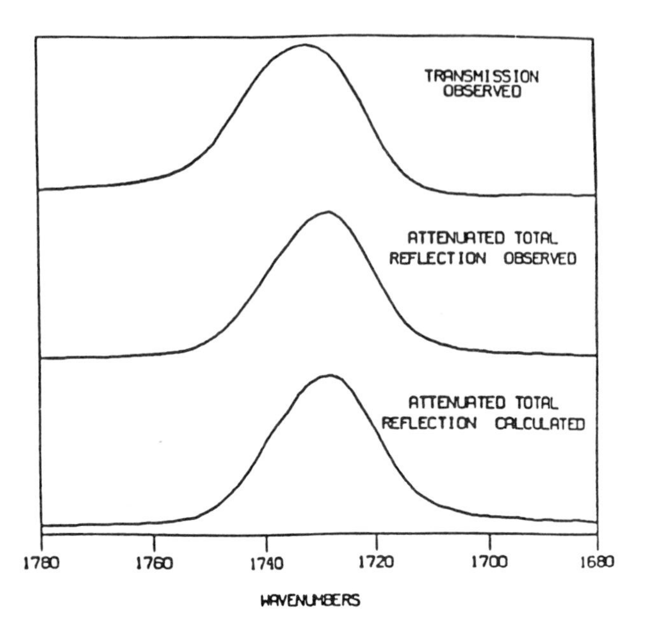

Figure 3.14. Comparison of the transformed ATR spectrum with the observed transmission and ATR spectra. (Reproduced with permission from reference 45. Copyright 1985 Society for Applied Spectroscopy.)

tance and ATR spectra show an asymmetric carbonyl band, while the transmission spectrum gives the correct profile. Yet, the same sample was used for all three experiments.

By using optical constants previously derived for PMMA, the expected band profiles for specular reflection and ATR experiments were calculated. For the ATR experiment, 13 reflections occur on the face of the Ge element covered by PMMA. For the specular reflectance experiment, all three interfaces (air – film, film – Ge, and Ge – air) must be included in the calculations to obtain the correct band intensities. The results of these calculations are shown in Figures 3.13 and 3.14. The peak shifts and shape changes evident in Figure 3.12 are completely accounted for by the Kramers – Kronig calculation.

Spectral Subtraction

Spectral subtraction, which has widespread utility, has become an important tool in the arsenal of the IR spectroscopist. Spectral subtraction allows the study of solid samples with high precision because corrections can be made for differences in sample thickness. Interfering absorbances can be removed, and the spectrum can be purified by removing artifacts, solvents, and other impurities, with the final result being the pure spectra of the components when they can be interconverted to each other or have a constant sum of concentrations.

There are many problems in using spectral subtraction, including the interaction of the individual components with each other to produce frequency shifts and changes in intensities. Problems in subtraction arise when the samples are too thick or are imperfect so that the absorbances are not linear. Also, spectral subtraction is applicable only to spectra obtained in the absorbance mode; that is, spectra plotted in the transmission mode are exponential rather than linear and should never be digitally subtracted. Optical subtraction (solution minus solvent) is possible if a truly double-beam instrument with perfectly matched cells is available.

Spectral subtraction requires the prior scaling of the spectra to correct for variations in the thicknesses of the samples in the IR beam. The spectroscopy of polymers usually involves working with solids rather than with solutions or liquids. With solids, it is almost impossible to match the thicknesses of the different samples in the beam, so a means must be found to correct for this mismatch in order to perform spectral subtraction. If the spectra obtained are in the linear optical range, a reference spectrum can be subtracted from the sample spectrum. The absorbances can be scaled to correct for differences in the amount of sample in the beam if an internal thickness band for the sample can be selected. An *internal thickness band* is a band whose intensity is a function of only the amount of sample in the beam. Generally, isolated bond stretching modes such as the C–H modes have this important property. The absorbance of the internal thickness band for the reference sample can be written

$$A_{ir} = a_{ir}C_i b_r \qquad (3.22)$$

where A_{ir} is the absorbance of the internal thickness band of reference sample; a_{ir} is the absorptivity coefficient; C_i is the concentration of the internal thickness band, i, that is, $C_i = 1$; and b_r is the thickness of the reference sample.

For the sample,

$$A_{is} = a_{is}C_i b_s \qquad (3.23)$$

where A_{is} is the absorbance of the internal thickness band of the analytical sample; a_{is} is the absorptivity coefficient for the analytical sample; C_i is the concentration of internal thickness band, that is, $C_i = 1$; and b_s is the thickness of the analytical sample.

The reference spectrum is scaled by using a scaling factor, k, such that the following equation is satisfied:

$$A_{is} - kA_{ir} = 0 = a_{is}C_i b_s - k(a_{ir}C_i b_r) \qquad (3.24)$$

so

$$k = b_s / b_r \qquad (3.25)$$

The scaled reference spectrum can then be subtracted from the spectrum of the analytical sample to produce a difference spectrum. Peak absorbances, integrated peak areas, or a least-squares curve-fitting method can be used to calculate the scaling factor, k. The method of choice will depend on the system being examined. Remember that the calculation of the scaling factor cannot be done entirely analytically, and the only test of the scaling factor is the resultant difference spectrum, which should be examined carefully before any further analysis is carried out.

Absorbance subtraction is based on two fundamental assumptions. First, it is assumed that the absorbance and shape of a band does not change with the optical thickness. Generally, the absorbance values must be less than 1 in order to meet this requirement. The assumption is tested with every subtraction; if the residual absorbance after a subtraction has a different shape than the original absorbance bands, the assumption has been violated, and the procedure or samples should be reexamined for the cause of the nonlinear effects (i.e., holes, nonuniform samples, sample orientation, vignetting, etc.). The second assumption is that the absorbance of a mixture is the linear sum of the absorbances of the components; that is, the components do not interact with each other differently at different relative concentrations. Concentration-dependent interactions will lead to frequency shifts and band shape changes, but rarely are such effects observed in solids.

Care must be exercised because the technique is also sensitive to optical spectroscopic errors. It is probably not a misstatement that absorbance subtraction has generated more artifacts than facts because of lack of attention to the proper sample preparation procedures and spectroscopic techniques.

For example, if the sample is nonuniform or wedge shaped, errors can occur. A false difference spectrum will be obtained from two spectra of the same sample with the same mean thickness if the samples are wedge shaped. The severity of the wedging effect depends on the method used to determine the scaling factor and on the optical thickness of the sample.

Another problem often encountered in absorbance subtraction is the preferential orientation of one of the samples relative to the other. Unfortunately, there is no general method or algorithm to eliminate this orientation problem because each band has its own characteristic dichroic behavior. The only solution is to prepare the samples with reproducible random orientation. Unless this is accomplished, three-dimensional tilted sample methods must be used. The three-dimensional tilted sample method removes the orientation of the sample and generates a pure "structure-factor" spectrum prior to absorbance subtraction.

If a properly calculated value of k is used, the spectral differences will be independent of sample thickness, and only structural differences will be observed. Additionally, with properly compensated thicknesses and the removal of strong interfering absorbances, the dynamic range of the difference spectrum is large enough to permit the differences in absorbances to be magnified through computer scale expansion, and reveal small details. By using spectral subtraction, the spectral changes that have occurred can be accentuated, and the spectral contributions that did not change can be minimized. Historically, spectral subtraction led to one of the first applications of FTIR spectroscopy to polymers, that is, the study of defects introduced during polymerization at different temperatures (*46*). In Figure 3.15, the FTIR spectra at 70 °C in the frequency range of 500 – 3200 cm^{-1} is shown for *trans*-1,4-polychloroprene polymerized at – 20 °C (spectrum a) and at – 40 °C (spectrum b). The b – a difference spectrum, which is also shown, reflects the increased presence of defects in the polychloroprene polymerized at – 40 °C relative to the sample polymerized at – 20 °C.

Absorbance subtraction can be considered a spectroscopic *separation* or *purification* technique for some problems inherent in polymers. An interesting application of FTIR difference spectroscopy is the spectral separation of a composite spectrum of a two-component system. However, complications can arise when there are intermolecular interactions that perturb the frequencies and intensities (*47*).

One example of the spectral separation approach using absorbance subtraction is a polymer containing two different geometric isomers, G_a and G_b. In general, the IR spectrum of each of these isomers will be different, but because there is a mixture of the two, the spectra of the individual species are difficult to observe. Proper use of subtraction allows the isolation of the spectra of the individual isomers.

The total absorbance, A_t, at a frequency, ν_i, of the polymer may be decomposed into the following contributions

$$A_t(\nu_i)_A = A_a(\nu_i)_A + A_b(\nu_i)_A \quad \text{(thickness compensated spectrum)} \quad (3.26)$$

where $A_a(\nu_i)$ and $A_b(\nu_i)$ are the contributions to the total absorbance at frequency ν_i resulting from the different geometric isomers, G_a and G_b, respectively, in sample A. For sample B, with a different relative amount of the two isomers, a similar equation can be written:

$$A_t(\nu_i)_B = A_a(\nu_i)_B + A_b(\nu_i)_B \quad \text{(thickness compensated spectrum)} \quad (3.27)$$

In order to proceed, isolated bands of both of the geometric isomers must be available for a second subtraction. For this second subtraction, the scale parameter, k_a, is chosen such that

$$A(\nu_{ia})_A - k_a A(\nu_{ia})_B = 0 \quad (3.28)$$

The resultant difference spectrum is the "purified" spectrum of the geometric isomer G_b. A similar set of equations holds for the spectrum of the geometric isomer, G_a. This technique can be applied to other separations where a variation in the relative amount of the structural components occurs.

Consider two components, a and b, which are interconverted by some external treatment. In the spectrum of any specimen, the total absorbance of one specimen, A_1, can be expressed as

$$A_1 = A_a + A_b \quad (3.29)$$

With the appropriate treatment (annealing and extraction) of this sample, a certain fraction x of the structure a is converted to b. The total absorbance of this new sample is

$$A_2 = (x) A_a + (1 - x) A_b \quad (3.30)$$

To remove the characteristic absorptions of structure a subtraction is performed:

$$A_s = A_2 - kA_1 \quad (3.31)$$

When $k = 1 - x$ or $k = x$,

$$A_s = A_a \quad \text{or} \quad A_s = A_b \quad (3.32)$$

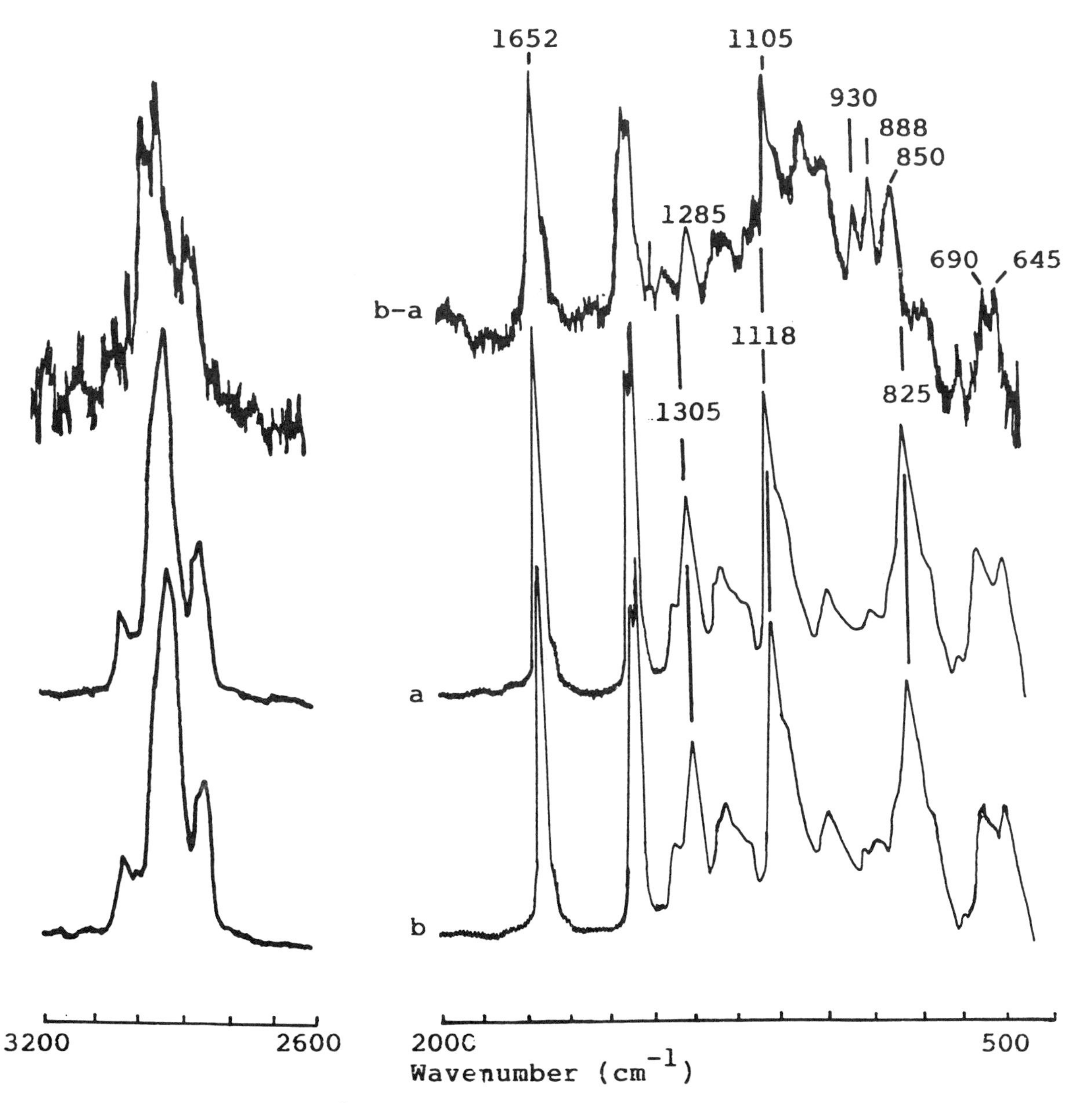

Figure 3.15. FTIR spectra at 70 °C in the range 500 – 3200 cm⁻¹. (a) trans-1,4-polychloroprene polymerized at −20 °C, and (b) at −40 °C; b−a, difference spectrum. (Reproduced with permission from reference 48. Copyright 1975 John Wiley Sons, Inc.)

and the remaining spectrum represents either component a or component b.

This technique was first applied to the isolation of the crystalline isomer vibrational bands of *trans*-1,4-polychloroprene (*48*). The spectrum of a cast film of predominately (>90%) *trans*-1,4-polychloroprene polymerized at −20 °C was compared with the spectrum of the same sample heated to 80 °C (above the melting point) for 15 min (Figure 3.16). Elimination of the amorphous contribution of the composite semicrystalline spectrum was accomplished by subtracting spectrum b from spectrum a until the bands at 602 and 1227 cm^{-1} were reduced to the base line. The "purified" crystalline isomer spectrum, spectrum c at the top of Figure 3.16, exhibits the sharp band structure expected for a regular crystalline

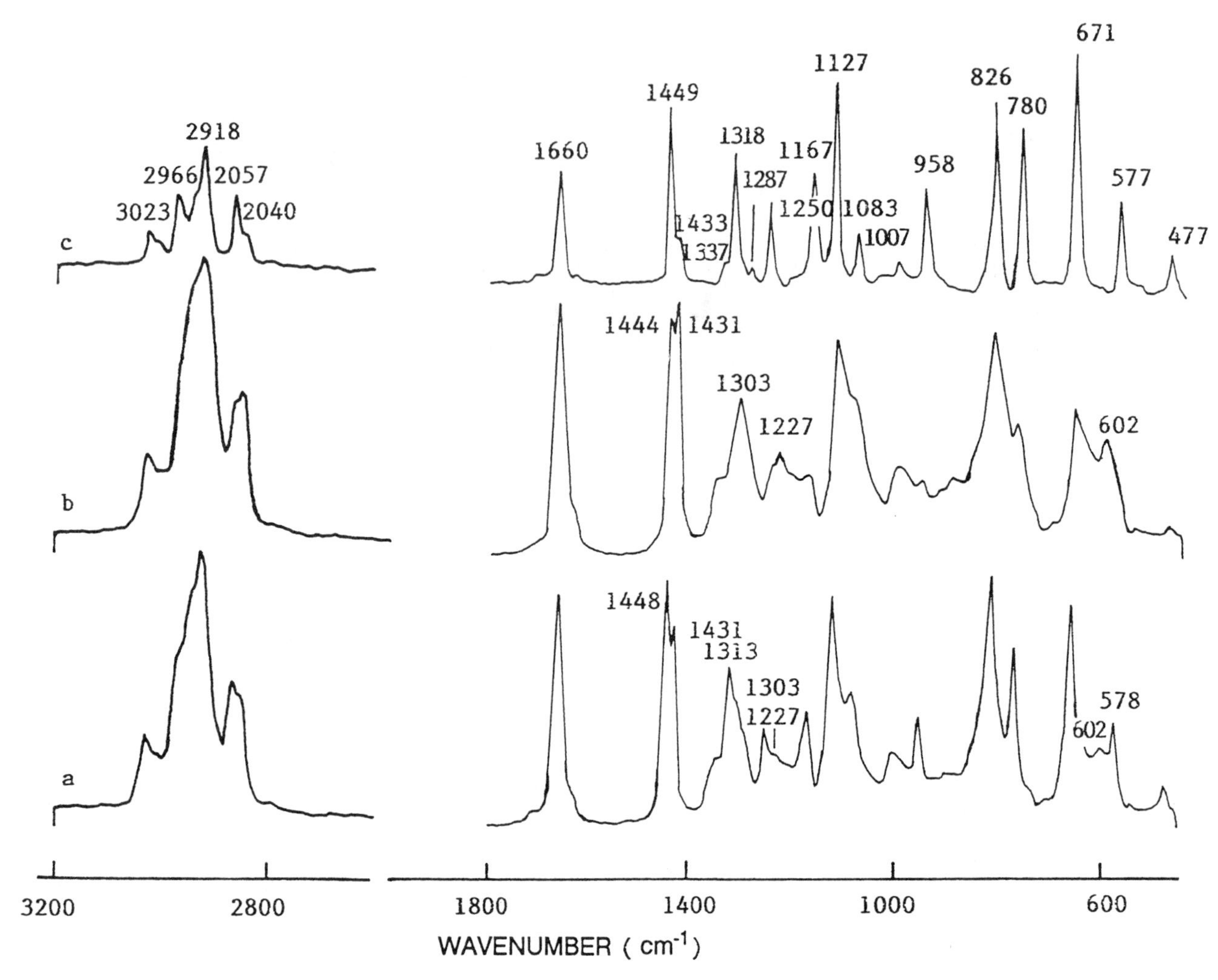

Figure 3.16. The spectrum of a cast film of predominately (>90%) trans-1,4-polychloroprene polymerized at −20 °C (spectrum a) is compared with the same sample heated to 80 °C (above the melting point) for 15 min (spectrum b). The "purified" crystalline isomer spectrum (spectrum c) exhibits the sharp band structure expected for a regular crystalline array. (Reproduced with permission from reference 48. Copyright 1974 John Wiley Sons, Inc.)

array. The interesting aspect of the crystalline isomer spectrum was that when the crystalline component spectra were obtained for samples polymerized at different temperatures through the same procedure, the crystalline isomer vibrational frequencies were different (*48*). This should not be the case if the crystalline isomer phase had the same isomeric structure. However, the spectra indicated that structural defects were incorporated into the crystalline domains, and the rate of defect formation increased as the polymerization temperature increased. As a result, the defects in the crystalline phase decoupled some of the longer sequences, and the spectrum of the crystalline isomer phase was different.

To date, this is the only crystalline isomeric spectrum showing this effect, yet it was the first observed!! As you can well imagine, this unique observation has generated a lot of concern on the part of the researchers involved.

The separation of the crystalline and amorphous isomeric phases into their respective spectra has been carried out for a number of polymers including poly(ethylene terephthalate) (*49*), polystyrene (*50*), poly(vinyl chloride) (*51*), polyethylene (*52, 53*), polypropylene (*54*), and poly(vinylidene fluoride) (PVDF) (*55*).

Digital subtraction of absorbance spectra is often

used to reveal or emphasize subtle differences between two samples. When a polymer is examined before and after a chemical or physical treatment, and the original spectrum is subtracted from the final spectrum, positive absorbances in the difference spectrum reflect the structures that were formed during the chemical or physical treatment, and negative absorbances reflect those structures that were lost.

An example of this type of application is shown in Figure 3.17 in which the difference spectra for polyethylene films irradiated in nitrogen and air for 50 and 100 h are shown (*56*). The positive peaks reflect the products of the reaction, including the carbonyl groups formed by oxidation, the methyl groups (1378 cm^{-1}) formed by by chain scission, and the isomerization of the vinyl end groups (965 cm^{-1}). The negative peaks are the reactants, including the amorphous methylene groups (1368, 1353, and 1308 cm^{-1}) and the loss of unsaturation of the end groups (991 and 909 cm^{-1}).

Figure 3.18 illustrates the use of subtraction to study blends. The FTIR spectra in the range 700 – 2000 cm^{-1} are shown for a compatible blend and an incompatible blend of PMMA – PVDF (*57*). The top spectrum is the difference spectrum of the compatible blend of PVDF – PMMA (39:61) minus PMMA, and the middle spectrum is the incompatible blend of PVDF – PMMA (75:25) minus PMMA. The bottom of the figure shows the spectrum of pure PVDF. Comparison of the spectra shows that the spectrum of PVDF is substantially different from that of the compatible blend but it is very similar to that of the incompatible blend, a result indicating phase separation in the incompatible blend.

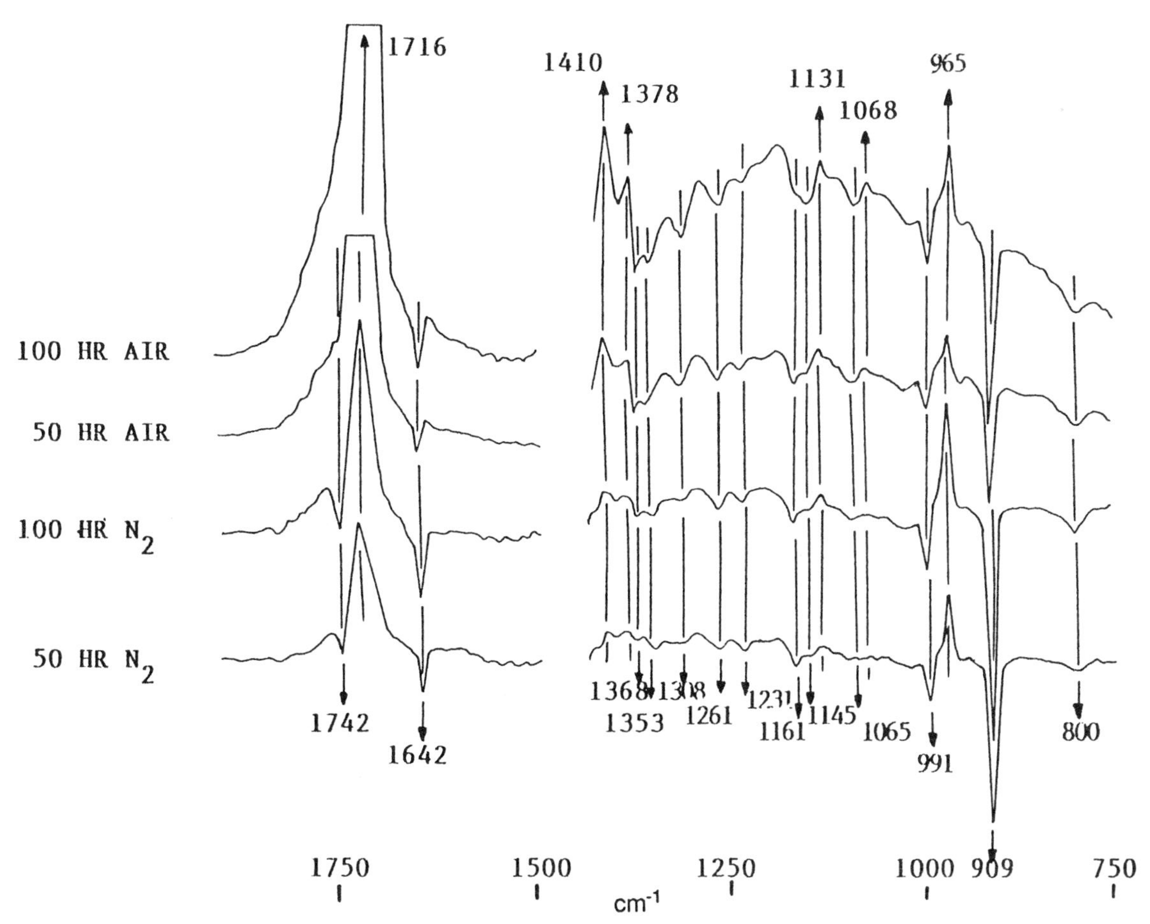

Figure 3.17. The difference spectra of polyethylene films irradiated in nitrogen and in air for 50 and 100 h are shown. (Reproduced with permission from reference 56. Copyright 1975 John Wiley & Sons, Inc.)

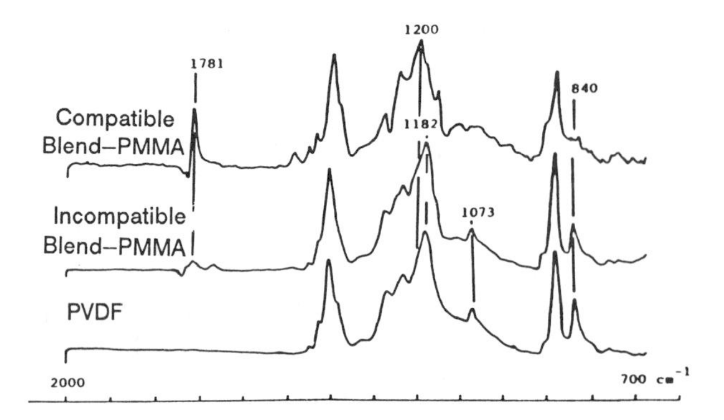

Figure 3.18. The FTIR spectrum of pure PVDF and the difference spectra of a compatible and an incompatible blend of PMMA – PVDF in the range 700 – 2000 cm^{-1} are shown. (Reproduced with permission from reference 57. Copyright 1984 Marcel Dekker, Inc.)

Resolution Enhancement Using Curve-Fitting Methods

One of the problems in polymer spectroscopy is the fact that the IR bands are inherently broad. This fact, coupled with the large number of bands occurring in the spectrum, usually leads to band overlap. Because the bandwidths are usually an order of magnitude greater than the resolution provided by the FTIR instrument, resolution enhancement techniques based on different mathematical procedures are often used (*58*). Peak finding methods such as derivative spectroscopy (*59*) and Fourier self-deconvolution (*60*) are helpful in narrowing the peaks and minimizing the overlap. The computer fitting methods employed generally are all iterative least-squares optimization processes (*58*). If the initial choice of parameters is close to the optimum, only two or three cycles of iteration are required.

The first option in fitting IR bands is the selection of the band shape (*61*). The common assumption is that vibrational peaks are Lorentzian in shape, although for polymers this is seldom the case. The approximate number of peaks is often determined by visual inspection of the spectral region under examination. Derivative spectra are helpful, as will be demonstrated later. It is recommended that a *minimum* number of peaks be used to fit the composite profile (*62*). As the number of peaks increases, so does the uncertainty.

The remaining problem is bandwidth selection. There appears to be no useful rule for solving this aspect of the curve-fitting problem because the individual bands in a spectrum can have a broad range of bandwidths. Therefore, the bandwidth is often used as one of the variables in the optimization process.

Derivative Spectroscopy. With digitized spectral data and the appropriate computer software, obtaining a derivative spectrum is a simple matter. The even derivatives have sharper peaks at the same frequency values as in the original spectrum. In the case of a Lorentzian profile with unit half-width, the second derivative (for which the peak is negative) has a value that is equal to 0.326 of the original bandwidth (Figure 3.19) (*63*), and for the fourth derivative, which has a positive peak, the value is equal to 0.208 of that of the original bandwidth (Figure 3.20). Band sharpening is therefore achieved with the derivative spectra. The resolution enhancement is 2.7 for the second derivative and 3.8 for the fourth derivative.

However, derivative spectroscopy does have some drawbacks. The amplitudes of the peaks in derivative spectra are a function of the widths of the original peaks, and this condition leads to a heavy bias toward the sharper peaks present in the original spectra. The noise peaks are sharper in the derivative spectra than in the spectra upon which they are superimposed. In addition, peaks located at the same frequency position as one of the lobes in the derivative spectrum can be lost. Also, the signal-to-noise ratio deteriorates as

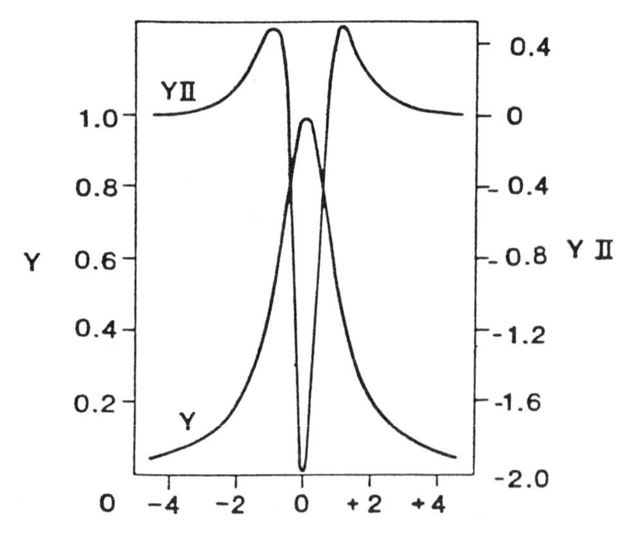

Figure 3.19. The second derivative of the Lorentzian profile. (Reproduced with permission from reference 63. Copyright 1986 Hüthig & Wepf Verlag.)

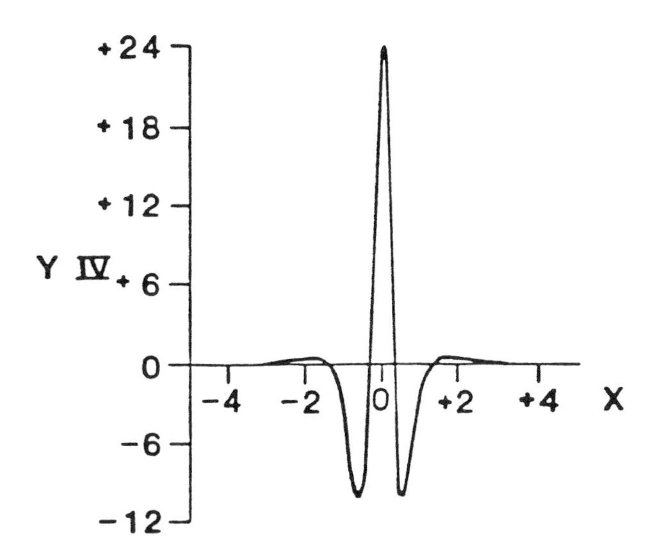

Figure 3.20. The fourth derivative of the Lorentzian profile. (Reproduced with permission from reference 63. Copyright 1986 Hüthig & Wepf Verlag.)

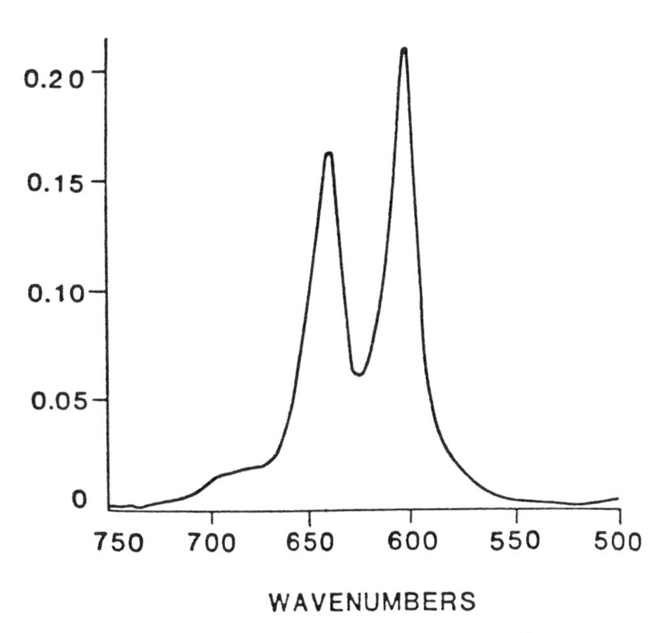

Figure 3.21. The spectrum of the carbon – chlorine region of the highly syndiotactic poly(vinyl chloride) made by the urea clathrate route has two peaks and a high frequency shoulder. (Reproduced with permission from reference 63. Copyright 1986 Hüthig & Wepf Verlag.)

the order, n, of the derivative increases. The rate of deterioration is approximately 2^n and is ultimately the limit to the number of derivatives that can be taken and to the maximum resolution enhancement achievable (*64*).

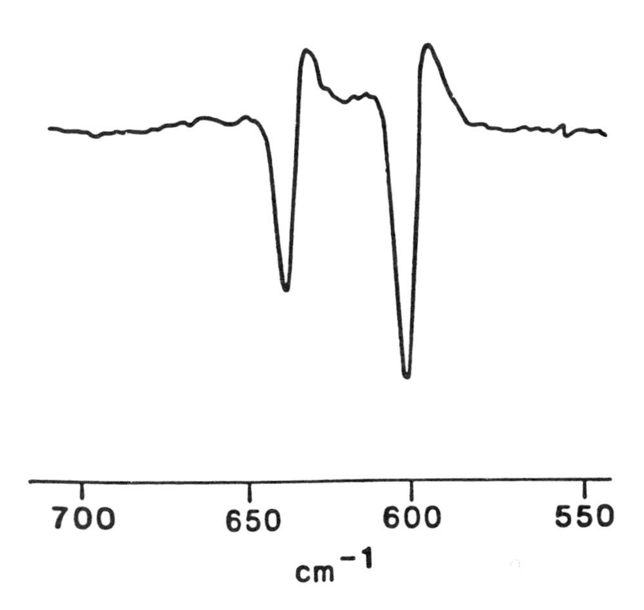

Figure 3.22. The second derivative spectrum of the carbon – chlorine region of highly syndiotactic poly(vinyl chloride) made by the urea clathrate route. (Reproduced with permission from reference 63. Copyright 1986 Hüthig & Wepf Verlag.)

An example is the spectrum of the carbon – chlorine region of the highly syndiotactic poly(vinyl chloride) (Figure 3.21) made by the urea clathrate route (*63*). It has two peaks and a high-frequency shoulder. The second derivative spectrum (Figure 3.22) shows two negative peaks at the positions of the two peaks in the original spectrum. There also appears to be a peak at 638 cm^{-1} in the second derivative spectrum; this peak is assigned to long planar syndiotactic sequences. The second derivative spectrum of a commercial poly(vinyl chloride) (PVC) reveals six peaks, but let us wait until later for an analysis.

Fourier Self-Deconvolution.

> *Deconvolution can be used to extract information about component elements of a data system that are partially or completely overlaid by stronger components so as to be undetectable to the eye in the original curve or other form of visual data plot.*
>
> – R. Norman Jones (*65*)

Another approach to resolution enhancement is the use of Fourier self-deconvolution, which is useful when the spectral features are broader than the instrument resolution (*66*). This technique is based on the fact that the breadths of the peaks are the result of the convolution of the intrinsic line shape with a broadening function. Deconvolution allows the removal of the broadening. The broadening function is Lorentzian, and deconvolution with a Lorentzian function will lead to a substantial, physically significant narrowing (*67*).

The data set is multiplied by the inverse transform of the Lorentzian function, which is the exponential function. The data set is then transformed into the frequency domain to give a conventional, sharpened spectrum. By using spectra from FTIR instruments, resolution enhancement factors of 3 can be obtained (*68*). However, no line can be narrower than the original instrument function.

The use of Fourier deconvolution requires a knowledge of the half-widths of the peaks to be examined, but in practice the half-widths of the component peaks of overlapping band systems are not known. Underestimation of the bandwidth results in a deconvoluted band that has a shape close to Lorentzian, but that has a lower resolution enhancement than would be achieved with the correct bandwidth. When overestimation occurs, the deconvoluted band is flanked by negative lobes, and the degree of sharpening achieved is greater than expected. Overconvolution also results in serious distortions of the intensities.

Fourier deconvolution is a valuable means for finding peaks in spectra of polymers with extensive overlapping of bands. The Fourier self-deconvoluted IR spectrum of the urea clathrate type of poly(vinyl chloride) in Figure 3.23 was obtained with a deconvolution function half-width of 11 cm^{-1} (*69*).

The resolution was increased by a factor of 2.5. The spectrum is similar in quality to the second derivative spectrum in Figure 3.22. There is, however, no clear indication of a peak in the 650-cm^{-1} region.

Fourier self-deconvolution degrades the signal-to-noise ratio of the spectrum. Another problem is that the same deconvolution width is used for the entire spectrum, but individual IR bands have different widths and therefore may not be optimized. In other words, individual bands may be overconvoluted, and overconvolution may lead to the appearance of negative peaks in the final spectrum in much the same manner as with derivative spectroscopy.

Maximum Entropy Method. The maximum entropy method approach to resolution enhancement treats the signal and the noise as separate entities (*70*). This method sets out to minimize random errors and noise on the basis of two criteria: (1) an assumption is made about the shapes of the peaks present, and (2) the best fit with respect to noise is obtained by maximizing the configurational entropy (negative information content), S, by using the equation

$$S = -\sum_i p_j \log p_j \qquad (3.33)$$

where p_j is the normalized intensity at the spectral frequency j. The maximum entropy method selects the solution that contains the smallest amount of false information from the large number of possible solutions (*71*). This method is not used often in IR spectroscopy, but it is used extensively in NMR spectroscopy and NMR imaging.

Quantitative IR Spectroscopy of Polymers

IR spectroscopy is routinely used to estimate the identity and concentration of absorbing species in multicomponent systems. The standard approach is to measure the spectra of a series of mixtures and use simultaneous equations to extract the desired concentration data.

In the analysis of multicomponent systems, there are four different classes of spectral problems (*72*). In

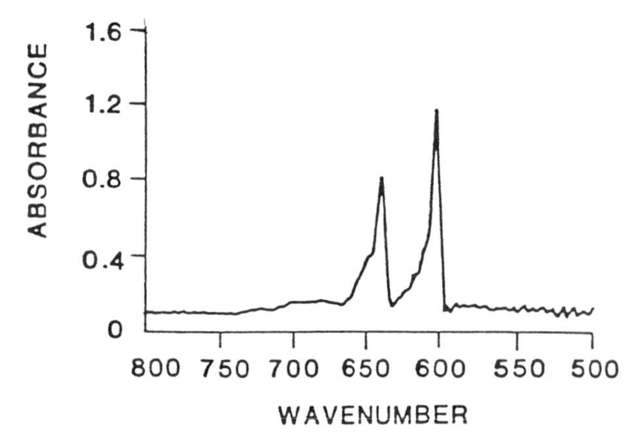

Figure 3.23. The Fourier self-deconvoluted IR spectrum of the urea clathrate-type poly(vinyl chloride) obtained with a deconvolution function half width of 11 cm^{-1}. (Reproduced with permission from reference 69. Copyright 1986 Society for Applied Spectroscopy.)

the first situation, all of the components and their spectra are known, and calibration data are available. In this case, the method of least-squares is appropriate for finding the quantity of each component. When proper calibration is carried out, this approach yields quantitative data for mixtures. In the second situation, the spectra of the components are not known, but the concentrations of the components of interest are known. This situation requires the use of a cross-correlation procedure. In the third situation, none of the components are known, and factor analysis is applied. The factor analysis method provides a lower limit to the number of linear-independent components present in the mixture and estimates the spectra when low numbers of components (fewer than three) are present. In the final and most difficult situation, attempting to quantify a number of known components in the presence of a variable background of unknowns, the method of rank annihilation has been developed. When there is only one known component, the amount of that component can be found by iteratively subtracting it from the observed data until the rank of the remaining matrix is reduced by one.

The quantitative relationship between the concentration, c, of a component in a sample and its absorbance, A, is given by the Bouguer–Beer law:

$$A = abc = \log \frac{I_0}{I} \qquad (3.34)$$

where the term I_0/I is termed the transmittance, and the percent transmittance is 100 log I_0/I. The constant a is the absorptivity, and b is the unit of thickness. Logarithms to the base 10 are ordinarily used, and a factor of 2.3 is incorporated into the constant a. This law assumes photometric linearity and accuracy of the spectrometer. Traditionally, quantitative analysis is carried out by constructing a calibration curve of absorbance of the analytical frequency vs. concentration, and then measuring the concentration of an unknown sample by using either the absorbance peak or the absorbance area. When two components are involved, absorbance measurements at two frequencies are necessary to estimate the individual concentrations. This method can be extended to more components as required.

Quantitative IR spectroscopy requires determination of the absorptivity by calibration or some other method (*73*). For polymers, this aspect of the problem is particularly difficult because usually there is no independent method of calibration. Low-molecular-weight analogs of the structure are often useful for calibration when the structures are similar and the spectral complications of vibrational coupling are not present. Often, however, there are no suitable model systems available (e.g., polymers that exhibit stereoregularity).

Least-Squares Analysis Using Spectra of Pure Components

The least squares method is applicable when the number of components and their pure spectra are known, and measurement of the relative concentrations of the components is desired (*74*). Beer's law can be applied in matrix form to allow the simultaneous determination of multiple components even when there is no unique set of spectral features for each of the components (i.e., when there is no single band of one of the components in the system that is not overlapped by one or more of the bands of the other components in the system).

The least-squares approach to quantitative analysis considers the complete spectrum with a large number of frequencies simultaneously and uses a least-squares fitting of the pure component spectra and the mixture spectrum to determine the concentrations of the components (*74*). This approach has the principle advantage of high precision because an increase in the number of data points improves the precision of the measurement. Other advantages of the method are

- using all frequencies in each of the spectral peaks (rather than only the peak maximum) (*75*)
- performing a weighted least-squares analysis to give greater emphasis to data with higher signal-to-noise ratios (*76*)
- fitting a linear or nonlinear base line under each spectral peak (*77*)

The computations can incorporate a variety of statistical methods to improve the analysis in the case of deviations from Beer's law (*77*).

The simultaneous determination of multiple components by IR spectroscopy depends on the fact that no two components have exactly the same spectrum. For a multicomponent system,

$$A_i = \Sigma\, k_{ij} c_j + e_i \qquad (3.35)$$

at each frequency, i, in the spectrum where k_{ij} represents the product of the path length and the absorptivity for component j at frequency i. The term c_j is the concentration of the jth component, and the term e_i is the random error in the measured spectrum. This error is assumed to be normally distributed with an expectation of zero. The full spectrum then yields a series of n equations, where n is the number of frequencies. These equations may be put in matrix form

$$\mathbf{A} = \mathbf{KC} + \mathbf{E} \qquad (3.36)$$

where **A** is the $n \times m$ matrix whose columns represent the spectrum of each of the m standard mixtures, **K** is the $n \times l$ matrix whose columns represent the l pure-component spectra at unit concentration and unit relative path length, **C** is the $l \times m$ matrix of the known component concentrations, and **E** is the $n \times m$ matrix of random measurement error in the spectra.

The least-squares estimate of the matrix of pure-component spectra, $\underline{\mathbf{K}}$, is given by

$$\underline{\mathbf{K}} = \mathbf{AC'(CC')^{-1}} \qquad (3.37)$$

where the primes indicate transposed matrices, and the negative exponent represents the inverse of the matrix. In the analysis step, the absorbances, A, of the unknown sample are measured at the sample frequencies, and the concentrations of the components are then calculated as

$$\mathbf{C} = \mathbf{(K'K)^{-1}K'A} \qquad (3.38)$$

The results can be used for analysis of unknown mixtures

$$\mathbf{C} = \mathbf{(KK')^{-1}K}a \qquad (3.39)$$

where a is the spectrum of the unknown sample. This method improves the precision by simultaneously using all of the measured frequencies. A major disadvantage of the method is that all interfering chemical components in the spectral region of interest need to be known and included in the calibration.

Inverse Least-Squares Method. The inverse least-squares method for quantitative analysis assumes that concentration is a function of absorbance:

$$\mathbf{C} = \mathbf{AP} + \mathbf{E_c} \qquad (3.40)$$

where **C** and **A** were defined previously, **P** is the $n \times l$ matrix of the unknown calibration coefficients relating the l component concentrations to the spectral intensities, and $\mathbf{E_c}$ is the $m \times l$ matrix of the random concentration errors. This representation has the advantage that the analysis is invariant with respect to the number of chemical components, l, included in the analysis (*78*). During the calibration,

$$\mathbf{P} = \mathbf{(A'A)^{-1}A'}c \qquad (3.41)$$

and the analysis of the unknown is simply

$$\mathbf{C} = \mathbf{A'P} \qquad (3.42)$$

This means a quantitative analysis can be performed even if the concentration of only one component in the calibration mixtures is known. The disadvantage of the inverse least-squares method is that the analysis is restricted to a small number of frequencies because the matrix that must be inverted has dimensions equal to the number of frequencies, and this number cannot exceed the number of calibration mixtures used in the analysis.

Partial Least-Squares Method. Partial least-squares analysis depends on the iterative estimation of a linear combination of spectral features in such a way that the concentrations are optimally predicted (*79, 80*). Partial least-squares is a method in which the calibration and prediction analyses are performed one component at a time. The concentration of only the chemical component of interest is used in the calibration. The calibration spectra can be represented as (*81*)

$$\mathbf{A} = \mathbf{TB} + \mathbf{E_a} \qquad (3.43)$$

where **B** is an $h \times n$ matrix, and the rows of **B** are the new partial least-squares basis set of h full-spectrum vectors, which are called the *loading spectra*. The term **T** is an $m \times h$ matrix of intensities in the new coordinate system of the h partial least-squares loading vectors for the m sample spectra. In partial least-squares, the rows of **B** are eigenvectors of **A′A**, and the columns of **T** are proportional to the eigenvectors of **AA′** (*82*).

By using partial least-squares, it is possible to model complex base lines and some types of nonlinearity in Beer's law (*82, 83*). In addition, this method can be used when molecular interactions or reactions are present.

The Ratio Method for Determination of Component Spectra

The ratio method is one of the techniques available for the extraction of pure-component spectra from mixture spectra (*84, 85*). The IR spectrum of a two-component mixture can be written as a linear combination of its constituent spectra:

$$[M_1] = [P_1] + [P_2] \qquad (3.44)$$

where $[M_1]$ is the spectrum of mixture 1, and $[P_1]$ and $[P_2]$ are the spectra of the pure components. Similarly, the spectrum of a second mixture made up of the same two components but with different relative concentrations can be written as

$$[M_2] = s_1[P_1] + s_2[P_2] \qquad (3.45)$$

where s_1 and s_2 are the scaling coefficients reflecting the different concentrations of $[P_1]$ and $[P_2]$ in $[M_2]$.

These two equations can be solved for $[P_1]$ and $[P_2]$ in terms of the spectra of the mixtures:

$$[P_1] = \left(\frac{1}{s_1 - s_2}\right)[M_2] - \left(\frac{s_2}{s_1 - s_2}\right)[M_1] \qquad (3.46)$$

$$[P_2] = -\left(\frac{1}{s_1 - s_2}\right)[M_2] + \left(\frac{s_1}{s_1 - s_2}\right)[M_1] \qquad (3.47)$$

The scaling coefficients s_1 and s_2 are derived by calculating the "ratio spectrum" from the two mixtures:

$$[R] = \frac{[M_2]}{[M_1]} = \frac{s_1[P_1] + s_2[P_2]}{[P_1] + [P_2]} \qquad (3.48)$$

In a spectral region where $[P_1] >> [P_2]$, $[R]$ is approximately equal to s_1. Conversely, if $[P_2] >> [P_1]$, then $[R]$ is approximately equal to s_2. An implicit assumption is that each pure-component spectrum contains a characteristic peak that is not overlapped by peaks appearing in the spectra of other pure components (*86*). With the introduction of relative concentrations in the preceding equations, this ratio approach allows quantitative analyses without external calibration (*87*).

The ratio method has been used to determine the pure spectra of the *trans* and *gauche* geometric isomers of PET (*88*). The IR spectra of PET films that have been annealed for different times at 230 °C are shown in Figure 3.24. The ratio spectrum of two of the films is shown in Figure 3.25, and the computed pure spectra of the *trans* and *gauche* isomers are shown in Figure 3.26. By using the ratio method

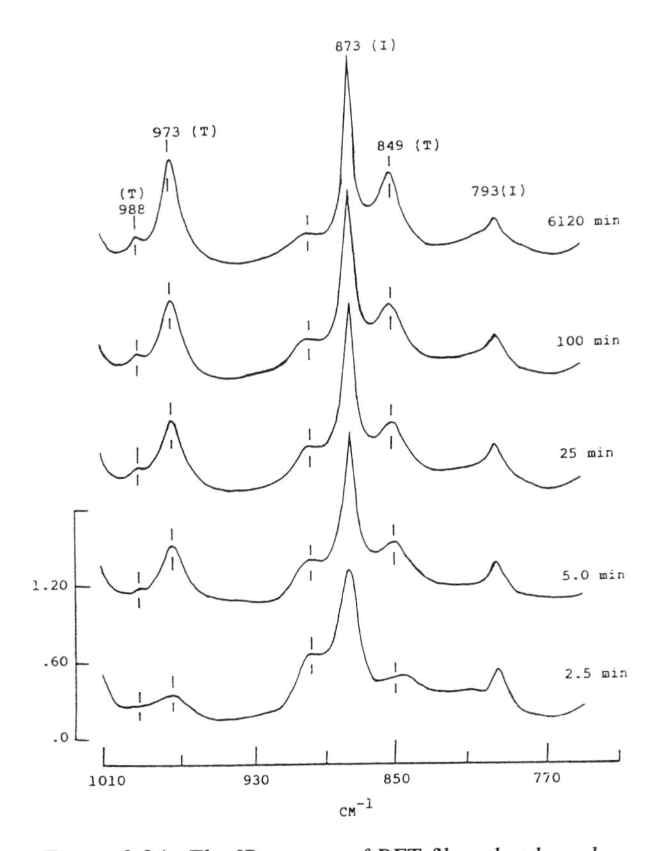

Figure 3.24. The IR spectra of PET films that have been annealed for different times at 230 °C. The letters T, G, and I correspond to trans, gauche, and internal thickness band, respectively. (Reproduced with permission from reference 88. Copyright 1982 John Wiley & Sons, Inc.)

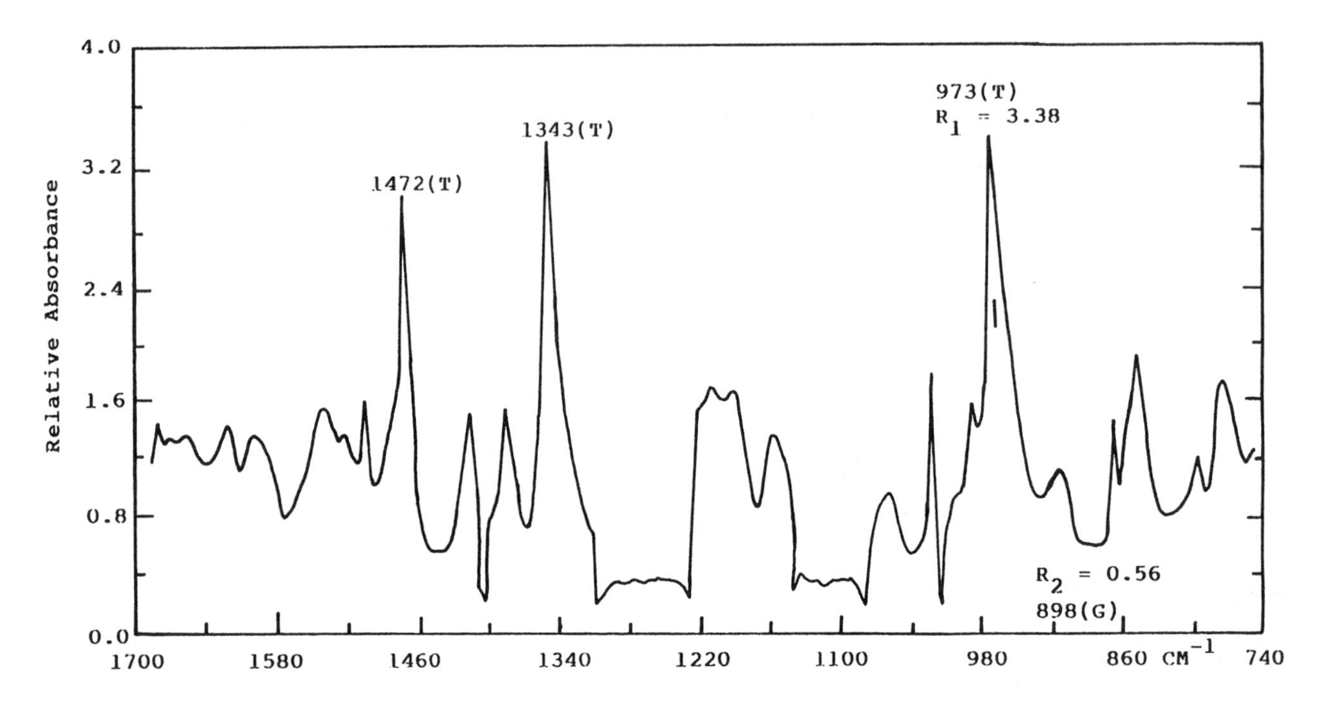

Figure 3.25. *The absorbance ratio spectra of PET for 1700 – 750-cm^{-1} region. The spectrum from the sample annealed at 200 °C for 102 h was divided by the spectrum from the sample annealed at 75 °C for 10 h. The letters T and G correspond to trans and gauche, respectively;* R_1 *is the maximum absorbance ratio; and* R_2 *is the minimum absorbance ratio. (Reproduced with permission from reference 88. Copyright 1982 John Wiley & Sons, Inc.)*

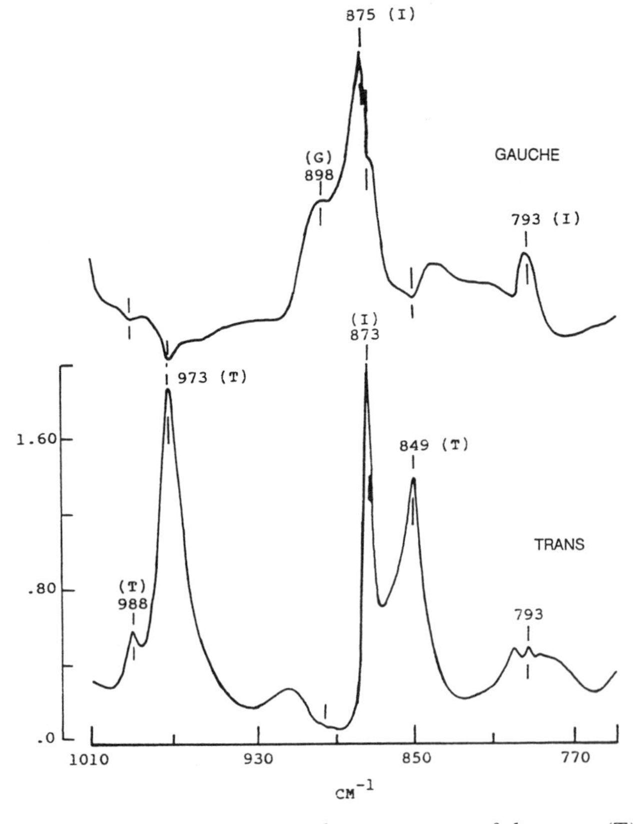

Figure 3.26. The computed pure spectra of the trans (T) and gauche (G) isomers of PET obtained by the ratio method. The letter I corresponds to an internal thickness band. (Reproduced with permission from reference 88. Copyright 1982 John Wiley & Sons, Inc.)

in this straightforward fashion, the pure spectra can be obtained.

Factor Analysis

> *Factor analysis (principal component analysis) is a multivariate statistical technique, which can be used to determine the number of independent components contributing to an experimentally observed quantity, such as an infrared absorption band.*
>
> – H. F. Shurvell *(89)*

Factor analysis is based on the expression of a property as a linear sum of terms, called factors. This method has wide applicability to a variety of multidimensional problems *(90)*. The Beer – Lambert law can be written for a number of components over a frequency range as

$$A_i = \sum k_j c_{ij} \tag{3.49}$$

where A_i is the absorbance spectrum of mixture i, k_j is the absorptivity path length product for the jth component, and c_{ij} is the concentration of component j in mixture i. Factor analysis is concerned with a matrix of data points. In matrix notation the absorbance spectra of a number of solutions can be written as

$$\mathbf{A} = \mathbf{KC} \tag{3.50}$$

where **A** is a normalized absorbance matrix that is rectangular in form. The columns in this matrix contain the absorbance at each wavenumber, and the rows correspond to the different mixtures being studied. The **A** matrix could thus be 400 by 10 corresponding to a measurement range of 400 wavenumbers at one wavenumber resolution for 10 different mixtures or solutions. The term **K** is the molar absorption coefficient matrix, and it conforms with the **A** matrix for the wavelength region but has only the number of rows corresponding to the number of absorbing components. The concentration matrix, **C**, has dimensions corresponding to the number of components and the number of mixtures or solutions being studied. Of course, **K** and **C** are unknown. In principle, factor analysis can be used to generate **K** and **C**, which will allow a complete analysis of a series of mixtures containing the same components in differing amounts.

There are two basic assumptions in factor analysis. First, the individual spectra of the components are not linear combinations of the other components, and second, the concentration of one or more species cannot be expressed as a constant ratio of another species. The different relative concentrations of the components in the mixtures provide the additional information necessary to deconvolute the spectra.

Determination of the Number of Components in a Mixture. Factor analysis initially allows a determination of the number of components required to reproduce the absorbance or data matrix, **A**. The rank of the matrix **A** can be found by using factor analysis, and this rank of **A** is considered equal to the number of absorbing components. To find the rank of **A**, the matrix $\mathbf{M} = \mathbf{A}'\mathbf{A}$ is formed where $\mathbf{A}'$ is the transpose of **A**. This matrix, termed the *covariance matrix*, has the same rank as **A** but has the advantage of being a square matrix with dimensions corresponding to the number of mixtures being examined. In the absence of noise, the rank of **A** is given by the number of nonzero eigenvalues of **M**. In the presence of noise, however, the problem of identifying the number of components is more difficult.

A common method for estimating k (the rank of **M**) is to plot the values of the eigenvalues versus the number of the eigenvalue and to inspect the singular values. If the noise is low compared to the signal, there should be a large drop between the k and $(k + 1)$ values and then much smaller drops between the $k + 1$ value and subsequent values. When the noise level is high, this method is likely to fail, especially when the mixture contains more than two or three individual components.

With the availability of proper software, the application of factor analysis is straightforward. The normalized spectra of the mixtures to be analyzed are entered into a computer, and the computer calculates the eigenvalues. A plot of the log of the eigenvalue versus the number of components is made. Often, a visual inspection will reveal the difference between the "real" eigenvalues and the "noise or error" eigenvalues. A plot of this type for a series of eight poly(ethylene terephthalate) films that have been annealed for various times is shown in Figure 3.27. The

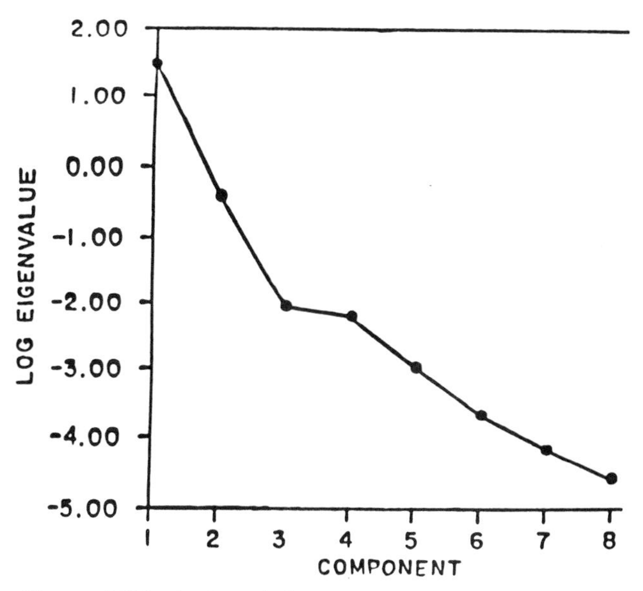

Figure 3.27. A plot of the magnitude of the eigenvalue versus the number of components for a series of eight PET films that have been annealed for various times. (Courtesy Lin, S. B. Ph.D. Thesis, Case Western Reserve University.)

plot shows the results using the spectra in the 3120- to 2850-cm^{-1} region. Because of the large drop in the eigenvalues, it is clear that there are only two real eigenvalues, and the remaining eigenvalues correspond to noise (*91*). When the eigenvalue differences are not sufficiently large for a visual evaluation, it is sometimes helpful to calculate an *indicator function* (*92*). This function reaches a minimum at the point where the principle contribution comes from the real eigenvalues and then increases as the error contribution takes over. An example of the use of the indicator function is shown in Figure 3.28 for compatible blends of polyoxyphenylene – polystyrene (PPO – PS) and the incompatible blend of polyoxyphenylene – poly(parachlorostyrene) (PPO – P_4ClS) (*93*). One more component would be expected (the interaction spectrum arising from mixing at the molecular level) for a compatible blend than for an incompatible blend. This is because the incompatible blend would be phase separated and its spectrum should be the simple sum of the spectra of the component homopolymers. The results in Figure 3.28 suggest such a difference exists for blends made up with different weight fractions.

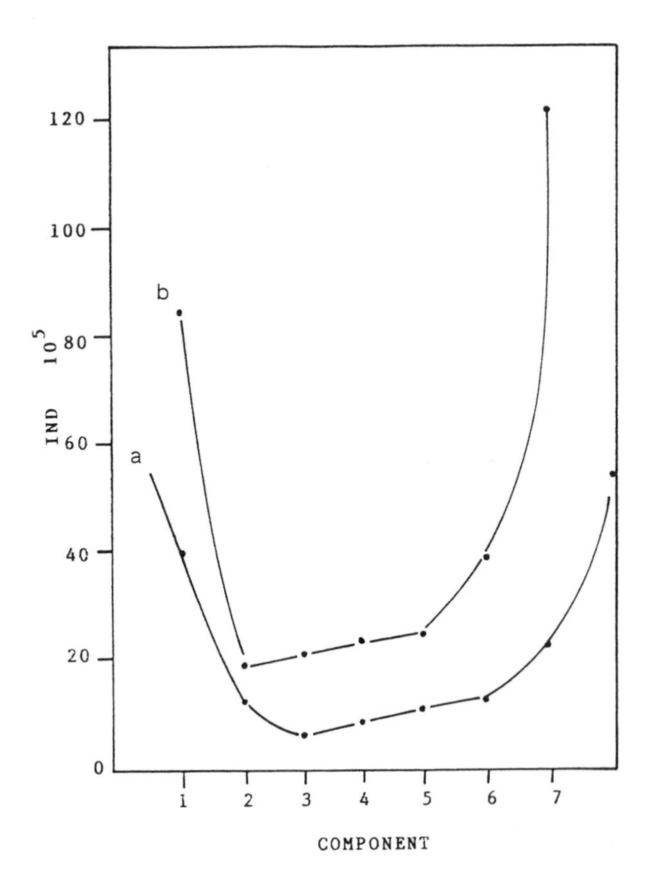

Figure 3.28. Plots of the indicator function versus the number of components for compatible blends of PPO – PS (a) and the incompatible blend of PPO – P_4ClS (b). (Reproduced with permission from reference 93. Copyright 1981 Society for Applied Spectroscopy.)

A new approach that provides additional information that is useful for estimating the rank of the covariance matrix has been suggested. This method is based on a *canonical correlation technique* (*94*). In this procedure, the correlation matrices, which are easily generated using the *multiscan technique*, are compared for two different levels of signal-to-noise ratio.

Determination of the Pure Spectra of Components. If the number of components, *n*, in a mixture is determined by factor analysis, then the dimension of the vector space in which all spectra of the mixtures can be represented is known, and a set of mutually orthogonal abstract eigenspectra are obtained. These eigenspectra define an *n*-dimensional coordinate system within which each data spectrum and each pure component spectrum are represented by points. The coordinate values of the points constitute a transformation vector by which the abstract eigenspectra can be rotated to physically real spectra. The magnitudes of the eigenvalues are a measure of the relative importance of corresponding eigenvectors in data reconstruction. The first eigenspectrum corresponds to the dominant eigenvalue, and hence it contains the maximum information. It represents the spectrum of the system if all the components were to coexist. The first eigenvector is the same line as that obtained from a linear least-squares fit of the data (*95*). The first and second eigenvectors are orthogonal by definition, so eigenvector 2 must be orthogonal in information relative to eigenvector 1. Factor analysis constructs eigenvector 2 in order to account for absorbance variations that are not accounted for by eigenvector 1. The information content of the corresponding eigenspectrum can have both positive and negative absorbances. As a result, eigenspectrum 2 resembles a difference spectrum (*95*).

Special considerations must be made, however, when determining the appropriate linear combinations of these orthogonal spectra to obtain the spectra of the pure components. In two-component systems, it is fairly easy to establish the range of a two-dimensional vector space in which the spectrum of each of the two pure components must fall. For a single spectrum in a two-component mixture, the following equation results:

$$\mathrm{PS} = a\mathbf{V}_1 + b\mathbf{V}_2 + e \tag{3.51}$$

where PS is the pure component spectrum; a and b are the coordinate values of the solution point; $\mathbf{V}_1$ and $\mathbf{V}_2$ are the first and second eigenvectors, respectively; and e is the error associated with the analysis. The coordinate values a and b, which transform the eigenspectra into the actual spectra, are unknown quantities that must be estimated. Criteria for obtaining an estimate for the location of the vector within the range of possible solutions must be developed. The two basic restrictions are (1) spectra must have positive intensities, and (2) the concentration of the components must be positive. Normalization of the mixture spectra to unit area results in their projections lying on a straight-line segment with the spectra of the purest mixtures located at each end of the line. The projections of the pure spectra also lie on the line and at the intersection of the limits determined by the requirement of positive absorbance.

The classical way of rotating the eigenspectra into physically meaningful spectra requires the restriction that each pure component contains a peak in a region where the other pure components do not absorb. Two- and three-component systems (*96, 97*) have been analyzed on this basis.

Determining the allowed ranges in higher dimensional vector spaces becomes increasingly difficult with increasing dimensions. For a three-component system, the projections lie on a plane, within a triangular boundary, on a three-axis coordinate system. The vertices of the triangular boundary are the projections of the pure underlying component spectra. Under certain circumstances, the operation that is carried out to obtain the rotation matrix needed to convert the orthogonal eigenspectra to real spectra can use the criterion that each pure component has a unique frequency, called a *pure point*. At the pure point only one component has a definite intensity, and the other components have zero intensity (*98*). By using pure points, it is possible to extract spectra and concentrations from a mixture data set of three or four components. One approach is to simplify the spectra by removing contributions from the system one at a time by using *minimum path* or *minimum entropy* techniques (*99*).

The factor analysis approach can be illustrated by using computer simulation. Simulated spectra containing two components of Lorentzian line shapes for four different mixtures are shown in Figure 3.29. Factor analysis of the covariance matrix indicates two pure components, as anticipated. The two primary eigenvectors for the two real eigenvalues were used to construct the abstract "eigenspectra" in Figure 3.30. In Figure 3.31, a plot of the four ordered pairs of the two primary eigenvector mixtures is shown on

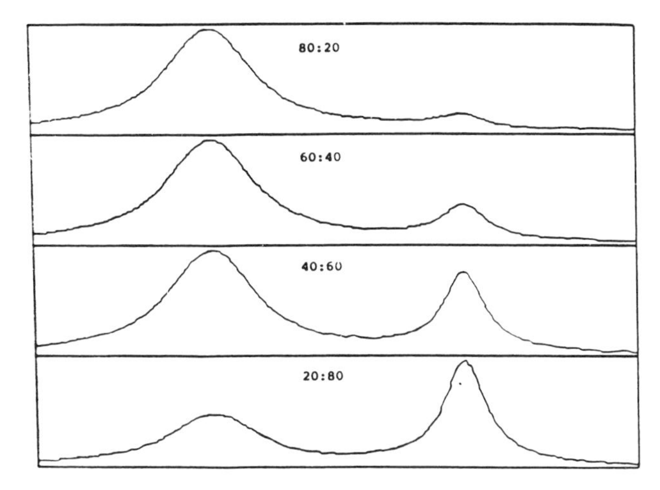

Figure 3.29. Computer simulated spectra with noise containing two components with Lorentzian peaks for four different mixtures. (Reproduced from reference 100. Copyright 1983 American Chemical Society.)

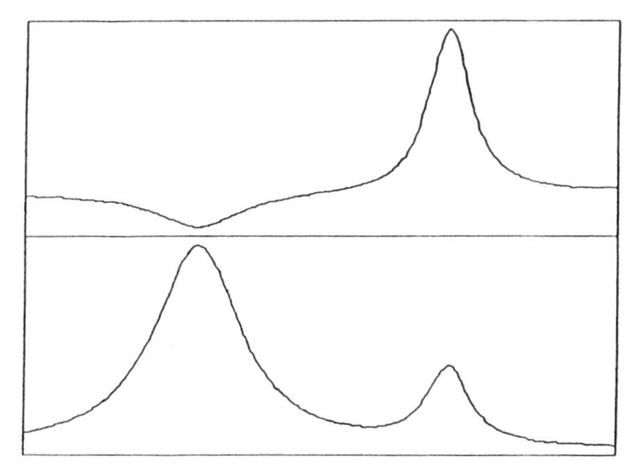

Figure 3.30. The abstract eigenspectra for the two real eigenvalues. (Reproduced from reference 100. Copyright 1983 American Chemical Society.)

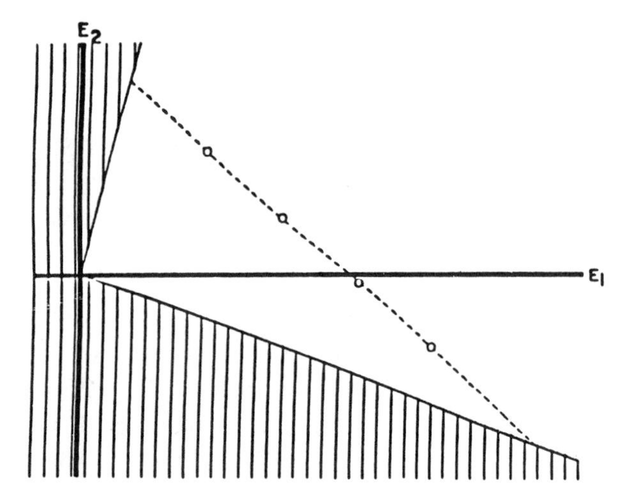

Figure 3.31. The plot of the four ordered pairs of the two primary eigenvector mixtures is shown on the eigenvector axis. (Reproduced from reference 100. Copyright 1983 American Chemical Society.)

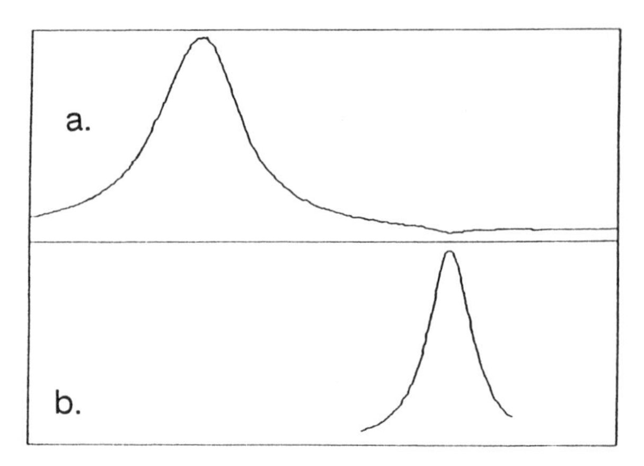

Figure 3.32. The pure component spectra obtained from the multiplication of the abstract eigenspectra by coefficients obtained from the intersection points of the linear line and the shaded area corresponding to zero absorbance from Figure 3.31. The spectrum in a is of component 1, and the spectrum in b is of component 2. (Reproduced from reference 100. Copyright 1983 American Chemical Society.)

the eigenvector axis, clearly indicating a linear relationship between the eigenvectors. The points of intersection of this linear line with the boundary region yield the coefficients by which the abstract eigenspectra are multiplied to obtain the pure component spectra, which are shown in Figure 3.32 (*100*).

Rank Annihilation

In many quantitative situations, the available sample contains a few known species of interest mixed with other unknowns. In such cases, it would be convenient to be able to obtain quantitative information for the known compounds without having to consider the other species present. When information is available about only one component in a mixture, the method of rank annihilation has been used successfully to determine the concentration (*101*). For a multicomponent system, the rank of the correlation matrix, **M**, should equal the number of components. If one of the components, N, is known, and the correct amount of N is subtracted from **M**, the original rank of **M** should be reduced by 1. In such an instance, the eigenvalue of **M** corresponding to N should become 0. Given a pure component that is known to be in the mixture, the process of rank annihilation subtracts an amount, β, of the standard from the mixture. The residual matrix **E**, obtained by using this process, can be expressed as

$$\mathbf{E} = \mathbf{M} - \beta N \tag{3.52}$$

where **M** ($\mathbf{A}^T\mathbf{A}$) is the correlation matrix. When the correct amount, β, of the pure component, N, is subtracted, the rank of the residual matrix will decrease by 1. This decrease can be monitored by observing the change in the eigenvalues of $\mathbf{E}^T\mathbf{E}$ with various choices of β (*102, 103*). At the appropriate choice of β, one of the eigenvalues will approach 0. Actually, because of noise, the eigenvalue will show only a minimum (*104*). The amount of N subtracted to achieve a minimum in the corresponding eigenvalue will correspond to the relative concentration of the known component in the mixture (*105*).

Cross-Correlation

Cross-correlation is the method of choice for the analysis of one known component in a complex, unknown mixture. The method can be proficient when the background, as well as the number and kinds of components, changes a great deal (*106*).

The absorbance of all mixtures at a particular frequency can be viewed as the absorbance attributable to the desired pure component multiplied by its fractional concentration plus the absorbances of all other species multiplied by their fractional concentrations. The approach of cross-correlation is to generate "reference spectra" from a knowledge of the sample's quantitative composition, rather than the more traditional method of multilinear regression, which uses the spectra to generate the composition of the mixture.

Cross-correlation is used to evaluate the similarity between the spectra of two different systems, for example, a sample spectrum and a reference spectrum. This technique can be used for samples where background fluctuations exceed the spectral differences caused by changes in composition. The cross-correlation technique also can be used to generate the spectra of the pure components from the mixture spectra when the pure component spectra are not available, or when the pure component spectra differ significantly from the isolated pure spectra because of interaction or matrix effects.

Cross-correlation can also be used for spectral reconstruction. If $a(x)$ is the absorbance or reflectance at a specific wavelength, and $c(x)$ is the concentration of the desired component in the xth sample, then the cross-correlation function, $C_{ac}(d)$, between signals $a(x)$ and $c(x)$ at zero displacement can be written as

$$C_{ac}(d) = (1/n)\Sigma_n a(x)c(x \pm d) \qquad d = 0, 1, 2 \cdots n - 1 \tag{3.53}$$

where n is the number of samples, x is the sample index, and d is the displacement from the current x index. For each d, the corresponding samples of the absorbance at a specific frequency and the corresponding known concentrations are multiplied, and the products are summed over the entire set to give the value of C_{ac}. When $d = 0$, C_{ac} represents a quantitative measure of the degree of similarity between the functions. Taking products of pairs of points produces a weighted measure. The correlation function has a minimum value at $d = 0$, when the desired component is absent, and increases linearly with increasing concentration of the desired component.

In the absence of noise or other absorbing components in the sample set, the value of $C_{ac}(d)$ at $d = 0$ depends on only the absorbance and concentration of the desired component, and all values of $C_{ac}(d)$ are zero when d does not equal zero. Under these circumstances, repeated application of the cross-correlation function equation at the various IR frequencies will extract the desired spectrum.

In real samples, the problem is complicated by experimental noise sources, the interference of other sample constituents, and a limited number of samples. The cross-correlation function will therefore contain undesired contributions. To overcome these errors, the average of the d values which are not equal to zero is subtracted from the value of $C_{ac}(d)$ at $d = 0$:

$$C_{ac} = (1/n)\Sigma_n[a(x) - \bar{a}][c(x) - \bar{c}] \tag{3.54}$$

When this calculation is repeated for a number of frequencies, the resulting spectrum shows the correlation between the sample absorbance and the concentration of the component and should be free of contributions from unwanted sources. The spectrum should be that of the desired component (*107*)

One of the advantages of the cross-correlation method is that the correlation is less sensitive to background fluctuations because the change may not affect all of the component peaks and will be attenuated accordingly. The correlation scales the noise in proportion to the information, tending to force a constant optimal signal-to-noise ratio over the entire frequency range (*108*).

References

1. Potts, W. J. *Chemical Infrared Spectroscopy;* Wiley: New York, 1963.; Vol 1, Techniques.
2. Griffiths, P. R.; de Haseth, J. A. *Fourier Transform Infrared Spectrometry;* Wiley: New York, 1986.
3. Cooley, J. W.; Tukey, J. W. *Math. Comput.* **1965,** *19,* 297.
4. Culler, S. R.; Ishida, H.; Koenig, J. L. *Ann. Rev. Mater. Sci.* **1983,** *13,* 363.
5. Anderson, R. J.; Griffiths, P. R. *Anal. Chem.* **1975,** *47,* 2339
6. Stimson, M. M.; O'Donnell, M. J. *J. Amer. Chem. Soc.* **1952,** *74,* 1805.
7. Carter, R. O., III; Carduner, K. R.; Peck, M. C.; Motry, D. H. *Appl. Spectrosc.* **1989,** *43,* 791.
8. Harrick, N. J. *Internal Reflection Spectroscopy;* Wiley-Interscience: New York, 1967.

9. Castillo, F. J.; Koenig, J. L.; Anderson, J. M.; Lo, J. *Biomaterials* **1984,** *5,* 319.

10. Jolley, J. G.; Geesey, G. G.; Hankins, M. R.; Wright, R. B.; Wichlacz, P. L. *Appl. Spectrosc.* **1989,** *43,* 1062.

11. IRIS Fiber Optics, Inc., British Telecom Laboratories, Martlesham Heath, Ipswich, England.

12. Compton, D. A. C.; Hill, S. L.; Wright, N. A.; Druy, M. A.; Piche, J.; Stevenson, W. A.; Vidrine, D. W. *Appl. Spectrosc.* **1988,** *42,* 972.

13. Kortum, G. *Reflectance Spectroscopy;* Springer-Verlag: New York, 1969.

14. Wendlandt, W. W.; Hecht, H. G. *Reflectance Spectroscopy;* Interscience: New York, 1969.

15. Fuller, M. P.; Griffiths, P. R. *Anal. Chem.* **1978,** *50,* 1906.

16. Kubelka, P. *J. Opt. Soc. Am.* **1948,** *38,* 448.

17. Brimmer, P. J.; Griffiths, P. R. *Appl. Spectrosc.* **1988,** *42,* 242.

18. Brimmer, P. J.; Griffiths, P. R.; Harrick, N. J. *Appl. Spectrosc.* **1986,** *40,* 258.

19. Chalmers, J. M.; Mackenzie, M. W. *Appl. Spectrosc.* **1985,** *36,* 634.

20. Chase, B.; Amey, R. L.; Holtje, W. G. *Appl. Spectrosc.* **1982,** *36,* 155.

21. Korte, H.; Otto, A. *Appl. Spectrosc.* **1988,** *42,* 38

22. McKenzie, M. T.; Culler, S. R.; Koenig, J. L. In *Fourier Transform Infrared Characterization of Polymers;* Ishida, H., Ed.; Plenum: New York, 1987.

23. McKenzie, M. T.; Koenig, J. L. *Appl. Spectrosc.* **1985,** *39,* 408.

24. Culler, S. R.; McKenzie, M. T.; Fina, L. J.; Ishida, H.; Koenig J. L. *Appl. Spectrosc.* **1984,** *38,* 791.

25. McKenzie, M. T.; Culler, S. R.; Koenig, J. L. *Appl. Spectrosc.* **1984,** *38,* 786.

26. Rosencwaig, A. *Photoacoustics and Photoacoustic Spectroscopy;* Wiley: New York, 1980.

27. Koenig, J. L. *Adv. Polym. Sci.* **1983,** *54,* 89.

28. Chatzi, E. G.; Urban, M. W.; Ishida, H.; Koenig, J. L. *Langmuir* **1988,** *4,* 846.

29. Urban, M. W.; Koenig, J. L. *Appl. Spectrosc.* **1985,** *39,* 1051.

30. Urban, M. W.; Koenig, J. L. *Anal. Chem.* **1985,** *60,* 2408.

31. Chatzi, E. G.; Urban, M. W.; Koenig, J. L. *Makromol. Chem. -Macromol. Symp.* **1986,** *5,* 99.

32. McDonald, W.; Goettler, H.; Urban, M. W. *Appl. Spectrosc.* **1989,** *43,* 1387.

33. Chase, D. B. *Appl. Spectrosc.* **1981,** *35,* 77.

34. Hvistendahl, J.; Rytter, E.; Oye, H. A. *Appl. Spectrosc.* **1983,** *37,* 182.

35. King, V. W.; Lauer, J. L. *J. Lubrication Tech.* **1981,** *103,* 65.

36. Jones, R. W.; McClelland, J. F. *Anal. Chem.* **1989,** *61,* 650.

37. Handke, M.; Harrick, N. J. *Appl. Spectrosc.* **1986,** *40,* 401.

38. Nagasawa, Y.; Ishitani, A. *Appl. Spectrosc.* **1984,** *38,* 168.

39. *Infrared Microscopy: Theory and Applications;* Messerschmidt, R. G.; Harthcock, M. A., Eds.; Dekker: New York, 1988.

40. Harthcock, M. A.; Atkin, S. C. *Appl. Spectrosc.* **1988,** *42,* 449.

41. Burns, D. H.; Callis, J. B.; Christian, G. D.; Davidson, E. R. *Appl. Opt.* **1985,** *24,* 154.

42. Liu, J.; Koenig, J. L. *Appl. Spectrosc.* **1987,** *41,* 447.

43. Hirschfeld, T.; Mantz, A. W. *Appl. Spectrosc.* **1976,** *30,* 552.

44. Clark, F. R. S.; Moffatt, D. J. *Appl. Spectrosc.* **1978,** *32,* 547.

45. Graf, R. T.; Koenig, J. L.; Ishida, H. *Appl. Spectrosc.* **1985,** *39,* 405.

46. Coleman, M. M.; Tabb, D. L.; Koenig, J. L. *Polym. Lett. Ed.* **1974,** *12,* 577.

47. Hirshfeld, T.; Kizer, K. *Appl. Spectrosc.* **1975,** *29,* 345.

48. Koenig, J. L.; Tabb, D. L.; Coleman, M. M. *J. Polym. Sci., Polym. Phys. Ed.* **1975,** *13,* 1145.

49. D'Esposito, L.; Koenig, J. L. *J. Polym. Sci., Polym. Phys. Ed.* **1976,** *14,* 1731.

50. Painter, P. C.; Koenig, J. L. *J. Polym. Sci., Polym. Phys. Ed.* **1977,** *15,* 1885.

51. Tabb, D. L.; Koenig, J. L. *Macromolecules* **1975,** *8,* 929.

52. Painter, P. C.; Koenig, J. L. *J. Polym. Sci., Polym. Phys. Ed.* **1977,** *15,* 1223.

53. Painter, P. C.; Koenig, J. L. *J. Polym. Sci., Polym. Phys. Ed.* **1977,** *15,* 1235.

54. Painter, P. C.; Watzek, M.; Koenig, J. L. *Polymer* **1977,** *18,* 1169.

55. Bachman, M. A.; Gordon, M.; Koenig, J. L. *J. Appl. Phys.* **1979,** *50,* 6106.

56. Tabb, D. L.; Sevcik, J. J.; Koenig, J. L. *J. Polym. Sci., Polym. Phys. Ed.* **1975,** *13,* 815.

57. Coleman, M. M.; Painter, P. C. *Appl. Spect. Rev.* **1984,** *20,* 255.

58. Maddams, W. F. *Appl. Spectrosc.* **1980,** *34,* 245.

59. Vandeginste, B. G. M.; De Galan, L. *Anal. Chem.* **1975,** *47,* 2124.

60. Goldman, A.; Alon, P. *Appl. Spectrosc.* **1973,** *27,* 50.

61. Baker, C.; Cockerill, I. P.; Kelsey, J. E.; Maddams; W. F. *Spectrochim. Acta,* **1978,** *34A,* 673.

62. Gillette, P. C.; Lando, J. B.; Koenig, J. L. *Appl. Spectrosc.* **1982,** *38,* 401.

63. Maddams, W. F. *Die Makromol. Chim. Macromol. Symp.* **1986,** *5,* 38.

64. Gans, P.; Gill, J. B. *Anal. Chem.* **1980,** *52,* 351.

65. Jones, R. N.

66. Kauppinen, J. K.; Moffatt, D. J.; Mantsch, H. H.; Cameron, D. G. *Appl. Spectrosc.* **1981,** *35,* 271.

67. Kauppinen, K.; Moffatt, D. J.; Mantsch, H. H.; Cameron, D. G. *Anal. Chem.* **1981,** *53,* 1454.

68. Kauppinen, K.; Moffatt, D. J.; Mantsch, H. H.; Cameron, D. G. *Appl. Opt.* **1982,** *21,* 1454.

69. Compton, D. A. C.; Maddams, W. F. *Appl. Spectrosc.* **1986,** *40,* 239.

70. Sasaki, K.; Kawata, S.; Minami, S. *Appl. Opt.* **1984,** *23,* 1955.

71. Friedrich, H. B.; Yu, J. -P. *Appl. Spectrosc.* **1987,** *41,* 227.

72. Gillette, P. C.; Lando, J. B.; Koenig, J. L. In *Fourier Transform Infrared Spectroscopy;* Ferraro, J. R.; Basile, L. J., Eds.; Academic: New York, 1985; Vol 4, p 1.

73. *Computerized Quantitative Infrared Analysis;* McClure, G. L., Ed.; ASTM: Philadelphia, 1987.

74. Antoon, M. K.; Koenig, J. H.; Koenig, J. L. *Appl. Spectrosc.* **1977,** *31,* 518-524.

75. Haaland, M.; Easterling, R. G. *Appl. Spectrosc.* **1980,** *34,* 539.

76. Haaland, M.; Easterling, R. G. *Appl. Spectrosc.* **1982,** *36,* 665.

77. Haaland, M.; Easterling, R. G.; Vopicka, D. A. *Appl. Spectrosc.* **1985,** *39,* 73.

78. Kisner, H. J.; Brown, C. W.; Kavarnos, G. J. *Anal. Chem.* **1983,** *55,* 643.

79. Fuller, M. P.; Ritter, G. L.; Draper, C. S. *Appl. Spectrosc.* **1988,** *42,* 217.

80. Fuller, M. P.; Ritter, G. L.; Draper, C. S. *Appl. Spectrosc.* **1988,** *42,* 228.

81. Haaland, D. M.; Thomas, E. V. *Anal. Chem.* **1988,** *60,* 1193.

82. Haaland, D. M.; Thomas, E. V. *Anal. Chem.* **1988,** *60,* 1202.

83. Haaland, D. M. *Anal. Chem.* **1988,** *60,* 1208.

84. Hirschfeld, T. *Anal. Chem.* **1976,** *48,* 721.

85. Koenig, J. L.; D'Esposito, L.; Antoon, M. K. *Appl. Spectrosc.* **1977,** *31,* 292.

86. Diem, H.; Krimm, S. *Appl. Spectrosc.* **1981,** *35,* 421.

87. Koenig, J. L.; Kormos, D. *Appl. Spectrosc.* **1979,** *33,* 349.

88. Lin, S. B.; Koenig, J. L. *J. Polym. Sci., Polym. Phys. Ed.* **1982,** *20,* 2277.

89. Shurvell, H. F.

90. Malinowski, E. R.; Howery, D. G. *Factor Analysis in Chemistry;* Wiley: New York, 1980.

91. Lin, S. B.; Koenig, J. L. *J. Polym. Sci., Polym. Phys. Ed.* **1982,** *20,* 2277.

92. Malinowski, E. R. *Anal. Chem.* **1978,** *49,* 612.

93. Koenig, J. L.; Tovar, M. J. M. *Appl. Spectrosc.* **1981,** *35,* 543.

94. Tu, X. M.; Burdick, D. S.; Millican, D. W.;

McGown, L. B. *Anal. Chem.* **1989,** *61,* 2219.

95. Rao, G. R.; Zerbi, G. *Appl. Spectrosc.* **1984,** *38,* 795.
96. Lawton, W. H.; Sylvestre, E. A. *Technometrics* **1971,** *13,* 617.
97. Chen, J. H.; Hwang, L.-P. *Anal. Chim. Acta.* **1981,** *133,* 271.
98. Malinowski, E. R. *Anal. Chim. Acta.* **1982,** *134,* 129.
99. Friedrich, H. B.; Yu, J. P. *Appl. Spectrosc.* **1987,** *41,* 227.
100. Gillette, P. C.; Lando, J. B.; Koenig, J. L. *Anal. Chem.* **1983,** *55,* 630.
101. Ho, D. -N.; Christian, G. D.; Davidson, E. R. *Anal. Chem.* **1978,** *50,* 1108.
102. Ho, D. -N.; Christian, G. D.; Davidson, E. R. *Anal. Chem.* **1980,** *52,* 1071.
103. Ho, D. -N.; Christian, G. D.; Davidson, E. R. *Anal. Chem.* **1981,** *53,* 92.
104. Gianelli, M.; Burns, D. H.; Callis, J. B.; Christian, G. D.; Anderson, N. H. *Anal. Chem.* **1983,** *55,* 1858.
105. Burns, D. H.; Callis, J. B.; Christian, G. D. *Anal. Chem.* **1986,** *58,* 2805.
106. Horlick, G. *Anal. Chem.* **1973,** *45,* 319.
107. Honigs, D. E.; Hieftje, G. M.; Hirschfeld, T. *Appl. Spectrosc.* **1984,** *38,* 317.
108. Mann, C. K.; Goleniewski, J. R.; Sismanidis, C. A. *Appl. Spectrosc.* **1982,** *36,* 223.

Suggested Reading

Chia, L.; Ricketts, S. *Basic Techniques and Experiments in Infrared and FT-IR Spectroscopy*; Gibbs and Soell: New York, 1988.

Gillette, P. C.; Lando, J. B.; Koenig, J. L. In *Fourier Transform Infrared Spectroscopy*; Ferraro, J. R.; Basile, L. J., Eds.; Academic: New York, 1985; Vol. 4, p. 1.

Griffiths, P. R.; de Haseth, J. A. *Fourier Transform Infrared Spectrometry*; John Wiley and Sons: New York, 1986.

Fourier Transform Infrared Characterization of Polymers; Ishida, H., Ed.; Plenum Press: New York, 1987.

Computerized Quantitative Infrared Analysis; McClure, G. L., Ed.; ASTM: Philadelphia, 1987.

Messerschmidt, R. G. *Infrared Microscopy, Theory and Applications*; Harthcock, M. A., Ed.; Dekker: New York, 1988.

Siesler, H. W.; Holland-Moritz, K. *Infrared and Raman Spectroscopy of Polymers*; Dekker: New York.

4

Applications of IR Spectroscopy to Polymers

Structural Applications of IR Spectroscopy

Many reviews have been published on the application of IR spectroscopy to polymers, and you should examine them if you are concerned with a specific polymer system or application (see the general references at the end of this chapter). The numerous applications of IR spectroscopy to the study of polymers far exceed our limited time and space. The basic IR spectroscopic methods of structural characterization of polymers will be discussed in this chapter, and a few pedagogical examples will be given to illustrate the applications of IR spectroscopy.

Number-Average Molecular Weight by Using End-Group Analysis

IR spectroscopy is a valuable technique for measuring the concentration of end groups in a polymer sample, particularly when the measurement of colligative solution properties is not possible because the sample is insoluble in ordinary solvents at room temperature. From IR measurements, a number-average molecular weight of the polymer can be calculated on the basis of the assumption that the chain is linear (with no branch points). This assumption limits the number of endgroups of a given polymer to either one or two per molecule. When polymers are prepared by addition polymerization, there is always the possibility of branching in the polymer, and in this case, the IR measurement is only a relative number-average molecular weight of the system. For polymers formed by condensation, branching is less likely to occur, and end-group analysis is quite accurate once the chemical nature of the end groups has been determined. Because the end groups are not spectroscopically coupled to the remainder of the chain, low-molecular-weight analogs can be used for calibration. IR spectroscopy is particularly useful for measuring the number-average molecular weight of insoluble systems; it is the standard method used to determine the molecular weights of fluorocarbons.

As an example of the use of IR spectroscopy to determine end-group concentrations and molecular weight, consider poly(butylene terephthalate) (PBT) (*1*). In this condensation polymer, the end groups are an alcohol group and an acid, so the molecular weight, M_n, of the polymer is

$$M_n = \frac{2}{(E_1 + E_2)} \tag{4.1}$$

where E_1 is the equivalents per gram of alcoholic end groups, and E_2 is the equivalents per gram of acid end groups. This equation presumes that no other end groups are present and that no grafting has occurred. The IR spectra of two different samples of PBT are shown in Figure 4.1. The $-$COH end groups absorb at 3535 cm^{-1}, and the $-$COOH end groups absorb at 3290 cm^{-1}.

The extinction coefficients were determined, and the results are as follows: $a_{OH} = 113 \pm 18$ 1/(g-equiv cm); $a_{COOH} = 150 \pm 18$ 1/(g-equiv cm). The baseline for the $-$COH is from 3480 to 3600 cm^{-1} tangent to the spectral curves, and for the $-$COOH peak, the baseline is from 3120 to 3480 cm^{-1}. These

1904—4/92/0077$10.50/1

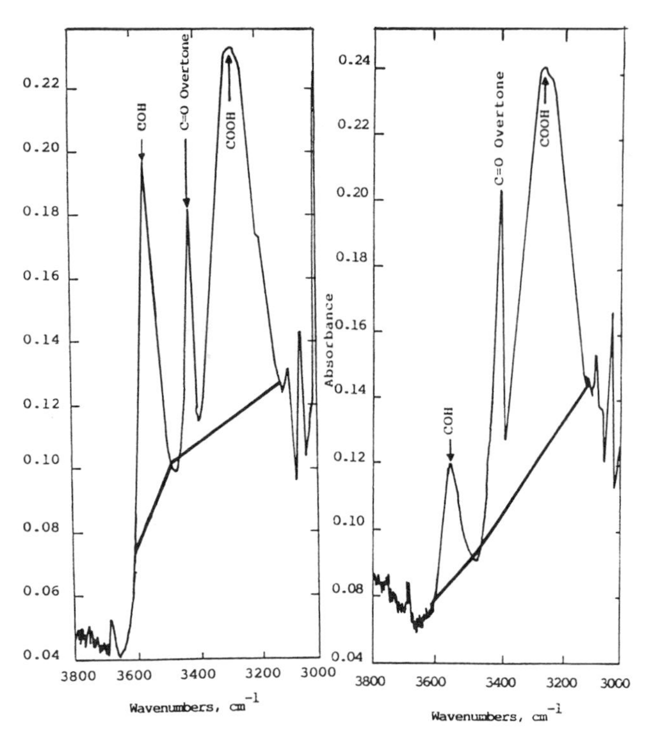

Figure 4.1. The spectra of two samples of PBT with different molecular weights. (Reproduced with permission from reference 1. Copyright 1985 Society of Plastics Engineers.)

IR measurements of molecular weight can be correlated with the results from viscosity measurements and can be used to study the changes in molecular weight that are induced by processing the PBT polymer (*1*).

Sometimes the end groups make only small contributions to the spectra, and in other cases their contributions are more profound. In the case of polyethylene, the bands at 909 and 990 cm^{-1}, which are associated with the terminal vinyl groups, are intense and readily observable in commercial polyethylene samples. These vinyl end groups are observable because the associated dipole moment is stronger than those of the methylene deformations, so the vinyl bands are inherently more intense than the weaker methylene wagging and twisting modes in the same frequency region. Therefore, although the vinyl end groups are very low in concentration, their absorptions appear clearly in the spectrum. On the other hand, the methyl end groups at 1378 cm^{-1} are inherently weak like the methylene modes and are difficult to view because they are badly overlapped by the larger number of amorphous methylene modes in this region. However, the methyl end-group absorptions can be used to measure branching in polyethylene.

IR Analysis of the Polymerization Process

IR spectroscopy is a convenient method for studying the curing of polymers. IR spectroscopy can be used to determine the reaction order and the chemical processes that occur in the reactant system. By using thermally controlled cells, the reactions can be studied at temperatures used in the industrial process. One system that has been studied extensively because of its industrial applications is the copolymerization of epoxide resin with cyclic anhydrides or amines. The reaction kinetics of methylbicyclo[2.2.1]heptene-2,3-dicarboxylic anhydride (Nadic methyl anhydride isomers or NMA) and diglycidyl ether of bisphenol A based epoxy (EPON 828, Shell) were followed by means of the 1858-cm^{-1} carbonyl band of the anhydride molecule (*2*). For a stoichiometric mixture of the reactants at 80°C, a zeroth-order kinetic plot was obtained up to 55% conversion (130 min). Beyond the gelation, the reaction slows and nearly stops at 71% conversion. A reduction in the polymerization rate is expected when the kinetic processes are slowed by an increase in the viscosity near the point of gelation. Arrhenius plots of the zeroth-order reaction rates in the temperature range of 70 – 120°C for a 1:1 anhydride: epoxy system are quite linear and yield an activation energy of 15.5 kcal/mol.

The extent of cross-linking in the NMA and EPON 828 (with 0.5 wt % benzyl dimethyl amine as a curative in each mixture) can be determined by using IR spectroscopy (*3*). The least-squares method was used to fit the spectra of the mixture of NMA and EPON 828. In order to determine the degree of cross-linking, a difference spectrum was calculated by subtracting the spectrum of a stoichiometric mixture of NMA and EPON 828 cross-linked for 37 min at 80°C from the spectrum of the same reactant mixture cross-linked for 83 min at 80°C. The procedure is illustrated in Figure 4.2 for the 600 – 2000-cm^{-1} region. The absorbances that are unaffected by the reaction, such as the aromatic ring modes at 1511 and 1608 cm^{-1}, are cancelled in the difference spectrum. Bands above and below the baseline represent the

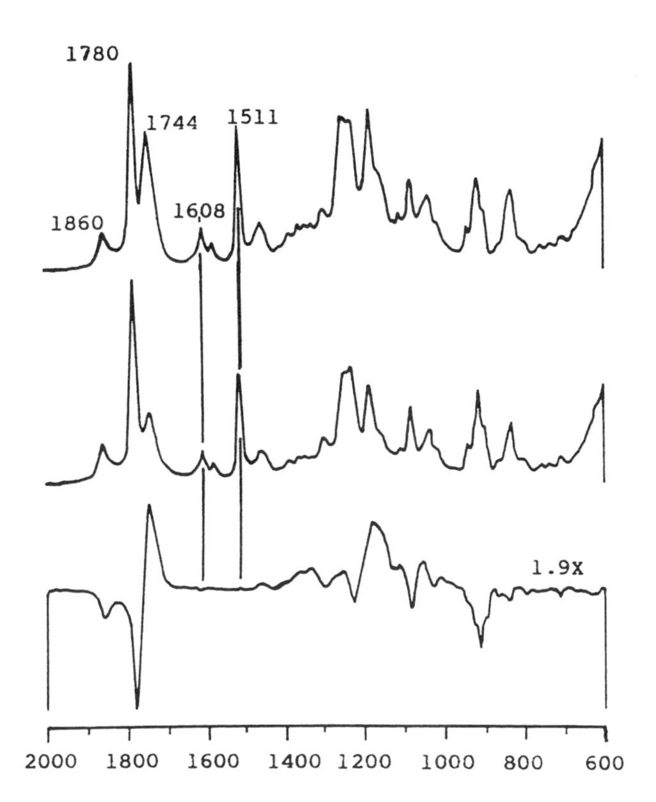

Figure 4.2. Generation of difference spectrum to characterize cross-linking at 80 °C of a stoichiometric mixture of NMA – EPON 828. The samples were cross-linked for 83 min (top spectrum) and 37 min middle spectrum). The bottom spectrum is the difference spectrum (top – middle). (Reproduced with permission from reference 2. Copyright 1981 Society of Plastics Engineers.)

ester cross-link formation and the disappearance of anhydride and epoxide groups, respectively. These measurements yielded the following results for different cure times at 80 °C.

Time (min)	*Degree of Cure (%)*
30	6.00
60	12.00
90	18.50
120	28.10
360	65.30

SOURCE: Reproduced with permission from reference 3. Copyright 1981.

Chemical Transformations in Polymers

We are interested in the nature of the structural changes produced by such chemical reactions as oxidation, reduction, and hydrolysis. Because the IR bands of polymers are inherently broad and weak, it is very difficult to detect minor chemical changes occurring on the polymer chain. It is necessary to eliminate from the observed spectrum the interfering absorptions of the unreacted portions of the polymer. This elimination step can be accomplished by using the absorbance subtraction of the spectrum of the control polymer from the spectrum of the reacted system. The resulting difference spectrum contains only those IR bands resulting from the chemical reactions that have occurred. By eliminating the interfering spectra of the unreacted components, the dynamic range is also enhanced, and scale expansion of the difference spectrum up to the limit of the signal-to-noise ratio is possible. By using both spectral subtraction and scale expansion of the difference spectrum, very small chemical changes can be detected. For example, in Figure 4.3 the spectra of polybutadiene are shown before and after a very mild oxidation, that is, storage in the instrument for 1 h. Visual inspection of the spectrum of the oxidized sample reveals no observable changes, but the subtraction of the spectrum of the unoxidized sample from that of the oxidized sample and scale expansion of the resulting difference spectrum reveals features

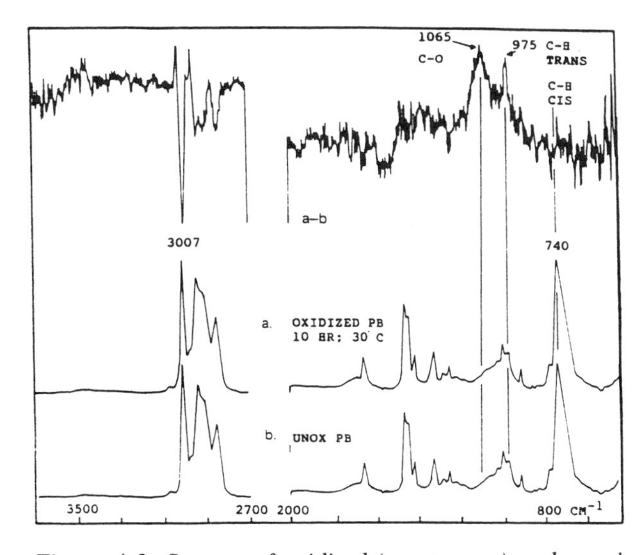

Figure 4.3. Spectra of oxidized (spectrum a) and unoxidized (spectrum b) samples of cis-*1,4-polybutadiene and the a – b difference spectrum. (Reproduced with permission from reference 4. Copyright 1976 Rubber Chemistry and Technology.)*

in the difference spectrum resulting from the onset of oxidation. Absorption in the 1080 – 1110-cm^{-1} region is characteristic of C – O groups. Loss of *cis*-methine groups is indicated by the negative bands at 740 and 3007 cm^{-1}. As the oxidation proceeds, the number of absorbing species increases, as does the complexity of the spectrum. By using absorbance subtraction, the spectral changes can be followed, and the concentration of the various species can be determined (*4*). The difference spectrum after 3 h shows formation of *trans*-methine at 975 cm^{-1} and specific carbonyl groups at 1700 and 1727 cm^{-1}. The 34-h difference spectrum shows a broadening of the C – O band. After 640 h, only 27% of the *cis*-methine remains, a result illustrating the need for oxidative stabilizers in rubber samples.

As another example, let us examine the effects of exposing an anhydride-cross-linked epoxy system to moisture (*5*). Spectra were obtained after exposure of the epoxy system to water at 80 °C for various time periods. The difference spectra obtained after 6.5 days (spectrum b) and 31.5 days (spectrum a) are shown in Figure 4.4. A decrease in the anhydride concentration is indicated by decreases in intensity of the vibrations at 1860, 1080, and 918 cm^{-1}. An increase in the concentration of carboxyl groups is indicated by the increases in the intensities of the bands at 1712 cm^{-1} (C=O), 1395 cm^{-1} (O – H), and 1195 cm^{-1} (C – O). These spectral results reflect the initial hydrolysis of the unreacted anhydride (approximately 5% of the initial anhydrides) to the diacid, which is in part leached from the sample by the water. As shown in Figure 4.4, the water induces some of the diacid molecules to react with the epoxy groups.

Figure 4.4. Hydrolysis and leaching of anhydride from epoxy resin exposed to water at 80 °C. Film thickness was 11 mm. (Reproduced with permission from reference 5. Copyright 1981 John Wiley & Sons, Inc.)

This general type of analysis can be used to study chemical reactions of polymers regardless of the specific type of reaction. The ultimate limitation is the sensitivity as determined by the experimental signal-to-noise ratio.

> The ability of the spectroscopist to interpret the spectral changes in terms of reactants or products is the principal limitation. With modern instrumentation, interpretation of the nature of the chemical changes is often the final barrier.

Copolymer Analysis

One of the more important applications of IR spectroscopy is the characterization of copolymers. An example of the analysis of the ethylene – propylene copolymer was given in Chapter 2. The steps needed to obtain the proper assignments of the absorbances to the copolymer sequences and the performance of the final analysis will be discussed in this chapter.

Vibrational coupling complicates the IR analysis of copolymers. The difficulty of the IR analysis is a function of the size of the comonomer repeat units. The magnitude of the vibrational coupling between units is a function of the intramolecular distance. When the repeat units are small, the coupling is large and is a function of the type of vibrational motion. When the repeat units are large, the coupling between units will be zero or very small. For example, the spectrum of the trimer of poly(ethylene terephthalate) (PET) is essentially the same as that of high-molecular-weight PET because the length of the PET repeat unit is so long that the intramolecular interactions between repeat units is essentially zero.

When the repeat units are small (i.e., contain only two carbons, as is the case for vinyl monomers), vibrational coupling can occur between the repeat units. This coupling results in spectral bands that have frequency and absorptivity depending on the length of the ordered sequence. This intensity perturbation by the coupling factor must be taken into consideration when developing a method of spectral analysis. For large comonomer repeating units, the IR analysis is simplified because the analysis of only the composition is possible, and the copolymer can be considered, from an IR point of view, to be simply a mixture of the two comonomers. Unfortunately, for the case of large repeating units, IR spectroscopy will not yield information about the copolymer sequence distribution.

Consider the case when vibrational coupling occurs; IR spectroscopy can yield connectivity information about the microstructure of the copolymer. The IR spectra of ethylene(E) – vinyl (V) chloride will be used as a specific example for this discussion. These copolymers were prepared by partial reduction of poly(vinyl chloride) (PVC) with tri-*n*-butyltin hydride (*6*). During the reduction of PVC, the reaction rate of tri-*n*-butyltin hydride with the V units depends on the dyad – triad environment of the V unit, with the central unit in the P_3(VVV) being more reactive than the central unit in the P_3(EVE). The mono-ad, dyad, and triad probabilities obtained from NMR data for the E – V copolymer are shown in Table 4.1. These results are plotted in Figure 4.5 for length sequences of 1, 2, 3, 4, 5, and 10.

The approach to interpreting the spectra of the copolymers is empirical. First, the copolymer spectra for a wide range of comonomer compositions are obtained, and then the spectra are examined by using the following steps.

Step 1: *Determine whether frequencies are due to the A or B component by determining at what copolymer composition the absorbance peaks disappear.*

The spectra of the copolymers for a range of compositions are shown in Figure 4.6. The frequencies of the various absorptions are shown as a function of composition in Figures 4.7 and 4.8. The IR bands resulting from long sequences of either monomer can be assigned by comparing them with the same frequencies in the spectra of the pure homopolymers and with those present predominantly at either end of the copolymer composition scale. These bands are indicated by the assignments given on the right sides of Figures 4.7 and 4.8. New bands appearing in the spectra most likely arise from short sequences in the copolymer.

Step 2: *Determine which peaks are sensitive to sequence length by determining which peak positions are dependent on composition.*

For the E – V copolymer, the 700 – 850-cm^{-1} region had three methylene rocking modes that were

Table 4.1. Monad, Dyad, and Triad Probabilities of E–V Copolymers

Copolymer	$P_V = 1 - P_E$	P_{VV}	$P_{VE} = P_{EV}$	P_{EE}	P_{EVE}	$P_{VVE} = P_{EVV}$	P_{VVV}	P_{VEV}	$P_{VEE} = P_{EEV}$	P_{EEE}
E–V–100	1.0	1.0	0.0	0.0	0.0	0.0	1.0	0.0	0.0	0.0
E–V–85.3	0.853	0.742	0.124	0.011	0.015	0.115	0.619	0.114	0.011	0.0
E–V–84.3	0.843	0.709	0.134	0.023	0.025	0.108	0.615	0.101	0.019	0.004
E–V–70.7	0.707	0.470	0.239	0.052	0.063	0.175	0.310	0.175	0.048	0.008
E–V–62.3	0.623	0.344	0.278	0.099	0.116	0.177	0.177	0.177	0.075	0.027
E–V–61.5	0.615	0.343	0.275	0.107	0.121	0.173	0.198	0.141	0.083	0.029
E–V–50.1	0.501	0.200	0.297	0.205	0.192	0.133	0.073	0.166	0.129	0.045
E–V–45.6	0.456	0.147	0.309	0.235	0.205	0.116	0.037	0.149	0.140	0.098
E–V–37.3	0.373	0.087	0.286	0.342	0.219	0.078	0.012	0.115	0.158	0.183
E–V–34.8	0.348	0.061	0.278	0.383	0.224	0.064	0.015	0.090	0.168	0.208
E–V–21.2	0.212	0.014	0.197	0.593	0.190	0.016	0.0	0.035	0.153	0.436
E–V–13.6	0.136	0.0	0.127	0.746	0.104	0.0	0.0	0.051	0.123	0.599
E–V–2.4	0.024	0.0	0.025	0.950	0.021	0.0	0.0	0.0	0.026	0.926
E–V–0	0.0	0.0	0.0	1.0	0.0	0.0	0.0	0.0	0.0	1.0

SOURCE: Reproduced with permission from reference 6. Copyright 1986 John Wiley & Sons, Inc.

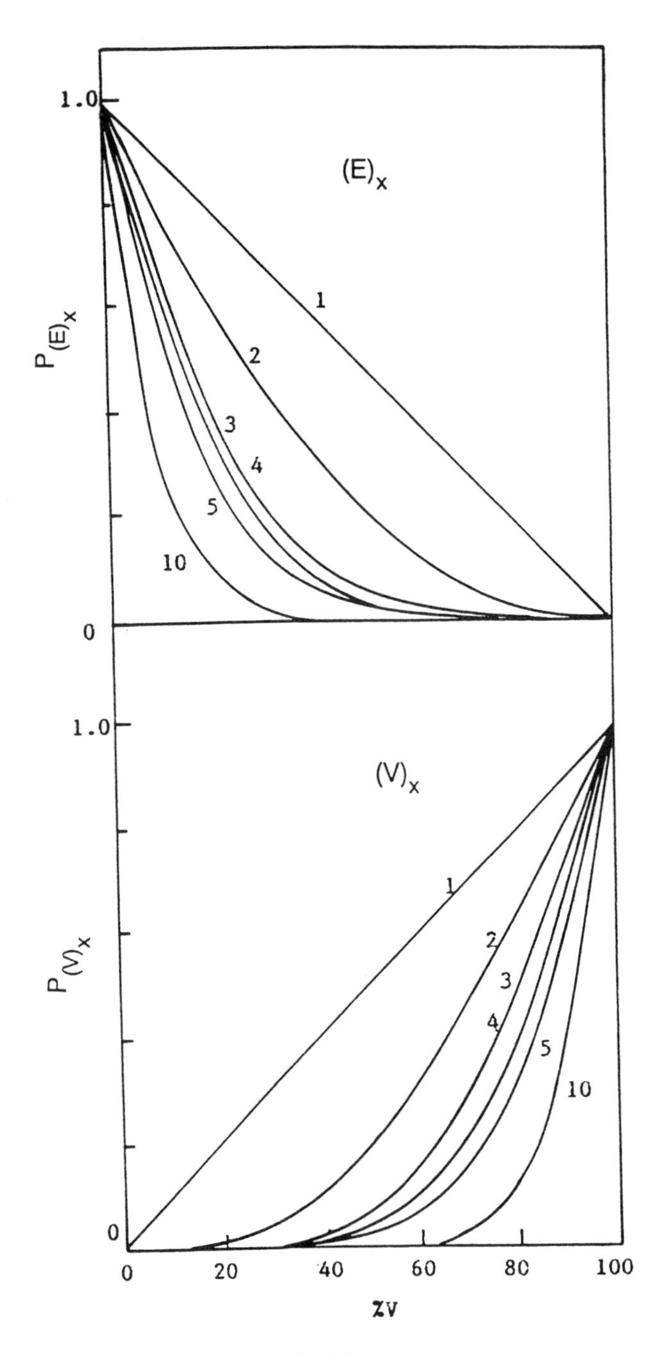

Figure 4.5. Sequence probability curves vs. concentration of V units for copolymer sequences of various lengths. (Reproduced with permission from reference 6. Copyright 1986 John Wiley Sons, Inc.)

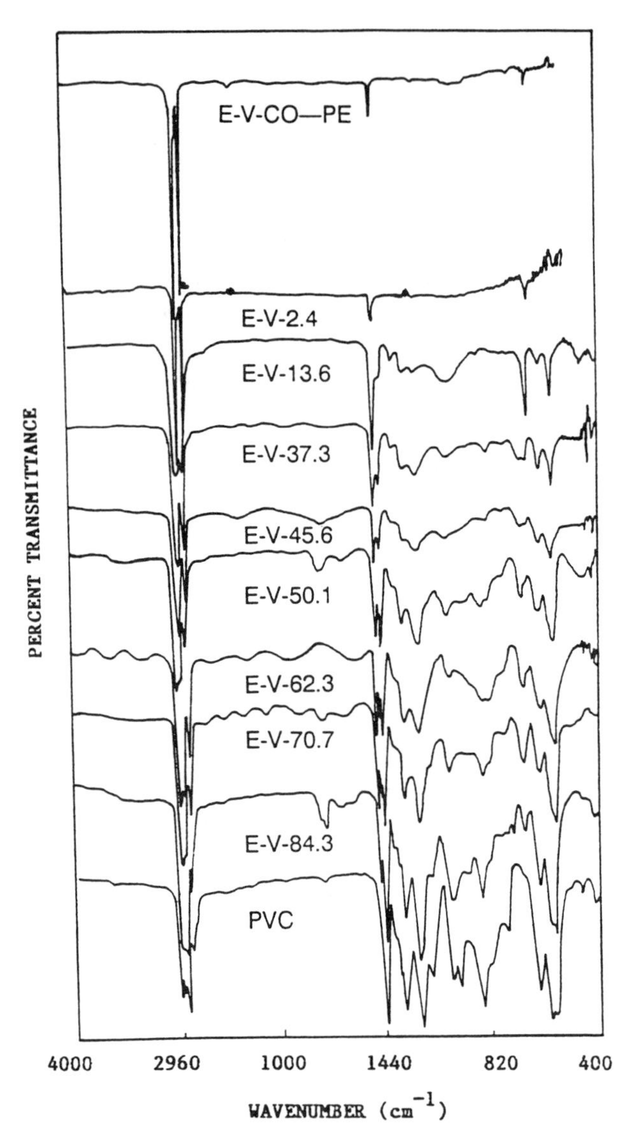

Figure 4.6. Transmission IR spectra of E – V copolymers. The number adjacent to the spectrum equals the concentration of V units present in the copolymer. (Reproduced with permission from reference 6. Copyright 1986 John Wiley Sons, Inc.)

found to be sequence dependent. The 850-cm^{-1} band of PVC results from the rocking motion of the methylene group between two methine groups and therefore is proportional to the concentration of VV dyads. A new absorbance band is observed at 750 cm^{-1} as the composition of the copolymer is changed. This band is assigned to the VEV triad and the VEEV tetrad. At the high end of the E concentration scale there are long sequences of E units (E_x, $x > 3$), and as a result, the 750-cm^{-1} band decreases and the 720 – 730-cm^{-1} band increases. The absorbance

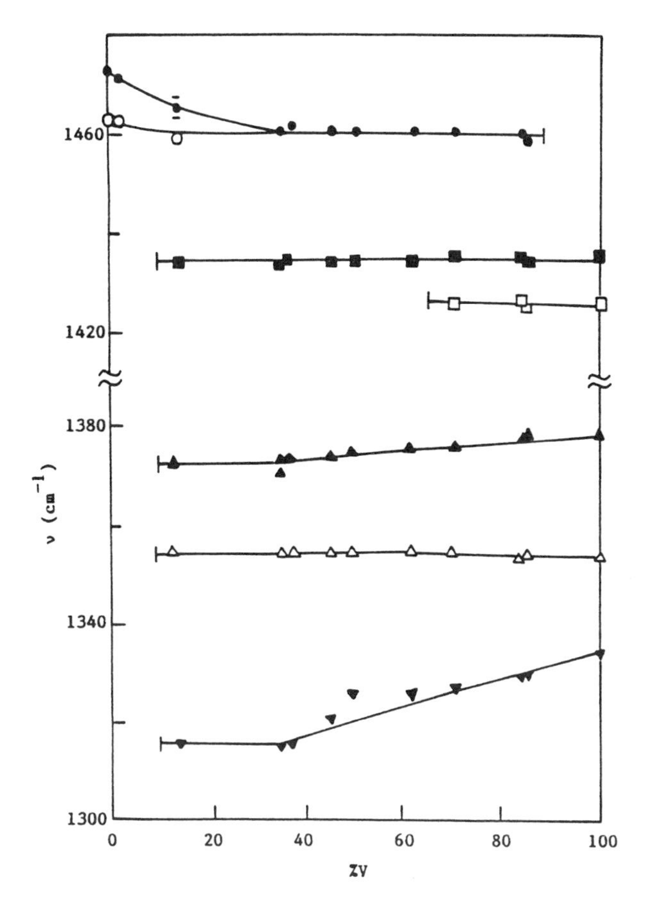

Figure 4.7. Frequency vs. concentration of V units for the 1300 – 1480-cm^{-1} absorbance range. The structural assignments are as follows: ● *and* ○ *1460 (E) CH_2b;* ■, *1434 (V) CH_2b;* □ *1426 (V) CH_2b;* ▲, *1379 (V) CH_2w;* Δ, *1355 (V) CH_2w; and* ▼, *1334 (V) CHb. (Reproduced with permission from reference 6. Copyright 1986 John Wiley Sons, Inc.)*

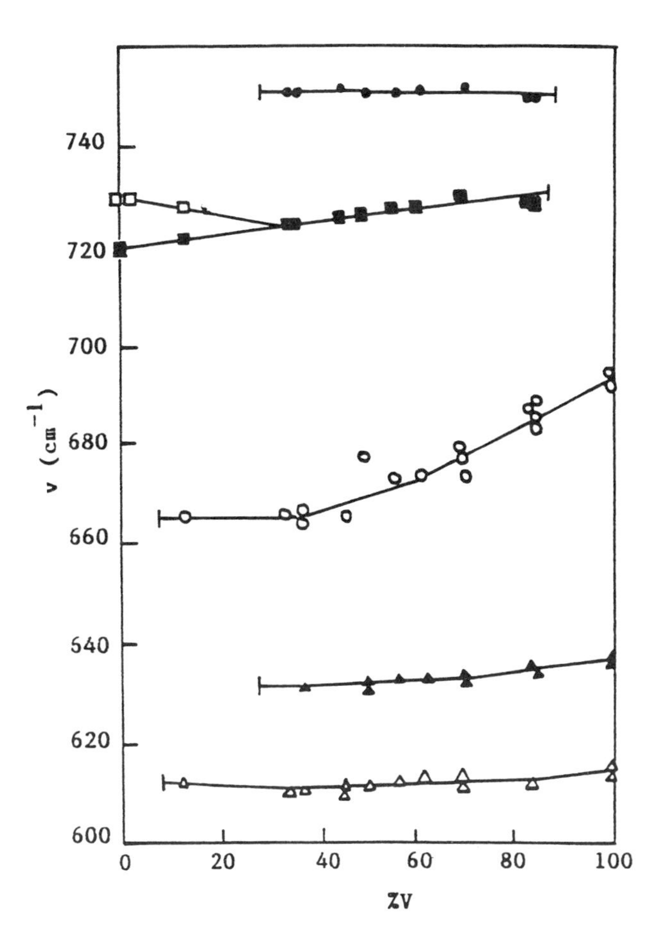

Figure 4.8. Frequency vs. concentration of V units for CH_2 rocking modes and C – Cl stretching modes. The assignments are as follows: ●, *750 (E);* □ *and* ■, *719 – 730 (E);* ○, *665 – 690 (V);* ▲, *630 (V); and* Δ, *611 (V). (Reproduced with permission from reference 6. Copyright 1986 John Wiley Sons, Inc.)*

intensity – sequence relationships for the various bands are listed in Table 4.2.

Step 3: *Determine the sequence sensitivity of the peaks by correlating the absorbances with the sequence data calculated from approximate reactivity coefficients.*

By using the absorbance areas for the various IR peaks vs. composition and a knowledge of the sequence probabilities, a correlation between the absorbance plots of the various bands and the specific sequences responsible for those absorptions can be established. The C – H stretching resonances at 2800 – 3000 cm^{-1}, which are proportional to the C – H concentration, are found to be proportional to the mono-ads of E (Figure 4.9). The peak at 750 cm^{-1} is sensitive to either P_3(VEV) or P_4(VEEV) (Figure 4.10). The peak at 728 cm^{-1} results from P_3(EEE) as determined by the comparison of the expected number of P_2(EE) with that of P_3(EEE) (Figure 4.10).

Step 4: *Determine the extinction coefficients of the bands as a calibration for quantitative analysis.*

Finally, it is necessary to determine the extinction coefficients so that quantitative analysis can be made.

Table 4.2. Absorbance Intensity–Sequence Relationship for E–V copolymers

ν (cm⁻¹)	*A = Absorbance area α (sequence probabilities)*[a]
2800–3000	P(E)
1473–1463	P(E) + P((E)$_x$), $x > 15$
1434–1426	P(V) + P((V)$_x$), $x = 15$
1379 → 1372	2P(EVE) + P(VVV)
1355	P((V)$_x$), $x = 7 \pm 2$
1334 → 1316	P(EVE) + P(VVV)
1255 multiplet	P(VVV) + ½ P(VVE) + ½ P(EVV) + ¼ P(EVE)
1100 multiplet	P(VVV) + P(EVV) + P(VVE)
960	P((V)$_x$), $x > 5$
915	either (a) P(EVE) + P(VVV) or (b) 2P(EVE) + P(VVV)
860 → 840	P(VV)
750	P(VEV) or P(VEEV)
728	either (a) P(EEE) or (b) P(EE) + P((E)$_x$); $x > 10$
693 → 666	P(V)
638,614	P(V) [+ possible contribution from P((V)$_x$, $x > 15$]

[a]Probabilities are set off by parentheses; e.g., P(EVE) = P_{EVE}.
SOURCE: Reproduced with permission from reference 6. Copyright 1986 Wiley.

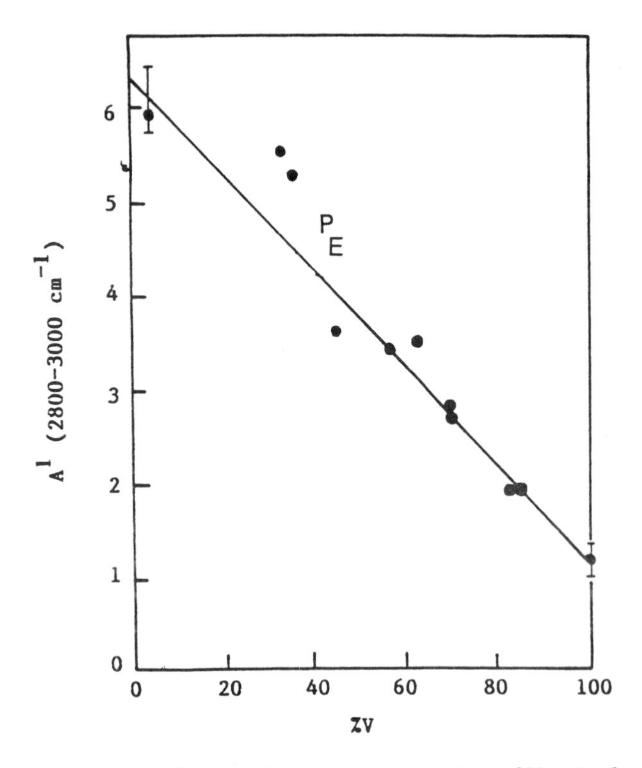

Figure 4.9. IR intensities vs. concentration of V units for the C–H stretching band from 2800 to 3000 cm⁻¹ showing that the band is sensitive to P(E) *monads. (Reproduced with permission from reference 6. Copyright 1986 John Wiley Sons, Inc.)*

In the case of the E–V copolymers, the NMR spectroscopic results can be used for calibration. The extinction coefficients vs. mole percent of V units for several of the absorption bands are shown in Figure 4.11. As expected, some composition dependence of the absorptivities is observed.

At this point, the assignments and calibrations are complete, and the IR spectra can be used to determine the compositions and microstructures of the copolymers prepared under a variety of conditions. The E–V copolymers under discussion have a "randomlike" comonomer sequence distribution.

Measurement of Stereoregularity

The presence of stereoisomers in polymers is indicated in IR spectra by the appearance of new absorption frequencies, shifting of absorption frequencies, and band broadening. Methods for measuring the stereoisomeric composition of a polymer chain depend on a knowledge of the relationship between these spectral properties and the stereoisomer structure.

The assignment of an IR frequency to a particular isomer may be established by (1) synthesis of a polymer containing only one of the stereoisomers, (2) correlation with the isomers of low-molecular-

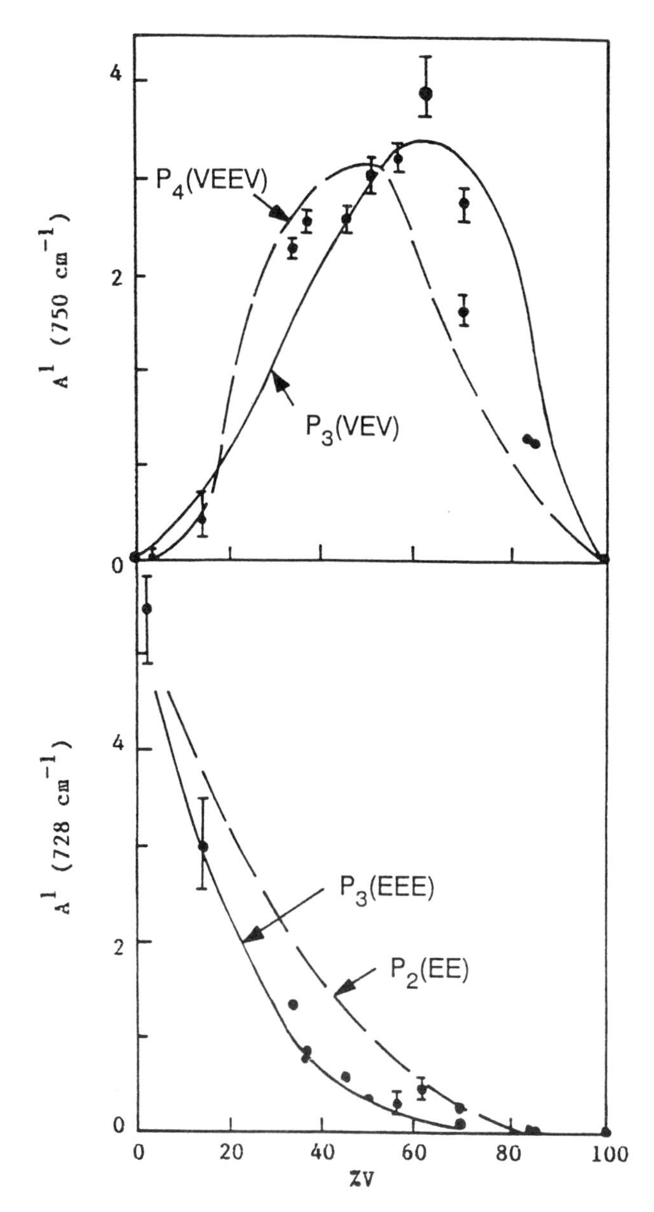

Figure 4.10. IR intensities vs. concentration of V units for the 750- and 728-cm^{-1} absorbances. (Reproduced with permission from reference 6. Copyright 1986 John Wiley Sons, Inc.)

weight analogs, (3) assignments based on normal coordinate analysis, and (4) comparison of absorbances with calculated values from other physical techniques such as NMR spectroscopy.

The ideal method of determining the spectral differences resulting from syndiotactic or isotactic structures is to prepare the two sterically pure polymers and compare their spectra. Unfortunately, there

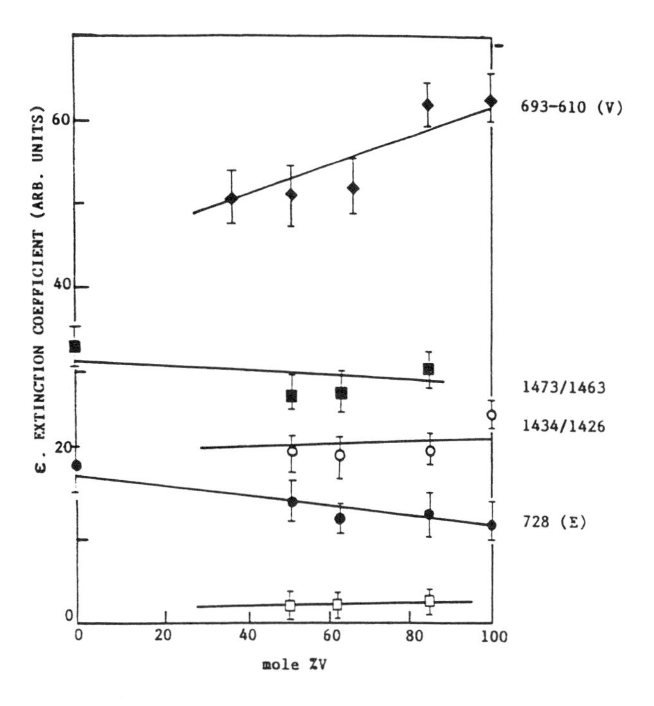

Figure 4.11. Extinction coefficients for several absorption bands of the E – V copolymer as a function of concentration of V. (Reproduced with permission from reference 6. Copyright 1986 John Wiley Sons, Inc.)

are few polymers for which the two pure stereoisomers have been prepared. For poly(methyl methacrylate), the spectra of the syndiotactic and isotactic polymers have been reported, and the spectra are shown in Figure 4.12 (7). The frequency differences

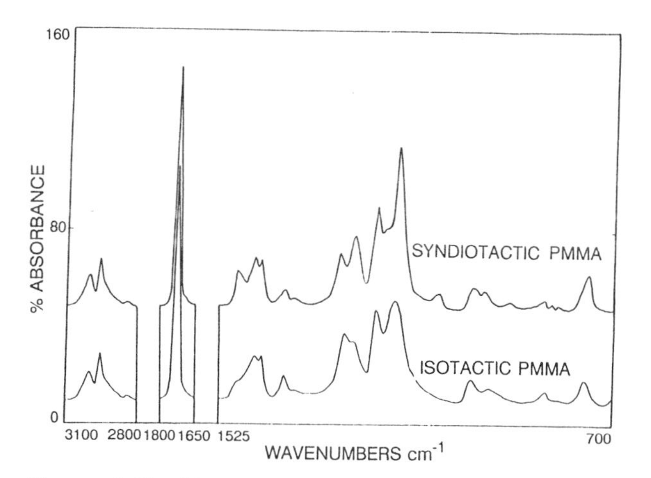

Figure 4.12. IR spectra of isotactic and syndiotactic poly(methyl methacrylate). (Reproduced from reference 7. Copyright 1981 American Chemical Society.)

and the structural assignments are listed in Table 4.3.

The sterically pure polymers of polystyrene have been prepared, and the spectra are shown in Figure 4.13 (*8*). The syndiotactic polymer can have several different conformations in the solid state, and these different conformations are reflected in the spectra.

Poly(vinyl chloride) has been studied frequently with FTIR spectroscopy because its IR spectrum contains a great deal of information about the configurational and conformational content of the sample (the next section gives definitions of these terms). The vibrations are usually studied by using the C – Cl bonds because they show the greatest sensitivity to the different structural features. A notation developed by Shipman et al. (*9*) directly specifies the local tacticity required for a given environment of a C – Cl bond. The notation specifies the environment by the symbol S (implying secondary chloride) with two subscripts, which may be either C or H, to specify the atom *trans* to the Cl atom across the two neighboring C – C bonds. The symbol S_{HH} is reserved for planar zig-zag structures at least five carbon atoms long, with the chlorine atom under consideration on carbon atom 3. The symbol S'_{HH} reflects nonplanar structures in which the chain bend occurs at carbon atom 2 or 4. From a study of model compounds, the following ranges of frequencies for C – Cl bonds were established : S_{HH}, 608 – 616 cm^{-1}; S'_{HH}, 627 – 637 cm^{-1}; $S_{CH'}$, 655 – 674 cm^{-1}.

Table 4.3. The Frequency Differences and Assignments for Poly(methyl methacrylate) for the Spectra of Syndiotactic and Isotactic Polymers

Isotactic	*Syndiotactic*	*Assignment*
2995		$\nu_a(C-H)$
2948		$\nu_s(C-H)$
1750		$\nu(C=O)$
	1485	$\delta_a(\alpha\text{-CH})$
1465	1450	$\delta(CH_2)$, $\delta_a(CH_3-O)$
	1438	$\delta_s(CH_3-O)$
	1388	$\delta_s(\alpha\text{-CH}_3)$
	1270	$\delta_s(\alpha\text{-CH}_3)$
1260		$\nu_a(C-C-O)$
1252		coupled with
	1240	$\nu(C-O)$
1190	1190	skeletal
1150	1150	
996	998	$\gamma_r(CH_3-O)$
950	967	$\gamma_r(\alpha\text{-CH}_3)$
759	749	$\gamma(CH_3)$ + skeletal

NOTE: Frequencies are in units of reciprocal centimeters.

SOURCE: Reproduced from reference 7.

The vibration of each C – Cl bond would be expected to depend on the configuration and conformation of the chain on either side of the bond. On this basis, the C – Cl bond can be considered the center of a triad which might be syndiotactic (s), isotactic (i), or heterotactic (h). Each of these stereoisomers may have more than one rotational isomeric form. Spectral assignments for PVC are given in the following table (*10*).

Conformational assignment	*Peak* (cm^{-1})	*Isomer*	*Conformation*
TTTT long sequences	602	s	S_{HH}
TTT short sequences	619	s	S_{HH}
TTTT long sequences	639	s	S_{HH}
TTTG syndiotactic	651	s	S'_{HH}
TG*G* syndiotactic	676	s	S_{HC}
TGTG isotactic	697	i	S_{HC}

SOURCE: Reproduced with permission from reference 10.

The assignment of bands to the stereoisomers encourages their use in semiquantitative measurements of differences in the stereoregularity of polymers. Although the validity of such measurements has been questioned, they are nevertheless playing an increasing role in the study of stereospecific polymerization.

Measurement of Conformation

The relative geometric arrangement of the chemical groups along the polymer chain, that is, the configuration or conformation of the chain is also of interest. In the polymer literature, the terms *configuration* and *conformation* are often used in quite different contexts, but sometimes they are used interchangeably.

> *The term "configuration" is used for describing an arrangement of atoms that cannot be altered by mere rotation of groups or atoms around single bonds. Configuration is determined during the polymerization process and cannot be altered except by breaking chemical bonds and forming new ones . . . On the other hand, "conformation" refers to the relative steric arrangement of atoms or groups that can be altered by rotation of the atoms or*

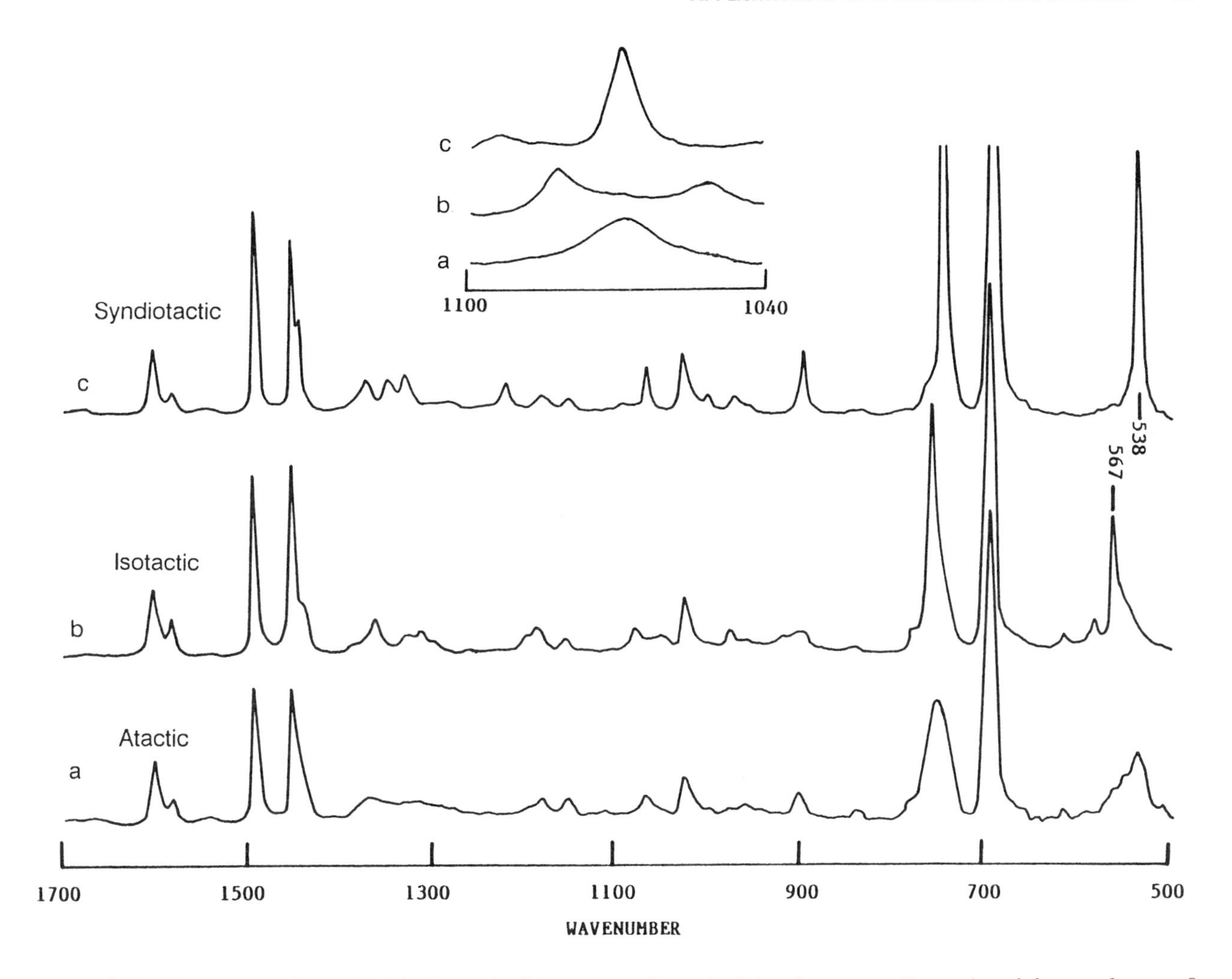

Figure 4.13. IR spectra of atactic (a), isotactic (b), and syndiotactic (c) polystyrene. (Reproduced from reference 8. Copyright 1989 American Chemical Society.)

> *groups around a single bond. For example, conformation includes the* trans *and* gauche *arrangements of consecutive C – C single bonds and the helical arrangement of several polymers in the crystalline state.*
>
> – H. Tadokoro (*11*)

A Nobel prize winning polymer scientist has the following to say about these two terms:

> *The class of configurations which are generated by executing rotations about single bonds of a molecule often are referred to as conformations. These two terms, configuration and conformation will be used somewhat interchangeably, in keeping with their dictionary definitions and broader usage, with only a minor difference in connotation.**
>
> – P. J. Flory (*12*)

Strong differences of opinion exist. Like Flory, we will use the terms *configuration* and *conformation* interchangeably and, as in the previous section, the

**The term conformation connotes form and a symmetrical arrangement of parts. The alternative term, configuration, is perhaps the more general of the two in referring to the disposition of the parts of the object in question without regard for shape and symmetry. Our usage of the latter term may at times violate conventions of organic chemists, who presume to designate the stereochemical arrangement of atoms or groups about the structural element of optical asymmetry in the molecule···. Confusion with configuration about an asymmetric center of the kind associated with optical rotation can easily be avoided by use of an appropriate prefix such as stereochemical.*

term *stereoconfiguration* will be used for stereochemical systems. The terminology can be made simpler if the isomers are termed *rotamers* when the system under consideration consists of one or more stable isomers resulting from the the rotation about a single chemical bond of one or more molecular units.

Polymer chains have a number of possible rotational isomers, depending on the temperature and thermal history of the sample. Rotational potentials for single bonds joining chemical groups such as methylene units are necessarily threefold and symmetric (Figure 4.14) (*11*). The energy minima occur when the substituents of the groups, hydrogens in this case, are in the staggered conformations, but maxima occur at the eclipsed conformations. In molecules possessing C–C bonds, the rotation angle values near the potential minima are strongly favored over those near the maxima. However, rapid interconversion is possible among the various forms.

The *trans* form is a staggered conformation in which the internal rotation angle X–C–C–X is 180°. There are two types of *gauche* isomers with an internal angle of 60°, as shown in Figure 4.14. The *cis* conformation of the *gauche* isomer corresponds to the internal rotation angle of 0° and is the least stable form. Several possible conformations of single-bonded carbon chains can result from these structures, as shown in Figure 4.15. The simplest conformation is the planar zig-zag (all *trans*) structure, as exhibited by polyethylene in the crystalline state. A chain of all *gauche* structures leads to left- or right-handed helices, as in poly(oxymethylene). This discussion has been limited to only two carbon atoms and, of course, more complicated systems give rise to more conformational possibilities. Polymers generally have several different conformations in the solid state depending on their thermal history.

In the crystalline state, the regular structure results in a repeating polymer conformation. This repeating conformation is usually determined by X-ray or electron diffraction methods, but for amorphous and glassy polymers, these diffraction methods are not applicable. Vibrational spectroscopic methods are

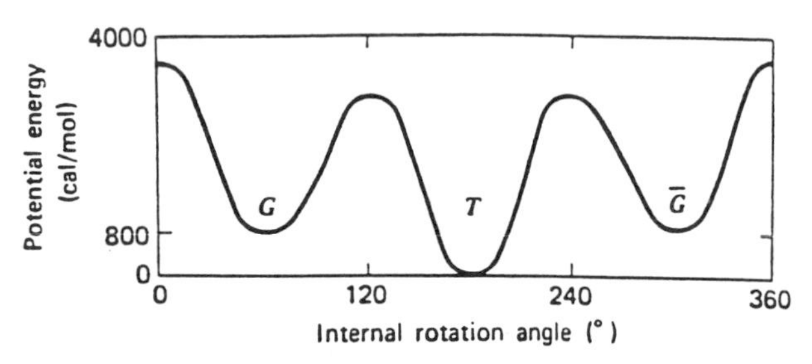

Figure 4.14. Potential energy curve for internal rotation about the C–C bond of butane. (Reproduced with permission from reference 11. Copyright 1979 J. L. Koenig.)

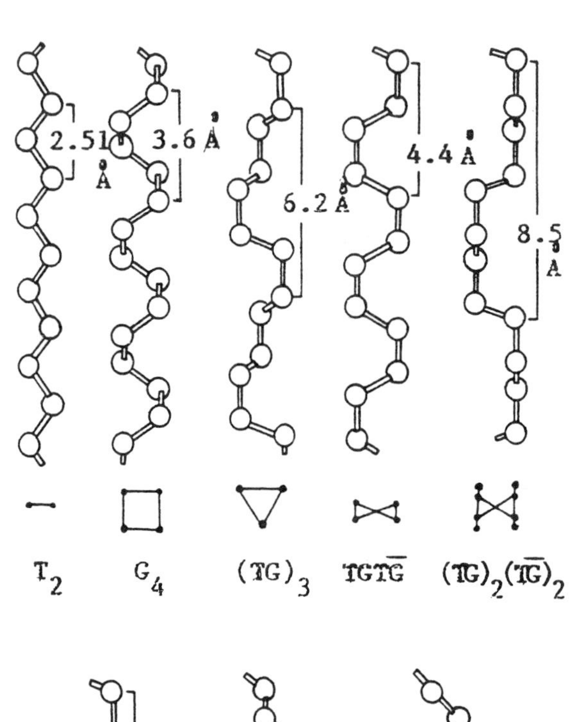

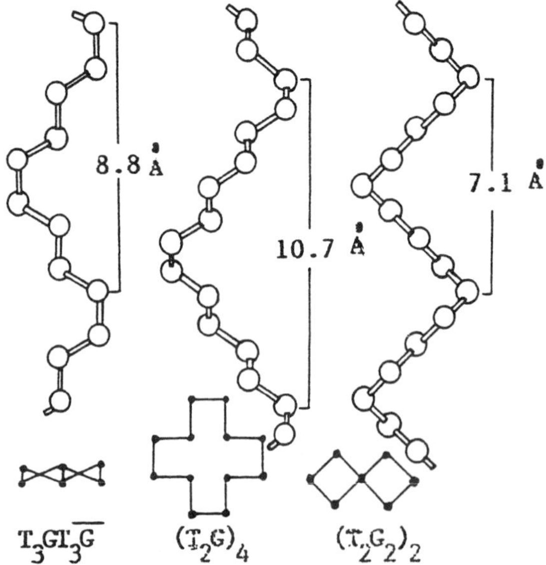

Figure 4.15. Several possible conformations of a single-bonded carbon chain. (Reproduced with permission from reference 11. Copyright 1979 by J. L. Koenig.)

useful because the vibrational modes are sensitive to differences in internal bond angles. When the polymer chains exhibit extended order, specific vibrational selection rules apply, and these rules can be used to determine the conformation of the isomer (*13*). These rules will be discussed in Chapter 5 because the Raman effect is particularly sensitive to the C – C – C bond angles.

In the disordered state, the polymer chain exists in a variety of conformations consisting of combinations of the various rotamers available to the chain. As stated in Chapter 1, it is not possible to determine the position of the atoms in space because there are far too many variations; rather, average values of the isomeric composition can be obtained.

To study rotational isomerism in polymers, the spectroscopic technique must be able to distinguish the spectral features of each identifiable conformer. The vibrational spectrum of a mixture of conformers will exhibit bands arising from the molecular vibrations of all conformers. In general, the conformer bands in a polymer system are observed in pairs that correspond to modes of similar form but slightly different frequencies in the high- and low-energy conformers. When the spectra are obtained as a function of temperature, pairs of bands with one band intensity increasing and the other band intensity decreasing will often be observed. For polymers, there is usually considerable overlap of these band pairs.

Polyethylene (PE) is the most-studied example of rotational isomerism. The crystalline domains are made up of the all-*trans* conformer structure. Bands that are characteristic of rotamers in the amorphous phase are also present. The most intense of the amorphous absorptions are the methylene wagging modes at 1303, 1353, and 1369 cm^{-1}. The TG conformation is correlated with the bands observed at 1303 and 1369-cm^{-1}. The 1353-cm^{-1} absorption band is assigned to the wagging of the GG structure. When PE is heated through its melting point, the concentration of the TG and GG conformations increases. However, the concentration of the TG conformation increases well below the melting point, and this increase indicates the formation of localized conformational defects in the crystalline polymer.

One method of studying conformations is to selectively substitute deuterium and examine the vibrations of the C – D bonds. When isolated CD_2 groups are chemically substituted in PE, bands at 622 cm^{-1}, which are associated with the rocking mode of an isolated CD_2 group with *trans* adjacent dihedral angles, are observed. These bands are TT bands. The band observed at 650 cm^{-1} is associated with the isolated CD_2 in the TG structure (*14*). Studies of "surface melting" of stearic acid and *n*-nonadecane have been made by using this topological deuteration approach (*15*). For stearic acid, a premelting process consisting of a conformational disordering of the molecule at either end of the alkyl chain takes place at approximately 7 °C below the melting point, and it generates a disorder at the interface between the molecular layers. Such surface melting coexists with surface domains of ordered material. The longitudinal diffusion process for the chain occurs by a "conformational soliton" or "twiston" process.

A problem with these IR techniques is the difficulty in isolating the spectra of each of the "pure" rotational isomers in the disordered phase. Because there will never be a model polymer of only one rotational isomer, the spectra of the pure isomers must be deconvoluted from the spectra of mixtures containing different amounts of the rotational isomers. Even the preferred isomer found in the crystalline state is contaminated because a polymer sample is never 100% crystalline. The usual procedure is to use spectral subtraction or the ratio method to isolate the spectra of the rotational isomers. Both of these methods have been used in studies of poly(ethylene terephthalate) (PET) (*16*). The results of the determination of the spectra of the *trans* and *gauche* isomers of PET by using the ratio method were demonstrated in Chapter 3, Figures 3.24 and 3.25.

Another requirement of the spectroscopic technique for the study of rotamers is the ability to quantitatively measure the concentration of each of the two or more conformers. After the "pure" spectra have been obtained, least-squares curve fitting can be applied to the mixture spectra to obtain the rotational isomeric composition of the sample. The results for PET that has been annealed at various temperatures are shown in Figure 4.16 (*17*).

The conformational structures of polymer melts and solutions are also of interest. Studies of the spectra of molten isotactic polypropylene (PP) have yielded some interesting results (*18*). Separation of

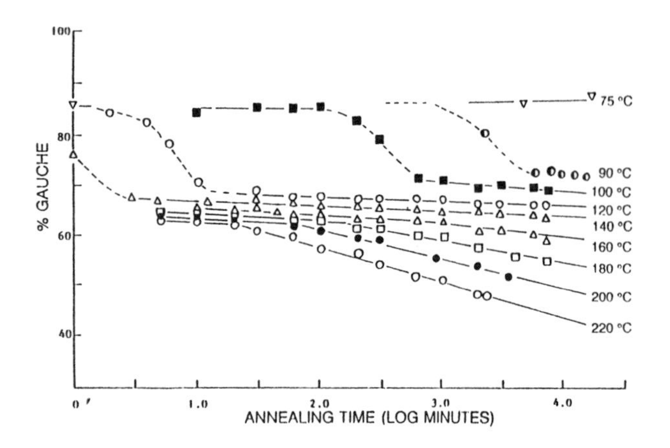

Figure 4.16. The percent of gauche *conformer in PET films as a function of annealing time. (Reproduced with permission from reference 88, Chapter 3. Copyright 1983 John Wiley & Sons, Inc.)*

the spectral components of the ordered and disordered phases permits an accurate determination of the frequencies of the vibrational modes of the polymer chain in the preferred conformation. The amorphous component is determined by obtaining a difference spectrum of two films, one of which is annealed. The "ordered" spectrum is obtained in a similar manner. The results are shown in Figure 4.17 (*18*). The ordered spectrum of isotactic PP shows sharp lines that are well resolved.

> This spectrum is precisely what is expected for long lengths of polypropylene chains in the preferred conformation, which is a 3_1 helix. This spectrum might be considered to be a "crystalline" spectrum because in most polymers such ordered chains are incorporated into a crystal lattice. However, we could detect no differences in the spectra resulting from chain packing. Therefore, we would prefer to refer to this spectrum as a "helical" spectrum and reserve the term "crystalline spectrum" for one that reveals intermolecular effects.

Examination of the amorphous spectrum reveals that the bands characteristic of irregular conformation sequences are broader, and a few are also considerably weaker in intensity than those in the ordered spectrum. The spectrum of PP in the melt is compared with the difference spectrum characteristic of the amorphous region in Figure 4.18. The bands in the molten spectrum resemble those in the spectrum

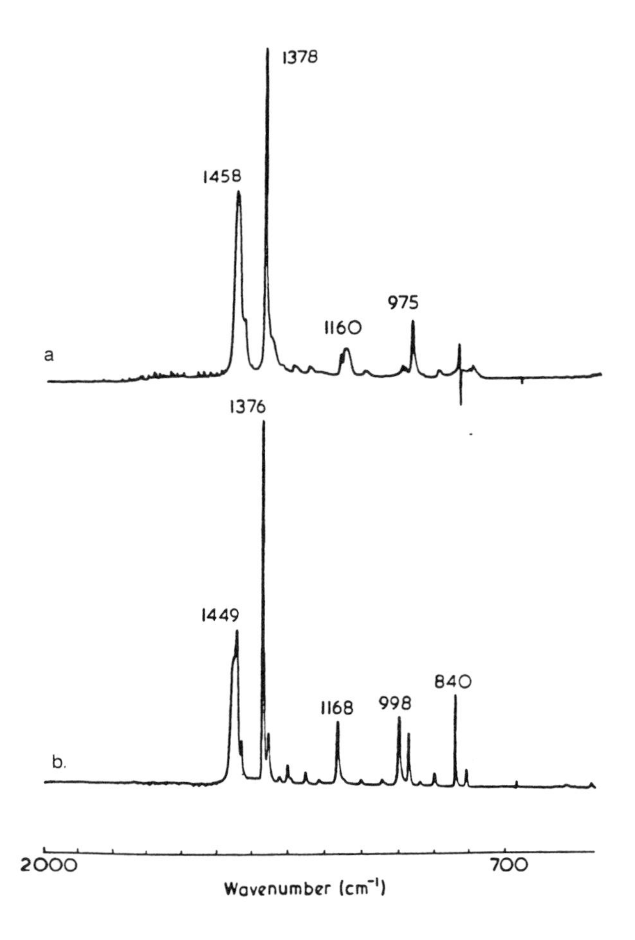

Figure 4.17. Difference spectra characteristic of the amorphous phase of quenched isotactic polypropylene (spectrum a) and of the ordered phase of the annealed sample (spectrum b). (Reproduced with permission from reference 18. Copyright 1977 Butterworth–Heinemann, Ltd.)

of the amorphous region, except for the two bands in the 1700-cm^{-1} region, which are characteristic of oxidation occurring in the molten polymer. In terms of the frequencies of many of the bands, the amorphous spectrum is intermediate between the spectrum of the melt and the spectrum of the helical chains. These results indicate that there are helical polymer chain segments in the amorphous phase. The molten spectrum can also be interpreted in terms of residual helical segments occurring in the melt.

Measurement of Branching in Polymers

One of the most important aspects of the study of polymers is the nature of the branches or side chains. This subject is particularly important for crystalline

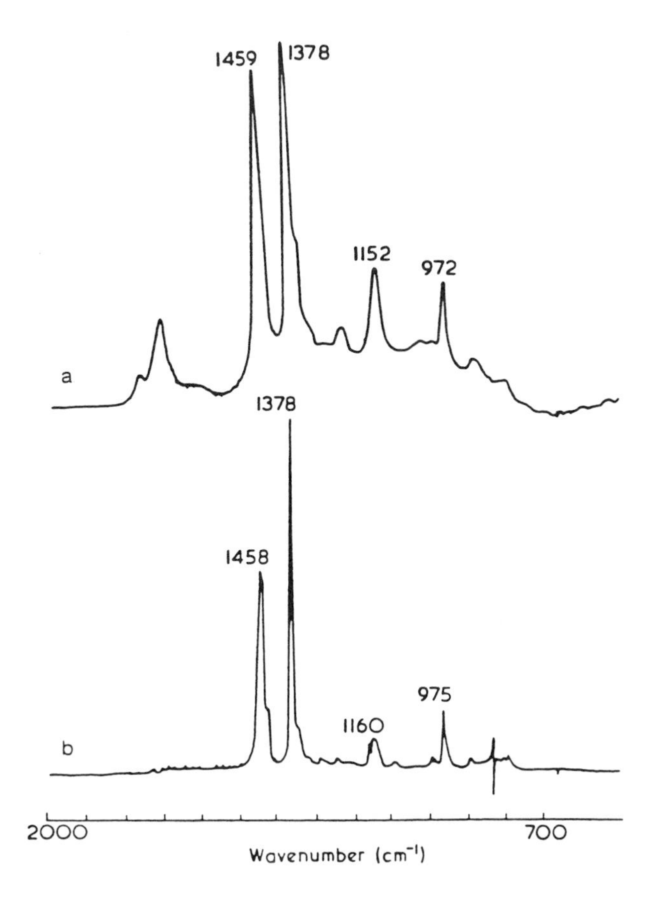

Figure 4.18. The IR spectrum of polypropylene in the melt (a) and the difference spectrum characteristic of the amorphous phase of polypropylene (b). (Reproduced with permission from reference 18. Copyright 1977 Butterworth–Heinemann, Ltd.)

polymers, because branches disrupt the crystalline order. It is important not only to establish the number and distribution of branches along the chain but also to determine their distribution between amorphous, interfacial, and crystalline phases and to characterize the degree of disorder introduced into the crystal by their incorporation.

Polyethylene (PE) can be made by either a low-pressure catalyzed polymerization or a high-pressure polymerization. The low-pressure process yields an essentially linear PE chain, and the high-pressure process yields a polymer with a number of short-chain branches, including ethyl and butyl branches, and occasional long-chain branches. PE has been studied extensively by using IR methods, and most commercial lots of PE have a reported branch content that was determined by IR analysis. The spectra of a low-density (branched) and a high-density (unbranched) PE in the region of interest are shown in Figure 4.19. The band at 1378 cm^{-1} arises from the methyl capped branches. Unfortunately, the methyl band at 1378 cm^{-1} is badly overlapped by three interfering bands at 1304, 1352, and 1368 cm^{-1} that result from amorphous methylenes. The interference band at 1368 cm^{-1} is particularly strong. By using the spectrum of a linear polymethylene standard, it is possible to obtain a difference spectrum that isolates the 1378-cm^{-1} band of the methyl branch for measurement of its absorbance. The intensity of the 1378-cm^{-1} band reflects the methyl group concentration.

A major complication arises with the IR technique because the absorbances of the methyl groups on the ends of the ethyl branches are greater than the absorbances of methyl groups attached to longer pendant alkyl groups. The differences in the absorptivities of the methyl groups in going from methyl to ethyl to longer branch lengths are 1.55:1.25:1, respectively (*19*). Unfortunately, with an ordinary PE sample it is not possible to separate these absorptivity differences, so the measured absorbance of the 1378-cm^{-1} band is a weighted average of the different branch lengths in PE. As a result, the calculated methyl branch content is not accurate, but it is a scale with which to compare PE samples from similar sources. When an absolute number of the different types of branches is sought, high-resolution solution NMR spectroscopy must be used, as will be discussed in Chapter 8. In practical terms, it is too expensive to use high-resolution NMR spectroscopy for every sample of PE, so the IR scale of branching in PE is accepted and is widely used by manufactures and consumers alike.

Characterization of Polymer Blends

Polymer blends are used for a variety of reasons. The principal motivation is to enhance the properties of the individual homopolymers in the blend. However, the polymers must be compatible, which means that they must form stable mixtures at the molecular level. The behavior of polymer blends depends, in general, on the degree of mixing of the components and their mutual interaction, as well as on the individual properties of the components. Most pairs of polymers are not miscible on a molecular level, and,

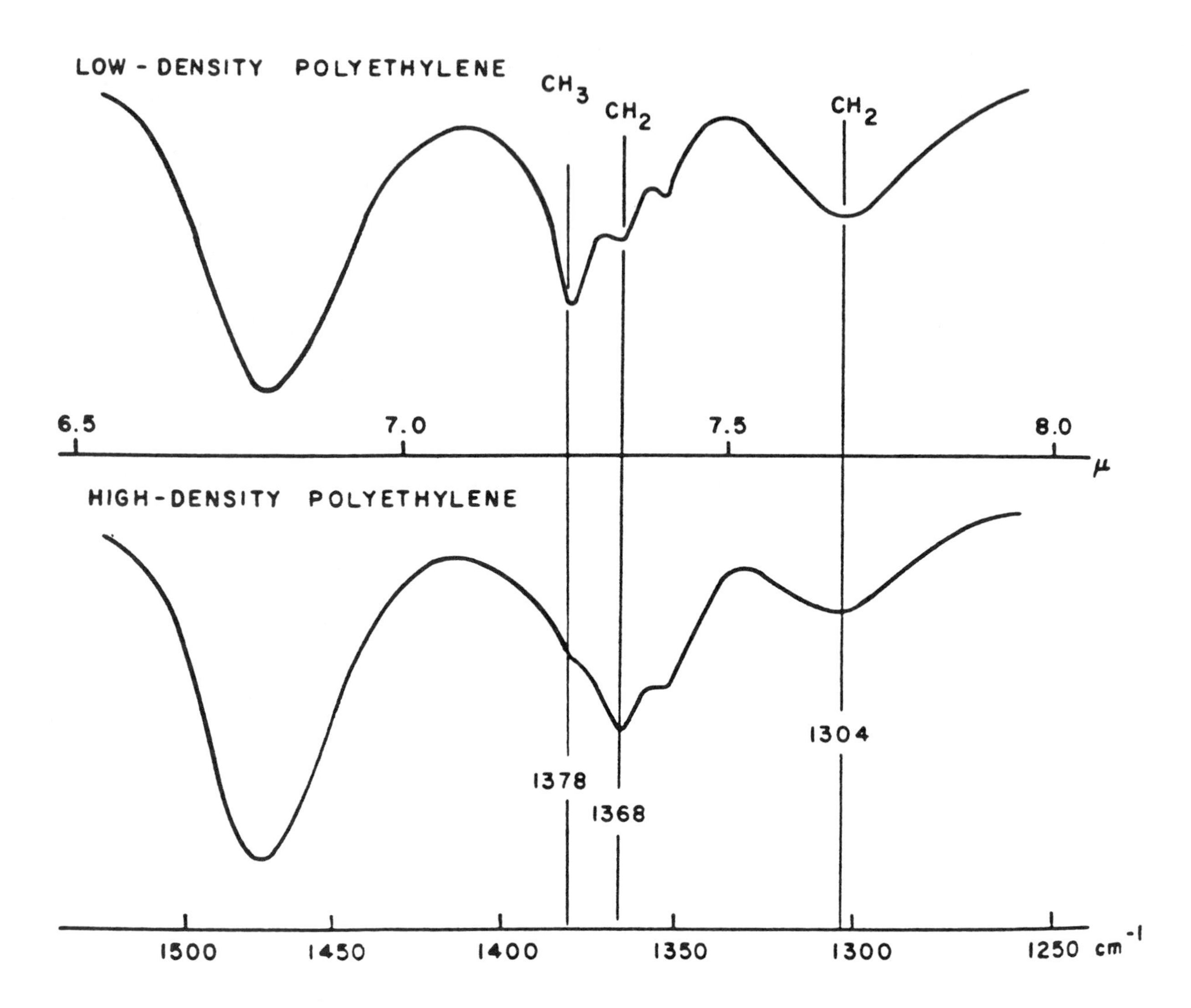

Figure 4.19. IR spectra of low- and high-density polyethylene in the 1500–1250-cm^{-1} region. (Reproduced with permission from reference 19. Copyright 1982 John Wiley & Sons, Inc.)

in the majority of cases, the mixing of two polymers results in phase separation. Such phase-separated systems exhibit poor mechanical properties. The presence of one polymer in the other polymer phase is commonly demonstrated by experimental observations of the glass transition temperature, T_g, of the coexisting phases. The T_g of one component will be displaced in the direction of the T_g of the second component. For amorphous polymer blends, only a single T_g is observed.

When compatible polymer blends are prepared, the important information is the nature of the interactions leading to compatibility, the reproducibility of the blend interactions, and the changes induced in the interactions with time and temperature after blending. Although a number of methods have been used to establish the compatibility of blends (e.g., thermoanalysis and NMR spectroscopy), none of the techniques are conducive to rapid analysis of the kinetics of the phase separation process. However, FTIR spectroscopy can be used to establish the nature and level of molecular interactions of blends and the changes in these interactions with aging. From an IR point of view, compatibility of a blend is defined in terms of the presence of a detectable "interaction" spectrum that arises when the spectrum of the blend is compared to the spectra of the two homopolymers. If the homopolymers are compatible, an interaction spectrum with frequency shifts and intensity modifications that are intrinsic to the system will be observed. If the homopolymers are incompatible, the spectrum of the blend is simply the spectral sum, within experimental error, of the spectra of the two homopolymers.

Factor analysis can be used as a quantitative method to establish the existence of a measurable interaction spectrum (*20*). To determine whether the interaction spectrum is a contributing factor to the spectrum of the blend, a series of polymer blends with different volume fractions of each homopolymer is prepared, and the spectrum of each blend is obtained. The number of components present in these blend spectra is then determined by factor analysis. In the case of compatible blends, three components are expected, but for incompatible blends, only two should be observed.

The blends of poly(vinylidene fluoride) (PVF_2) – poly(vinyl acetate) (PVAc) have been studied (*21*) as a function of composition and temperature during solvent evaporation. Factor analysis was performed on PVF_2 – PVAc blends that were heat treated at 75 and 175 °C. The minimum indicator function values for both blends correspond to a three-component system, a result indicating that both blends are compatible.

The interaction spectrum was generated by a double subtraction of the spectra of the two homopolymers from the spectrum of the blend. An example of such a subtraction for a 50:50 PVF_2 – PVAc blend that was heat treated at 75 °C is illustrated in Figure 4.20. The interaction spectrum shows substantial interaction between the two homopolymers in the blend, as reflected by the shifts in frequencies and intensities. The interaction spectra of several compositions of the PVF_2 – PVAc blends that were heat treated at 75 and at 175 °C are shown in Figures 4.21 and 4.22, respectively.

Quantification of changes in the degree of interaction in the blends as a function of blend composition and heat-treatment temperature was determined by the least-squares curve fitting method. The three spectral contributions used in this curve-fitting technique were the spectra of the two homopolymers and the interaction spectrum. In Figure 4.23, the contribution of the interaction spectrum is shown as a function of weight percent PVF_2 for PVF_2 – PVAc blends that were heat treated at 75 and 175 °C. As expected, the magnitude of the interaction changes with composition.

Aging studies were performed to monitor changes in the compatibility of these blends with time. As shown in Tables 4.4 and 4.5, the largest changes in the magnitude of the interaction occur during the early stages of aging.

The most IR-observable effect of molecular interactions in blends occurs when the two components' hydrogens bond with each other. An example is given in Figure 4.24, which shows the IR spectra of polyvinylphenol (PVPh) – PVAc blends that were cast from tetrahydrofuran (THF) and that contain different amounts of PVAc in the regions from 3800 to 3000 cm^{-1} and from 1800 to 1650 cm^{-1} recorded at room temperature (*20*). In this case, the carbonyl and hydrogen bonding IR regions have large spectral differences. Frequency shifts and changes in band intensities show molecular interactions resulting from hydrogen bonding between the chains.

An interesting method of forcing two polymers that do not exhibit hydrogen bonding between each other to be compatible is to introduce into one of the systems a small amount of comonomer that can hydrogen bond and that can act as the chemical link between the polymer chains (*23*). The spectra of the styrene (92%) – acrylic acid copolymer, the 89% styrene – acrylic acid (SAAS) – poly(methyl methacrylate) (PMMA) blend (4:1), and the interaction spectrum obtained after double subtraction are shown in Figure 4.25. Hydrogen bonding has been introduced in the blends to make these two polymers compatible.

A note of caution must be inserted here. Allara (*24*) theoretically investigated the effects of the refractive indices of the component polymers on the IR band shapes associated with strongly absorbing groups. Residual peaks can arise purely from optical dispersion effects. These residual peaks shift from the absorption peak center in both positive and negative directions with varying intensities. Consider a 10:90 blend of two polymers, A (with a strong absorption band) and B, with refractive indices of 1.5 and 1.4, respectively. Optical dispersion effects will result in the presence of a large positive peak that has an intensity of about 4% of that of the main blend peak and that is located 12 cm^{-1} from the main blend

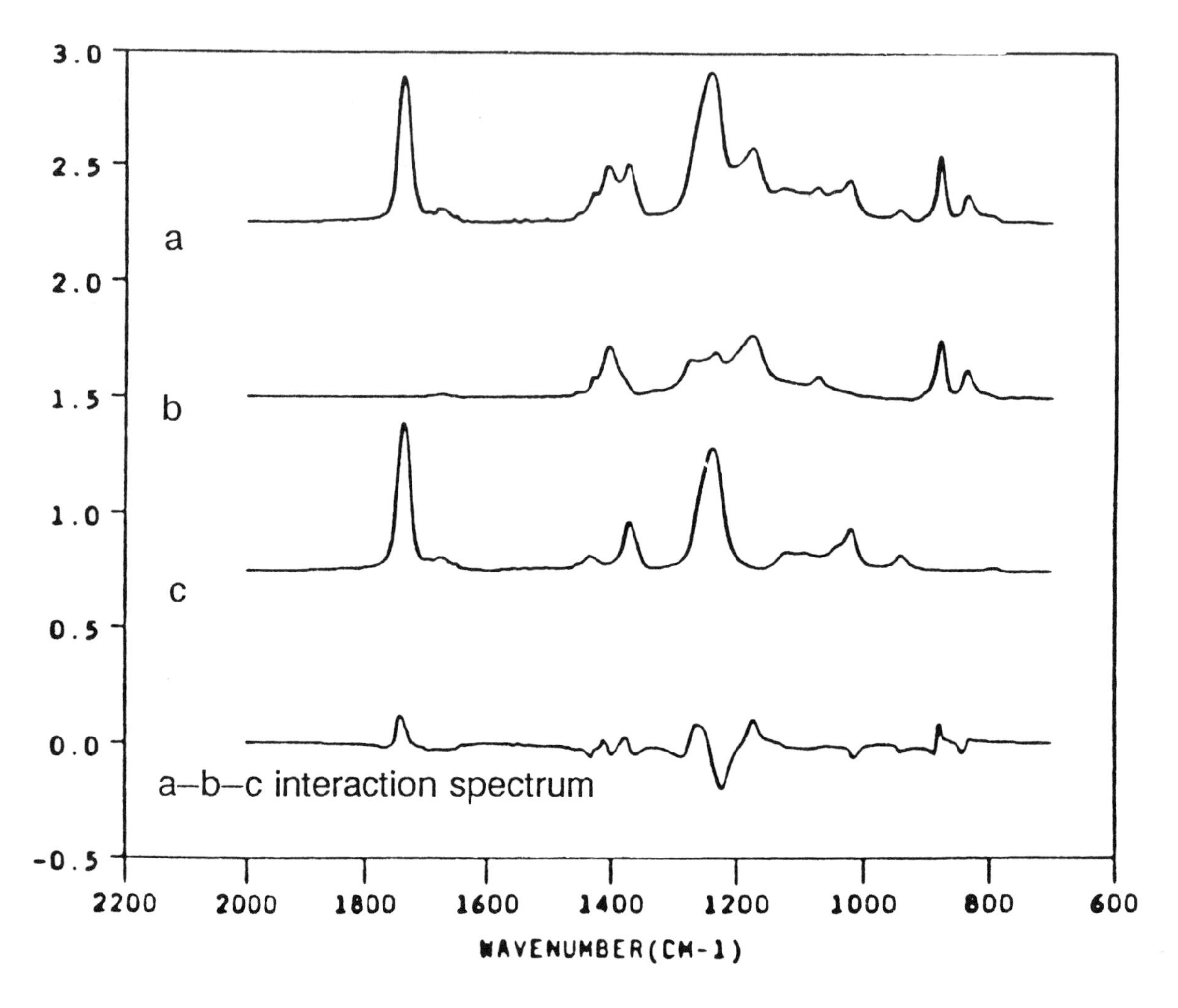

Figure 4.20. FTIR transmission spectra of 50:50 PVF_2-PVAc (a), PVF_2 (b), PVAc (c), and the interaction spectrum of the 50:50 PVF_2-PVAC blend obtained from the subtraction a − b − c. The spectra were collected immediately after heat treatment at 75 °C for 1 h.

Table 4.4. Interaction Contribution (Aging Time) of PVF_2–PVAc Blends Heat Treated at 75 °C

PVF_2–PVAC wt %	0 h	8 h	16 h	24 h	2 days	7 days	24 days
10:90	5.29	4.43	5.02	4.34	3.87	4.19	4.12
20:80	4.92	3.41	4.11	3.19	2.85	3.21	2.89
30:70	5.04	5.25	6.64	5.40	4.89	5.95	4.96
40:60	9.55	8.35	9.91	7.55	7.48	7.52	6.43
50:50	9.67	8.70	9.56	9.89	8.28	7.93	8.57
60:40	13.89	14.01	12.89	14.88	13.08	12.35	11.44
70:30	12.94	11.90	12.05	13.48	10.98	11.13	9.57
80:20	13.06	12.20	12.25	11.34	11.44	11.59	10.61
90:10	10.36	10.23	9.86	9.12	11.08	9.95	11.37

NOTE: Experimental error ± 2%.

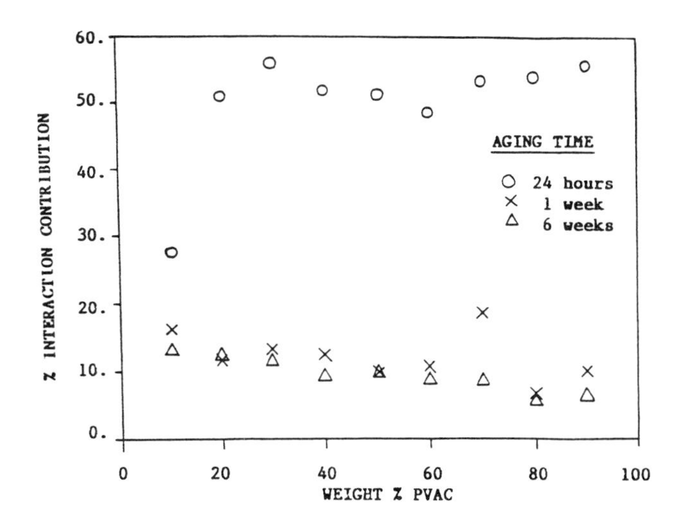

Figure 4.21. Interaction spectra of PVF_2-PVAc blends obtained immediately after heat treatment at 75 °C for 1 h. Spectra were collected for blends having weight percent ratios of 10:90 to 90:10 PVF_2-PVAc.

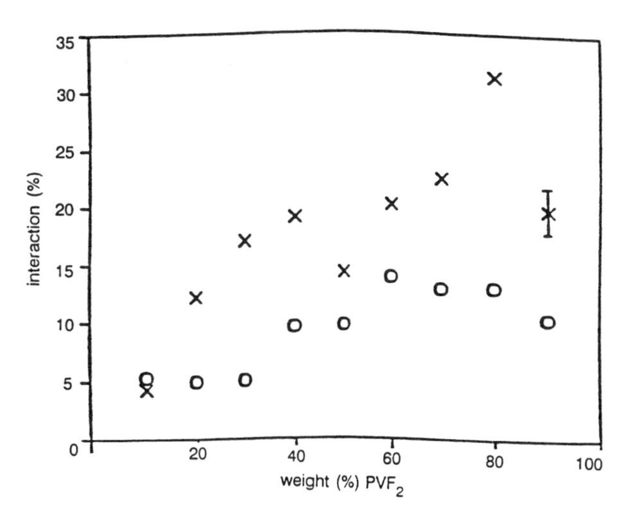

Figure 4.23. Plot of the percent contribution of the interaction spectrum vs. weight percent PVF_2 for PVF_2-PVAc blends heat treated at 75 °C (○) and 175 °C (×).

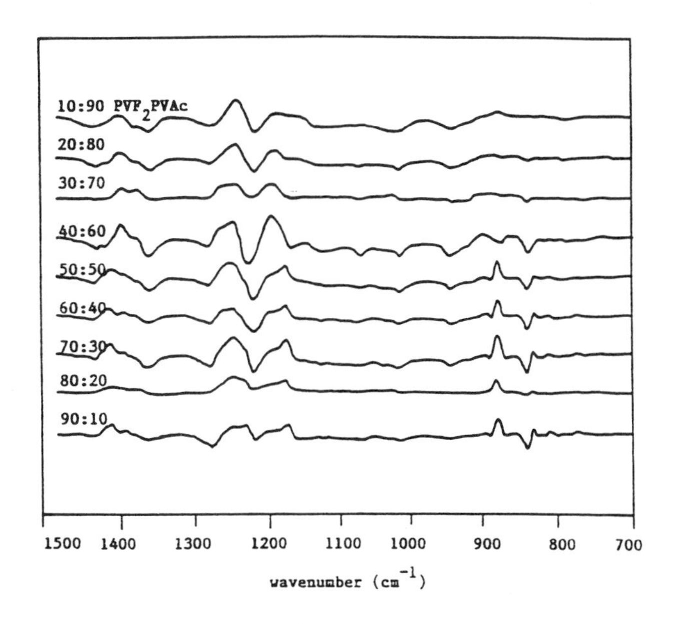

Figure 4.22. Interaction spectra of PVF_2-PVAc blends heat treated at 175 °C.

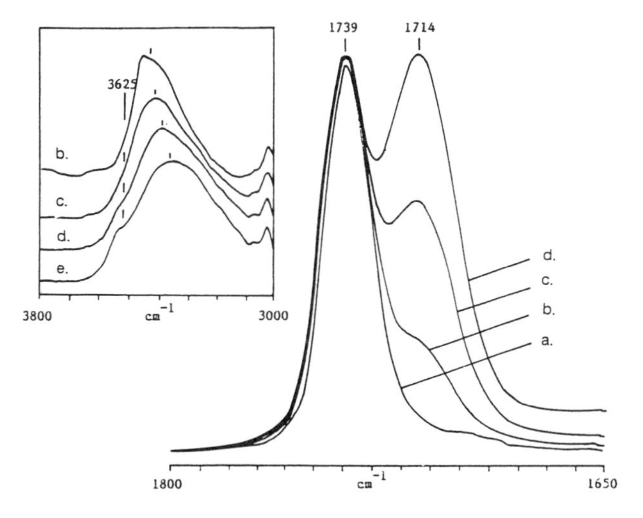

Figure 4.24. The IR spectra in the regions of 3800 and 1800 – 1650 cm^{-1} recorded at room temperature of PVPh – PVAc blends cast from THF containing 100 (a), 80 (b), 50 (c), 20 (d), and 0 (e) wt % PVAc. (Reproduced with permission from reference 22. Copyright 1984 Society for Applied Spectroscopy.)

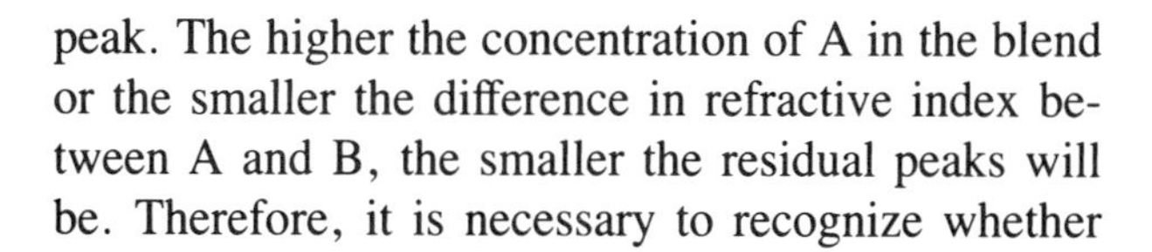

peak. The higher the concentration of A in the blend or the smaller the difference in refractive index between A and B, the smaller the residual peaks will be. Therefore, it is necessary to recognize whether the observed spectral differences are purely optical artifacts or whether they instead reflect the true intermolecular interactions between the component polymers.

Table 4.5. Interaction Contribution (Aging Time) of PVF_2– PVA_C Blends Heat Treated at 175 °C

PVF_2–PVA_C wt %	*0 h*	*8 h*	*16 h*	*24 h*	*2 days*	*7 days*	*24 days*
10:90	4.27	6.81	6.58	4.83	4.49	4.60	5.87
20:80	12.20	14.63	12.65	11.91	12.03	13.01	10.03
30:70	17.05	18.56	15.76	16.88	15.16	13.73	17.12
40:60	19.05	22.27	20.43	20.36	18.84	16.92	19.01
50:50	14.27	14.64	15.79	15.83	14.49	15.35	15.22
60:40	20.10	20.57	20.77	20.81	19.45	19.54	23.00
70:30	22.51	25.00	23.04	22.98	18.76	21.17	21.05
80:20	31.33	28.72	24.94	28.41	29.21	28.95	29.18
90:10	19.80	21.96	16.00	22.52	24.16	24.10	22.58

NOTE: Experimental error ± 2%.

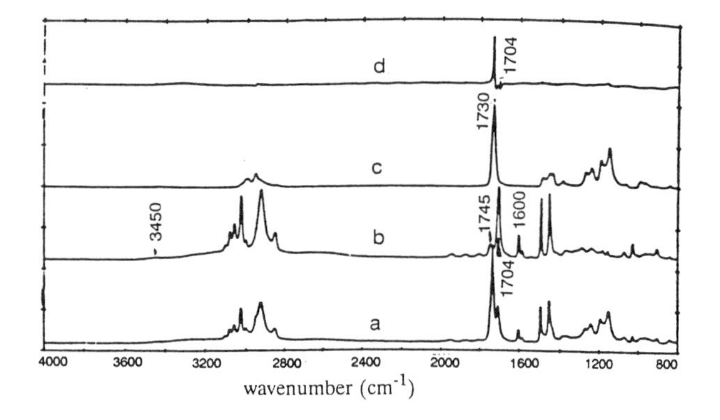

Figure 4.25. The IR spectra of SAAS – PMMA (4:1) blend (a), pure SAAS (b), pure PMMA (c), and (d) the interaction spectrum (a – b – c). (Reproduced with permission from reference 23. Copyright 1989 John Wiley & Sons, Inc.)

Deformation of Polymer Systems

The strength properties of synthetic polymers are considerably enhanced by molecular orientation. Therefore, it is desirable to determine the relationship between the molecular orientation and the strength properties. The characterization of orientation effects is difficult because of the multiphase nature of most oriented engineering thermoplastics. For example, X-ray diffraction is an excellent method for determining the orientation of the crystalline regions of the polymer, but it contributes nothing to the knowledge of the disordered or amorphous regions. IR spectroscopy is a particularly useful tool for orientation studies when specific bands from each phase are observed. IR spectroscopy is also useful for determining "internal" orientation, that is, the differences in orientation of the flexible functional groups attached to the polymer chain.

Dichroic IR Measurements of Orientation in Polymers. The dichroic ratio may be considered to be characteristic of the directional orientation of the segments of the molecule. For a polymer whose molecular axis is oriented parallel to the spectrometer sampling plane, the dichroic ratio, R, is defined as

$$R = \frac{A_{\parallel}}{A_{\perp}} \qquad (4.2)$$

where $A_{\parallel}$ is the absorbance parallel to the chain axis, and $A_{\perp}$ is absorbance perpendicular to the chain axis. For highly oriented samples, the dichroic ratio may approach either infinity or zero, depending on the alignment of the transition-moment vector with respect to the molecular chain axis. The alignment of the chain segments can be determined from dichroic ratio measurements if the inherent polarizations are known (*25*). In general, $A_{\parallel}$ and $A_{\perp}$ are determined successively by using a polarizer that is aligned parallel and perpendicular to the stretching direction. For samples that have a low level of orientation, the magnitude of the dichroic ratio is close to 1. In these cases of minimal orientation, it is better to measure the quantity $A_{\parallel}$ and $A_{\perp}$ because it is a more sensitive measurement under these conditions.

Differences in internal orientation have been demonstrated by studies of poly(acrylonitrile) (*26*). The dichroic behaviors of the nitrile band at 2241 cm^{-1} and the chain methylene band at 1452 cm^{-1}, which represent the orientation of the side groups and the polymer chains, respectively, are shown in Figure 4.26. Although there is only a small degree of orientation of the groups, a greater perpendicular alignment of the nitrile transition moments with respect to the chain axis occurs with draw.

IR dichroic measurements have also been made for the compatible blends of atactic polystyrene (PS) – poly(2,6-dimethyl-1,4-phenylene oxide) (PPO) (*27*). The PPO and PS chains orient in a different manner when subjected to a uniaxial strain on the blend. The PPO orientational behavior does not depend on the PPO concentration, but the PS orientation increases regularly up to 25% PPO and then remains constant. The sensitivity of the orientation of PS to the concentration of PPO is thought to be a result of the hindrance by the PPO of the stress relaxation behavior of PS.

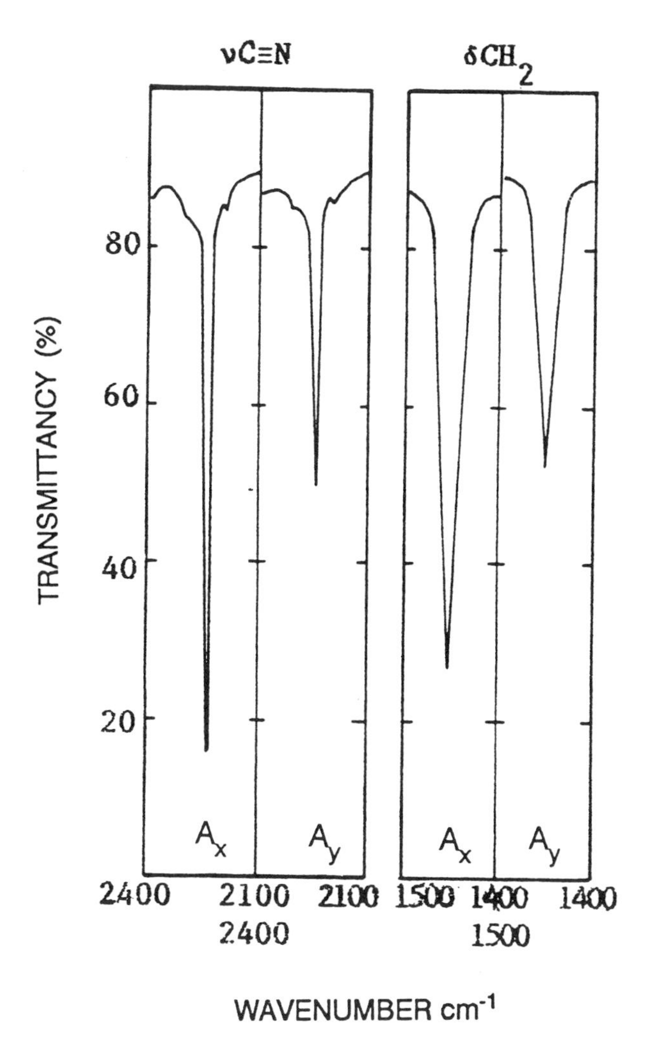

Figure 4.26. IR dichroism of the nitrile and CH_2 absorption bands of a 3 × drawn poly(acrylonitrile) film. (Reproduced with permission from reference 26. Copyright 1970 Marcel Dekker, Inc.)

Because IR spectra can be obtained rapidly by using FTIR equipment, the deformation process can be studied while the polymer is being stretched (*28*). The dichroic ratio and stress as a function of strain and time for the CH_2 rocking bands between 736 and 726 cm^{-1} and between 726 and 710 cm^{-1} are shown in Figure 4.27. The CH_2 rocking band between 736 and 726 cm^{-1} has a transition moment that is parallel to the *a* axis, and the CH_2 rocking band between 726 and 710 cm^{-1} has a transition moment that is perpendicular to the *a* axis. Before the stretching process, the 730-cm^{-1} band exhibits parallel dichroism, indicating a preferred initial orientation of the *a* axis along the stretching direction. For small strains, the

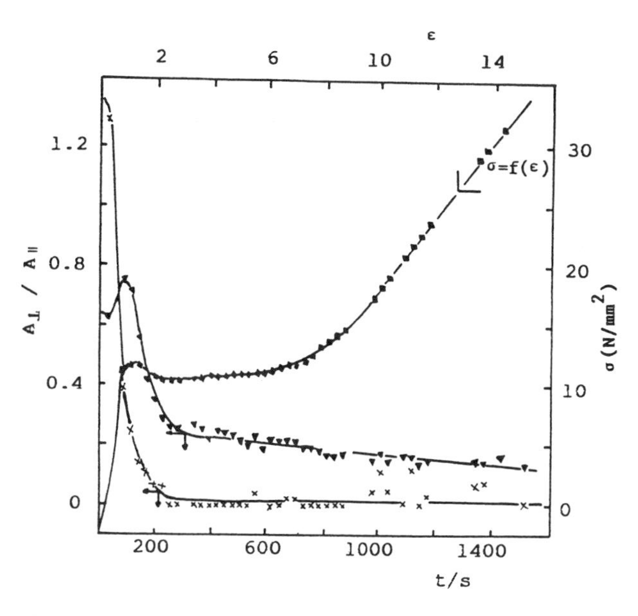

Figure 4.27. Dichroic ratios and stress as a function of strain and time (stretching velocity) for the CH_2 rocking modes at 730 cm^{-1} (×) and at 720 cm^{-1} (▼). The stress – strain diagram is given by (■). (Reproduced with permission from reference 28. Copyright 1981 Hüthig & Wepf Verlag.)

original orientation increases slightly. With the beginning of the necking process, the dichroic ratio shows the occurrence of a small rotation of the *b* axis into the stretching direction and an opposite movement of the *a* axis. There is a rotation of the *a* and *b* axes into directions perpendicular to the stretching direction when the strain is equal to draw ratios between 1 and 3.

The actual spectra from the various portions of the stress – strain curve of PE are shown in Figure 4.28, along with the deconvolution of the observed spectra into their individual components. During the stretching process, a new band appears at 716 cm^{-1}. This shoulder is associated with the conversion of the orthorhombic PE modification to the monoclinic structure. As the draw rate is increased, the band at 716 cm^{-1} also increases. From the curve resolution, the amount of the monoclinic form produced is approximately 20% of the total integrated absorbance of the rocking bands in PE (*28*).

Trichroic IR Measurements of Orientation. As indicated in Chapter 2, many polymers have orientation with a three-dimensional character. It is therefore necessary to obtain a structure factor

$$A_0 = \frac{1}{3}\left[A_x + A_y + A_z \right] \qquad (4.3)$$

in order to determine the relative amounts of the structural components in the sample (*A* is absorbance). The three-dimensional orientation function can be determined by using the ratios A_x/A_0, A_y/A_0, and A_z/A_0. The method of performing such measurements has been described (*29*). In Figure 4.29, the trichroic spectra of uniaxially drawn PET illustrate the differences in the spectra for the directions parallel, perpendicular, and through the thickness direction of the films (*30*). The isotropic IR spectra of the PET films are obtained from a linear combination of these three spectra. These calculated isotropic spectra for one-way drawn PET are shown as a function of percent elongation in Figure 4.30 for the frequency region 800 – 1000 cm^{-1}. By using the spectra of the pure *gauche* and *trans* isomers and the least-squares curve-fitting technique, the conformer composition can be determined as a function of draw ratio. The results are shown in Figure 4.31. The orientation parameters for the three different directions can also be calculated as a function of the percent elongation, as shown in Figure 4.32.

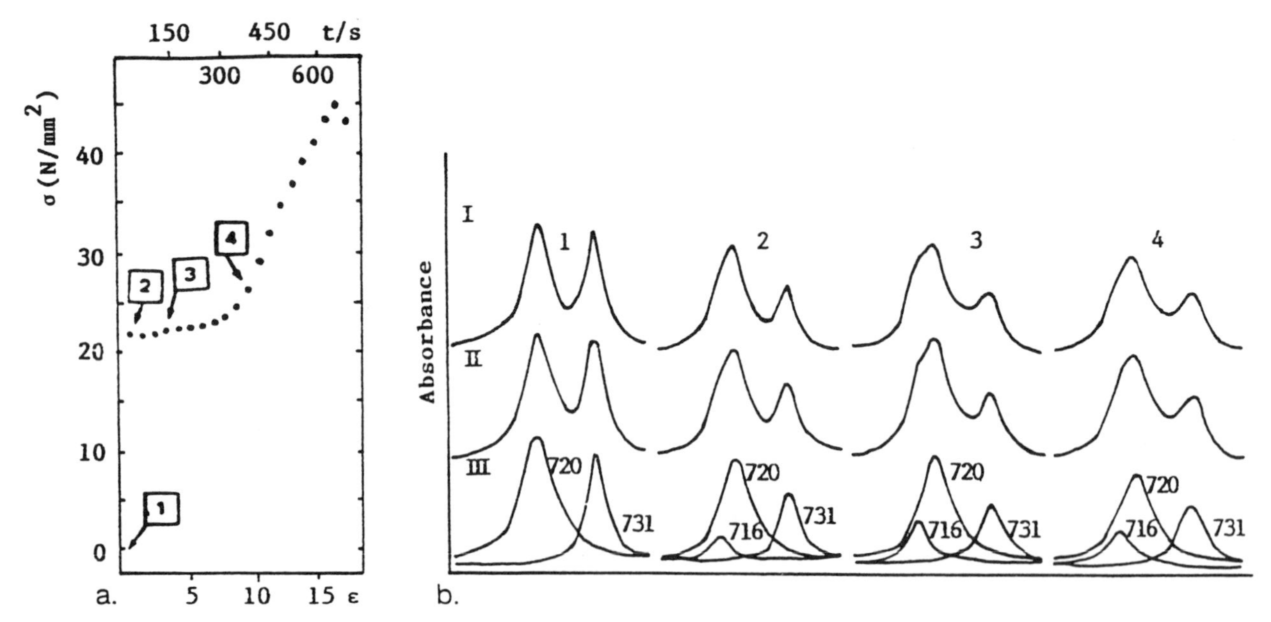

Figure 4.28. (a) Stress – strain diagram of polyethylene with a draw rate of 0.26 mm/s. (b) Experimental (row I), synthesized (row II), and resolved (row III) bands of the CH_2 rocking modes. The experimental spectra were scanned at the indicated positions (numbers in squares) of the stress – strain diagram. (Reproduced with permission from reference 28. Copyright 1981 Hüthig & Wepf Verlag.)

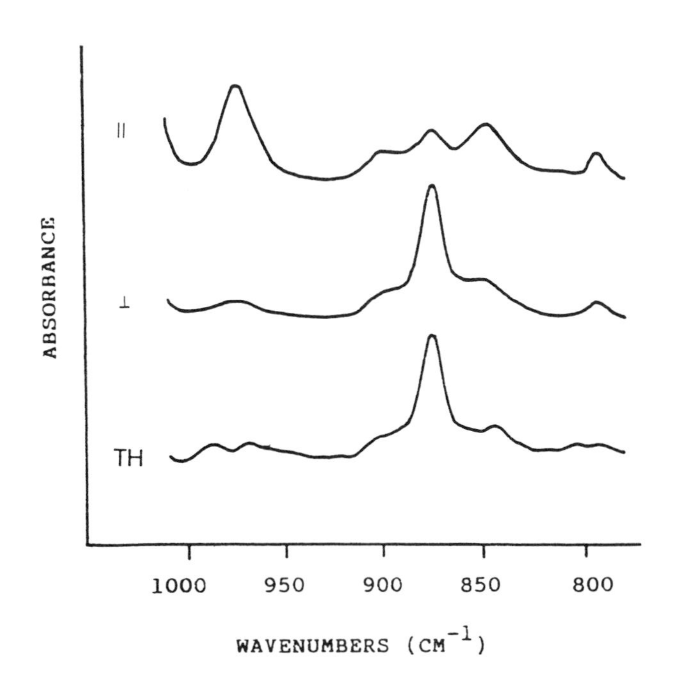

Figure 4.29. The spectra of one-way drawn PET film in the directions parallel and perpendicular to the direction of draw and through the thickness direction of the film. (Reproduced with permission from reference 30. Copyright 1986 John Wiley & Sons, Inc.)

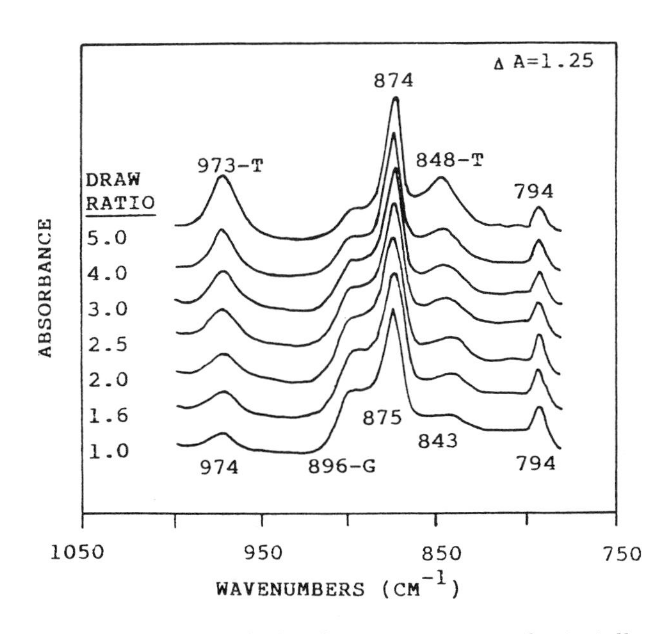

Figure 4.30. The calculated isotropic spectra of uniaxially drawn PET as a function of percent elongation. ΔA is the absorbance range for each spectrum. (Reproduced with permission from reference 30. Copyright 1986 John Wiley & Sons, Inc.)

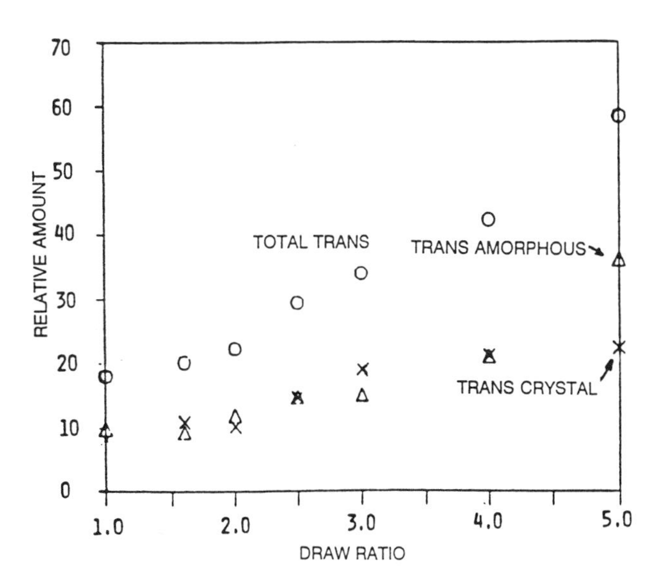

Figure 4.31. The conformer composition of uniaxially drawn PET as a function of draw ratio. (Reproduced with permission from reference 30. Copyright 1986 John Wiley Sons, Inc.)

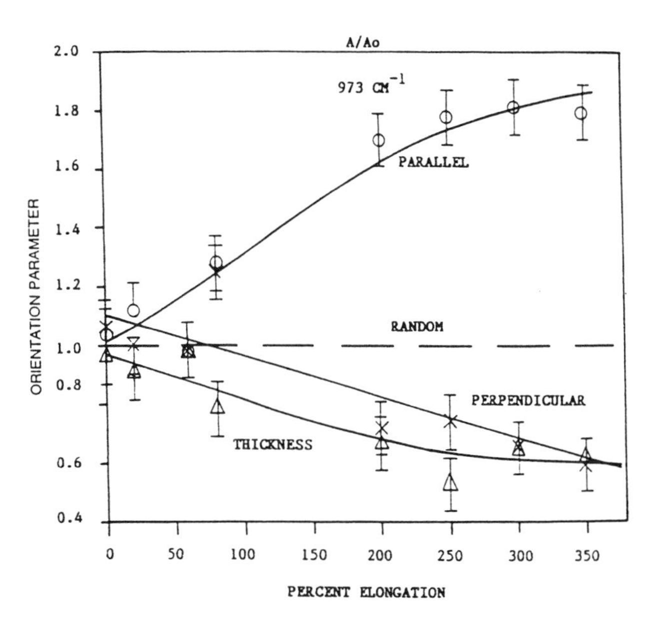

Figure 4.32. The trichroic orientational parameters for one-way drawn PET as a function of percent elongation. (Reproduced with permission from reference 30. Copyright 1986 John Wiley Sons, Inc.)

When structural changes occur, the trichroic measurements will reveal frequency shifts as well as changes in relative intensities. This is the case when phase I of poly(vinylidene fluoride) (PVDF) is oriented and annealed both under stress (Figure 4.33) and without stress (Figure 4.34) (*31*). Frequency differences can be observed at 842 cm^{-1} in the perpendicular direction and at 845 cm^{-1} in the thickness direction in the samples annealed under stress and then relaxed. Similarly, there is a difference in the frequencies at 891 cm^{-1} in the thickness direction and at 882 cm^{-1} in the perpendicular direction. The origin of these shifts has been attributed to defect structures in the phase-I lattice. Defects in the crystal lattice will occur through the introduction of *gauche* (+) or *gauche* (−) conformations. Nucleation of a transforming chain onto a preexisting defect-containing chain can propagate a kink in a direction-specific manner. Molecular modeling studies suggest a steric feasibility of phase-I crystals with defects aligned approximately perpendicular to the *trans* lattice. In essence, the vibrations of the defect regions are in register with and at a fixed and defined angle from the defect-free crystal regions. Hence, the orientation dependent frequencies of these defect bands are observed (*31*).

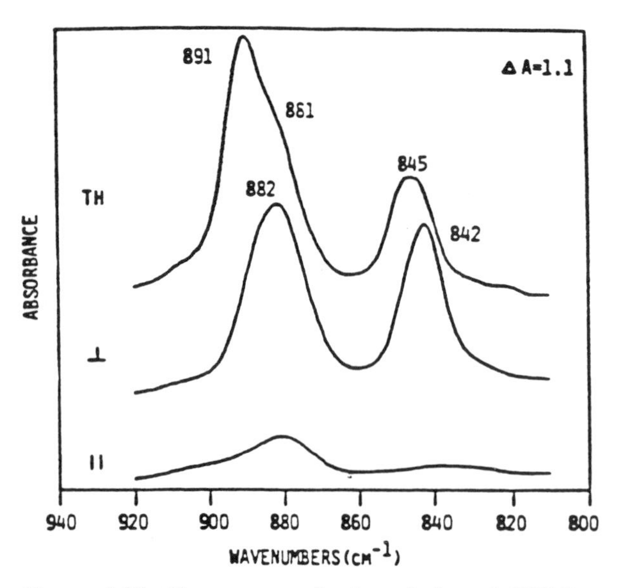

Figure 4.33. The spectra of oriented phase-I PVDF annealed under tension. Parallel, perpendicular, and calculated thickness-direction spectra are shown. (Reproduced with permission from reference 31. Copyright 1986 John Wiley & Sons, Inc.)

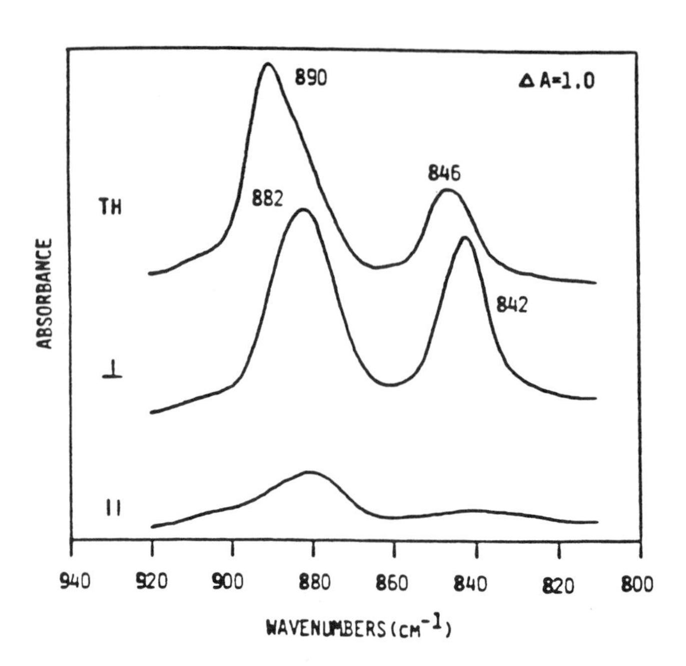

Figure 4.34. The spectra of oriented phase-I PVDF annealed under strain relaxation. Parallel, perpendicular, and calculated thickness-direction spectra are shown. (Reproduced with permission from reference 31. Copyright 1986 John Wiley & Sons, Inc.)

Mechanically Stressed Polymer Systems. IR spectroscopy can be used to study the effects of applied mechanical stress on highly oriented samples. The goal is to obtain the *molecular stress distribution function*, which is an important quantity for determining the stress relaxation moduli or creep compliances. Shifts in the peak frequencies are observed and an attempt is made to determine the molecular stress distribution by deconvolution (*32*). The shifts as a function of stress for the 1168-cm^{-1} band of oriented isotactic polypropylene are shown in Figure 4.35 (*33*). Stress sensitivity typically involves a small shift (0 − 5 cm^{-1}) to a lower frequency of the band's maximum intensity along with an asymmetric shape change of the entire band. The results are often interpreted in terms of a nonuniform distribution of external load, resulting in a nonsymmetrical displacement of the individual bond absorption frequencies. The shift appears as a linear function of stress and can be expressed by

$$\Delta\nu_\sigma = \nu(\sigma) - \nu(0) = \alpha_x\sigma \qquad (4.4)$$

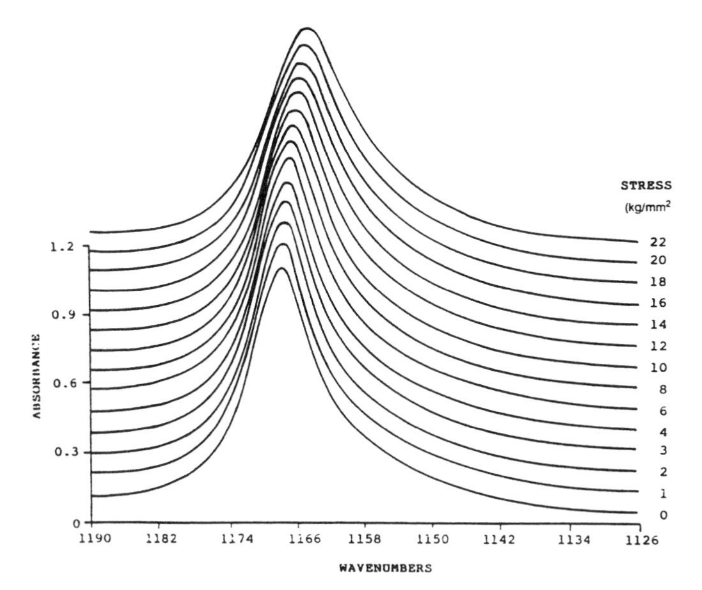

Figure 4.35. Parallel-polarized 1168-cm^{-1} band of oriented isotactic polypropylene at successive stress levels. The average stress application rate was 0.5 (kg/mm^2)/min. The spectra were obtained at 4-cm^{-1} resolution, and the profiles of successive levels were vertically offset by 0.09 absorbance units. The peak frequency varies from 1167.72 cm^{-1} at s = 0, to 1164.90 cm^{-1} at s = 22 kg/mm^2. (Reproduced with permission from reference 33. Copyright 1984 John Wiley & Sons, Inc.)

where $\Delta\nu_\sigma$ is the mechanically induced peak frequency shift; $\nu(\sigma)$ is the frequency of the bond under a stress, σ; $\nu(0)$ is the frequency of the bond without stress; α_x is the mechanically induced frequency shifting coefficient at constant temperature, T; and σ is the applied uniaxial stress. The value of α_x is an approximate measure of the stress distribution asymmetry for a band whose IR intensity is independent of conformation and molecular environment. If the band is associated with a particular conformational isomer, variations in α_x can occur. The magnitude of a frequency shift resulting from stress depends on the draw ratio, annealing treatments, and thermal history of the sample. This dependence has been demonstrated particularly for PET (*34*). There is also a threshold stress below which no shift in frequency will occur. The threshold stress effect may be related to the ability of the disordered amorphous chains to sustain the load before appreciable strains are transferred to the crystalline chains.

The vibrational frequency shift and α_x are significantly affected by temperature. An increase in temperature is accompanied by an increase in the magnitude of the frequency shift.

For some polymers, poly(butylene terephthalate) (PBT) in particular, the application of stress leads to a nearly reversible crystalline-phase transformation. The crystal structure of the unstressed PBT has a crumpled *gauche – trans – gauche* conformation of the aliphatic chain segments. Under stress, the crystal structure changes to a strained form in which the polymer chains are essentially fully extended with an all-*trans* sequence of aliphatic segments. The sensitivity of IR spectroscopy to this phase change arises primarily from the conformational changes of the aliphatic segments. The simplest representation of the dependence of the transformation on strain and the reversible nature of the transformation is a plot of the absorbance variations of the CH_2 modes at 1485 cm^{-1} (stressed form) and bands at 1460 cm^{-1} (relaxed form) that are shown in Figure 4.36 (*35*). The decrease in absorption that occurs at about 2 – 3% strain coincides with the onset of the crystalline-phase transition. The PBT system has been studied extensively because of its interesting phase transformation as well as its industrial value as a stretchable fiber for blue jeans.

Dynamic IR Linear Dichroism. Dynamic mechanical techniques are important in the characterization of the rheological properties of polymers. In

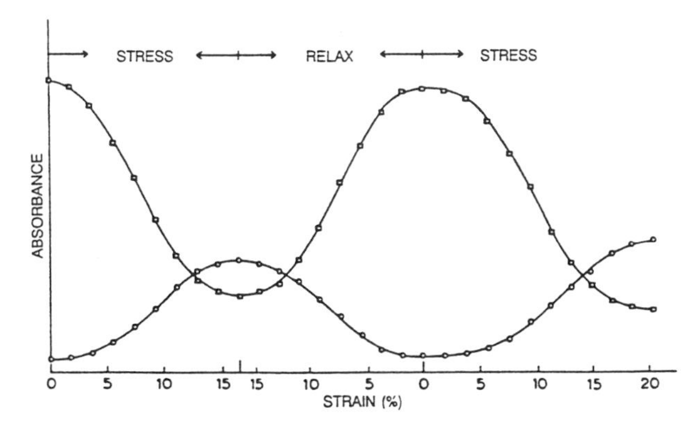

Figure 4.36. Absorbance of the methylene absorption bands at 1460 (□) (relaxed form) and 1485 (○) cm^{-1} (stressed form) as a function of reversible stress and strain. The absorbances have been corrected for changes in sample thickness by using the aromatic ring band at 1510 cm^{-1} as a reference internal-thickness band. (Reproduced with permission from reference 35. Copyright 1979 Steinkopff Verlag Darmstadt.)

dynamic mechanical analysis, a small-amplitude oscillatory strain is applied to a sample, and the resulting dynamic stress is measured as a function of time. The dynamic mechanical technique allows the simultaneous measurement of both the elastic and the viscous components of the stress response. Typically, the temperature and deformation frequency are changed in order to determine the mechanical relaxation spectrum of the system. In order to ensure a linear viscoelastic response, the oscillatory strain amplitude is typically well below 1.0%. The sinusoidal strain, $\epsilon(t)$, can be written:

$$\epsilon(t) = \epsilon + \epsilon' \sin \omega_s t \tag{4.5}$$

where ϵ is the static strain, ϵ' is the strain coefficient, ω_s is the frequency, and t is the time. The time-dependent stress response to the strain, $\sigma(t)$, is given by

$$\sigma(t) = \sigma + \sigma' \sin(\omega_s t + \delta) \tag{4.6}$$

where δ is the loss angle. This dynamic stress response, $\sigma(t)$, can be separated into two orthogonal components that are in phase and quadrature ($\pi/2$ out of phase) with the dynamic strain, $\epsilon(t)$:

$$\sigma(t) = \mathrm{E}'\epsilon' \sin \omega_s t + \mathrm{E}''\epsilon' \cos \omega_s t \tag{4.7}$$

The coefficients E' and E'' are the in-phase and quadrature components of the time-dependent stress and are referred to as the *dynamic tensile storage modulus* and *loss modulus*, respectively. They are related to the loss angle, δ, and the amplitudes of the dynamic stress and strain by

$$E' = \left(\frac{\sigma'}{\epsilon'}\right) \cos \delta \tag{4.8}$$

and

$$E'' = \left(\frac{\sigma'}{\epsilon'}\right) \sin \delta \tag{4.9}$$

The storage modulus, E', represents the ability of the material to elastically store the absorbed mechanical energy as potential energy. The loss modulus, E'', represents the ability to dissipate the absorbed energy as heat. The ratio between the dissipated and stored mechanical energy is referred to as the *mechanical dissipation factor*.

$$\tan \delta = \frac{E''}{E'} \tag{4.10}$$

This quantity is a convenient index for characterizing the viscoelastic state of the material because the onset of a new type of dissipation process for mechanical energy often results in a change in tan δ. This dissipation factor is used to detect various transition phenomena in polymers.

A dynamic IR linear dichroism (DIRLD) study was performed (*36*). The DIRLD spectrometer was constructed around a Dynastat dynamic mechanical analyzer and a dispersion IR instrument. The polarization modulation requires a photoelastic modulator (PEM) so that the plane of the polarization of the light can alternate rapidly between two directions that are parallel and perpendicular to a fixed reference axis taken as the direction of strain. Like the stress response discussed previously, the time-dependent dichroic difference signal, $\Delta A(t)$, is written

$$\Delta A(t) = \Delta A_0 + \Delta A' \sin(\omega_s t + \beta) \tag{4.11}$$

Because the molecular orientation is a rate-dependent process, there is an optical phase-loss angle, β, between the dynamic dichroism and the strain signals.

Like the dynamic loss, the dynamic dichroism signal can be separated into two orthogonal components:

$$\Delta A(t) = \Delta A' \sin \omega_s t + \Delta A'' \cos \omega_s t \tag{4.12}$$

Here the terms $\Delta A'$ and $\Delta A''$ are referred to as the in-phase spectrum and the quadrature spectrum of the dynamic IR linear dichroism, respectively. The two orthogonal spectra are related to the amplitude, ΔA, and loss angle, β, by

$$\Delta A' = \Delta A \cos \beta \tag{4.13}$$

$$\Delta A'' = \Delta A \sin \beta \tag{4.14}$$

The in-phase spectrum is proportional to the extent of instantaneous strain. The quadrature spectrum represents the component of reorientation proportional to the rate of strain that is out of phase with the strain by $\pi/2$. A *dichroic dissipation factor*

$$\tan \beta = \frac{\Delta A''}{\Delta A'} \quad (4.15)$$

is also defined in a way analogous to the mechanical dissipation factor, tan δ. Because $\Delta A'$ and $\Delta A''$ are measured as a function of frequency, it is possible to compare the responses of specific functional groups to each other as well as to the macroscopic strain signal.

The DIRLD spectra, $\Delta A'$ and $\Delta A''$, and the absorbance spectrum of an atactic polystyrene film in the region between 1425 and 1525 cm^{-1} are shown in Figure 4.37 (*36*). The 1490-cm^{-1} absorption band of polystyrene is assigned to the coupling of the aromatic ring and the aromatic CH deformation and is polarized in the plane of the ring perpendicular to the bond between the phenyl group and the backbone aliphatic chain. This band has a significant signal in the quadrature component that is shifted to a higher frequency, but the rest of the spectrum is nearly in phase with the applied strain. This variation in dynamic dichroism for different absorption bands suggests that some fraction of the aromatic side chains in the sample may be responding to the applied strain at a rate that is different from that of the polymer backbone.

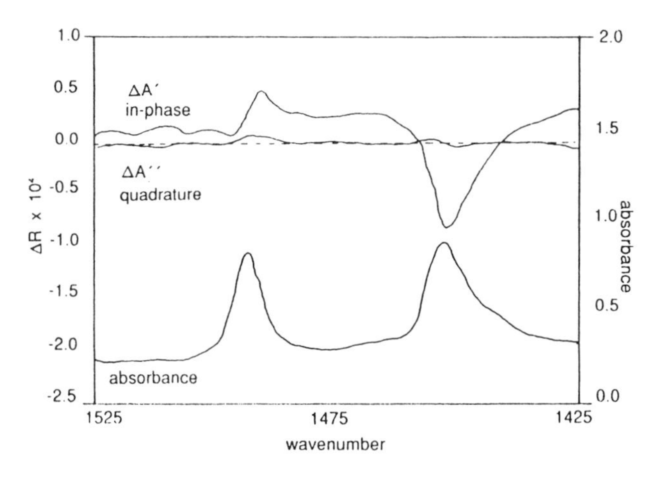

Figure 4.37. The DIRLD spectra, $\Delta A'$ (in-phase) and $\Delta A''$ (quadrature), and the absorbance spectrum of an atactic polystyrene film in the region between 1525 and 1425 cm^{-1}. (Reproduced with permission from reference 36. Copyright 1988 Society for Applied Spectroscopy.)

Two-Dimensional IR Spectroscopy. Linear dynamic perturbations of a polymer sample will generate directional changes in the transition moments of functional groups. The relaxation of this perturbation depends on the intra- or intermolecular couplings of the functional groups. By using this concept, it is possible to generate two-dimensional IR correlation spectra (*37*). The correlations of the specific fluctuation rates of the individual dipole transition moments can be used as spectroscopic labels to differentiate the relationships between highly overlapped IR bands.

A time-dependent IR absorbance band under a small dynamic strain has both a static absorbance and a dynamic absorbance resulting from the perturbation-induced fluctuation. The dynamic absorbance is influenced by the reorientation rates of the electric dipole transition moments, and differences in the rates are determined by the local environment of the functional group, that is, its intra- and intermolecular couplings with other groups. Directional absorbances, $A_{\parallel}$ and $A_{\perp}$, are measured with IR beams polarized in directions parallel and perpendicular to a reference axis, and these directional absorbances contain both the fixed and dynamic components of the absorbances.

The dynamic dichroic difference can be expressed in terms of an in-phase and a quadrature spectrum. The in-phase spectrum represents the reorientational motions that are occurring simultaneously with the applied dynamic strain and is determined primarily by the extent of strain. The quadrature spectrum, on the other hand, characterizes the motions that are out of phase with the dynamic strain by $\pi/2$ and is determined primarily by the rate of strain. When the in-phase and quadrature spectra have different shapes, there is an indication that the time dependencies of the reorientation rates vary among the measured absorbance bands.

A *synchronous correlation intensity* that characterizes the correlation between two IR bands undergoing dynamic oscillation can be calculated. This intensity reaches its maximum value when the dynamic IR correlation signals are totally in phase with each other, and it reaches its minimum value when they are out of phase with each other. The *asynchronous correlation intensity* becomes either maximum or minimum when the signals are orthogonal to each other and vanishes when the pair of signals are exactly in phase or antiphase with each other. Two-dimensional (2-D) IR spectra are obtained by plotting these correlation intensities as functions of two independent wavenumbers, ν_1 and ν_2. In the 2-D synchronous correlation spectra, diagonal peaks are always positive, and they indicate the local reorientational motions of functional groups. They also reveal the presence of additional peaks that are not observed in ordinary IR spectra. Synchronous cross peaks occur when the two dipole transition moments associated with the molecular vibrations of different functional groups orient simultaneously. The presence of cross peaks suggests the possibility of the existence of inter- or intramolecular interactions among functional groups. These cross peaks can be used in a synchronous 2-D IR spectrum to map out correlations among IR bands in a manner analogous to 2-D NMR correlation spectroscopy.

In an asynchronous 2-D IR correlation spectrum, there are only cross peaks and no diagonal peaks. Cross peaks occur when the electric dipole transition moments reorient independently and at different rates. On this basis it is possible to deconvolute highly overlapped multiplet bands.

The contour map of a synchronous 2-D IR correlation spectrum of an atactic polystyrene film is shown in Figure 4.38a. There are diagonal and cross peaks at 1450 and 1491 cm^{-1} and a weak autopeak at 1430 cm^{-1}. The asynchronous 2-D IR spectrum for the same region is shown in Figure 4.38b. The normal spectrum has only one band at 1450 cm^{-1}, but the 2-D IR spectrum has three, at 1443, 1450, and 1456 cm^{-1}, because they are spread out in the second dimension. Six IR bands have been located by using the 2-D IR spectrum, and the band locations on a conventional absorbance spectrum of polystyrene are shown in Figure 4.38c. When the sample is perturbed by a dynamic strain, the transition moments at 1443, 1450, and 1456 cm^{-1} reorient perpendicular to the strain direction, and those at 1465, 1490, and 1495 cm^{-1} reorient parallel to the strain direction. The functional groups corresponding to the bands at 1443 and 1495 cm^{-1} reorient the fastest and are coupled. The pair of bands at 1450 and 1490 cm^{-1} are also coupled, but they orient 4 ms later. The last pair, at 1456 and 1465 cm^{-1}, move about 5 ms behind the first pair.

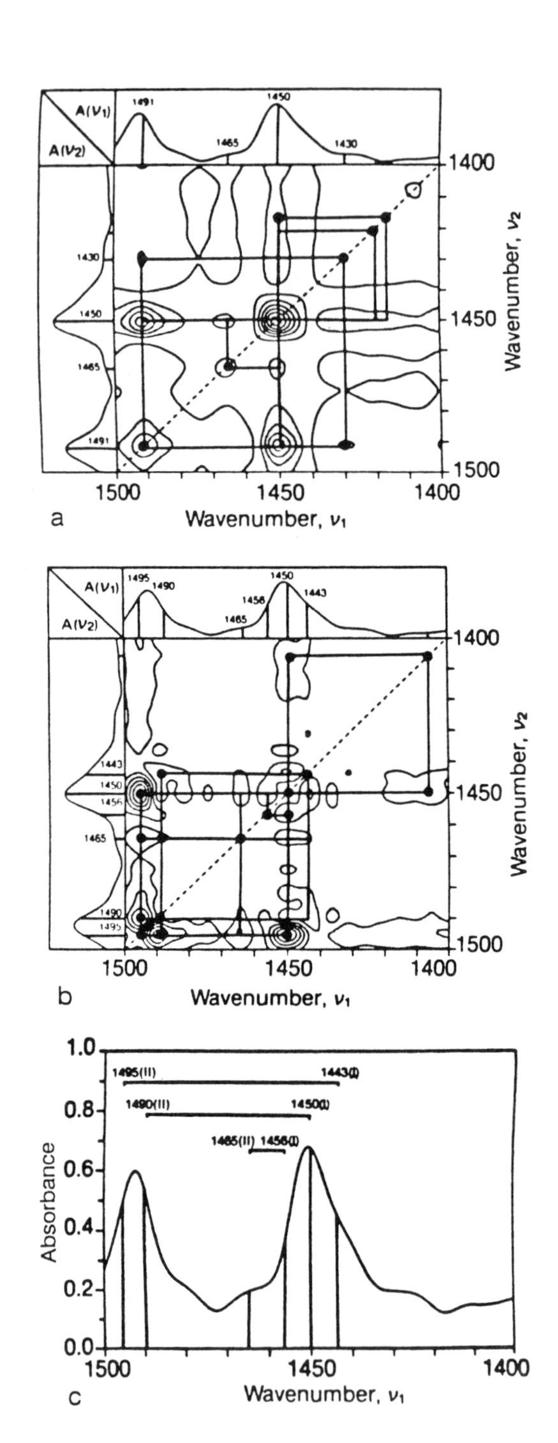

Figure 4.38. Contour map of synchronous (a) and asynchronous (b) 2-D IR spectra of atactic polystyrene, and the six IR bands in the spectral region between 1400 and 1500 cm^{-1} of atactic polystyrene resolved by 2-D IR correlation analysis (c). (Reproduced with permission from reference 37. Copyright 1990 Society for Applied Spectroscopy.)

Measurement of Morphological Units in Polymers

The mechanical, physical, and chemical properties of polymers are ultimately determined by the structure and organization of the macromolecules. In the crystalline regions, the polymer chains are highly ordered and immobile. These regions give the thermoplastic material its mechanical strength. In the amorphous layers, the polymer chains are mainly disordered and mobile. These disordered regions allow plastic deformation of the material. Thus the precise characterization of molecular order is a primary prerequisite to understanding the macroscopic properties of polymeric materials.

The crystallization conditions directly affect the structure and properties of the polymer. A knowledge of the arrangement of molecular chains within the lamellae of a semicrystalline polymer should be valuable in understanding the properties. Most polymers crystallize into folded chain crystals. Semicrystalline polymers exhibit a lamellar morphology when crystallized both from the melt and from dilute solution. The lamellae are typically 100 – 500 Å thick, with the amorphous polymer interspersed between the crystalline regions. The molecular chains are normal to the lamellae and have lengths that are much greater than the lamellar thickness. Thus the molecular chains can traverse one or more lamellae several times. A considerable fraction of the chains must return to the same crystal; that is, they are folded. Whether the molecule returns with a reentry that is either predominantly adjacent or random to the crystallite stem of origin is a matter of debate. Vibrational spectroscopy can contribute to our knowledge of the nature of the order in polymers and is useful in such determinations.

This problem has been studied in polyethylene (PE) by using deuterium isotopic labelling (*38, 39*). The orthorhombic unit cell of crystalline PE contains two symmetrically nonequivalent chains. These two chains give rise to both in-phase and out-of-phase modes whose frequency differences are the correlation splitting. When deuterated PE and normal PE are crystallized together, bands from both species give rise to correlation splittings. Tasumi and Krimm (*38, 39*) observed resolved double splittings as evidence of adjacent reentry along a (110) fold plane because the magnitude of the splitting was found to agree with calculated values for highly extended (110) fold planes.

The chain fold in PE must contain TG (*trans—gauche*) sequences, and probably six TG bond sequences are required per fold. Although the number of TG bond sequences in a solution-crystallized sample can be calculated directly, the number of folds is not available because the TG conformation can be found in cilia and other defects as well as in folds. Relative changes in the number of TG units can be observed in a sample that has been annealed because annealing reduces the number of folds and therefore increases the stem length (*40, 41*).

Intermolecular Interactions in Polymers

Vibrational spectroscopy can be used to study intermolecular effects in the solid state and the changes produced by temperature effects. If the intermolecular forces are sufficient, the fundamental modes of the single chain are split into different spectral components in the crystal.

> The splitting is also called *correlation field splitting* by the purist. You will often find the term in the literature, although not very often in polymer science, as will become apparent momentarily. This effect is also sometimes referred to as *Davydov* or *factor group splitting.*

The number of theoretically expected bands depends on the number of molecules in the unit cell. Polyethylene has a planar conformation and an orthorhombic unit cell containing two molecules. Each group mode of the isolated PE molecule is predicted to be split into two components for crystalline PE.

> There is an old rule about theory and experiment. Theory can always predict the results after the experimental results are in. However, the theory may not always be verified by experiment.

This crystal field splitting has been observed for the methylene rocking mode at 720 cm^{-1} and for the methylene bending mode at 1460 cm^{-1} in spectra of crystalline PE. Although other modes should also exhibit such splitting, their inherent band width prevents the observation of separate components. When PE is melted, the crystal field splitting disappears.

Consequently, a measure of the relative intensities of the 720- to 730-cm^{-1} bands can be used to rank the relative crystallinity of PE samples.

> This crystalline result for PE is known to nearly everyone in the polymer field. It has caused untold misery for practicing polymer spectroscopists because they have been asked to perform IR crystallinity measurements on all of the semicrystalline polymers of interest to their colleagues. Crystallinity plays a major role in the determination of the physical and mechanical properties of semicrystalline polymers, and a simple, rapid, and cheap method to measure crystallinity is desired.
>
> My belief is that the density of the sample is such a measure of crystallinity. But such a simple measurement gets little respect and requires a knowledge of the density of the crystalline and amorphous phases, and therefore, one is often reluctant to use it. Crystallinity should probably be measured by X-ray diffractometry whenever possible because this is the final test of crystallinity. IR analysis is usually required of most samples for reasons of chemistry or impurities, so the IR spectroscopist is asked, "Why don't you measure the crystallinity while you are at it?" After all, we can measure a lot of things, so why not do it all!

The problem is that very few polymers exhibit the crystal field splittings observed for PE. The explanation is that the crystal field splitting is very sensitive to the separation of the polymer chains. The intermolecular interaction forces fall off at a rate of r^6, where r is the distance between chains. A very small difference in separation results in a large difference in the magnitude of the interactions, and consequently, in the frequency separation of the bands. In fact, the PE chains are closer together than chains of any other polymer, so the crystal field splitting is the largest for PE.

For polypropylene (PP), on the other hand, it has not been possible to observe any IR splittings resulting from intermolecular or crystalline packing. In fact, there are many features in the PP spectrum that look like pairs of bands, and indeed they are. However, the band pairs arise from *intramolecular helical splitting*, not from *intermolecular crystalline splitting*. Actually, the IR spectrum of PP is influenced very little by intermolecular packing. In Figure 4.39, the X-ray patterns of the monoclinic α phase and the smectic δ phase are shown. The monoclinic α phase consists of well-ordered crystalline 3_1 helices, while the smectic δ phase consists of 3_1 helices that are out of register with each other. The smectic phase will convert with time to the monoclinic α phase at a rate that is a function of temperature. The IR spectra of these two phases are also shown in Figure 4.39, and the spectra are strikingly similar. Only minor differences in intensities are observed in the IR spectra, yet from the X-ray point of view, the smectic phase is noncrystalline. The reason for these small differences is that the helices are far apart (relative to the distance in PE), and consequently, the intermolecular forces are much lower, and the intermolecular splitting is below the resolution of the instrument.

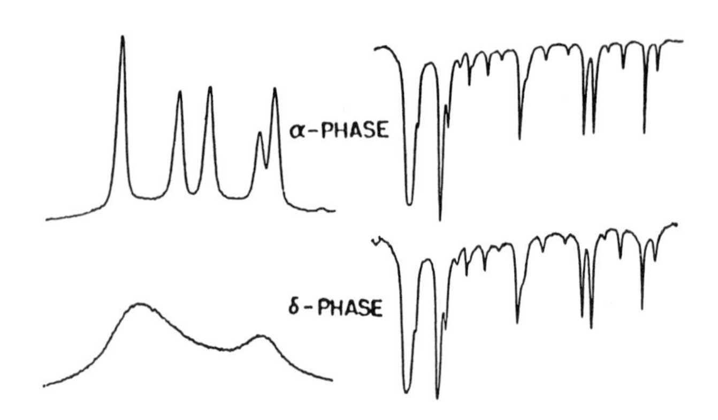

Figure 4.39. A comparison of the X-ray diffraction patterns (left) *and the IR spectra* (right) *of the monoclinic α phase and smectic δ phase of isotactic polypropylene.*

Unfortunately, the spectral results for PP are typical for most semicrystalline polymers. *Generally, the chains are simply too far apart and the intermolecular forces are so small that crystal field splitting is not observed.*

Even the experienced IR spectroscopist may be confused at this point. There are a number of reports in the polymer literature of correlations of spectral features with crystallinity or density. However, these IR correlations are based on the measurement of the amount of the rotational isomer preferred in the crystalline state, which, on the average, may correlate with the crystallinity or density. Two different measurements of crystallinity are being considered here. X-ray diffraction measures the *long-range order* or *inter*molecular order as a result of chain packing. IR spectroscopy measures the concentration of the crystalline isomer, which is a *short-range order* or *intramolecular* phenomenon. These two physical methods do not necessarily determine the same crystallinity. The short-range intramolecular order is a necessary condition for the occurrence of long-range intermolecular order, but the short-range order can exist without the presence of the long-range order. For example, in the cilia connecting one lamella to another, the short-range order can be high. However, there is little possibility of intermolecular order because the cilia can be completely isolated from each other. IR spectroscopy would detect these repeat units in the crystalline domain, but in X-ray diffraction they would go undetected. Consequently, the IR spectroscopic and X-ray diffraction measurements of crystallinity can be different, and this result is often the case. This does not mean either measurement is wrong. The two methods simply measure different phenomena that are generally lumped into the term "crystallinity". IR spectroscopy measures short-range crystallinity, and X-ray diffraction measures long-range crystallinity.

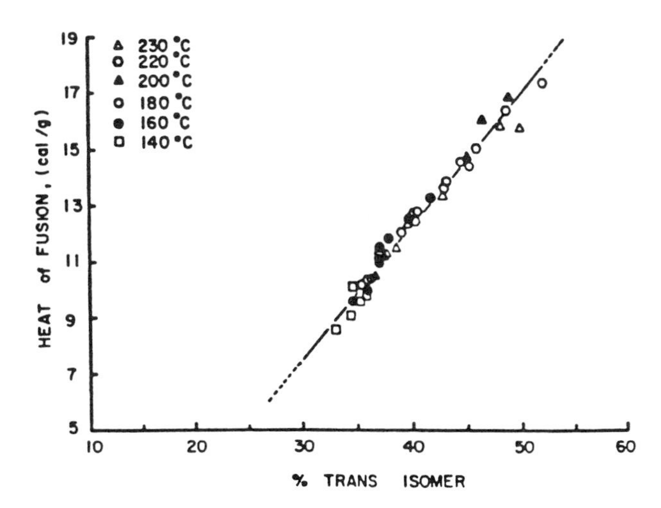

Figure 4.40. Correlations between the heat of fusion and the percent of trans *isomer in semicrystalline PET films.* $\Delta H_f = -6.710 \pm 0.473$ *cal/mol. (Reproduced with permission from reference 42. Copyright 1984 John Wiley & Sons, Inc.)*

Another technique for measuring crystallinity is the differential scanning calorimeter (DSC) measurement of the heat of fusion. The DSC measures the amount of heat required to melt a sample, and by using a knowledge of the heat of fusion of a 100% crystalline sample, the percent crystallinity is calculated. DSC measures yet another type of crystallinity–the *meltable* portion. The meltable crystalline portion represents a degree of order less than that of the long-range X-ray order because some crystallites can melt but are too small to be observed with X-rays. Yet, DSC does not measure the contributions of cilia and other isolated ordered chains to short-range order. Again, correlations can be made, as shown in Figure 4.40 for PET (*42*), but careful annealing conditions are required to establish such correlations.

Therefore, the method used to measure crystallinity, whether by density, IR spectroscopy, X-ray diffraction, or DSC, needs to be specified in order for the results to be meaningful.

Structural Changes and Transitions as a Function of Temperature

The intensity, band shape, and frequency of an IR absorption band can change as a result of a change in the temperature of the sample. There are two extremes of interpretation of the effects of temperature on spectroscopic results. Either thermal expansion affects the inherent nature of the dipole moment change as a result of changes in the intermolecular forces or the concentration of the absorbing species changes with temperature. These two effects can occur simultaneously, and this event often precludes a straightforward analysis.

With increasing temperature, intermolecular expansion reduces the electric dipole moment of the interaction and therefore reduces the intensity of the absorption band. The temperature effect on band intensity is most pronounced for bands that arise from very polar chemical groups. Changes in band shape can also result from changes in temperature. These changes are the result of additional thermal energy being imparted to the multiplicity of vibrational energy levels available to the polymer. The effects of thermal broadening are most easily observed in the IR spectra of molten polymers.

At cryogenic temperatures, no concentration changes are expected; thus IR spectroscopy can be used to detect transitions in polymers by recording abrupt or discontinuous changes in intensities as a function of temperature. This IR type of molecular dilatometry should indicate the same transitions that are observed in bulk thermal expansion measurements. If certain IR absorptions are related to the various components or morphological structures present, then a probe of the thermal responses of these structures is available. Multiphase and copolymer systems can be easily studied by using IR spectroscopic analysis.

IR spectroscopy can be used to identify the different conformers in a polymer. The criteria for identifying the absorption bands of the conformers is based on the following observations. As the temperature of the sample is increased, the relative intensities of the absorption bands of the high-energy conformers increase, and those of the lower-energy conformers decrease. When the sample is cooled rapidly to form a stable crystalline form, the bands of the stable conformers dominate and those of the high-energy conformers essentially disappear. Conformer bands should be observed in pairs corresponding to the same vibrational modes in both the high-energy conformer and the low-energy conformer.

The equilibrium between two conformers A and B, where A = B, is dependent on temperature. The concentration of the high-energy A conformer, C_A, relative to the concentration of the low-energy form, C_B, is given by the van't Hoff expression

$$\frac{C_B}{C_A} = \exp\left[-\frac{\Delta G}{RT}\right] \tag{4.16}$$

where R is the gas constant, T is the temperature, and ΔG is the Gibbs free energy difference between conformers, given by the expression

$$\Delta G = \Delta H - T\Delta S \tag{4.17}$$

The standard enthalpy difference, ΔH, is the difference in enthalpy of the two conformers, and ΔS is the difference in entropy.

The vibrational band intensities of the conformer bands, A_A and A_B, are related to the concentration of each conformer as follows:

$$A_A = a_A C_A b \tag{4.18}$$

$$A_B = a_B C_B b \tag{4.19}$$

where a is absorptivity and b is path length.

Substitution of these relationships into the van't Hoff relation yields

$$\frac{A_B a_A}{A_A a_B} = \exp\left[-\frac{\Delta H}{RT_e} + \frac{\Delta S}{R}\right] \tag{4.20}$$

This may be expressed as

$$\ln\left[\frac{A_B}{A_A}\right] = -\frac{\Delta H}{RT} + \frac{\Delta S}{R} - \ln\frac{a_A}{a_B} \tag{4.21}$$

If it is assumed that ΔS and the ratio a_A/a_B are constant with temperature, it is possible to determine ΔH by making a series of measurements of A_B/A_A at various temperatures. On this basis, the van't Hoff energy is defined as

$$\Delta H = h_+ - h_- \tag{4.22}$$

with

$$h_\pm = \frac{-R\partial \ln A_\pm(\nu)}{\partial(1/T)} \tag{4.23}$$

where the bands that increase in intensity, $A_+(\nu)$, or decrease in intensity, $A_-(\nu)$, yield the energies h_+ and h_-, respectively. Then

$$\Delta H = \frac{R\partial\left[\frac{\ln A_+(\nu)}{A_-(\nu)}\right]}{\partial(1/T)} = R\partial\left[\frac{\ln K}{\partial(1/T)}\right] \tag{4.24}$$

where the equilibrium constant, K, is defined by

$$K = \frac{\frac{A_+(\nu)}{a_+}}{\frac{A_-(\nu)}{a_-}} \tag{4.25}$$

where $A_\pm$ and $a_\pm$ are the peak absorbances and extinction coefficients, respectively, of the increasing and decreasing bands.

The FTIR spectrum of syndiotactic PMMA was obtained at 20-degree intervals starting at 30 °C (*43*). The difference spectra of syndiotactic PMMA are shown in Figure 4.41. These differential absorbances were analyzed by using the van't Hoff equation and

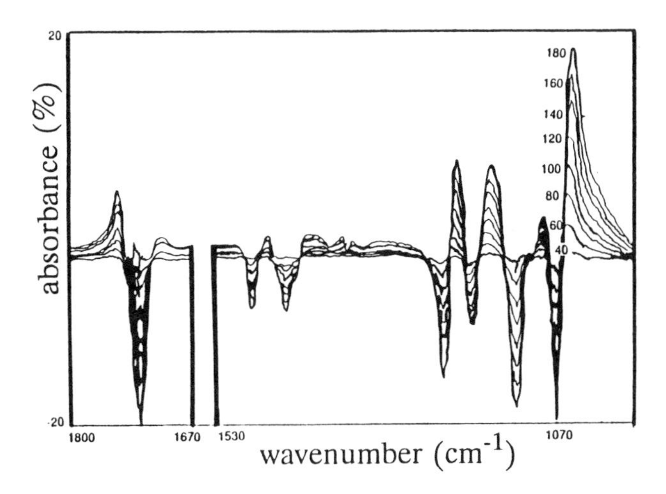

Figure 4.41. The FTIR difference spectra of syndiotactic PMMA obtained at 20-degree intervals starting at 30 °C. (Reproduced from reference 43. Copyright 1981 American Chemical Society.)

were plotted as shown in Figure 4.42. The conformational energies are largest (2000 cal/mol) for the 1276 – 1264-cm^{-1} bands and range from 1080 to 1283 cal/mol for the 1195 – 1168- and 1152 – 1140-cm^{-1} bands. These data have been interpreted in terms of the rotational isomeric state theory. The IR spectroscopic approach is straightforward and is useful for obtaining information of this type.

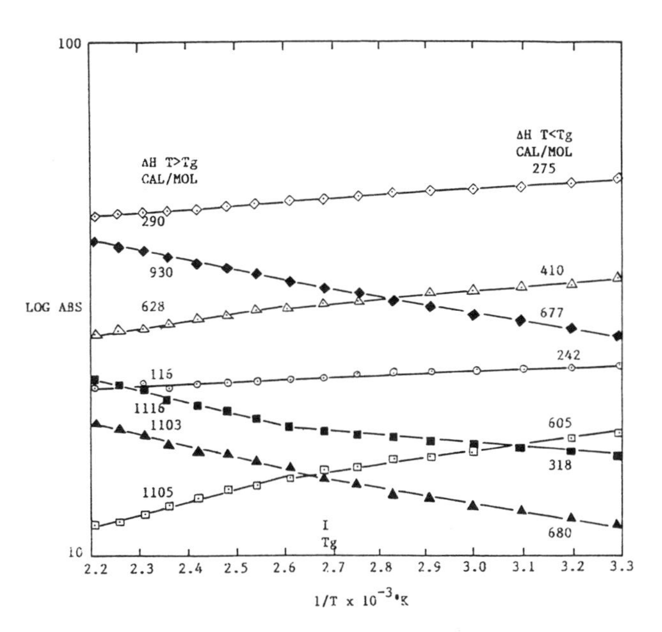

Figure 4.42. The differential absorbances of PMMA analyzed by using the van't Hoff equation. The activation energy, Δ H, is given in calories per mole. The IR band frequencies (cm^{-1}) are as follows: ◇, 1152; ◆, 1136; Δ, 1195; ■, 1260; ○, 1246; ▲, 1224; and □, 1276. (Reproduced from reference 43. Copyright 1981 American Chemical Society.)

Time-Dependent Phenomena in Polymers

> *The use of the inherent advantages of Fourier transform infrared (FTIR) spectroscopy for the observation of time-varying phenomena has so far been limited to some rather exotic methods of acquiring time-resolved interferograms. The difficulty of these methods and the pitfalls into which the unwary might stumble have precluded their use by the general chemical community. This is quite unfortunate, since detailed knowledge of the course of a chemical reaction and the structure of intermediates can be garnered from an examination of the infrared spectrum as a function of time for a system which is undergoing a physical or chemical change.*
>
> – S I. Yaniger and D. W. Vidrine (*44*)

One of the major advantages of FTIR spectroscopy is its rapid scanning capability, which has opened new applications of IR spectroscopy. These new applications require continuous monitoring in short time intervals and extremely small differences between spectra. With currently available commercial instrumentation, interferograms may be collected and stored at a rate of 50 to 85 scans per second to give a time resolution of 12 – 20 ms. If the time frame of the transient phenomenon is long with respect to the mirror scan time, and if the signal is sufficiently large, single interferograms may be transformed in a rapid-scan instrument to yield essentially instantaneous incremental spectra.

Kinetic Studies of Polymerization Reactions

From the study of time-dependent intensity changes of absorption bands characteristic of the polymerization reactants and reaction products, the order and specific rate, *k*, of a reaction can be derived. Also, the activation energy can be calculated when studies of absorption are made as a function of temperature.

With the rapid-scanning capability of FTIR spectroscopy, the chemical changes of a sample in a heated cell can be monitored, provided that the spectral sampling interval is shorter than the interval during which the change occurs. Conversion curves as a function of time and temperature for each species involved in the cure process can be generated from the IR absorbances. The determination of the kinetic parameters from a single dynamic scan is based on the general *n*th-order rate expression in which the conversion, F, as a function of time, t, and temperature, T, is given by

$$\frac{\mathrm{d}\,F(t,\,T)}{\mathrm{d}\,t} = k\,[1 - F(t,\,T)]^{n} \qquad (4.26)$$

where n is the reaction order. Assuming that the temperature dependence of the rate constant is described by the Arrhenius equation

$$k = A \exp\left[-\frac{E}{RT}\right] \qquad (4.27)$$

where E (kcal/mol) is the activation energy, and A ($\sec^{-1}$) is the *Arrhenius frequency factor*. The conversion curve can be used to determine the reaction kinetic parameters, E, n, and $\ln A$.

The rapid-scan FTIR technique has been applied to the study of the cure kinetics of isocyanate coatings (*45*). The generalized dynamic FTIR spectra for absorbance vs. time – temperature, as well as the fractional conversion vs. time – temperature, are shown in Figure 4.43. The reaction can be followed by monitoring the absorbance of the peak at 2256 cm^{-1}, which results from the isocyanate functionality, as a function of time. Film thickness can be monitored using the band at 1446 cm^{-1}. The fractional conversion as a function of temperature and time is calculated by

$$F(t,\,T) = \frac{[A_f - A(t,T)]}{(A_f - A_0)} \qquad (4.28)$$

where A_0 is the initial absorbance, $A(t,\,T)$ is the absorbance at temperature T and time t, and A_f is the final absorbance. The kinetic parameters were determined by using a simplex fitting procedure (*45*).

Dynamic Deformation Measurements

Vibrational spectroscopic studies of short-time phenomena in polymers during elongation, mechanical load, stress relaxation, and creep, and after fracture

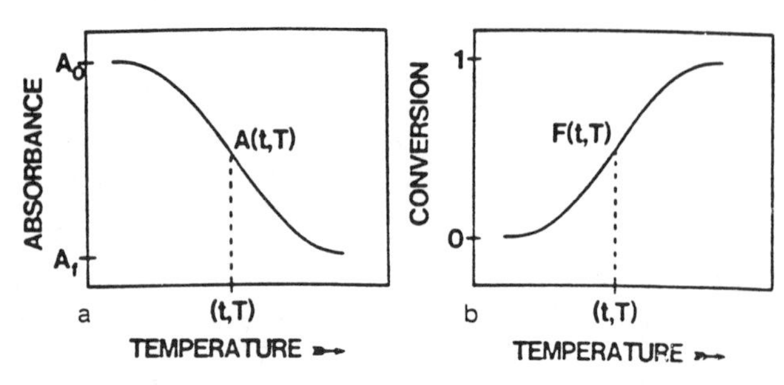

Figure 4.43. Generalized dynamic FTIR spectra for absorbance vs. time – temperature (a) and fractional conversion vs. time – temperature (b). (Reproduced with permission from reference 45. Copyright 1984 Plenum.)

are now possible with the rapid scanning capabilities of FTIR spectroscopy.

To follow the spectral changes that occur during the application of stress and subsequent relaxation, a stretching device is placed in the spectrometer chamber. Synchronous recording of the stress – strain diagrams and the IR spectra are made. An example of the results for the deformation of an amorphous PBT sample is shown in Figure 4.44 (*46*). Applied strain forces the approximately *gauche – trans – gauche* conformation (α form) of PBT into an approximately all-*trans* conformation (β form) that exists only under stress. During relaxation, the interactions of the terephthalate residues drive the chains back to the *gauche – trans – gauche* conformation of the α form. Upon drawing the isotropic amorphous sample, the absorption bands at 1460 – 1455 and 917 cm^{-1} resulting from the α form decrease, and the bands at 1475 – 1470 cm^{-1} resulting from the β form increase.

A number of other applications of rapid-scanning FTIR spectroscopy have been reported, and an excellent review of these studies is available (*47*).

Fatigue Measurements Using IR Time-Resolved Spectroscopy

When fatigue measurements are made, repetitive oscillatory strains must be applied to a preloaded polymer. Under these circumstances, time-resolved FTIR spectroscopy is used to follow the structural changes. The object is to construct a curve of response vs. log time or temperature. The time scale involved is often from seconds to milliseconds. By using time-resolved spectroscopic techniques, it is possible to make such measurements.

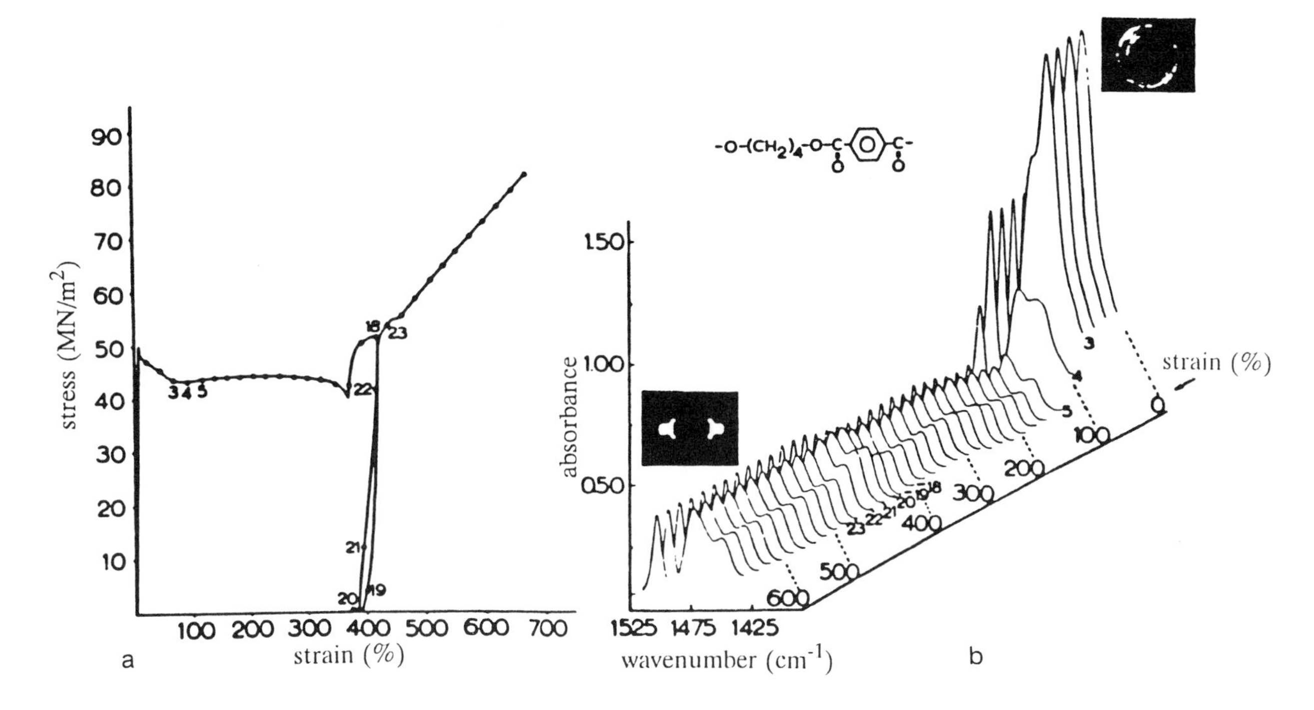

Figure 4.44. Simultaneous stress – strain and FTIR measurements during uniaxial deformation and recovery of a primarily amorphous PBT: (a) stress – strain curve of the mechanical treatment (elongation rate: 1.66%/s), (b) FTIR spectra in the 1500 – 1400-cm^{-1} (CH_2) region with the wide-angle X-ray diagrams (b, inset) of the original sample and a 600% drawn sample. (Reproduced with permission from reference 46. Copyright 1981 Springer-Verlag.)

> *If the material being studied is linearly viscoelastic, a condition usually satisfied at low cyclic strain amplitudes, theoretically we are capable of capturing events occurring in time intervals as short as 50 μs.*
>
> – J. E. Lasch et al. (*48*)

When time-dependent repetitive processes such as fatigue measurements in polymers are under investigation, it is possible to do stroboscopic measurements in such a fashion as to obtain the complete spectrum of the sample in microseconds (*48, 49*). The technique is based on the existence, at any given time, of a small interval of time during which the event appears stationary. During this time interval, the interferogram can be collected at a specific path difference. The complete interferogram from zero to the maximum retardation is obtained in this fashion and can be transformed to give a spectrum at time t. In fact, n interferograms can be obtained, each corresponding to a specific time. If the process is repeated a sufficient number of times, and if the timing is accurate and the sampling of the data are properly sorted, a spectrum with a time resolution of microseconds can be obtained after Fourier transformation. For an experiment involving an oscillatory strain, the spectrum at the initial time with no stress or strain and at intervals of time t can be obtained

$$t = t_o + \left(\frac{x}{\nu}\right) \tag{4.29}$$

where x is the retardation, and v is the velocity of the mirror corresponding to different strain levels as the sample is deformed. In eq 4.29, t is the nth data point, and t_o is a constant time interval at the start of the chosen number of zero crossings. With a suitable offset and with proper time sorting of the data, the complete interferogram can be obtained. It is crucial to the success of the experiment that the sample does not undergo irreversible changes during the cyclic deformation for the time period of the data collection. Otherwise, the data will not be representative of the portion of the cyclic deformation process. Much of the early work in time-resolved IR spectroscopy with FTIR instruments suffered from a time drift of the sample during the accumulation period. Because polymers can undergo thousands of cycles of fatigue if the strain level is not high, it appears that the fatigue experiment represents a nearly ideal problem for investigation with time-resolved spectroscopy (TRS).

The first reported work in this area was an investigation of isotactic polypropylene that was fatigued at a rate of 10 Hz with an elongation of 1 – 5% (*49*). The spectra showed no shifts in frequency, but reversible intensity changes were observed. The data were interpreted in terms of the presence of a "smecticlike" structure resulting from cyclic stress.

The response of ion-containing ethylene – methacrylic acid copolymers has also been investigated (*50*). The strain amplitude of the samples was 2%, and the period of external stress was 50,000 μs. Changes in the dichroic ratio of the 2673-cm^{-1} band were observed as a function of strain in times as short as 200 μs. Other spectroscopic changes were observed but were not interpreted in terms of structural changes.

Experiments of this type are extremely tedious and are difficult to perform because of the precision required to maintain synchronization between the interferometer movement and the external event. A separate microprocessor has been utilized to overcome this difficulty for dynamic deformation studies (*51*) and for triggering chemical reaction parameters (*52*). The mating of in situ FTIR spectroelectrochemistry with rapid-scan techniques allows the study of electrochemical reactions that, unlike the other time-resolved techniques just discussed, do not require reaction reversibility (*44*).

A rather new approach to time-resolved spectroscopy that uses modern optoelectronic-controlled step-scan methods of Fourier transform interferometry has been suggested (*53*). When using step-scan interferometry, the initiation of a reversible event must occur at each mirror position. The time evolution at each mirror position subsequent to initiation is then measured at discrete time intervals. Data processing involves sorting by time in order to produce an interferogram for each sampling time. These interferograms are subsequently Fourier transformed to yield a set of time-resolved spectra. To quote the authors :

> *It is anticipated that, given sufficient source intensity, and with the availability of fast mercury cadmium telluride (MCT) detection and transient digitizing electronics, time intervals down to* ~10 ns should be accessible.
>
> – R. A. Palmer et al. (*53*)

Characterization of Surfaces

IR spectroscopy has been shown to be an effective means for the study of surface species and reactions occurring at interfaces. These results can be reflected in the nature of interfacial bonding and the role of adhesion in property determination. A review on this subject has been published (*54*).

Where Do We Go from Here?

The applications of IR spectroscopy seem to be limited by only the imagination of the spectroscopist. The quality of the instrumentation is continuing to improve, and the cost is decreasing so that quality instrumentation is available to laboratories of all sizes. Therefore, the number of applications of IR spectroscopy will continue to grow. IR instrumentation is moving into the quality control laboratory because of its sensitivity to small differences in samples. It is also being used in the chemical plant as a process monitoring device because of its capability of continuously monitoring a large variety of structural aspects in a process stream. Growth in these areas will continue as instrumentation is specifically designed for these applications.

> *Our modern instruments are capable of giving signal/noise ratios greater than one part in 105 which means if your chart paper is over 60 m tall, it would display the data appropriately.*
>
> – Tomas Hirshfeld (*55*)

References

1. Kosky, P. G.; McDonald, R. S.; Guggenheim, E. A. *Polym. Eng. Sci.* **1985,** *25,* 389.
2. Antoon, M. K.; Koenig, J. L. *J. Polym. Sci., Part A: Polym. Chem.* **1981,** *19,* 549.

3. Antoon, M. K.; Zehner, B. E.; Koenig, J. L. *Polym. Compos.* **1981,** *2,* 81.

4. Pecsok, R. L.; Painter, P. C.; Shelton, J. R.; Koenig, J. L. *Rubber Chem. Technol.* **1976,** *49,* 1010.

5. Antoon, M. K.; Koenig, J. L. *J. Polym. Sci., Part B: Polym. Phys.* **1981,** *19,* 197.

6. Bowmer, T. N.; Tonelli, A. E. *J. Polym. Sci., Part B: Polym. Phys.* **1986,** *24B,* 1681.

7. O'Reilly, J. M.; Mosher, R. A. *Macromolecules* **1981,** *14,* 602.

8. Zimba, C. G.; Rabolt, J. F.; English, A. D. *Macromolecules* **1989,** *32,* 2867.

9. Shipman, J. J.; Folt, V. L.; Krimm, S. *Spectrochim. Acta,* **1962,** *18,* 1603.

10. Compton, D. A. C.; Maddams, W. F. *Appl. Spectrosc.* **1986,** *40,* 239.

11. Tadokoro, H. *Structure of Crystalline Polymers;* Wiley-Interscience: New York, 1979; p 9-10.

12. Flory, P. J. *Statistical Mechanics of Chain Molecules;* Wiley-Interscience: New York, 1969; p 15.

13. Koenig, J. L. *Appl. Spectrosc. Rev.* **1971,** *4,* 233.

14. Snyder, R. G.; Poore, M. W. *Macromolecules* **1973,** *6,* 708.

15. Zoppo, M. D.; Zerbi, G. *Polymer* **1990,** *31,* 658.

16. Lin, S. B.; Koenig, J. L. *J. Polym. Sci., Polym. Symp.* **1984,** *71,* 121.

17. Lin, S. B.; Koenig, J. L. *J. Polym. Sci., Part B: Polym. Phys.* **1983,** *21,* 2365.

18. Painter, P. C.; Watzek, M.; Koenig, J. L. *Polymer* **1977,** *18,* 1169.

19. Koenig, J. L. *Chemical Microstructure of Polymer Chains;* Wiley: New York, 1982; p 63.

20. Koenig, J. L.; Tovar, M. J. M. *Appl. Spectrosc.* **1981,** *35,* 543.

21. Sargent, M.; Koenig, J. L. Unpublished results.

22. Coleman, M. M.; Painter, P. C. *Appl. Spectrosc. Rev.* **1984,** *20,* 255.

23. Jo, W. H.; Cruz, C. A.; Paul, D. R. *J. Polym. Sci., Part B: Polymer Phys.* **1989,** *27,* 1057.

24. Allara, D. L. *Appl. Spectrosc.* **1979,** *33,* 358.

25. Jasse, B.; Koenig, J. L. *J. Macromol. Sci. --Rev. Macromol. Chem.* **1979,** *C17(1),* 61-135.

26. Wolfram, L. E.; Koenig, J. L.; Grasselli, J. G. *Appl. Spectrosc.* **1970,** *24,* 263.

27. Lefebvre, D.; Jasse, B.; Monnerie, L. *Polymer* **1981,** *22,* 1616.

28. Holland-Moritz, K.; van Werden, K. *Makromol. Chem.* **1981,** *182,* 651.

29. Fina, L. J.; Koenig, J. L. *J. Polym. Sci., Part B: Polym. Phys.* **1986,** *24,* 2509.

30. Fina, L. J.; Koenig, J. L. *J. Polym. Sci., Part B: Polym. Phys.* **1986,** *24,* 2525.

31. Fina, L. J.; Koenig, J. L. *J. Polym. Sci., Part B: Polym. Phys.* **1986,** *24,* 2541.

32. Bretzlaff, R. S.; Wool, R. P. *Macromolecules* **1983,** *16,* 1907.

33. Lee, Y. -L.; Bretzlaff, R. S.; Wool, R. P. *J. Polym. Sci., Part B: Polym. Phys.* **1984,** *22,* 681.

34. Wool, R. P. *Polym. Eng. Sci.* **1980,** *20,* 805.

35. Siesler, H. W. In *Proceedings of the 5th European Symposium on Polymer Spectroscopy;* Hummell, D., Ed.; Verlag Chemie: Weinheim, 1979.

36. Noda, I.; Dowrey, A. E.; Marcott, C. *Appl. Spectrosc.* **1988,** *42,* 203.

37. Noda, I. *Appl. Spectrosc.* **1990,** *44,* 550.

38. Tasumi, M.; Krimm, S. *J. Chem. Phys.* **1967,** *46,* 755.

39. Tasumi, M.; Krimm, S. *J. Polym. Sci., A-2* **1968,** *6,* 995.

40. Painter, P. C.; Havens, J.; Hart, W. W.; Koenig, J. L. *J. Polym. Sci.* **1977,** *15,* 1223-1236.

41. Painter, P. C.; Havens, J.; Hart, W. W.; Koenig, J. L. *J. Polym. Sci.* **1977,** *15,* 1237-1249.

42. Lin, S. -B.; Koenig, J. L. *J. Polym. Sci.* **1984,** *71C,* 121.

43. O'Reilly, J. M.; Mosher, R. A. *Macromolecules* **1981,** *14,* 602.

44. Yaniger, S. I.; Vidrine, D. W. *Appl. Spectrosc.* **1986,** *40,* 174.

45. Provder, T.; Neag, C. M.; Carlson, G.; Kuo, C.; Holsworth, R. M. "Cure Reaction Kinetics Characterization of some Model Organic Coatings Systems by FT-IR and Thermal Mechanical Analysis." In *Analyt-*

ical Calorimetry; Johnson, J. F.; Gill, P. S., Eds.; Plenum: New York, 1984; p 377.

46. Holland-Moritz, K.; Siesler, H. W. *Polym. Bull.* **1981,** *4,* 165.

47. Siesler, H. W. *Adv. Polym. Sci.* **1984,** *62,* 2.

48. Lasch, J. E.; Burchell, D. J.; Masoaka, T.; Hsu, S. L. *Appl. Spectrosc.* **1984,** *38,* 351.

49. Fateley, W. G.; Koenig, J. L. *J. Polymer Sci., Polym. Lett. Ed.* **1982,** *20,* 445.

50. Burchell, D. J.; Lasch, J. E.; Dobrovolny, E.; Page, N.; Domian, J.; Farris, R. J.; Hsu, S. L. *Appl. Spectrosc.* **1984,** *38,* 343.

51. Molis, S. E.; MacKnight, W. J.; Hsu, S. L. *Appl. Spectrosc.* **1984,** *38,* 529.

52. Daschbach, J.; Heisler, D.; Pons, S. *Appl. Spectrosc.* **1986,** *40,* 489.

53. Palmer, R. A.; Manning, C. J.; Rzepiela, J. A.; Widder, J. M.; Chao, J. L. *Appl. Spectrosc.* **1989,** *43,* 193.

54. Ishida, H. *Rubber Chem. Technol.* **1987,** *60,* 497.

55. Hirshfeld, T. ''Quantitative FT-IR: A Detailed Look at the Problems Involved.'' In *Fourier Transform Infrared Spectroscopy: Applications to Chemical Systems;* Ferraro, J. R.; Basile, L. J., Eds.; Academic Press: New York, 1979; Vol. 2.

Suggested Reading

Culler, S. R.; Ishida, H.; Koenig, J. L. *Annu. Rev. Mater. Sci.* **1983**, *13,* 363.

Henniker, J. C. *Infrared Spectrometry of Industry Polymers*; Academic: 1967.

Fourier Transform Infrared Characterization of Polymers; Ishida, H., Ed.; Plenum: New York, 1987.

Koenig, J. L. *Adv. Polym. Sci.* **1983**, *54,* 87.

Koenig, J. L. *Appl. Spectrosc.* **1975**, 29, 293

Krimm, S. *Fortschr. Hochpolym. Forsch.*, **1960**, 2, 51.

Siesler, H. W.; Holland-Moritz, K. *Infrared and Raman Spectroscopy of Polymers;* Dekker: New York, 1980.

Zbinden, R. *Infrared Spectrometry of High Polymers*; Academic: New York, 1964.

5

Raman Spectroscopy of Polymers

One can say that anyone presently using an IR instrument will be likely to benefit from a Raman spectrometer. Anyone presently using only IR techniques is obtaining less than half of the spectral information. With Raman spectroscopy, one can measure the other portion of the spectrum, and for many molecular systems, the Raman effect constitutes the "richer" portion of the spectral data (*1*).

Many chemists and physicists, even those who have recently finished school, have never seen a Raman spectrometer, let alone measured a Raman spectrum.

– Carl Zimba (*2*)

In the previous chapters, IR spectroscopy of polymers was discussed. In this chapter, a complementary technique that also measures the vibrational energy levels, Raman spectroscopy, will be discussed. There is a fundamental difference in the IR and Raman techniques. IR spectroscopy depends on a change in the permanent dipole moment of the chemical bond or molecule with the vibrational normal mode in order to produce absorption. Raman spectroscopy, on the other hand, depends on a change in the induced dipole moment or polarization to produce Raman scattering. This difference in the physical nature of the selection rules may seem minor, but it plays a major role in the manner in which the two vibrational techniques are used for the study of polymers.

The Nature of Raman Scattering Spectroscopy

Raman spectroscopy ··· is viewed by many people as an exotic technique.

– Carl Zimba (*2*)

In the first paragraph of the paper by C. V. Raman and K. S. Krishnan (*3*) in which they introduce this technique, they state

If we assume that the X-ray scattering of the "unmodified" type observed by Professor Compton corresponds to the normal or average state of the atoms and molecules, while the "modified" scattering of altered wavelength corresponds to their fluctuations from that state, it would follow that we should expect also in the case of ordinary light two types of scattering, one determined by the normal optical properties of the atoms or molecules, and another representing the effect of their fluctuations from their normal state. It accordingly becomes necessary to test whether this is actually the case . . .

The results of these tests led to the following statement later in the paper:

Some sixty different common liquids have been examined in this way, and every one of them showed the effect in greater or less degree.

Thus the technique we now call the Raman effect was first observed (*3*).

When a beam of light is incident upon a molecule, it can be either absorbed or scattered. Scattering can be either *elastic* or *inelastic*. The electric field of the incident light induces a dipole moment, P, in the molecule, given by

$$P = \alpha E \tag{5.1}$$

where E is the electric field, and α is the polarizability of the molecule. Because the electric field oscillates as it passes through the molecule, the induced

1904—4/92/0115$06.25/1

dipole moment in the molecule also oscillates. This oscillating dipole moment radiates light at the frequency of oscillation in all directions except along the line of action of the dipole. The electric field is an oscillating function that depends upon the frequency of the light, ν_o, according to

$$E = E_o \cos 2\pi\nu_o t \tag{5.2}$$

where E_o is the impinging electric field, and t is time.

Substitution in eq 5.1 gives

$$P = \alpha E_o \cos 2\pi\nu_o t \tag{5.3}$$

The polarizability, α, depends on the motion of the nuclei in the molecule. The motion of the nuclei of a diatomic molecule can be expressed in terms of the normal coordinate of the vibration, x_1, and the dependence of α on x (the change in internuclear separation with vibration) can be approximated by a series expansion

$$\alpha = \alpha_o + (\partial\alpha/\partial x)x + \cdots \tag{5.4}$$

The normal mode is a time-dependent vibration with a frequency, ν_1. This dependence can be expressed as

$$x_1 = x^o{}_1 \cos 2\pi\nu_1 t \tag{5.5}$$

where $x^o{}_1$ is the equilibrium position. Substitution gives

$$P = (E_o \cos 2\pi\nu_o t)\,[\alpha_o + (\partial\alpha/\partial x)x]\,(x^o{}_1 \cos 2\pi\nu_1 t) \tag{5.6}$$

From basic trigonometry

$$\cos\theta\cos\phi = \frac{\cos(\theta + \phi) + \cos(\theta - \phi)}{2} \tag{5.7}$$

Substitution gives

$$P = (\alpha_o E_o \cos 2\pi\nu_o t) + \left(\frac{1}{2}\right) E_o x^o{}_1 \left(\frac{\partial\alpha}{\partial x}\right) [\cos 2\pi(\nu_o + \nu_1)t + \cos 2\pi(\nu_o - \nu_1)t] \tag{5.8}$$

This complex equation demonstrates that three lines are predicted in the light scattered by a diatomic molecule. The α_o term represents the light that is not shifted in frequency (*Rayleigh scattering*). If $(\partial\alpha/\partial x)$ does not equal 0, Raman lines are shifted higher and lower in frequency than ν_o by ν, the frequency of vibration of the molecule. If $(\partial\alpha/\partial x)$ equals 0, the second term is 0, and no Raman lines are observed. This change in polarization with the vibrational motion of the nuclei, $\partial\alpha/\partial x$, is the basis for the selection rule governing the Raman activity of a vibrational mode.

The molecule will scatter light at the incident frequency. However, the molecule vibrates with its own unique frequencies. If these molecular motions produce changes in the polarizability, α, the molecule will further interact with the light by superimposing its vibrational frequencies on the scattered light at either higher or lower frequencies.

When a beam of photons strikes a molecule, most of the photons are scattered elastically, and this *elastic scattering* is termed *Rayleigh scattering*. Rayleigh scattering is responsible for the sky appearing blue because scattering is more efficient at shorter wavelengths. A few photons (1 in 10^8) undergo *inelastic scattering* or *Raman scattering* (*see* Figure 5.1). These Raman or inelastically scattered photons have different frequencies and produce a spectrum of frequencies in the scattered beam that constitute the Raman spectrum of a molecule (Figure 5.2). The photons that lose energy appear on the lower fre-

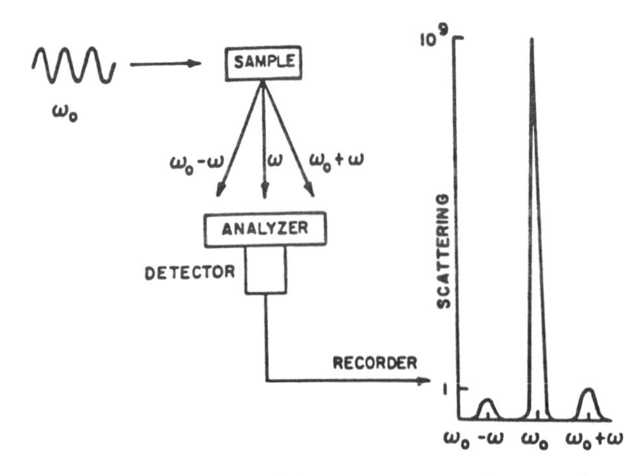

Figure 5.1. Diagram of the Raman scattering experiment showing the large effect of Rayleigh scattering compared to the Raman effect.

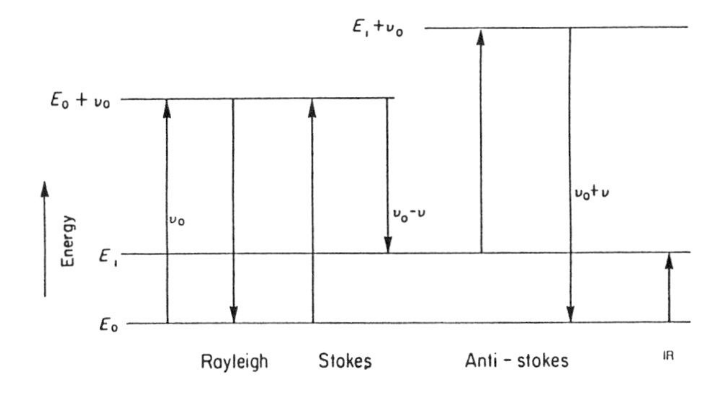

Figure 5.2. The energy levels of the Raman spectroscopic experiment showing the Stokes and anti-Stokes lines and the IR energy levels.

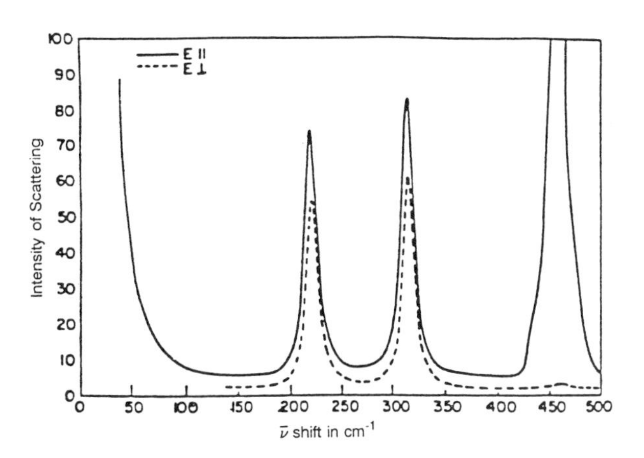

Figure 5.3. The Raman spectrum of CCl_4 .

quency side of the exciting line and are called *Stoke's lines*, and the photons that gain energy appear at higher frequencies and are called *anti-Stokes lines*. The IR frequency is the difference between the two levels, so Raman spectroscopy arises from the same energy levels as IR spectroscopy, but by a quite different route.

Polarization Effects in Raman Spectroscopy

In general, the induced polarizability is not necessarily in the direction of the incident beam, and P_X, P_Y, and P_Z, which are the X, Y, and Z components of the dipole moment, can be calculated by

$$P_X = \alpha_{XX}E_X + \alpha_{XY}E_Y + \alpha_{XZ}E_Z \quad (5.9)$$

$$P_Y = \alpha_{YX}E_X + \alpha_{YY}E_Y + \alpha_{YZ}E_Z \quad (5.10)$$

$$P_Z = \alpha_{ZX}E_X + \alpha_{ZY}E_Y + \alpha_{ZZ}E_Z \quad (5.11)$$

where α_{XX}, α_{YY}, and α_{ZZ} are the components of the principle axes of the polarizability ellipsoid, and α_{XY}, α_{YZ}, and α_{XZ} are the other components. Consequently, the Raman scattered light emanating from even a random sample is polarized to a greater or lesser extent. For randomly oriented systems, the polarization properties are determined by the two tensor invariants of the polarization tensor, that is, the trace and anisotropy tensors. The depolarization ratio is always $\leq 3/4$. For a specific scattering geometry, this polarization is dependent upon the symmetry of the molecular vibration giving rise to the line. As shown in Figure 5.3, the line of CCl_4 at 457 cm^{-1} is completely polarized. The partial polarization of the other two modes indicates that the motions of the CCl_4 are antisymmetric. The unique polarization properties of this "new radiation" led C. V. Raman to believe that he was observing a new phenomenon rather than normal emission such as fluorescence.

For solids, the problem of polarization is more complicated, but the results are more rewarding. In solids, the molecular species are oriented with respect to each other. Therefore, the molecular polarizability ellipsoids are also oriented along definite directions in the crystal. Because the electric vector of the incident laser beam is polarized, the directionality in the crystal can be used to obtain Raman data from each element of the polarizability ellipsoid. When the laser polarization and collection are along the z axis, a spectrum from the α_{zz} component of the tensor is obtained. Rotating the analyzer by 90° allows collection of the x-polarized light while still exciting along the z axis, and in this manner α_{zx} is obtained.

However, some polymeric samples contain crystallites or voids comparable in size to the visible wavelengths of the laser. These crystallites or voids scramble the incident laser polarization and thereby prevent any useful measurement of depolarization ratios. Some error in Raman polarization measurements arises because the incident light and Raman scattered light are multiply reflected at the surface of the sample and are also refracted upon entering or leaving the sample. The light-polarization directions are therefore poorly defined. Immersing the sample in a liquid that has a refractive index close to that of the polymer helps to minimize this problem (*4*).

In the usual Raman experiment, the observations are made perpendicular to the direction of the incident beam, which is plane polarized. The *depolarization ratio* is defined as the intensity ratio of the two polarized components of the scattered light that are parallel and perpendicular to the direction of propagation of the polarized incident light. The polarization of the incident beam is perpendicular to the planes of propagation and observation. For this type of geometry, the depolarization ratio is defined as the intensity ratio:

$$\rho = \frac{VH}{VV} \tag{5.12}$$

where for the right-angle scattering experiment, V is perpendicular to the scattering plane, and H is in the scattering plane. An alternate notation expressed in terms of the laboratory coordinate system is $A(BC)D$, where A is the direction of travel of the incident beam; B and C are the polarization of the incident and scattered light, respectively; and D is the direction in which the Raman scattered light is observed. Generally, the incoming beam travels along the X axis, the scattered beam travels along the Z axis, and the Y axis is perpendicular to the scattering plane (*see* Figure 5.4).

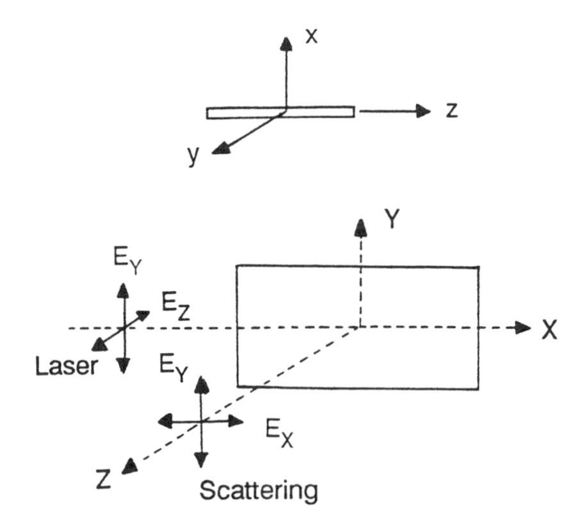

Figure 5.4. Coordinate system for the Raman scattering experiment. The laser beam path is along X, *with vertical polarization along* Y *or horizontal polarization along* Z. *The Raman-scattered radiation is collected along* Z *and analyzed either horizontally along* X *or vertically along* Y. *The rod models a local chain segment and shows the attached molecular coordinate system. (Reproduced from reference 6. Copyright 1984 American Chemical Society.)*

To study the orientation of the sample, the aforementioned experiments must be related to the molecular polarizabilities. By using upper-case letters for the laboratory coordinate system and lower-case letters for the molecule-fixed coordinate system, the derived polarizability tensor can be found. For excitation polarized along F, the scattered radiation, which is polarized along F', will have an intensity that is proportional to $(\alpha_k)FF'^2$ (F and F' may be either vertical or horizontal). The appropriate relationships have been determined (*5, 6*). In an isotropic system, there are orientationally invariant terms, δ and γ, that are defined as

$$\delta = \left(\frac{1}{3}\right) \sum \alpha_{ii} \tag{5.13}$$

and

$$\gamma^2 = \frac{1}{3} \sum_{i<j} [(\alpha_{ii} - \alpha_{jj})^2 - 6\alpha_{ij}^2] \tag{5.14}$$

The contributions of the various geometries of three-dimensionally isotropic systems to the Raman spectrum are shown in Table 5.1.

These results show that the intensities of the VH, HV, and HH experiments should be identical. These results have been confirmed, as shown in Figure 5.5, which shows the polarized Raman spectra of a thin film of an unoriented isotactic poly(methyl methacrylate) (PMMA) sample.

Table 5.1. Contribution to Spectrum for Three-Dimensionally Isotropic Systems

Experiment	*Geometry*	*Symmetric*	*Anisotropic*
VV	X(YY)Z	δ^2	$4\gamma^2/45$
VH	X(YZ)Z	0	$\gamma^2/15$
HV	X(ZY)Z	0	$\gamma^2/15$
HH	X(ZX)Z	0	$\gamma^2/15$

SOURCE: Reproduced from reference 6. Copyright 1984 American Chemical Society.

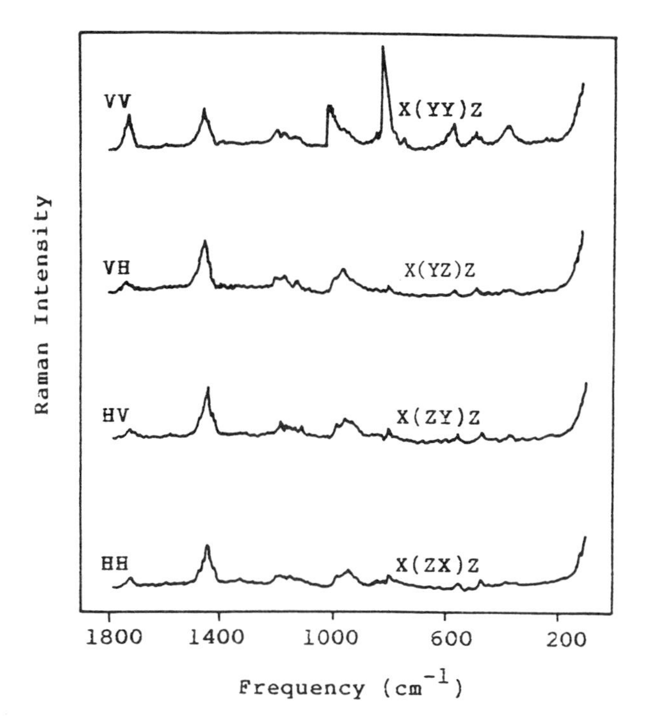

Figure 5.5. Polarized waveguide Raman spectra of an unoriented thin film of isotactic PMMA. The TE spectra (VV and VH) are at the top, and the TM spectra (HV and HH) are at the bottom. (Reproduced from reference 6. Copyright 1984 American Chemical Society.)

Theoretically, the depolarization ratio can have values ranging from 0 to 3/4, depending on the nature and symmetry of the vibrations. Nonsymmetric vibrations give depolarization ratios of 3/4. Symmetric vibrations have depolarization ratios ranging from 0 to 3/4, depending on the polarizability changes and symmetry of the bonds in the molecule. Accurate values of the depolarization ratio are valuable for determining the assignments of Raman lines, and, in conjunction with dichroic measurements from IR spectroscopy, the two measurements represent a powerful tool for the structural determination of polymers. Because the laser beam is inherently polarized and highly directional, polarization measurements can be made easily.

Spectral Differences Between IR and Raman Spectroscopy

The Raman effect is produced by the exchange of energy between the incident photons and the vibrational energy levels of the molecule and is unique to each molecule. In many instances, the magnitudes of the Raman shifts in frequency correspond exactly to the frequencies of IR absorption. As a result, the information obtained from IR measurements can also be found in the Raman spectra. Some vibrational modes appear only in the IR spectrum, and other modes appear only in the Raman spectrum. The differences in the vibrational patterns of IR and Raman spectroscopy can be used to a great advantage in the determination of the structure of molecules, and these differences represent one of the prime reasons for interest in Raman spectroscopy. Generally, the more symmetric the molecule is, the greater the differences will be between the IR and Raman spectra. Strong Raman scattering arises from nonpolar molecular groupings, and strong IR absorption occurs when the molecule contains polar groups (*7*). In chainlike polymer molecules, the vibrations of the substituents on the carbon chain are most easily studied by using IR spectroscopy (as indicated in Chapter 2), and the vibrations of the carbon chain can be studied by using Raman spectroscopy.

These differences between Raman and IR spectroscopy for polymers can be illustrated. The Raman and IR spectra of linear polyethylene are shown in Figure 5.6 (*8*). The polyethylene molecule has a center of symmetry, and the IR and Raman spectra should exhibit entirely different vibrational modes. Examination of the spectra reveals that this is the case. In the IR spectrum, the CH_2 modes are clearly the strongest, but the C – C modes dominate the Raman spectrum. The IR and Raman spectra of poly(ethylene terephthalate) (PET) are shown in Figure 5.7. The C – C stretching modes of the aromatic ring clearly dominate the Raman spectrum, and the C – O modes are strongest in the IR spectrum. The Raman spectrum has a fluorescence background that makes the dynamic range of the spectrum less than is desirable. The Raman lines are weak when superimposed on the broad background. The IR and Raman spectra of poly(methyl methacrylate) (PMMA) are compared in Figure 5.8. The Raman spectrum is particularly rich in the lower frequency range, where there is little absorbance in the IR spectrum. The C = O and C – O bands dominate the IR spectrum, and the C – C modes dominate the Raman spectrum.

Perhaps the greatest reason for the limited application of Raman spectroscopy is the low-perceived value of Raman spectroscopy to the industrial labo-

ratory. The analytical chemist must decide what additional information can be obtained by having a Raman spectrometer and the requisite trained personnel.

— Carl Zimba (2)

Compared to IR spectroscopy, Raman spectroscopy has a number of advantages, and these advantages are as follows:

- Raman spectroscopy is a scattering process, so samples of any size or shape can be examined.
- Very small amounts of materials can be examined without modification (sometimes localized degradation can occur with high laser power).
- Fiber optics can be used for remote sampling.
- Aqueous solutions can be analyzed.
- The low frequency (10 – 500 cm^{-1}) region is available on the same instrument.
- Less restrictive selection rules apply, so richer spectra are obtained.

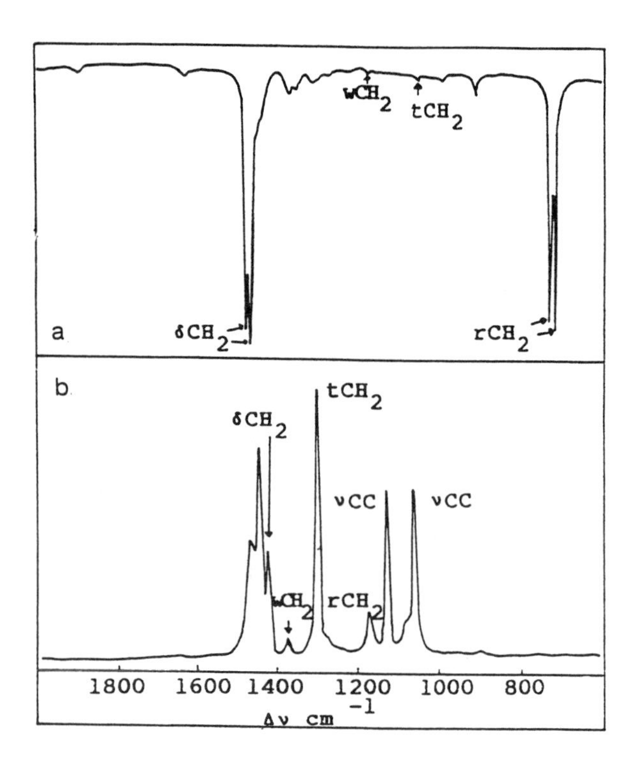

Figure 5.6. IR (a) and Raman (b) spectra of linear polyethylene. (Reproduced with permission from reference 8. Copyright 1979 Steinkopff Verlag Darmstadt.)

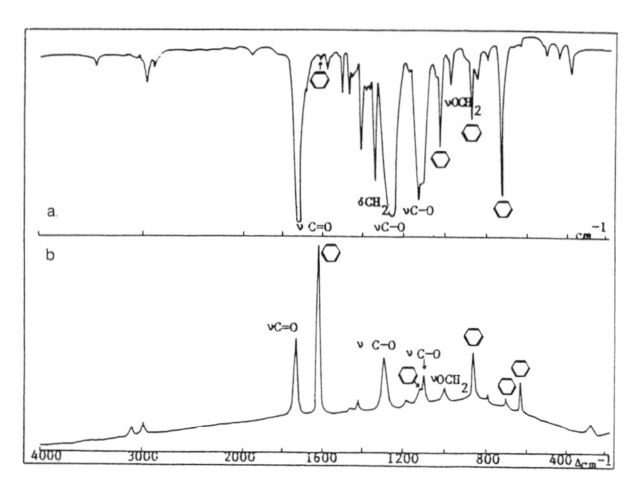

Figure 5.7. IR (a) and Raman (b) spectra of poly(ethylene terephthalate). (Reproduced with permission from reference 8. Copyright 1979 Steinkopff Verlag Darmstadt.)

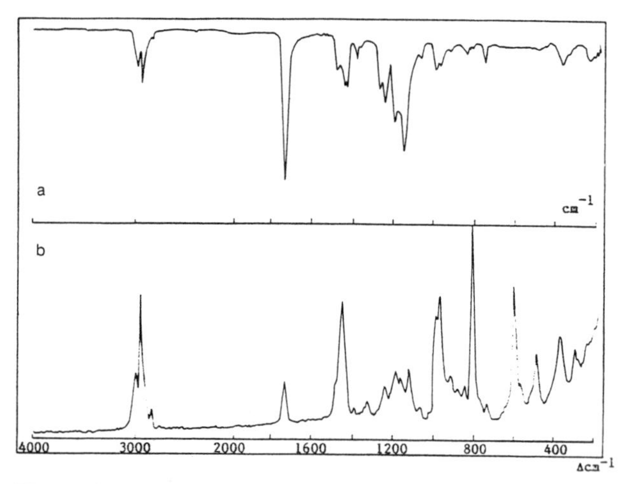

Figure 5.8. IR (a) and Raman (b) spectra of poly(methyl methacrylate). (Reproduced with permission from reference 8. Copyright 1979 Steinkopff Verlag Darmstadt.)

Because the Raman effect is a scattering process, it does not have the problems associated with the requirement of light transmission. Raman front-surface reflection allows the examination of a sample of any size or shape. If the sample can be hit with the laser beam, a Raman spectrum can usually be obtained.

Raman spectroscopy has an inherent microsampling capability because the laser beam has a small diameter, and the beam can be focused to an even smaller size. By using a commercial Raman instrument (called the MOLE) with a computerized microscopic sampling device, a sample can be scanned, and chemical heterogeneities in the composition can be determined. By using this instrument, two-dimensional images can be obtained, as shown in Figure 5.9 (*9*). Selected Raman band images that show on the microscopic scale the contributions of the different components in the sample can be obtained by selecting a specific Raman frequency. This technique is not widely used for two reasons: The commercial instrument is very expensive (and presently may not be available), and fluorescence in polymers is still a major problem.

Because the Raman effect can be successfully integrated with fiber optics, remote sampling is possible. IR spectroscopy can also be interfaced with fiber optics, but the loss of light in the IR region is much higher than the loss in the visible region, and the maximum distance between the sample and the instrument is much smaller for IR spectroscopy than for Raman spectroscopy.

> *The advantage of working with Raman rather than IR spectra was that the Raman spectra could be recorded in great detail for substances in aqueous solution, with little interference by the frequencies arising from water, whereas the IR spectra of the solutes would have been almost completely blotted out by the intense absorption of water.*
>
> — J. T. Edsall (*10*)

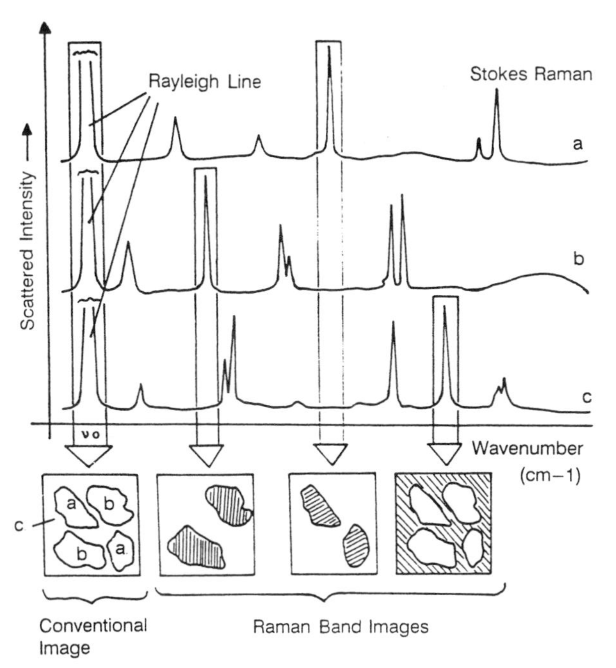

Figure 5.9. The principle of selective imaging using specific spectral "bands" of the Raman-scattered radiation. The matrix, c, contains two inclusions, a and b. Each material possesses a characteristic Raman spectrum, shown as a, b, and c, respectively. (Reproduced from reference 9. Copyright 1990 American Chemical Society.)

Because water is a highly polar molecule, it is a strong IR absorber. By the same token, it is a weak Raman scatterer. Therefore, the Raman effect can be used to study aqueous solutions, and this fact has been beneficial to the biological field.

Studies of water-soluble polymers such as surfactants can be successfully carried out with Raman spectroscopy.

As will be demonstrated later in our discussion of Raman instrumentation, a single instrument can be used to scan the region from 10 cm^{-1} (depending somewhat on the scattering power of the sample) to 4000 cm^{-1}. FTIR spectroscopy, on the other hand, requires changing the beam splitter and detector to reach this low frequency, or *far-IR*, region of the spectrum. Raman spectroscopy has been used to detect the low-frequency acoustic modes of polymers, and the frequencies of these modes have been used to measure the dimensions of crystals.

For polymers and most other molecules, the Raman selection rules are less restrictive than the IR selection rules, so the Raman spectra are richer in vibrational information than the IR spectra. Ideally, a spectroscopist would like to have both the IR and Raman spectra of a sample. Recent developments in Fourier transform instrumentation now make this a real possibility.

There are differences in the instrumentation needed to make IR and Raman measurements. These differences in instrumentation are compared in Table 5.2 (*6*). Rapid changes are taking place in Raman instrumentation, but the general character remains the same. The major difference between the two techniques is the need for a laser light source for Raman spectroscopy compared to a blackbody source (usually a heated wire) for IR spectroscopy. The Raman detector for the visible light range is a photomultiplier tube, which has a much higher level of sensitivity than the thermal or pyroelectric detectors used in IR spectroscopy. Although energy is the limitation for both techniques (as it is eventually for all forms of spectroscopy), inefficient scattering (1

Table 5.2. Comparison of Raman and IR Spectroscopic Methods and Instrumentation

Feature	*Raman*	*IR*
Relative complexity	Moderate	Slightly greater
Source	Laser	Blackbody or diode laser
Detector	Photomultiplier tube	Thermal, pyroelectric, bolometers
Resolution	ca. 0.25 cm^{-1}	ca. 0.05 cm^{-1}
Principal limitation	Energy	Energy
Wavenumber range	10 – 4000 cm^{-1}	180 – 4000 cm^{-1} (one instrument) 10 – 40 cm^{-1} (second instrument or new beamsplitter, source, and detector)
Purge requirement	No	Yes
Photometry	Scattering single beam	Absorption double beam

photon in 10^8) is the limiting factor in Raman spectroscopy, and the low power level of the blackbody source is the limiting factor in IR spectroscopy.

These differences between Raman and IR spectroscopy lead to different sample handling techniques (Table 5.3).

Instrumentation and sample handling considerations are reflected in the applications of IR and Raman spectroscopy. The applicability of Raman spectroscopy to various types of polymer samples is rated in Table 5.4. Raman spectroscopy is effective for aqueous solutions and in low-frequency ranges.

Raman and IR Spectroscopy in Combination To Determine Polymer Conformation

When IR and Raman spectroscopic techniques are used in combination, the results are much greater than with the use of either technique individually. The combined use of IR and Raman spectroscopy extracts most of the obtainable information ("silent", or optically inactive, modes and extremely weak modes are not detected). The complementary nature of the IR and Raman data has important practical applications. This complementary nature arises from the differences in selection rules governing the vibra-

Table 5.3. Comparison of Sample Handling for Raman and IR Experiments

Feature	*Raman*	*IR*
General applicability	95%	99%
Sample limitations	Color; fluorescence	Single crystals; metals; aqueous solutions
Ease of sample preparation	Very simple	Variable
Liquids	Very simple	Very simple
Powders	Very simple	More difficult
Single crystals	Very simple	Very difficult
Polymers	Very simple, but see sample limitations	More difficult
Single fibers	Possible	Difficult
Gases and vapors	Now possible	Simple
Cells	Very simple (glass)	More complex (alkali halide)
Micro work	Good (<1 μg)	Good (<1 μg)
Trace work	Sometimes	Sometimes
High and low temperature	Moderately simple	Moderately simple

Table 5.4. Applicability of Raman Spectroscopy to Polymer Samples

Sample	*Excellent*	*Very Good*	*Good*	*Poor*
Polymer				
Homonuclear backbone	×			
Polar substituents			×	
End groups				×
Multicomponent systems				
Additives (<1%)				×
Fillers (>5%)				
Glass	×			
Carbon black				×
Inorganic (TiO_2)		×		
Pigments			Variable	
Properties				
Variable size and shape	×			
Limited solubility	×			
Colors with aging				×
Sensitivity to thermal history	×			

tional energy levels. For molecules with a center of symmetry (there are identical atoms on either side of the center of symmetry), no vibrational frequencies are common to the IR and Raman spectra. This principle is called the *mutual exclusion principle*. Although symmetry might be considered important for low-molecular-weight substances like ethylene and benzene (both of which have a center of symmetry), polymers are not usually expected to have a center of symmetry. Polyethylene has a center of symmetry, and the observed IR and Raman lines do not coincide in frequency (*see* Figure 5.6).

Consider this result from a point of view of structural determination; that is, if the Raman and IR lines are mutually exclusive in frequency, a center of symmetry must exist in the polymer. Consider polyethylene sulfide (PES), which contains a succession of $(-CH_2CH_2SCH_2CH_2S-)$ repeat units. The conformations about the CC, CS, SC, CC, CS, and SC bonds, are *trans*, *gauche* (right), *gauche* (right), *trans*, *gauche* (left), and *gauche* (left), respectively. This structural model of PES has a center of symmetry, and theory predicts that the observed IR and Raman frequencies are mutually exclusive.

If PES exists in a helical form like polyethylene oxide (PEO), it does not have a center of symmetry, and theory would suggest that there are many frequencies that would be coincident in the Raman and IR spectra. PEO has 20 coincident Raman and IR bands, but PES has only two coincident frequencies, and these coincident bands are the result of "accidental" degeneracies. Accidental degeneracies arise when the frequency differences between modes are too small to be resolved within the instrument. Thus, the configuration of PES is opposite to that of PEO. In PEO, the C – C bond takes the *gauche* form, and the C – O bond takes the *trans* form; but in PES, the C – C bond takes the *trans* form, and the C – S bond takes the *gauche* form (*7*).

This mutual exclusion rule for molecules with a center of symmetry is one example of the manner in which symmetry within a molecule can influence the selection rules for vibrational spectroscopy. All elements of symmetry influence the selection rules. The selection rules can be evaluated by using the structural models of the molecules, and the proper structural model for a molecule can be determined by comparing the experimental results with the theoretical results. This approach to structure determination has been a valuable tool for the study of the various geometric structures of low-molecular-weight substances. This approach can also be used in the study of polymer conformation.

For polymers, the additional measurements of polarization in Raman spectroscopy and dichroic behavior in IR spectroscopy can be used to aid in the classification of the vibrational properties. In combination, the two techniques can be used to measure the vibrational modes and to classify these vibrational modes into types that are unique to the molecular symmetry of the polymer chain.

A general example is the monosubstituted vinyl polymers $(CH_2-CHX)_n$, which can exist in two possible stereoregular forms–*isotactic* (substitution on the alternate carbons on the same sides of the chain) and *syndiotactic* (substitution on the alternate carbons on the opposite sides of the chain). These ordered stereoregular forms can assume either planar or helical conformations. Each of these ordered structures has well-defined, unique selection rules for IR and Raman activity. These various structures can be distinguished on the basis of spectral properties without a detailed knowledge of the molecular motions or energies, that is, without normal coordinate analysis.

Any observed vibrational spectra can be classified into sets of modes that are defined as follows:

- [R, IR] Frequencies coincident in both Raman and IR spectroscopy
- [R, 0] Frequencies active in Raman spectroscopy, but inactive in IR spectroscopy
- [0, IR] Frequencies inactive in Raman spectroscopy, but active in IR spectroscopy

In addition to these three general classifications, each set of IR and Raman polymeric frequencies can be subclassified according to their individual Raman polarizations and IR dichroic behaviors. The Raman lines may be polarized or depolarized, and the IR bands may have parallel or perpendicular dichroism with respect to the polymer chain axis. The terminology for describing the observed frequencies in the IR and Raman spectra is based on the general expression [a,b], where a indicates the polarization properties of the Raman lines, and b indicates the dichroic behavior of the IR bands. On this basis, a may be p (polarized), d (depolarized), or 0 (inactive), and b may be σ (perpendicular), π (parallel), or 0 (inactive). Therefore, the expression [p, σ] indicates that the observed frequency is found in both the Raman and IR spectra and is polarized in the Raman spectrum and has perpendicular dichroism in the IR spectrum. On this basis, all observed frequencies in the IR and Raman spectra must fall into one of the following classifications: [p, 0], [d, 0], [0, π], [0, σ], [p, π], [p, σ], [d, π], and [d, σ].

Any spectrum can consist of any or all combinations of the aforementioned mode types. The range of possibilities for a given ordered polymer structure is determined solely by the symmetry properties of the polymer. However, the picture may be distorted by the limited spectral sensitivity of the Raman and IR instrumentation (in other words, a Raman or IR vibrational mode can be expected theoretically, but the intensity of the mode is so low that it is not observed experimentally). Instruments yielding high signal-to-noise ratios will minimize this problem. Another complication is the presence of multiple or disordered structures (which have no specific selection rules) that cause deviations from the pure mode selection rule requirements of the perfect chain. The spectral bands that are associated with minor amounts of irregular structures are weak and variable in intensity from sample to sample, which makes their detection possible in some cases.

The band classification for each type of polymer chain stereoconfiguration and conformation for monosubstituted vinyl polymers can be calculated, as shown in Figure 5.10 (*7*). Every polymer structure possesses a unique collection of spectroscopic vibrations, and on this basis, the stereoconfiguration and conformation of the polymer chain in the solid state

Structure	Symmetry	Optical Activity								Example
	R IR	p π	p σ	d π	d σ	p 0	d 0	0 π	0 σ	
Center of Symmetry	D_{2h} C_{2h}					•	•	•	•	PE, PES
Atactic		•	•	•	•			•		PVF
Syndiotactic										
Helix >3_1	D_R				•	•	•	•		PEO
Helix 3_1	D_3				•	•				
Helix 2_1	D_2			•	•	•				
Planar	C_{2v}		•	•	•		•			PVC
Isotactic										
Helix >3_1	C_r	•			•		•			Polybutene
Helix 3_1	C_3	•			•					PP
Planar	C_s		•	•						

Figure 5.10. Selection rules for monosubstituted polyvinyl polymers.

can be determined by measuring of the spectroscopic vibrations.

A specific example is poly(vinyl chloride) (PVC), which is a monosubstituted vinyl polymer that has a syndiotactic-rich character and a conformation that can be either an extended all-*trans* structure or a folded syndiotactic structure. The vibrational modes of these conformational models obey different selection rules and have different dichroic properties that can be used to spectroscopically test these structures (*11*). The folded syndiotactic model of PVC has the [p, 0] classification that requires unique Raman lines (no coincident IR frequency) that are polarized. The extended syndiotactic model has the two unique classifications of [d, 0] and [p, σ], which means that the unique Raman lines are depolarized, and the Raman lines that are polarized have perpendicular dichroism in the IR spectrum. In the Raman spectrum of PVC (*12*), polarized lines are observed at 363, 638, 694, 1172, 1335, 1430, and 1914 cm^{-1}, and IR bands are also observed at each of these frequencies. This result rejects the folded syndiotactic structure, because this structure requires the polarized lines to be unique. In addition, each of these frequencies is perpendicularly dichroic in the IR spectrum, a fact that supports the planar syndiotactic structure.

This structural determination method, which is based on differences in the IR and Raman frequencies, is a general method for ordered polymer structures in the solid state.

Limitations of Raman Spectroscopy Resulting from Fluorescence

The principle limitation of Raman spectroscopy in the visible region is that of fluorescence, and Raman spectra cannot be obtained from many samples because of the curtain of fluorescence that hides them. Even if the fluorescence is assumed to arise from an impurity that is present at the parts-per-million level, with a fluorescent quantum yield of only 0.1, 10 fluorescent photons will be produced for every Raman photon. In the visible excitation region, 95% of polymers are not amenable to Raman analysis because of unquenchable fluorescence.

> It is also safe to say that the more desperately that you need the Raman spectrum, the more likely it is that unquenchable fluorescence will raise its ugly head.

A number of methods to remove fluorescence resulting from impurities have been attempted.

> If the fluorescence is not the result of an impurity but rather from the sample itself, the sample will usually burn up very quickly in the laser beam, and a neat looking hole in the sample will result. Under these circumstances, you have removed the fluorescence along with the sample, and you can't obtain a Raman spectrum.

The most convenient and often-used method of fluorescence removal is photobleaching, that is, exposure of the sample to the laser beam until the fluorescence decays. In many cases, a few minutes of exposure to the laser beam will cause the fluorescence to decay sufficiently that a Raman spectrum can be observed. Sometimes, exposure of the sample to the laser beam for several hours is required. The decay of the fluorescence is easily detected by driving the recorder as a function of time without scanning the monochromator. A perceptible "bleaching" of the sample at the spot of the laser focus can sometimes be observed.

Removal of the colored impurities by exposure to activated charcoal columns (the old sugar chemist's trick) or by extraction has proven to be a successful purification method for some samples.

Sometimes, it is possible to change the laser frequency so that excitation occurs outside the absorption envelope of the fluorescent impurity. Occasionally, changing the frequency from blue to red excitation improves the Raman signal by decreasing the fluorescence level. The laser excitation frequency can be decreased below the critical value by using a krypton laser at 6471 Å rather than an argon laser at 4880 Å. A more drastic approach is to use an exciting frequency in the IR region that is well below the absorbing electronic frequencies. This technique has been termed the Fourier transform Raman technique, and it will be discussed later in this chapter.

Fluorescence and Raman scattering can be distinguished on the basis of the duration of the two processes. In fluorescence, the system absorbs radiation and is excited to a higher electronic state. After approximately 10^{-9} s, the system re-emits radiation. Raman scattering involves no absorption of radiation. The scattering system is never in an excited electronic state, and the whole process takes place in less than 10^{-12} s. Therefore, another approach may be used to obtain the Raman spectra. *Time-resolved gated-detection* recognizes that the fluorescence has a finite delay time and the Raman signal does not, and so the detection is gated to observe the immediate Raman signal and to suppress the latent fluorescence signal. This technique is very sophisticated and demanding experimentally, and unfortunately, does not completely eliminate the fluorescence signal. Consequently, it is not widely used.

Resonance enhancement can be used to remove fluorescence by making the Raman signal much more intense than the fluorescence signal. The UV-resonance Raman method can be used to enhance the Raman signal above the fluorescence signal if the sample has the proper electronic structure for resonance enhancement and if the proper UV frequency source is available. The resonance Raman technique is discussed in the section "Resonance Raman Spectroscopy" in this chapter.

In summary, there is no universally adequate solution to the fluorescence problem in Raman spectroscopy. All polymer samples will fluoresce, and the concern is a matter of how much fluorescence and for how long. However, with modern techniques such as FT Raman and resonance Raman spectroscopy, there is considerable optimism that this problem can be overcome.

Experimental Raman Spectroscopy

The three general types of Raman spectrometers are (1) conventional single-channel Raman spectrometers, (2) optical multichannel-detector spectrometers, and (3) Fourier transform Raman spectrometers. Two types of Raman spectroscopy involve resonance enhancement: resonance Raman spectroscopy and surface-enhanced Raman spectroscopy. These latter two techniques use conventional instrumentation and exhibit unconventional enhancement effects.

Each type of Raman spectrometer has been designed to optimize certain aspects of Raman instrumentation. Conventional Raman spectrometers are designed to eliminate the unwanted stronger Rayleigh-scattered light from the weaker Raman signal. These conventional Raman spectrometers are the simplest and are readily available for routine work when interfering fluorescence is not a major problem. Optical multichannel detectors for conventional Raman measurements use the multiplex advantage to increase the sensitivity and speed of the acquisition of Raman spectra. Fourier transform Raman spectrometers are designed to deal with interfering fluorescence by using low-energy laser excitation. Modifications have been made to the FTIR instruments to convert them to Fourier transform Raman spectrometers. All of these instruments will be discussed briefly.

Experimental Conventional Raman Spectroscopy

The experimental apparatus for conventional Raman spectroscopy, outlined in block form in Figure 5.11, includes (1) a powerful laser irradiating in the visible-wavelength region, (2) an illuminating chamber for the sample, (3) a high-performance light-dispersion system to resolve the more intense, elastically scattered light from the weak, inelastically scattered Raman signal, (4) a light detection and amplification system capable of detecting weak light levels, and (5) a recorder.

The first source for Raman spectroscopy was the sun, which is not a very intense spectroscopic source. A helical mercury arc (the Toronto arc) that surrounded a cylindrical sample was developed later and

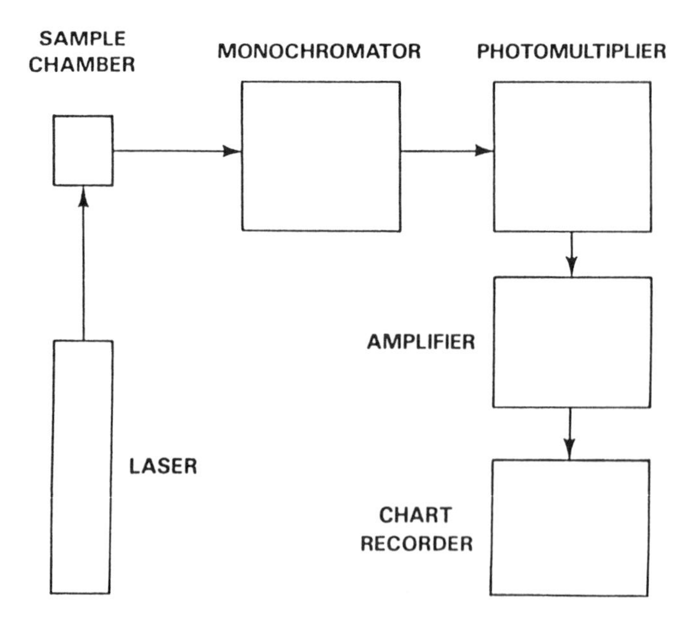

Figure 5.11. Block diagram of a conventional Raman spectrometer.

was the only source used for many years. In order to obtain a Raman spectrum with a Toronto arc, the samples must be colorless and free of dust and fluorescence. This requirement was a major restriction for the study of polymers. Finally, with the advent of the laser light source, Raman spectroscopy underwent major growth.

> *However, despite the fact that lasers are present in supermarket checkout counters throughout the country, it is the very use of a laser that tends to symbolize the perception of Raman spectroscopy as an exotic technique of limited utility.*
>
> – Carl Zimba (2)

The main advantages of a laser light source for Raman spectroscopy are directionality, coherency, intensity, monochromaticity, and polarization.

A monochromatic laser light beam is characterized by its wavelength (λ), power, and polarization. The frequency, ν, is given by

$$\nu = \frac{c}{\lambda} \tag{5.15}$$

where c is the speed of light (2.998×10^{10} cm/s in a vacuum), λ is expressed in centimeters, and ν is expressed in Hertz, or cycles per second.

The wavelength of light is commonly expressed in angstroms (Å), and 10 Å = 1 nm = 10^{-9} m. The spectral regions of light and the corresponding wavelengths, frequencies, and wavenumbers are listed in Table 5.5.

The wavelength (or frequency) of light is related to the energy, E, by

$$E = h\nu \tag{5.16}$$

Table 5.5. The Wavelengths of Light and the Corresponding Colors, Frequencies, and Wavenumbers

Color	*Wavelength (nm)*	*Frequency (10^{14} Hz)*	*Wavenumber (cm^{-1})*
IR	10,000	0.300	1000
Near IR	1000	3.00	10,000
Blue	470	6.38	21,300
Near ultraviolet	300	10.00	33,300

where h is Planck's constant. The energy of a single photon in joules, E_p, is

$$E_p = \frac{(1.986 \times 10^{-16})}{\lambda\ (\text{nm})} \tag{5.17}$$

where λ is the frequency.

The photon energy available from 1 mol of photons can be calculated as joules × Avogadro's number (in kilocalories per mole), where 1 cal (thermochemical) = 4.184 J. Because Avogadro's number is equal to 6.022×10^{23}, the energy of 1 mol of photons at 500 nm can be calculated as

$$\frac{(3.973 \times 10^{-19}) \times (6.022 \times 10^{23})}{4.184 \times 10^{3}} = 57.18 \text{ kcal/mol}$$

Double and triple monochromators can be used to reduce the intense elastically scattered light to a low enough level that the extremely weak Raman signal can be detected. Because the Raman signal is only $10^{-9} - 10^{-6}$ as strong as the Rayleigh line, two and three grating monochromators are used in tandem to reduce the Rayleigh-scattered light to an acceptable level. These monochromators are scanned, and the scattered light is passed over the exit slits to the detector.

The detector for conventional Raman spectroscopy is a photomultiplier tube that is coupled with photon-counting electronics that can detect single photons.

The illuminating chamber is simple and is built to facilitate the focusing of the beam on the sample. The beam can be passed through a liquid or solution and reflected several times to enhance the signal. The scattered light is condensed by using a lens system and is focused on the slits of the light-dispersion system.

Optical Multichannel Raman Spectrometers

Multichannel Raman spectrometers are similar to the conventional systems just described except that the photomultiplier tube is replaced with an optical multichannel detector (OMD). The OMD consists of many (500 – 1000) photosensitive elements, each monitoring a different wavelength at the focal plane. Instead of scanning the wavelengths across the exit slit, the dispersed spectrum is focused onto an OMD at the focal plane. An intensified photodiode array can monitor 1000 wavelengths simultaneously.

Multichannel Raman spectrometers have a number of advantages compared to conventional Raman spectrometers (*13*). First, OMDs allow a substantial segment of the Raman spectrum to be recorded simultaneously without scanning the spectrometer's wavelength setting. This capability of simultaneous analysis of multiple data points makes it possible to record an image (10^3 channels) rather than an average light intensity of a single frequency. Thus, the instrument can be operated as a spectrograph rather than as a scanning monochromator. Typically a frequency range of 600 cm^{-1} can be monitored.

Sensitivity is another advantage of the OMD. The sensitivity of the photodiode for an area of 25 $\times$ 2.5 μm is similar to that of a good photomultiplier tube. The OMD also has a short response time that permits the storage of all the spectral elements in a "single-exposure" duration of as short as a few milliseconds. Each element, or pixel, of the photodiode array stores a charge proportional to the integral number of photons striking the element. At intervals of 10 ms or more, the stored charge from all pixels is read, and the pixel is reset. With this repetitive integration and readout, a series of spectra can be generated, and each of these spectra represents a Raman scattering during a 10-ms segment of time. In this manner, time-resolved Raman spectra can be obtained.

The limitations of multichannel Raman spectrometers arise primarily from the fact that the maximum spectral region that can be measured at one time is limited by the active area of the detector and the available dispersion of the grating. The connection of two or more separately measured spectra is a prerequisite for obtaining a complete Raman spectrum. There are also the inevitable channel-to-channel differences in sensitivity, particularly at the edges of the array. Because detector sensitivity falls off at each end of the active area, it is necessary to make a sensitivity calibration to generate a smoothly connected spectrum (*14*).

Fourier Transform Raman Spectroscopy

Fourier transform Raman spectroscopy is designed to eliminate the fluorescence problem encountered in conventional Raman spectroscopy (*15*). Fluorescence can be avoided by using an excitation frequency below the threshold for any fluorescence process. The most common excitation frequency for FT Raman spectroscopy is the Nd:YAG laser source at 1.064 μm (9398 cm^{-1}). Consequently, the Raman scattering occurs in the 5398 – 9398-cm^{-1} (near-IR) region, and conventional FTIR instrumentation can be used with slight modification. Rejection of the Rayleigh line is accomplished by using absorption filters (dielectric interference filters) that are designed to pass wavelengths that are longer than the laser wavelength. The sharpness of the cutoff determines the low-frequency limit (200 cm^{-1}) of the FT Raman spectrometer.

The FT Raman instrument has the following components: (1) a laser excitation source, (2) a Fourier transform interferometer equipped with the appropriate beam splitter and detector for the near-IR region, (3) a sample chamber with scattering optics that match the input port of the Fourier transform instrument, and (4) an optical filter for rejection of the Rayleigh-scattered light. A diagram of such an instrument is shown in Figure 5.12 (*16*). Typically, a He:Ne laser beam is coaligned with the Nd:YAG laser beam to make it possible to observe the Nd:YAG beam for alignment and focusing.

The greatest advantage of FT Raman spectroscopy is the removal of fluorescence. The FT Raman technique is specifically designed to eliminate fluorescence, and in some cases, the effects are quite dramatic. The spectra of a highly fluorescing dye shown in Figure 5.13 were obtained with conventional Raman and FT Raman spectroscopy. The resulting spectra suggest that the effort required to accomplish the FT Raman method is rewarding (*16*).

> *It is apparent after considering the signal-to-noise ratio obtained for these measurements that the sensitivity of FT Raman is competitive with scanning measurements.*
>
> – John Rabolt et al. (*17*)

The advantages of collecting Raman spectra by using Fourier-transform techniques are counterbalanced by potential difficulties that arise primarily from the distinction between inherently weak Raman signals and extraordinarily strong, undesired Rayleigh scattering. Because the noise associated with an intense Rayleigh line is redistributed across the entire spectrum by the Fourier transform process, the multi-

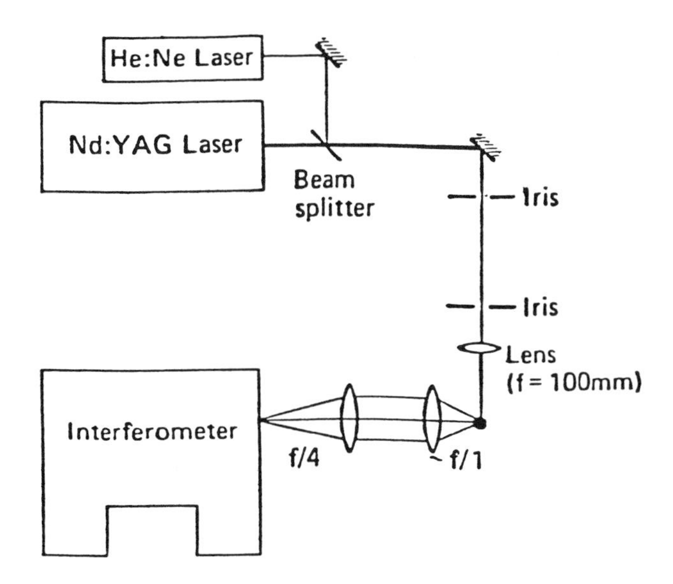

Figure 5.12. Schematic diagram of the FT Raman spectrometer. The He:Ne and Nd:YAG laser beams are made colinear by using a dichroic beam splitter. The 632.8-nm line is used to align the sample in the beam and to image the collected light on the entrance aperture of the interferometer. The Rayleigh line rejection-filter assembly, which is placed inside the sample compartment of the interferometer, is not shown. (Reproduced with permission from reference 15. Copyright 1987 Society for Applied Spectroscopy.)

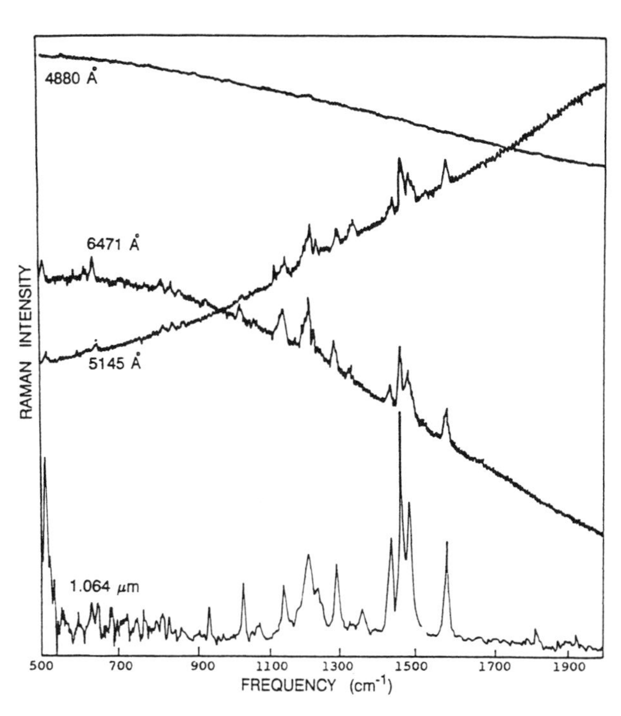

Figure 5.13. Raman spectra of a cyanine dye, 3,3-dioctadecyl-2-2′-methylenebis(benzothiazole) perchlorate, recorded at different wavelengths throughout the visible and into the near-IR region. (Reproduced with permission from reference 15. Copyright 1987 Society for Applied Spectroscopy.)

plex advantage, which is attractive in collecting FTIR spectra, significantly affects the integrity of the Raman-scattered light. Thus, effective use of the FT Raman technique requires a reduction in the intensity of the Rayleigh line so that the energy reaching the detector is primarily Raman-scattered radiation. Improvements in FT Raman spectroscopy will depend on improvements in the cut-off frequencies of the filters (*18*). Notch filters and colloidal-gel diffracting filters are currently being developed, and the laser line is Bragg diffracted so that its transmission through the filter is less than 10^{-10}. Raman-shifted lines are transmitted by the notch filter with greater than 50% efficiency.

FT Raman spectroscopy using FTIR instrumentation solves another problem frequently encountered in conventional Raman spectrometers, which is the lack of sufficient frequency precision required to perform spectral subtractions.

FT Raman spectroscopy does have some disadvantages, as might be expected. The cross section for Raman scattering at the IR excitation wavelength of 1.06 μm is reduced by a factor of 16 from that of the visible excitation wavelength of 5145 Å. The result is a considerable loss in sensitivity because the Raman scattering cross sections of molecules are inherently small. Improved detectors (i.e., near shot noise limited) for the IR region are necessary. Because the process requires IR radiation, absorption can result in local heating, particularly for black samples.

> Bruce Chase, one of the coinventors of the FT Raman method, perhaps summarized it best when comparing conventional and FT Raman spectroscopy. He says: "If you can do the experiment by conventional methods, do it. If not, you have no alternative except FT Raman if you want the Raman spectrum."

Another unforeseen problem is the invisibility of the IR laser beam, which makes alignment of the sample difficult. One option is to colinearly align a visible laser, such as a He:Ne laser, with the laser light source. Another option is to use fiber optics (*19*). With fiber-optic components, optical alignment is virtually eliminated to allow rapid switching from one sample to another.

> *It is worth noting that the advent of detectors with a noise performance that is superior by an order of magnitude (10^{-15} water noise-equivalent power) would largely negate any multiplex gain in FT Raman spectroscopy. At present, the use of pulsed YAG lasers and time-resolved detection may prove advantageous with scanning instruments. Furthermore, if near-IR-sensitive multichannel detectors (intensified diode arrays or CCDs) become commercially available, the use of a spectrograph/multichannel detector would probably become the preferred route for obtaining Raman spectra in the near-IR, since Rayleigh-line rejection would be very efficient and the low-frequency vibration region (<200 cm^{-1}) would become easily accessible.*
>
> – N. J. Everall and J. Howard (*20*)

Resonance Raman Spectroscopy

When the laser wavelength used to excite the Raman effect lies under an intense electronic absorption band of a chromophore, this condition will lead to a considerable resonance enhancement of the Raman signal by a factor of 10^3 to 10^6. A tunable pulsed-UV laser source (frequency-doubled Nd:YAG laser) that pumps a dye laser can produce radiation from 217 to 450 nm and will allow the observation of the resonance Raman effect. Contrary to the high power required by conventional Raman and FT Raman spectroscopy, the typical power requirement for resonance Raman spectroscopy is only a few milliwatts. The resonance Raman spectrometer and detectors are the same as for the conventional or multichannel Raman systems.

The most obvious advantage of resonance Raman spectroscopy is the enhancement of the observed UV excitation by 6 to 7 orders of magnitude. These high intensity enhancements allow trace analysis at the 20-ppb level.

The enhanced vibrational modes are generally totally symmetric and are associated only with the electronic chromophore being excited. This condition results in a considerable simplification of the Raman spectrum and allows a selected chromophore to be used as a probe in the molecule.

Exciting higher electronic states sometimes eliminates fluorescence interference and makes resonance Raman spectroscopy a viable method when fluorescence is a problem in the conventional Raman experiment (*21*).

There are some inherent disadvantages in resonance Raman spectroscopy. The observed modes are associated only with the excited chromophore, and in some cases, this may not be the chromophore of interest. Because irradiation occurs at an absorption frequency of the sample, undesirable photochemistry may destroy the species under investigation. For quantitative analysis, there is the possibility of nonlinear variations in intensities that depend in a complex way on the proximity of the laser excitation wavelength to the electronic maximum of the sample (*22*).

At this stage it may be desirable to compare resonance Raman spectroscopy with FT Raman spectroscopy. FT Raman spectroscopy allows vibrations that are not associated with the electronic chromophore to be observed. Additionally, FT Raman spectroscopy exhibits linear scattering intensities that allow quantitative analysis. On the other hand, resonance Raman spectroscopy is much more sensitive than FT Raman spectroscopy (*23*).

Surface-Enhanced Raman Scattering

Surface-enhanced Raman spectroscopy (SERS) offers considerable promise for the study of polymers for several reasons (*24*). The enhancement effect can increase Raman scattering by a factor of $10^3 - 10^6$ (*25*). Adsorption of molecules on the SERS-active metal surface causes fluorescence quenching in highly fluorescent compounds. In addition, surface-enhanced resonance Raman scattering can further enhance the Raman scattering efficiency by a factor of $10^3 - 10^5$ above that observed under resonance or surface-enhanced conditions alone.

Local electromagnetic effects on certain roughened materials (e.g., Ag, Au, and Cu) enhance the Raman signal by factors of up to 10^6 and make detection of monolayers straightforward (*26*). Enhanced Raman spectra that are $10^5 - 10^6$ as strong as normal Raman scattering were reported for pyridine (*27*).

Sampling Techniques in Raman Spectroscopy

The sampling techniques used in Raman spectroscopy are shown in Figure 5.14. A sample in any state can be examined without difficulty by using Raman spectroscopy. The laser beam is narrow, collimated, and unidirectional, so it can be manipulated in a variety of ways depending on the configuration of the sample.

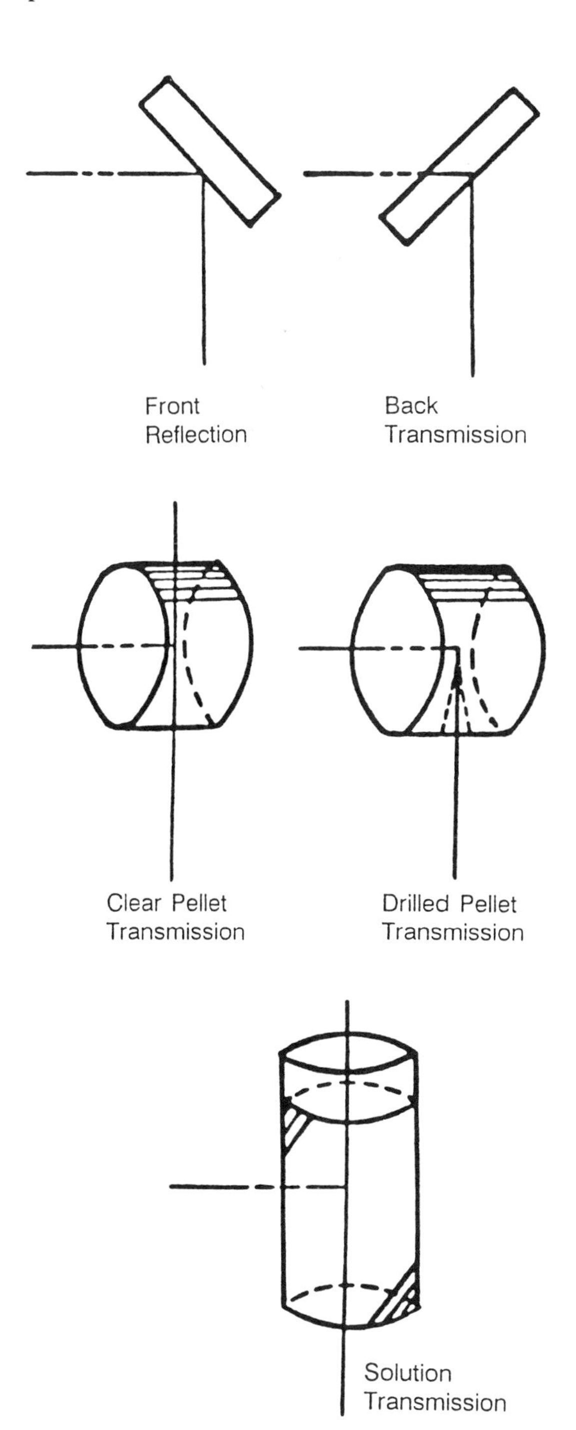

Figure 5.14. Raman sampling techniques for polymers.

For liquids, a cylindrical cell of glass or quartz with an optically flat bottom is positioned vertically in the laser beam. For solids, the particular method used depends on the transparency of the sample. For clear pellets or samples, right-angle scattering is used. With translucent samples, it is helpful to drill a hole in the sample pellet. Powdered samples can be analyzed by using front-surface reflection from a sample holder consisting of a hole in the surface of a metal block inclined at 60° with respect to the beam. Injection-molded pieces, pipes, and tubing; blown films; cast sheets; and monofilaments can be examined directly.

Filled polymers can be studied by using Raman spectroscopy because the fillers such as glass, clay, and silica are weak Raman scatterers and do not interfere with the Raman spectrum of the matrix.

> Carbon-black-filled samples present a problem for Raman spectroscopy because the carbon black will totally absorb the laser beam. The subsequent heating of the sample can lead to destructive results, including loss of sample. But it is fun watching it smoke!

It is desirable to use Raman spectroscopy to make quantitative measurements of functional groups. The relationship between concentration (g/ml) and intensity of the Raman scattering is linear, and changes in IR transmission are logarithmic with concentration. In Raman spectroscopy, the "internal standard" technique must be used because the number of scattering sites in the laser beam cannot be determined. For solutions, a known amount of a standard can be added to determine the relative amount of the unknown material in solution.

Application of Raman Spectroscopy to Polymer Structure Determination

Chemical Structure and Composition

The choice of Raman spectroscopy for analysis of chemical composition and structure is based on the high sensitivity of the Raman effect for certain nonpolar chemical groups. In polymers, these groups are primarily the nearly homonuclear single and multiple C – C bonds that are weak or absent in the IR spectra. The characteristic group frequencies for Raman spectroscopy have been tabulated (*28*).

Raman spectroscopy can differentiate between internal and external bonds as well as *cis* and *trans* isomerism and conjugation in compounds with ethylenic linkages. The type of unsaturation in butadiene and isoprene rubbers can be determined from the intense Raman scattering of the C=C stretching modes (*29*). The *trans*- and *cis*-1,4-polybutadiene structures scatter at 1664 and 1650 cm^{-1}, respectively. The 1,2-vinyl structure of polybutadiene scatters at 1639 cm^{-1}, and this scattering is well-resolved from that of the 1,4-polybutadiene structures. For polyisoprene, a slightly different situation prevails. The *cis*- and *trans*-1,4-polyisoprene structures are not resolved, and they scatter at 1662 cm^{-1}, but the 3,4-polyisoprene structure scatters at 1641 cm^{-1}, and the 1,2-vinyl structure scatters at 1639 cm^{-1} (*30*).

With its unique sensitivity to highly polarizable bonds, Raman spectroscopy offers considerable potential for the study of accelerated sulfur vulcanization (*31*). Although conventional Raman spectroscopy is limited by the fluorescence problem, FT Raman spectroscopy should be particularly useful in this situation. The Raman spectrum of a mixture of thiuram sulfides and sulfur indicates that tetrasulfides and disulfides are easily identified. As a result, it becomes feasible to detect and measure the amount of mono-, di-, and polysulfidic cross-links without resorting to chemical modification of the network. The detection of cyclic sulfides, conjugation, and pendant side groups along the main chain is also possible. For the vulcanizates of *cis*-polybutadiene, the lines at 1633, 1187, 734, and 720 cm^{-1} are associated with dialkenyl sulfide cross-links. The lines occurring at 440 and 505 cm^{-1} are assigned to polysulfidic and disulfidic structures. The 635-, 690-, and 708-cm^{-1} lines are associated with cyclic sulfides. Pendant side groups derived from the accelerator thiuram (TMTD) produce lines at 1142 and 577 cm^{-1} (*32*).

Raman spectroscopy has also been used to study urea–formaldehyde resins (*33*). The important moieties in the cured resins are $-CH_2OH$ and $-CH_2OCH_2-$. The Raman line occurs at 1450 cm^{-1} for the $-CH_2OH$ moiety and at 1435 cm^{-1} for the $-CH_2OCH_2-$ moiety. The concentration of $-NCH_2N-$ groups increases with cure relative to the concentration of $-CH_2O-$ groups. These results support the acid-catalyzed mechanism in which methyl groups are lost by reaction with $-NH_2$ to form $-NCH_2OCH_2N-$ linkages. This latter type of $-CH_2O-$ moiety is also removed by the loss of formaldehyde to yield additional $-NCH_2N-$.

Conformation of Polymer Chains in the Solid State

One of the important applications of Raman spectroscopy (with IR spectroscopy) is the determination of the configuration and conformation of polymer chains in the solid state. As will be discussed in Chapter 6, high-resolution NMR spectroscopy of polymers in solution is a powerful technique for determining stereoregularity, but it is not useful for solids. The applications of NMR spectroscopy to solids will be discussed in Chapter 7.

When molecules possess symmetry, this symmetry restricts the types of vibrational modes that can be observed in the IR and Raman spectra (*see* Figure 5.10). For polymers with C–C backbones, the Raman spectra are dominated by the strong lines arising from the C–C skeletal modes. These skeletal modes are sensitive to the conformation because they are highly coupled, and any change in the conformation will vary the coupling and shift the frequencies accordingly. Additionally, for the planar 2_1 and 3_1 helices, differences in selection rules for Raman and IR spectra allow a determination of the conformation. For helical conformations with a pitch higher than that of a 3_1 helix, the selection rules do not change.

When a polymer chain coils into a helix, a characteristic splitting of nearly all of the IR and Raman modes is observed. Theory offers an explanation of these observations. All monosubstituted vinyl helical polymers have [p, π] vibrational modes, which are termed the A modes, and [d, σ] modes, which are termed the E modes. Theoretically there are two different E modes for each helix, but they are degenerate in frequency and do not appear separately. The frequencies of the helical A and E modes depend on the helix angle. As was demonstrated in Chapter 2, the normal modes of a helical chain can be represented by a dispersion curve of frequency vs.. a phase angle, θ. Each pitch of the helix will have a characteristic dispersion curve, and similar motions will have similar energies. On the dispersion curve, the A-mode vibrations occur at $\theta = 0$, and the

E-mode vibrations occur at $\theta = \psi$, where ψ is the helix angle. Thus for a polymer with the same chemical repeat units, differences in conformations will be reflected in the A modes because the different helical conformations will depend on only energy considerations, and the phase-angle difference is the same. The E-mode shifts from one helical form to another depend on the energy differences and on ψ differences corresponding to the different helix angles. The helical modes should be slightly more sensitive to the changes in conformation.

As indicated in Chapter 2, some branches of the dispersion curve of the helical conformations will be insensitive to the environment of the chemical groups. These are referred to as the *characteristic* modes, and they correspond to the uncoupled vibrational modes. These characteristic modes have a very flat dispersion curve, and their frequencies do not depend on the phase angle (i.e., the conformation). Other branches of the dispersion curve are sensitive to the phase angle, and the frequencies of the modes depend on the pitch of the helix of the repeating units. Generally, the observed spectra will have modes that have the same frequency (characteristic modes) regardless of the helix type, as well as modes that have different frequency positions because of the helix form. These latter modes are useful for characterizing the helical conformation of a polymer in the solid state.

Raman spectroscopy will not supplant X-ray diffraction for the determination of the conformation of a polymer in the solid state, but Raman spectroscopy can be useful for those systems that are unstable and cannot be satisfactorily oriented to give a proper X-ray pattern. Polybutene is an example where Raman spectroscopy has been of value for the determination of the conformation of the polymer in the solid state.

1-Polybutene exists in at least three crystalline forms. Form I has a hexagonal unit cell with 3_1 helices. Form II is prepared by casting a film from different solvents, but it will transform slowly and irreversibly into form I at room temperature, so X-ray diffraction patterns of form II are difficult to obtain. Form III of 1-polybutene transforms to form II upon heating, and then spontaneously transforms to form I. The Raman spectra of forms I, II, and III are substantially different because of differences in the conformations (*34*). Detailed analysis (including the dreaded use of normal coordinate analysis) can establish the conformation of 1-polybutene from the Raman spectra.

Conformation of Polymers in Solution and in the Melt

Spectroscopic studies of polymers in solution are of interest primarily to relate the structure in solution to other solution properties. In many cases, the conformation of the polymer changes upon dissolution or melting, or undergoes transformation with changes in the pH, ionic strength, or salt content of the solution. The preferred solvent for Raman spectroscopy is water because the scattering of water is very weak except for the regions of 1650 and 3600 cm^{-1}. As a consequence, Raman spectroscopy is quite useful for studying the secondary and tertiary structures of biological molecules, including carbohydrates, proteins, and nucleic acids. For synthetic polymers, the spectral results are less dramatic but nevertheless revealing.

The Raman spectra of polyethylene glycol (PEG) in aqueous and chloroform solutions are shown in Figure 5.15. Comparison of the Raman spectrum of PEG in aqueous solution with that of PEG in the melt indicates that the changes that occur on dissolution in water are considerably less dramatic than the changes

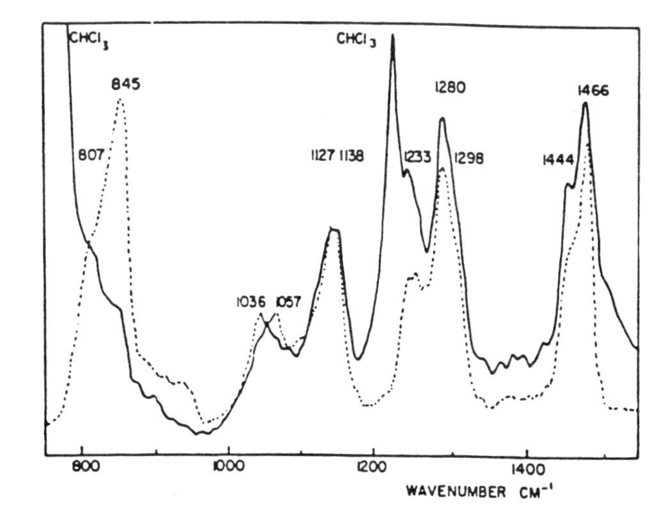

Figure 5.15. Raman spectra of PEG in aqueous solution (dashed line) and in chloroform (solid line). (Reproduced with permission from reference 35. Copyright 1970 John Wiley & Sons, Inc.)

observed upon melting. On the other hand, the spectrum of PEG in chloroform is very similar to the spectrum of PEG in the melt. The half-widths of the Raman lines of the spectrum of the aqueous solution are considerably narrower than those of the spectrum of the molten state, a result indicating fewer energy states are available to the molecule in the aqueous solution than in the melt (*35*).

References

1. Koenig, J. L. *Appl. Spectrosc. Rev.* **1971,** *4,* 233.
2. Zimba, C. *Spectroscopy* **1988,** *3,* 8.
3. Raman, C. V.; Krishnan, K. S. *Nature* **1928,** *121,* 501.
4. Bower, D. I.; Ward, I. M. *Polymer* **1982,** *23,* 645.
5. Snyder, R. G. *J. Mol. Spectrosc.* **1971,** *37,* 353.
6. Schlotter, N. E.; Rabolt, J. F. *Polymer* **1984,** *25,* 165.
7. Koenig, J. L.; Angood, A. C. *J. Polym. Sci.* **1970,** *A8,* 1787.
8. Willis, H. A. In *Proceedings of the 5th European Symposium on Polymer Spectroscopy;* Hummel, D. O., Ed.; Verlag Chemie: Weinheim, 1979; p 15.
9. Carey, P. R.; *Biological Applications of Raman and Resonance Raman Spectroscopies;* Academic: New York, 1982; p 51.
10. Edsall, J. T. *Reflections by an Emminent Chemist;*
11. Angood, A. C.; Koenig, J. L. *J. Macromol. Sci. (Phys.)* **1969,** *B3,* 323.
12. Koenig, J. L.; Druesdow, D. *J. Polym. Sci.* **1969,** *A-2,* 1075.
13. Champion, A.; Woodruff, W. H. *Anal. Chem.* **1987,** *59,* 1299A.
14. Iwata, K.; Hamaguchi, H.; Tasumi, M. *Appl. Spectrosc.* **1988,** *42,* 12.
15. Hirschfeld, T.; Chase, B. *Appl. Spectrosc.* **1986,** *40,* 133.
16. Zimba, C. G.; Hallmark, V. M.; Swalen, J. D.; Rabolt, J. F. *Appl. Spectrosc.* **1987,** *41,* 722.
17. Rabolt, J.; et al *Appl Spectrosc.* **1987,** *41,* 722.
18. Lewis, E. N.; Kalasinsky, V. F.; Levin, I. W. *Appl. Spectrosc.* **1989,** *43,* 156.
19. Lewis, E. N.; Kalasinsky, V. F.; Levin, I. W. *Anal. Chem.* **1988,** *60,* 2658.
20. Everall, N. J.; Howard, J. *Appl. Spectrosc.* **1984,** *43,* 778.
21. Dudik, J. M.; Johnson, C. R.; Asher, S. A. *J. Phys. Chem.* **1985,** *89,* 3805.
22. Asher, S. A.; Johnson, C. R. *Science* **1984,** *225,* 311.
23. Asher, S. A. *Anal. Chem.* **1984,** *56,* 720.
24. Fleischmann, M.; Hendra, P.; Mcquillan, A. *Chem. Phys. Lett.* **1974,** *26,* 163.
25. *Surface Enhanced Raman Scattering;* Chang, R.; Furtak, F., Eds.; Plenum Press: New York, 1982.
26. Angel, S. M.; Katz, L. F.; Archibal, D. D.; Lin, L. T.; Honigs, D. E. *Appl. Spectrosc.* **1988,** *42,* 1327.
27. Angel, S. M.; Myrick, M. L. *Anal. Chem.* **1989,** *61,* 1648.
28. Dollish, F. R.; Fateley, W. G.; Bentley, F. F. *Characteristic Raman Frequencies of Organic Compounds;* Wiley-Interscience: New York, 1974.
29. Cornell, S. W.; Koenig, J. L. *Rubber Chem. Technol.* **1970,** *43,* 313.
30. Cornell, S. W.; Koenig, J. L. *Rubber Chem. Technol.* **1970,** *43,* 322.
31. Coleman, M. M.; Shelton, J. R.; Koenig, J. L. *Rubber Chem. Technol.* **1973,** *46,* 938.
32. Koenig, J. L.; Coleman, M. M.; Shelton, J. R.; Starmer, P. H. *Rubber Chem. Technol.* **1971,** *44,* 938.
33. Hill, C. G., Jr.; Hedren, A. M.; Meyers, G. E.; Koutsky, J. A. *J. Appl. Polym. Sci.* **1984,** *28,* 2749.
34. Cornell, S. W.; Koenig, J. L. *J. Polym. Sci.* **1969,** *A7,* 1965.
35. Koenig, J. L.; Angood, A. C. *J. Polym. Sci.* **1970,** *A8,* 1787.

Suggested Reading

Bower, D. I.; Maddams, W. F. *The Vibrational Spectroscopy of Polymers*; Cambridge University Press: Cambridge, 1989.

Carey, P. R. *Biological Applications of Raman and Resonance Raman Spectroscopies*; Academic: New York, 1982.

Dollish, F. R.; Fateley, W. G.; Bentley, F. F. *Characteristic Raman Frequencies of Organic Compounds*; Wiley-Interscience: New York, 1974.

Fateley, W. G.; Dollish, F. R.; McDevitt, N. T.; Bentley, F. F. *Infrared and Raman Selection Rules for Molecular and Lattice Vibrations: The Correlation Method*; Wiley: New York, 1972.

Freeman, S. K. *Applications of Laser Raman Spectroscopy*; Wiley-Interscience: New York, 1974.

Grasselli, J. G.; Snavely, M. K.; Bulkin, B. J. *Chemical Applications of Raman Spectroscopy*; John Wiley and Sons: New York, 1981.

Koenig, J. L. "Raman Scattering of Synthetic Polymers: A Review," *Appl. Spectrosc. Rev.*, **1971**, *4(2)*, 233-306.

Koenig, J. L. "Raman Spectroscopy of Biological Molecules: A Review," *J. Polym. Sci., Part D.*, **1971**, *4(2)*, 233-306.

Koenig, J. L. "Raman Spectroscopy of Biological Molecules: A Review," *J. Polym. Sci., Part D.*, **1972**, *6*, 59-177.

Sushchinskii, M. M. *Raman Spectra of Molecules and Crystals*; Keter: New York, 1972.

6

High-Resolution NMR Spectroscopy of Polymers in Solution

We can think of NMR spectroscopy as a means of interrogating a large number of spies that have been stationed within a chemical system. They are not foreign intruders. They belong there. But they can be made to reveal the nature of their surroundings. NMR messages are complicated by the fact that the language spoken by nuclei is an alien language. We have to learn how to decode these alien signals.

—James N. Shoolery (*1*)

Elements of NMR Spectroscopy

Nuclear magnetic resonance (NMR) spectroscopy is the experimental study of the energy levels of certain atomic nuclei of molecules in a magnetic field (*2, 3*). In a magnetic field, the magnetic properties of the nuclei dominate. All atomic nuclei possess a characteristic known as *nuclear spin*. Certain atomic nuclei that have an odd number of either protons or neutrons possess a nonzero spin. Examples are the principal isotopes of hydrogen, sodium, and phosphorus. As the positively charged nucleus spins on its axis, the moving charge creates a magnetic moment, μ, that tends to align in a magnetic field. Actually, the thermal motion of the molecule makes the magnetic moment wobble. The torque makes the magnetic moment act like a child's top, and it precesses about the axis of the external magnetic field at a frequency that depends on the strength of the field. Quantum mechanics indicates that a nucleus with spin I_s is characterized by an angular momentum with a spin quantum number I and is related to the magnetic moment, μ, by:

$$\mu = \gamma \bar{h} I \tag{6.1}$$

where γ is the gyromagnetic ratio (rad/G-s), and $\bar{h}$ is Planck's constant divided by 2π.

Subatomic particles spin about a theoretical kind of axis like a spinning top. One big difference between a spinning top and a spinning particle, however, is that a top can spin either faster or slower, but a subatomic particle always spins at exactly the same rate.

—Gary Zukak (*4*)

Nuclei with a spin number of zero ($I = 0$) are not observable in NMR experiments. There are 118 nuclei that have been studied by NMR (*5*), but the nuclei of primary interest to polymer chemists are the proton (^{1}H, $I = 1/2$), deuteron (^{2}H, $I = 1$), ^{13}C (I = 1/2) (for the more common ^{12}C isotope, $I = 0$), and ^{19}F ($I = 1/2$).

The two factors that determine the sensitivity and utility of a nucleus for NMR spectroscopy are the natural abundance and the gyromagnetic ratio. For example, the ^{1}H isotope is 100% naturally abundant and has a high gyromagnetic ratio, and this makes ^{1}H the most sensitive nucleus for NMR study. On the other hand, the ^{13}C nucleus is present in natural abundance at a level of only 1.1% and has a gyromagnetic ratio that is 1/4 that of hydrogen to make it 1.6×10^{-2} less sensitive than hydrogen for NMR study. Properties of nuclei that are useful for the study of polymers by NMR spectroscopy are listed in Table 6.1.

The nuclear spin generates a small magnetic field,

1904—4/92/0137$07.50/1

Table 6.1. NMR Properties of Nuclei of Interest in Polymers

Isotope	*Natural Abundance (%)*	*Resonance Frequency (MHz) for Field of 10 kG (1 T)*	*Relative Sensitivity for Equal Number of Nuclei at Constant* B_0	*Magnetic Moment (units of nuclear magnetons)*	*Spin I in Multiples of* $\hbar$
^{1}H	99.9844	42.577	1.000	2.79270	1/2
^{2}H (D)	0.0156	6.536	0.00964	0.85738	1
^{13}C	1.108	10.705	0.0159	0.70216	1/2
^{14}N	99.635	3.076	0.00101	0.40357	1
^{15}N	0.365	4.315	0.00104	− 0.28304	1/2
^{17}O	0.037	5.772	0.0291	− 1.8930	5/2
^{19}F	100	40.055	0.834	2.6273	1/2
^{29}Si	4.70	8.460	0.0785	− 0.55477	1/2
^{31}P	100	17.235	0.0664	1.1305	1/2
^{35}Cl	75.4	4.172	0.00471	0.82089	3/2
^{37}Cl	24.6	3.472	0.00272	0.68329	3/2

SOURCE: Reproduced from reference 40. Copyright 1980 American Chemical Society.

and in the absence of an applied magnetic field, the orientation of these dipoles is random (Figure 6.1a). However, when a sample is placed in a homogenous magnetic field, the dipoles will align with the lines of induction or the force of the applied magnetic field (Figure 6.1b). Because molecules are constantly in thermal motion, and because the molecules interact with each other, the thermal motion will cause the magnetic moments of most of the protons to point randomly. However, the average or net magnetization will be preferentially aligned along the magnetic field (Figure 6.1c). The average of all these magnetic moments is called the *thermal equilibrium magnetization*, M_o, and is given by

$$M_o = \left(\frac{N_1}{3kT}\right) \gamma^2\hbar^2 I(I+1) H_o = \chi_o H_o \quad (6.2)$$

where N_1 is the number of spins per unit volume, k is Boltzmann's constant, T is temperature, χ_o is the static magnetic susceptibility, and H_o is the applied magnetic field. The net magnetization is approximately 1.4 ppm, which means that there are only 1.4 excess aligned protons per million protons. The intensity of the NMR signal is proportional to M_o and is directly related to the magnitude of the applied field. The *macroscopic nuclear magnetization*, **M**, is the vector sum along H_o of the individual spin moments.

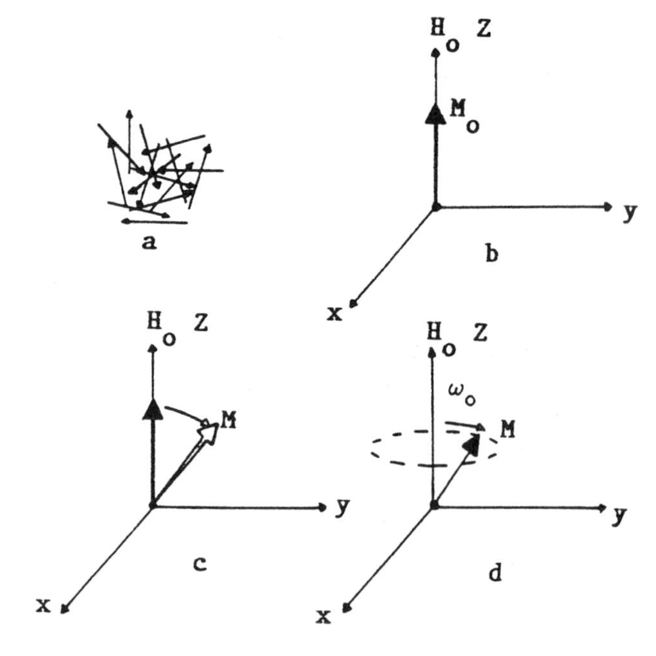

Figure 6.1. The nuclear spin generates a small magnetic field, and in the absence of an applied magnetic field, the orientation of these dipoles is random (a). When a sample is placed in a homogenous magnetic field, the dipoles will align themselves with the lines of induction or force of the applied magnetic field (b). However, thermal motion does not allow the alignment to remain, and the average alignment is at a small angle to the magnetic field (c). The magnetic moment precesses about the magnetic field at the Larmor frequency (d).

Every subatomic particle has a fixed, definite, and known angular momentum, but nothing is spinning! If you don't understand it, don't worry. Physicists don't understand these words either. They just use them.

– Gary Zukav (*4*)

Because the nuclei have angular momenta and thermal motions, they will not align completely parallel to the field, and the torque from the applied field will cause the magnetic moments to precess about the field direction with a characteristic angular frequency. The precession frequency is known as the *Larmor frequency* (Figure 6.1d) and is proportional to the applied magnetic field, H_o. The Larmor frequency, ω_o, is given by

$$\omega_o = \gamma H_o \qquad (6.3)$$

in radians per second.

The nuclei with a spin of 1/2 have two energy states: aligned with the field (lower energy) and aligned against the field (higher energy). The energy levels of the nuclei with a spin of 1/2 in the presence of a magnetic field are given by

$$\Delta E = \gamma \hbar H_o \qquad (6.4)$$

The energy levels that occur in the presence of a static magnetic field appear as shown in Figure 6.2.

Spin is quantized just like energy and charge. It comes in chunks. Like charge, all of the chunks are the same size. In other words, when a spinning top slows down, its rotation does not diminish smoothly and continuously, but in a series of tiny steps.

– Gary Zukav (*4*)

All nuclei that have spins greater than 1/2 have asymmetrical charge distributions and multiple energy levels. In general, there are $2I + 1$ nuclear energy states that correspond to magnetic quantum number (m) values of I, $(I - 1)$, $(I - 2)$, $\cdots$, $-I$. The energy difference falls in the radiofrequency range ($10^7 - 10^8$ Hz). The nuclei can be induced into a higher energy state by absorption of a radiofrequency pulse of the appropriate frequency and strength. This radiofrequency, rf, is generated by using an alternating current of variable frequency that is passed through a coil whose axis lies in the xy plane perpendicular to the applied magnetic field.

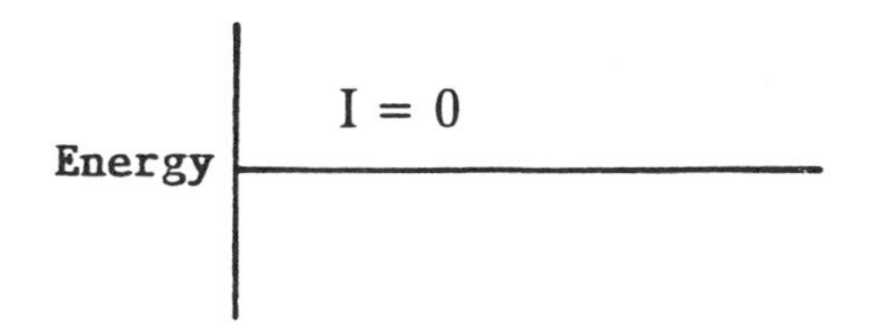

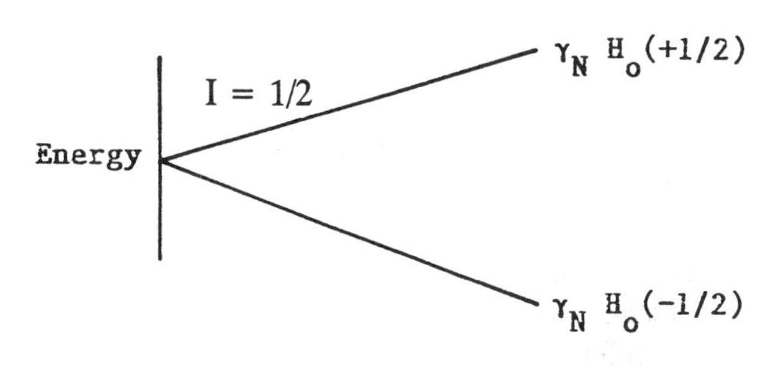

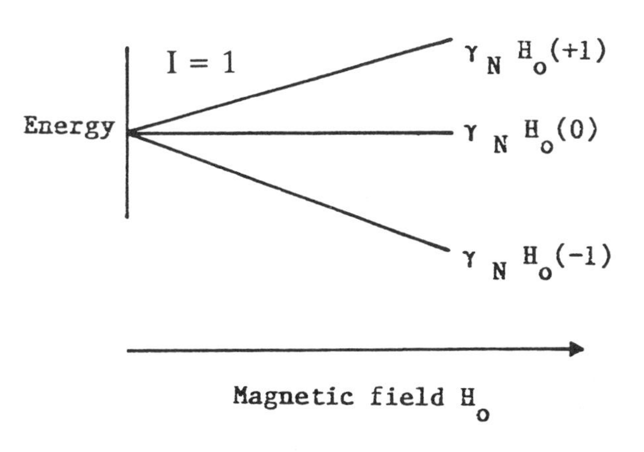

Figure 6.2. Quantized energies of nuclei in a magnetic field.

This current gives rise to an oscillating magnetic field that is also perpendicular to the applied field. As the frequency of the oscillating field is varied, there is a point at which it exactly matches the precessional frequency of the nuclei, and energy from the radiofrequency is absorbed by the nuclei. When this absorption occurs, the system is said to be *in resonance*. This absorption of energy is the resonance phenomenon. In resonance, energy is transferred from the rf radiation to the nuclei, which causes a change in the spin orientation of the nuclei or, in other words, a change in the spin populations in the various energy levels. The rf field must oscillate with a frequency, ν_o, given by

$$\Delta E = \bar{h}\nu_o \tag{6.5}$$

Expressed in terms of the applied magnetic field, the frequency of the rf field in Hertz is

$$\nu_o = \gamma H_o \tag{6.6}$$

This equation describes the resonance condition for the NMR experiment and shows that the "resonant" radiofrequency must correspond to the Larmor frequency for the applied magnetic field. Because the gyromagnetic ratio is different for each nuclear isotope, different nuclei resonate at widely different frequencies in a given applied magnetic field. In a magnetic field of 1 T, ^{1}H has a frequency of 42.5759 MHz, and ^{13}C has a frequency of 10.705 MHz.

The rf field, H_1, provides the nuclei with the quanta of energy necessary to move to the higher energy levels. The need for a time-dependent H_1 results from the fact that H_1 is active only if it can "chase" the spins at the frequency at which they are rotating. When the proper rf pulse is applied, the nuclei begin to precess in phase. As the rf pulse continues, more and more nuclei fall in line and precess in phase with the rf magnetic field. After a certain period of rf exposure, the magnetization becomes *coherent*. The protons then precess as a coherent group in the *xy* plane rather than as randomly phased individuals in the *z* direction. Just as the synchronization of randomly phased light waves forms a coherent laser, synchronization of radiowaves produces a coherent rf signal that is detectable over the random background.

Transfer of rf energy to the sample through resonance is indicated by the flipping of nuclei to the high-energy (antiparallel) cone. As coherence is established by the rf pulse, the net magnetization vector ($\mathbf{M_o}$) angles away from the $+z$ axis and precesses about this axis at the resonant frequency (Figure 6.3a). The amount of rf energy required to cause $\mathbf{M_o}$ to rotate from the *z* axis to the *xy* plane is called the 90° rf pulse.

After the rf field is turned off, the magnetization returns to its original equilibrium value by emitting energy or by transferring energy to the surrounding molecules. However, the probability of spontaneous emission depends on the frequency to the third power, and at radio frequencies this term is too small to be significant. Thus, all NMR transitions are *stimulated*. The stimulated process by which the energy is lost to the environment is called *relaxation*.

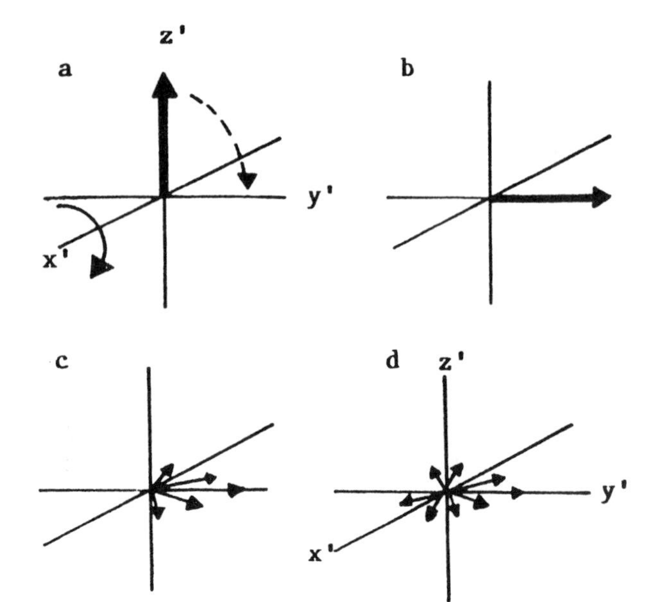

Figure 6.3. Effect of the 90° rf pulse on the magnetization vector.

The Pulsed NMR Experiment

In order to understand the NMR experiment, the magnetization vector, $\mathbf{M_o}$, must be resolved into a component, M_z, along the *z* axis, and a component, M_{xy}, perpendicular to the field. The M_z component is called the *longitudinal magnetization*, and the M_{xy} component is termed the *transverse magnetization*. At thermal equilibrium, M_z is equal to M_o (the initial magnetization), and M_{xy} is zero. When the sample is irradiated with the rf pulse, M_z decreases, and M_{xy} increases. The NMR signal is detected in the *xy* plane. The 90° rf pulse, also known as the *read pulse*, is generated by passing an oscillating electric current through the transmitting coil that surrounds the sample. The field generated by the rf pulse is perpendicular to the applied magnetic field. A short pulse of rf radiation of t seconds duration is equivalent to the simultaneous excitation of all the frequencies in the range $rf_o \pm t-1$, where rf_o is termed the *carrier frequency* of the nuclei under study. Because the pulse duration is only microseconds, the bandwidth is sufficiently large to excite all of the resonant nuclei in the sample.

As the nuclei receive the pulse, they come into phase with each other and become a bundle of precessing nuclei that has a net magnetization in the xy plane (Figure 6.3b). Following the pulse, the nuclear spins initially precess about the static field and are coherently in phase. Because of the inhomogeneities in the static magnetic field and the local magnetic fields of adjacent molecules, the nuclei precess at different frequencies and quickly lose their coherence and become out of phase (Figure 6.3c, d). The precession phase of the nuclei is not totally random, and a net component of magnetization is generated in the xy plane. This rotating M_{xy} component induces a voltage in the receiving coil that surrounds the sample.

The observed signal in the time domain is called a *free induction decay* (FID) because it is measured "free" of a driving rf field. It is a "decaying" voltage because the net nuclear magnetic field decays as the nuclei return to their equilibrium value (Figure 6.4a). The term "induction" is used because a current is induced in the receiver coil. This signal is collected as a function of time, amplified, and processed to give the detected NMR signal.

The FID signal for the nuclei can be expressed as

$$\text{FID} = A \cos \omega_o t \exp(-t/T_2^*) \qquad (6.7)$$

where A is the amplitude. The term T_2^* is the experimental *transverse relaxation time* and includes contributions resulting from the inhomogeneity of the static magnetic field as well as from the T_2, which is the *spin–spin relaxation time*. The FID observed in pulsed NMR spectroscopy can be described as a linear combination of damped oscillations characterized by a complex frequency. FID signals in NMR data usually consist of several groups of wave elements distributed within narrow-frequency bands. The spectrum in the frequency domain is obtained by taking the Fourier transform of this response:

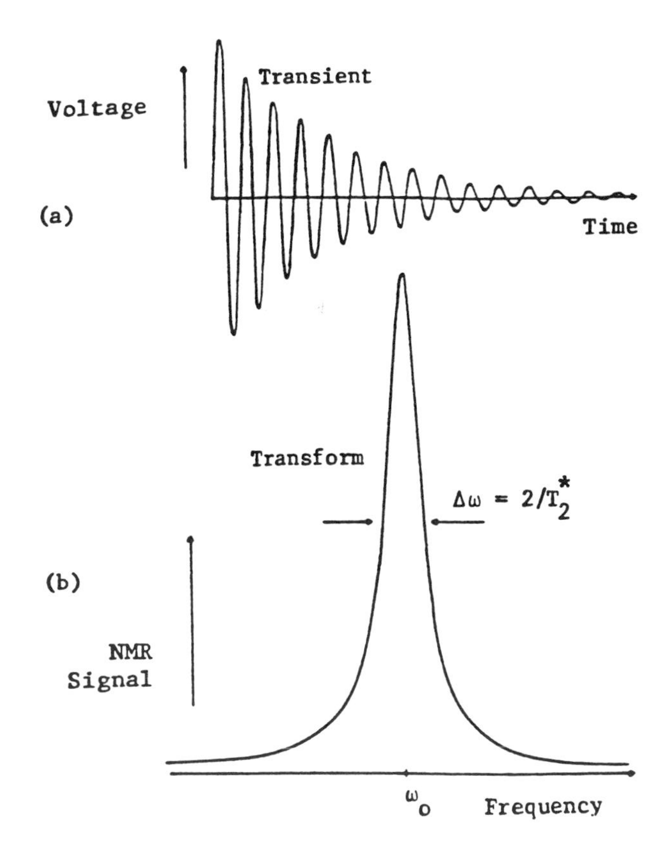

Figure 6.4. Free induction decay as a function of time (a) and the Lorentzian line obtained by the Fourier transform of the FID (b).

$$F(\omega) = \gamma_o A \cos \omega_o t \exp(-t/T_2^*) \exp(i\omega t)\, dt \qquad (6.8)$$

where ω is the Larmor frequency and i is the square root of -1. Integration yields

$$F(\omega) = \left(\frac{A}{2}\right) \frac{T_2^*}{1 + T_2^{*2}(\omega - \omega_o)^2} \qquad (6.9)$$

This function is termed a *Lorentzian line shape* (Figure 6.4b) and is the NMR spectral result for a single nuclear type. For actual systems, a number of different nuclear types will be present, and the FID will be quite complicated (*6*). The Fourier transform process separates the different contributions and produces the proper spectrum in the frequency domain (*7*).

To increase the signal-to-noise ratio, the spectra are signal-averaged by collecting a large number of pulses. The NMR signal adds coherently, whereas the random noise will add only as the square root of the number of scans accumulated. Thus, the improvement in the signal-to-noise ratio is proportional to the square root of the number of transients accumulated. During this summation process, the interval between pulses (the PD or pulse delay) must be sufficient for the system to return to equilibrium; otherwise, the signal can become saturated. The time required to return to equilibrium is the *longitudinal relaxation time* (which will be discussed later) and the general practice is to make PD $= 5T_1$ to ensure a return (~98%) to equilibrium (*7*).

Nuclear Spin Relaxation

After the resonance rf pulse has been applied, a higher energy excited state exists. This higher energy level corresponds to the nuclei that are elevated to the antiparallel position. To return to the equilibrium or ground state, this excess energy is passed to the surroundings by stimulated emission, and some of the antiparallel nuclei return to the parallel or low-energy state. This decay of the magnetization proceeds to equilibrium in an exponential manner. The rate of this process is determined by two external factors or relaxations: the spin – lattice relaxation time, T_1, and the spin – spin relaxation time, T_2.

The Spin – Lattice Relaxation Time

The rate of return to equilibrium along the static field (z axis) depends on the rate of exchange of energy between the nuclei and the environment or, *lattice* in NMR terminology (Figure 6.1). The longitudinal magnetic relaxation time, T_1, is often called the *spin – lattice relaxation time* and depends on the effectiveness of energy transfer from the excited nuclei to the lattice. Just as resonance requires stimulation by rf energy at a particular frequency, return to the ground state is enhanced by the presence of rf energy in the lattice at the resonant frequency (*stimulated emission*). This rf energy comes from fluctuating magnetic nuclei and electrons in the lattice. Magnetic fluctuations that are at or near the resonant Larmor frequency generate rf magnetic fields that stimulate the transition from the high-energy to the low-energy states. This coupling mechanism allows the energy initially added by the rf pulse to be dissipated to the lattice.

The T_1 is specific for each molecule and its environment and gives information about the molecular dynamics of the magnetic moments and the molecules around them. When the nuclear magnetization rotates at the Larmor frequency, local fluctuating magnetic fields (perpendicular to the rf axis) lead to a decay of the longitudinal spin magnetization. The local fluctuating magnetic fields are generated primarily by molecular motion. Thus when the lattice exhibits considerable molecular motion at the appropriate frequency, the energy coupling is effective, and the T_1 is short. On this basis, the T_1 is short for liquids where the molecular motion is extensive over a broad range of frequencies but is quite long for solids where the motion is restricted and occurs over a narrow range of frequencies. When a solid is heated to a viscous liquid, the T_1 shortens. Initially, thermal energy transfer improves as the molecular motion increases. An optimum is reached in a viscous liquid state, and T_1 is minimized. Additional heating results in molecular motions that are too rapid for efficient thermal energy transfer, and the T_1 value again increases.

The Spin – Spin Relaxation Time

The transverse magnetization decays by a different relaxation process than the spin – lattice process and involves the return to equilibrium by the loss of coherence of the transverse magnetization of the nuclei. The magnetization dephases in the xy plane because of differences in the individual precession frequencies of adjacent nuclei. The energy is transferred adiabatically between the nuclei as they are jostled between high- and low-energy positions. These internal differences introduce a "flip – flop" process and a loss of coherence. Because this internal adiabatic exchange occurs between two different spins, the transverse magnetic relaxation time, T_2, is also called the *spin – spin relaxation time*.

The T_2 gives information about the distribution of resonant frequencies and about the local fields experienced by the magnetic moments of the nuclei. Local fields are related to the structure and to the chemical nature of the environment around the nuclei. Because the local magnetic fields in liquids fluctuate very rapidly and can average to zero, the internal local fields are weak and yield long T_2s or narrow resonance lines. The atoms in solids are in nearly fixed positions, and the internal fields are significant and contribute to the rapid loss of coherence. Therefore, the T_2 in solids is very short (microseconds), and the resonance lines are very broad.

There is an additional contribution to T_2 that is not molecular in origin. The rate of decay of transverse magnetization is influenced by the homogeneity of the external static magnetic field. The inhomogeneity of the magnetic field across the sample causes the nuclei in different parts of the sample (called *isochromats*) to precess at different rates, which leads to additional irreversible destruction of the coherence. The experimentally observed T_2^* is the sum of the actual internal molecular T_2 and the contribution resulting from the nonuniformity of the magnetic field.

The effects of inhomogeneity of the magnetic field can be minimized by performing *spin–echo* experiments. Application of an rf pulse along the x axis rotates $\mathbf{M}_o$ about the x axis through an angle Θ given by

$$\Theta = \gamma H_1 t_p \quad (6.10)$$

where H_1 is the intensity of the rf pulse, and t_p is the length of the pulse (generally in the range of microseconds). Thus, the appropriate values of H_1 and t_p can be combined to make $\Theta = 90°$ and to rotate $\mathbf{M}_o$ through 90°. For ^{1}H, γ is 6.2577 kHz G^{-1}, and for an H_1 field of 50 G, t_p is 1.17 μs. The magnetization becomes colinear with the y axis and thus gives rise to a signal. Such a pulse is called a *90° pulse*. In similar fashion, a pulse that will invert the magnetization vector can be generated, and this pulse is called a *180°* or *inversion pulse*.

The principal source of external relaxation or coherence in the transverse plane for liquids and solutions is inhomogeneity in the static magnetic field. This source of loss can be eliminated by using a spin–echo sequence (Figure 6.5). After the 90° pulse, the individual nuclei precess around the z axis at different rates because of the inhomogeneity of the magnetic field and lose phase coherence (Figure 6.5c). A 180° rf pulse reverses the y components of the individual vectors so that after the pulse they start to come back into coherence (Figure 6.5d) to form an echo. Coherence is achieved when the individual vectors realign to produce maximum magnetization (Figure 6.5e). The echo is then detected as the FID with an increased signal.

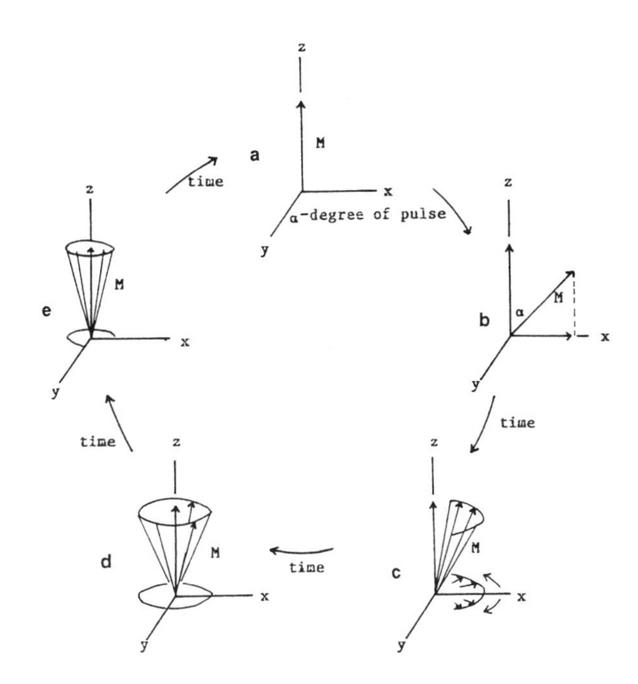

Figure 6.5. The spin–echo sequence.

Magnetic Interactions Between Nuclei

One of the most striking and characteristic features of nuclear magnetic resonance (n.m.r.) is that the spectra from solids are so very much broader than those from liquids. An excellent example is provided by water, whose proton n.m.r. linewidth at room temperature is about 0.1 Hz, while for ice at low temperatures it is about 105 Hz, six orders of magnitude broader.

–E. R. Andrew (*8*)

All nuclei are surrounded by magnetic dipoles that are associated with neighboring nuclei. The effective field, H_{eff}, at a particular nucleus may be either larger or smaller, depending on the relative orientation of these neighboring nuclei with respect to the static magnetic field. For solids, these orientations are fixed, and the nuclei may come into resonance over a broad range of frequencies (~1 – 100 kHz). In liquids or solutions, the effects of the magnetic dipoles are mutually canceled if the Brownian motion is sufficient to rapidly change the relative orientations of the molecules with respect to each other in a time shorter than the lifetime of a spin state, and thus to average the local field of the magnetic dipoles to zero. The width of the resonance line for samples of this kind decreases to a value (0.1 – 1 Hz) that depends on the inhomogeneity of the applied magnetic field. For polymers, the rotational and translational motions of the chain segments may be hindered because the motions of neighboring segments are correlated with each other (*9*). The result is a broadening of the resonance peaks up to 10 – 50 Hz, particularly if the chains are stiff and the motion is anisotropic.

Interactions Between Nuclei and Their Environments

When the NMR spectra of molecules in liquids and solutions are obtained, multiple resonances are observed (*10*). The first observed spectrum of ethyl alcohol, showing three different lines resulting from the three different types of protons, is shown in Figure 6.6. The ratio of the intensities is 3:2:1, suggesting the assignment on the basis of the number of protons to the CH_3, CH_2, and hydroxyl protons, respectively.

These multiple resonances arise from differences in the electronic environments of the nuclei. The source of the change in resonance is the shielding effect of the electron orbitals. The external field, H_o, induces orbital currents that produce a small local magnetic field, H_{loc}, which is opposed to H_o. The nucleus inside the electron cloud experiences an effective field, H_{eff}, that is slightly smaller than H_o, and therefore comes to resonance at a lower Larmor frequency. This effect is known as *electronic shielding*.

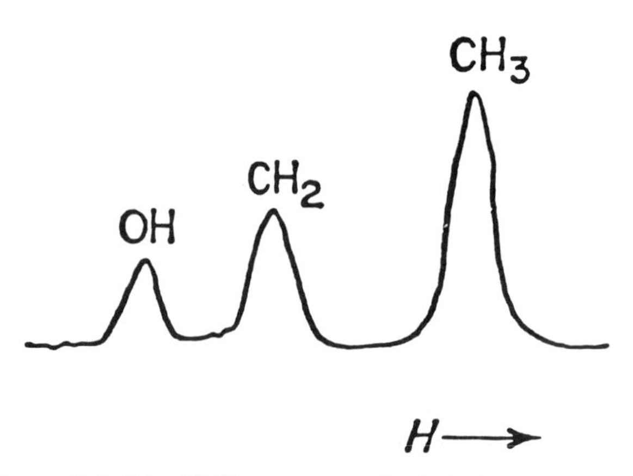

Figure 6.6. The NMR spectrum of ethyl alcohol at low magnetic field. This was the first NMR observation of different chemical shifts in the same molecule. (Reproduced with permission from reference 10. Copyright 1951 American Institute of Physics.)

The measured value of this electronic shielding depends on the relative motions of the molecules. When rapid molecular tumbling occurs, the induced secondary field is averaged with respect to the neighboring nuclei, and an isotropic-average chemical shift is observed. When motion is restricted, as with solids, the shielding depends not only on the electronic configuration but also on the orientation of the chemical bonds with respect to the applied field. In this case, the resonance peaks will shift with the alignment of the solid with the magnetic field. For powders, in which all possible orientations occur, broad resonance lines, often termed *powder patterns*, occur for each of the nuclei.

In the case of motional averaging in solutions or liquids, sharp, narrow resonances are observed in the NMR spectra. The appearance of different resonances for the observed nuclei in a given molecule arises from differences in the shielding effect, which is modified by the electron-withdrawing or electron-donating effects of neighboring groups. The electron-withdrawing or donating effect is proportional to the field strength, H_o, and the shielding field, H_s, is

$$H_s = -\sigma H_o \qquad (6.11)$$

or

$$H_{eff} = H_o + H_s = H_o(1 - \sigma) \qquad (6.12)$$

where σ is a proportionality factor called the *screening constant* and is characteristic of the shielding effect caused by the chemical surroundings of a given isotope. The absolute value of the effective magnetic field cannot be determined, so in order to obtain a point of reference for the different effective fields, a standard substance is used as an internal reference. Generally tetramethylsilane (TMS) is chosen as the standard reference for proton NMR spectroscopy. To compare H_{eff} of a proton in a specific chemical environment with H_{eff}(TMS), a dimensionless number, δ, is defined as a measure of the chemical shift:

$$\delta_{sample} = (\sigma_{TMS} - \sigma_{sample}) \times 10^6 \qquad (6.13)$$

The screening constants are found to be approximately 10^{-5} or less, and the factor 10^6 transforms δ to the units of parts per million (ppm). NMR spectra are plotted with decreasing frequency to the right. If one nucleus is more shielded than another, its signal will be shifted to lower frequency (or upfield). By convention, it will have a more negative chemical shift and will appear further toward the right side of the spectrum.

The benefit of chemical shifts is that nuclei exhibit specific resonances that depend on the chemical nature of the nuclei (*11*). Like characteristic group frequencies in IR and Raman spectroscopy, similar molecular groups have similar chemical shifts. Measurements of chemical shifts have generated catalogs of correlations between the magnitudes of these chemical shifts and the chemical nature of the resonating nuclei (*12*). Consequently, measurements of the chemical shifts of nuclei in different molecules allow a determination of the chemical nature of the nuclei involved. This method of chemical structure determination using ^{1}H, ^{19}F, ^{29}Si, or ^{13}C NMR spectroscopy is widely used to determine the chemical structures of small molecules and polymers (*13, 14*).

> *With the observation of three magnetically nonequivalent types of protons in ethyl alcohol, nuclear magnetic resonance began to be primarily the province of the chemist, as it is today.*
>
> —F. A. Bovey (*9*)

Through-Bond Interactions with Other Nuclei

> *Complications of proton and carbon NMR chemical shifts are a means of drawing correlations between different compounds. Spin coupling data, whether homo- or heteronuclear, extend the elucidation process a step further, providing connectivity information for structural fragments rather than just plausible identities of isolated atoms.*
>
> —Gary E. Martin (*15*)

At higher resolution (i.e., at higher magnetic fields), the proton resonances split into patterns reflecting the environments of the nuclei. When the resonating homonuclear nuclei are in the same molecule, spin–spin interactions via bonding electrons will occur. To first order, these scalar interactions are normally observed through three chemical bonds. The scalar interactions occur only when the interacting protons are magnetically and chemically distinguishable.

There is also an orientation-dependent dipole–dipole interaction that occurs through space between both bonded and nonbonded nuclei. When rapid motion occurs, as in liquids and solutions, the effects of these through-space dipolar nuclear interactions average zero. The *scalar* through-bond interactions are independent of the orientation of the chemical bonds with the magnetic field and are not averaged to zero by molecular motion. Thus, the scalar spin–spin interactions persist even in the presence of molecular motion.

The scalar bond-coupling pathway is used to transmit information concerning the spin states through the electron orbitals of the bonds to the neighboring nuclei. The result is a splitting of the energy levels to produce new resonances that appear as perturbations of the original energy levels.

Early ^{1}H NMR spectra of ethyl alcohol are shown in Figure 6.7. At a higher magnetic field, the resonances split into interesting patterns. The interpretation of the new lines is given in Figure 6.8, which shows the various orientations and energy levels of the methyl and methylene protons. The spectra obtained are a direct result of these different energy levels being available to the nuclei. The intensities of

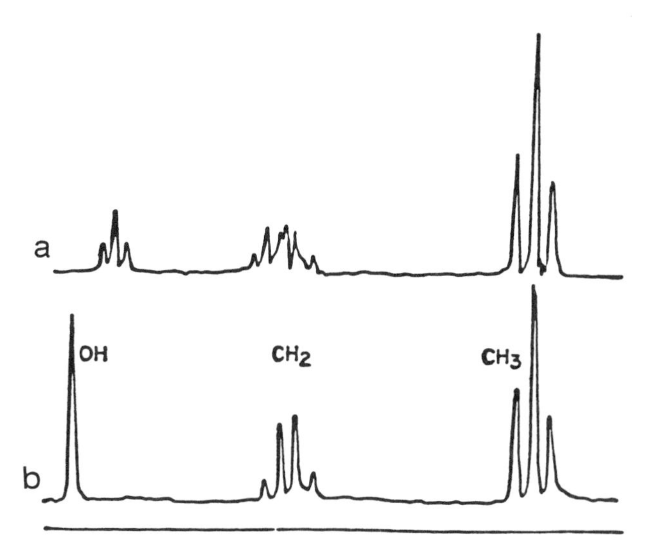

Figure 6.7. NMR spectra of pure dry ethyl alcohol (a) and slightly acidic alcohol (b). The slightly acid environment causes rapid exchange of the hydroxyl group between neighboring molecules. This exchange is sufficiently rapid to "average" the electronic environment. (Reproduced with permission from reference 10. Copyright 1951 American Institute of Physics.)

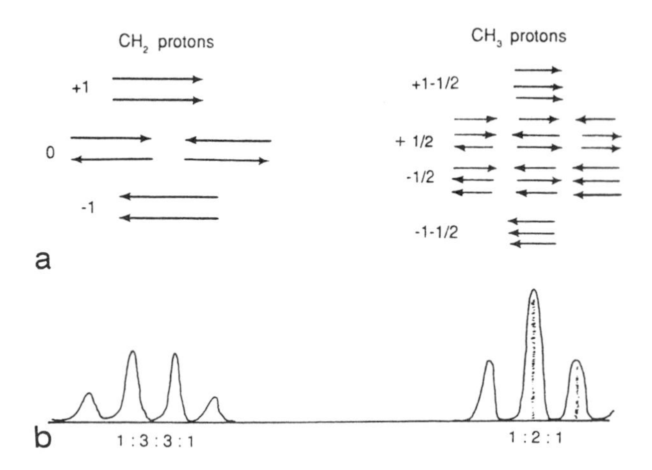

Figure 6.8. Spin – spin coupling diagrams showing the possible spin orientations of the methyl and methylene protons (a) and the corresponding spectrum with these functionalities as neighboring groups (b). (Reproduced with permission from reference 10. Copyright 1951 American Institute of Physics.)

the lines are a function of the number of available states for each energy level. The magnitude of the splitting for the methylene protons is the same as for the methyl protons, and the magnitude of the separation is termed the *J coupling constant*. When the ethyl alcohol is neat, there is no exchange of the slightly acidic protons on the hydroxyl group, and the methylene resonances are further split as a result of the presence of this proton on the hydroxyl group. The methyl protons are not affected (to first order) because they are more than three chemical bonds away from the hydroxyl protons, and the interaction effect is too small to observe at this level of resolution.

These results can be summarized in the following fashion. A nucleus, X, with a spin of 1/2 has spin states of either +1/2 or −1/2. A second adjacent homonucleus, A, in the same molecule will be able to recognize the two different spin states of the nucleus X through the scalar coupling of intervening bonds. Consequently, the coupled nucleus will resonate at two different frequencies corresponding to the upper- and lower-energy levels of the interacting nucleus. Because of the large number of nuclei present and the small differences in the populations of the two levels, the two coupled resonances of nucleus A will have nearly equal intensities, that is, 1:1. Similarly, two resonances of equal intensity, with a frequency separation corresponding to J_{AX}, will be observed for nucleus X. The magnitude of the frequency difference is determined by the strength of the scalar coupling and is specific for the chemical nature of the two interacting nuclei. Consequently, the magnitude of the coupling is given by a specific coupling constant J_{AX} (Hz), where A and X are the homonuclear coupled nuclei. Measurement of J_{AX} from the NMR spectrum is useful for determining the presence of the coupled A and X nuclei, that is, the chemical environment of the A and X nuclei (*16*).

When there are two nuclei (A_2) with magnetically equivalent surroundings in a molecule, the nuclei will influence a scalar-interacting nucleus, X, in this molecule via bonding electrons. The following spin-state combinations will be possible: (+1/2, +1/2), (+1/2, −1/2), (−1/2, +1/2), and (−1/2, −1/2). The net effect of the two energy-state combinations (+1/2, −1/2) and (−1/2, +1/2) is energetically the same. Hence, the resonance of the neighboring nucleus is split into three equally spaced peaks with a separation of J_{AX} and with an intensity ratio of 1:2:1 (Figure 6.8). The spin – spin coupling effect is illustrated for ethyl alcohol in Figure 6.8.

As the number of equivalent nuclei increases, the number of lines increases. In general, the multiplicity of the splitting is given by $2NI_s + 1$, where N is the number of magnetically equivalent nuclei with spin I_s (Figure 6.9). The intensity distributions of these lines

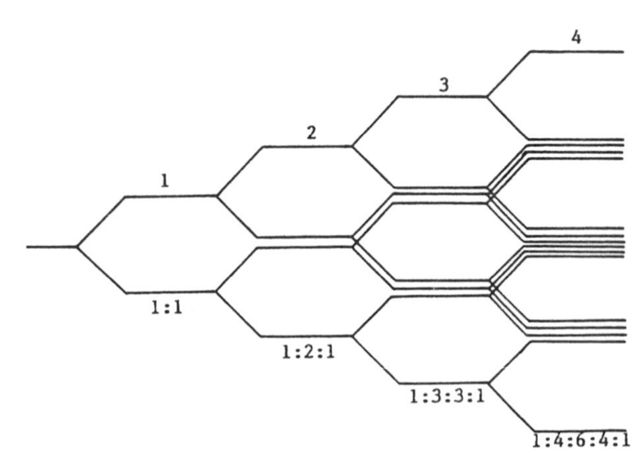

Figure 6.9. The spin – spin coupling effect for increasing numbers of equivalent nuclei showing the number of lines and their relative intensities.

are specific, and a general prediction for the intensities can be made on the basis of the coefficients of the binomial expansion $(a + b)^n$. The intensity ratio of the multiplet is given by the corresponding row of Pascal's triangle, with the first row corresponding to $N = 0$ (no interacting nucleus). These intensity ratios extend to five equivalent nuclei as represented in the sixth row in the triangle.

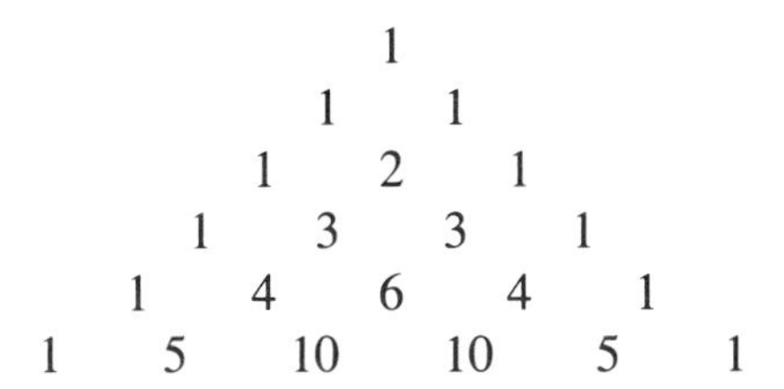

These rules are valid only when the chemical-shift differences between the coupled protons are much greater than the coupling constants. The spectra are then said to obey the *first-order approximation*. When the chemical shift differences become smaller, more complex spectra are obtained. Such spectra are said to be *second order* (*17*).

Vicinal coupling constants can vary in both sign and magnitude depending on a number of structural parameters, and particularly on the dihedral angle between the C – H protons. If one looks down the C – C bond between the carbon atoms to which two coupled vicinal protons are attached, the angle the protons make with one another is known as the *dihedral angle*. The dependence of the coupling constants of vicinal protons on the dihedral angle, ϕ, is given by the Karplus equation (Figure 6.10):

$$J = (8.5 \cos^2 \phi) - 0.28 \quad \text{for } \phi = 0-90° \qquad (6.14a)$$

$$J = (9.5 \cos^2 \phi) - 0.28 \quad \text{for } \phi = 90°-180° \qquad (6.14b)$$

This dependence of the coupling constants on bond angle is particularly useful for studying the stereochemistry in biopolymer systems (*18*).

The preceding discussion has particular relevance for ^{1}H NMR spectra, but similar effects are observed for other nuclei. For example, heteronuclear ^{1}H – ^{13}C scalar decouplings can be observed. However, high-resolution proton-decoupled ^{13}C spectra of liquids

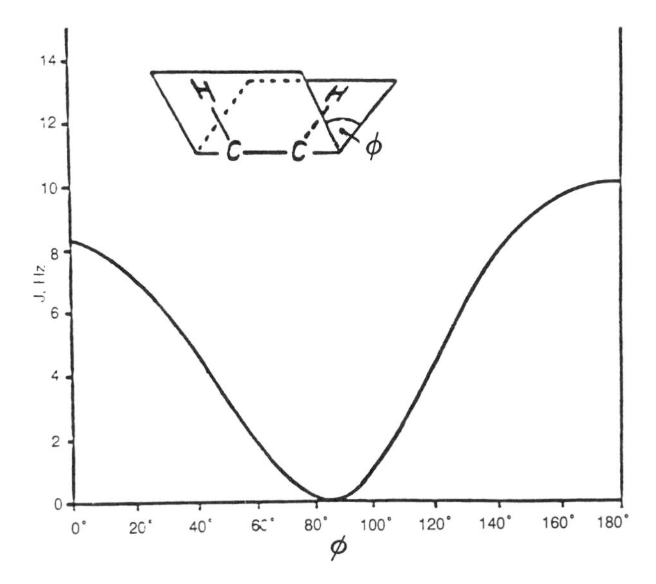

Figure 6.10. The dependence of the scalar coupling constants, J, *of vicinal protons on the dihedral angle* ϕ.

exhibit only single resonances for each magnetically distinct carbon because in this case the ^{1}H – ^{13}C scalar coupling is suppressed (decoupled) by a double-resonance experiment. In a double-resonance experiment, in addition to the carrier frequency of the ^{13}C, there is resonant irradiation of the protons that coherently signal-averages the scalar couplings to 0. This is called *scalar decoupling*. The ^{13}C resonance is measured at the carbon resonance frequency. Because the ^{1}H – ^{13}C scalar couplings are less than 200 Hz, an rf decoupling strength of ~0.7 G ($\gamma H_1/2\pi$ = 3 kHz) is sufficient.

In a manner similar to the spin – spin couplings apparent in the ^{1}H NMR spectra, ^{13}C – ^{13}C scalar couplings are present in ^{13}C NMR spectra. The one-bond carbon – carbon couplings ($^1J_{CC}$) are typically 30 – 40 Hz for C – C single bonds and 50 – 60 Hz for C=C double bonds. The one-bond C – C couplings are about an order of magnitude larger than typical two- and three-bond C – C couplings ($^2J_{CC}$, $^3J_{CC}$) (*16*). Unfortunately, because of a low natural abundance, most of the ^{13}C are surrounded by ^{12}C, and no coupling occurs. However, a small fraction (~10^{-4}) of ^{13}C – ^{13}C pairs exists at natural abundance for each C – C bond in the polymer. Moreover, these pairs are isolated from all other pairs and exhibit simple AX or AB patterns. However, the intensity of these ^{13}C – ^{13}C scalar coupling resonances is very weak, and it is generally necessary to enrich the sample with ^{13}C isotopes in order for these C – C couplings to be observed.

By using chemical shifts and spin coupling patterns, neighboring atoms in a molecule can be determined unambiguously. The coupling constant depends on the chemical structure of the moiety that contains the coupling nuclei. A system of nuclei coupling with each other is characterized by capital letters. Each set of magnetically equivalent nuclei is represented by one capital letter, with the number of the nuclei in the set as a subscript. Sets of magnetically equivalent nuclei whose differences in chemical shifts are comparable to the coupling constants to yield higher-order spectra are designated by A, B, C, ···. Other sets of magnetically equivalent nuclei with chemical shifts comparable to the coupling constants, but differing greatly from A, B, or C in chemical shift, are written X, Y, Z. Thus, the three aliphatic protons in styrene and vinyl chloride yield ABC spectra, and the four protons of 1,1-dichloroethane yield an A_3X spectrum.

Through-Space Interactions with Other Nuclei

In addition to the scalar couplings, there are also through-space interactions arising from dipolar interactions. These interactions are particularly important in the NMR spectroscopy of solids and will be discussed in Chapter 9.

Experimental Proton NMR Spectroscopy

Experimentally, proton (^1H) NMR spectroscopy involves the least demanding nuclei to study because of the nearly 100% natural abundance and high gyromagnetic ratio of the proton. Proton NMR spectroscopy suffers from the disadvantage of a small chemical-shift range of only 10 ppm. The proton spin – spin coupling constants yield useful information about the chemical environment of the resonanting nuclei. High-resolution proton NMR spectroscopy can furnish valuable information about the chemical structure, regiochemistry, stereochemistry, and conformation of a polymer (*9*). Model compounds are helpful for assigning peaks in the NMR spectra of polymers. Although data from the monomer and dimer models can be presumed to be transferable to the polymers, some unique features in the proton spectra of the polymers can give insight into the polymer structure.

Experimental ^{13}C NMR Spectroscopy

> *Thanks to the development of Fourier transform instruments with spectrum accumulation, carbon spectroscopy, despite the low natural abundance of ^{13}C (1.1%), has become a method of fairly high sensitivity, able to establish the presence of structural features at a level of less than one carbon per 10,000.*
>
> – Frank Bovey (*9*)

There are several aspects of ^{13}C NMR spectroscopy that make the sensitivity low. The low natural abundance of the ^{13}C nucleus (1.1%), the relatively small magnetic moment (1.59×10^{-2} for ^{13}C compared to 1.0 for ^1H), and the long ^{13}C relaxation times (limiting the frequency of pulsing or signal averaging) all contribute to a low NMR signal. The ^{13}C signal is 1/5700 as strong as the ^1H NMR signal, but modern FT NMR machines allow the detection of the ^{13}C in natural abundance.

The advantage of using ^{13}C NMR spectroscopy is that the ^{13}C nucleus is sensitive to subtle changes in its immediate electronic environment and insensitive to long-range effects, such as solvent effects and diamagnetic anisotropy of neighboring groups (*14*). The resonances of ^{13}C nuclei occur over a broad range of approximately 250 ppm (compared to 10 ppm for protons); generally the result is a separate resonance for every carbon atom in the molecule. Within a series of compounds (e.g., alkanes), the chemical shifts can be predicted with a high degree of accuracy. In compounds with high C/H ratios, much of the molecule is invisible in proton NMR spectroscopy. Quaternary carbons (those not bearing protons) are readily observed in ^{13}C spectra.

The low concentration of adjacent ^{13}C nuclei has the advantage that no homonuclear dipolar coupling occurs because the probability of two ^{13}C nuclei being adjacent in natural abundance is 1.1% of 1.1%, or 1 in 8264.

When protons are bonded to carbon, ^{13}C spectra show multiple lines resulting from spin – spin coupling to ^1H. This coupling is often of the same order (120 – 250 Hz) as the chemical-shift differences between the carbon nuclei, and complex spectra occur. As a result of spin – spin C – H interactions, a CH_3

signal is split into a quartet of peaks with intensities of 1:3:3:1, a CH_2 signal is split into a triplet with intensities of 1:2:1, a CH signal is split into a doublet with intensities of 1: 1, and the quaternary carbon appears as a singlet. The frequency separation in Hertz between the multiplets is the *spin – spin heteronuclear coupling constant*, J_{CH}.

This multiplet structure weakens the signal for the ^{13}C spectra and, in the case of many resonances, complicates the spectra beyond recognition. For ^{13}C solution NMR spectroscopy it is desirable to eliminate the multiplet fine structure arising from the J coupling of the carbons to the protons. Spin decoupling causes the vectors of a multiplet to be static in the rotating frame because their Larmor frequencies become identical if $\delta_A = \delta_1$. This scalar decoupling is accomplished by the use of broad-band proton decoupling. The magnitude of the scalar decoupling proton rf field is such that $\gamma_H H_1 >> J_{CH}$, where J_{CH} is the spin – spin coupling constant for the carbon and proton. The magnitude of the field is typically 4 kHz or 1 G. The pulse duration time, t_p, must meet the condition that $t_p << (1/4)\Delta - 1$, where Δ is the total range of frequencies from the rf pulse. For ^{13}C at 2.3 T, $\Delta = 4$ kHz, so $t_p << 60$ μs. In order to irradiate all the protons as one regardless of their chemical shifts, the irradiating rf field is modulated with random noise.

The application of this scalar broad-band rf proton field causes the lifetime of the particular state to be short relative to $1/J_{CH}$ and thus collapses the coupling to produce a singlet for every CH_x signal. The signal-to-noise ratio in such spectra is enhanced because the individual peaks of a multiplet collapse into a single line at the center of the multiplet. Additionally, the line widths in broad-band decoupled spectra are narrower because the broadening resulting from long range C – H interactions is also destroyed.

The continuous irradiation or decoupling of the protons in a sample disturbs the Boltzmann distribution of the upper and lower ^1H energy levels. The carbon nuclei react to the change in ^1H-energy-level populations by changing their own energy-level populations. This change results in an equilibrium excess of nuclei in the lower ^{13}C energy level relative to that specified by a Boltzmann distribution. Therefore more energy will be absorbed because of the excess population in the lower energy level. This enhancement is termed the *nuclear Overhauser effect* (NOE). The maximum signal gain or enhancement, E_s, that is obtained is

$$\Delta E_s = 1 + \frac{\gamma_H}{2\gamma_C} \qquad (6.15)$$

so the maximum NOE for ^{13}C NMR spectroscopy is 2.99. However, this NOE enhancement can be a mixed blessing, particularly if quantitative measurements are made. The NOE may be variable for the different lines, which makes quantitative measurements difficult unless due consideration is given to this effect.

For a methyl carbon with scalar-proton decoupling and a maximum NOE, the intensity of the resulting single line is 24 times greater than the outer lines in the 1:3:3:1 quartet that would be observed without decoupling. For ^{13}C NMR spectra, which are recorded for the low natural abundance of carbon, scalar decoupling can be used to obtain better signal-to-noise ratios and to improve the sensitivity.

Experimental NMR Spectroscopy of Polymers

> *Advances in both liquid and solid NMR techniques have so changed the picture that it is now possible to obtain detailed information about the mobilities of specific chain units, domain structures, end groups, branches, run numbers, number-average molecular weights, and minor structural aberrations in many synthetic and natural products at a level of 1 unit per 10,000 carbon atoms and below.*
>
> – J. C. Randall (*19*)

The ^1H and ^{13}C NMR spectra of a polymer can be obtained either in the solid state or in solution. If the polymer is soluble, a high-resolution solution spectrum can be obtained. The high-resolution NMR techniques used for polymers are similar to those used for low-molecular-weight compounds.

The most obvious differences between high-resolution solution NMR spectra of low-molecular-weight compounds and polymers are the greater line widths of the signals in the polymer spectrum. The line widths of the NMR signals depend on the T_2 relaxation time of the nucleus, which in turn depends on the rate and nature of motion of the nucleus. For low-molecular-weight molecules, the rates of motion are rapid, the relaxation times are longer, and the line widths are narrow. Because of the high viscosity of the polymer solutions, the rates of motion of the polymers are slow, and hence the relaxation times are short, and the line widths are broad.

The overall rotational mobility of the polymer rather than segmental motions is primarily responsible for the motional averaging of the dipolar and chemical-shift interactions to their isotropic values. The viscosity can be decreased and the motion can be increased by making more dilute polymer solutions and by increasing the measurement temperature. In this manner, narrower lines and high-resolution spectra can be obtained for polymers in solution. Special techniques are required to obtain high-resolution spectra of polymeric solids (*20, 21*) (*see* Chapter 8).

Another approach to wrestling the chemical-shift information from the polymer systems and to improving the NMR spectra of polymers is to use high-field magnets. The most important advantage of a high static magnetic field is that the chemical-shift dispersion is proportional to the field strength, and potentially more chemical-shift information is available at a higher field strength if the line widths do not increase at the same time. For studies of homopolymer configuration and conformation and of sequencing in copolymers, the higher field strengths allow the observation of the chemical shifts of longer sequences, which yield more structural information. The second advantage of a high-field magnet is that the sensitivity is proportional to the magnetic field to the 7/4 power. For polymers that must be studied in very dilute solutions in order to obtain high-resolution spectra, the increased sensitivity decreases the required measurement time. Finally, for quadrupolar nuclei, the lines tend to become narrower at higher fields because the second-order quadrupolar broadening is inversely proportional to the magnetic-field strength.

NMR Method of Structure Determination for Polymers

The first step in any NMR study of a polymer is to assign the NMR resonances to specific structural features of the polymer. Most structural assignments are made by using one or more of the methods in the following list.

- comparison of the observed chemical shifts with those observed for analogous low-molecular-weight model compounds
- calculation of chemical shifts by using derived additivity relationships, particularly for ^{13}C
- synthesis of polymers with known specific structural or compositional features to establish resonance–structure relationships
- synthesis of polymers with selectively enriched ^{13}C sites or deuterium substitution for protons
- comparison of the intensities of structural sequences with predicted intensities calculated on the basis of assumed polymerization kinetics and statistical models
- one-dimensional spectral-editing techniques such as selective-spin-decoupling experiments for the determination of the proton bonding of the carbons
- two-dimensional techniques

However, all these methods have limitations, and it is generally necessary to use a combination of these methods, as well as other physical and chemical techniques.

Structural elucidation is dependent upon establishing the nature of the chemical bonds between the various atoms in the polymer. Sometimes, bonding can be inferred from the ^{1}H and ^{13}C chemical shifts and from the spin – spin couplings between nuclei. Recently developed two-dimensional (2-D) NMR techniques can be used to determine the coupling between nuclei and to reveal the chemical shifts of these nuclei (*22, 23*). These new techniques largely replace a wide variety of selective spin-decoupling experiments.

The Use of Chemical Shifts To Determine Polymer Structure

Electronic shielding consists of two different contributions, termed *diamagnetic* and *paramagnetic effects*. The overall effect is the sum of both diamagnetic and paramagnetic contributions to the electron circulation. The local shielding fields that are induced by circulations resulting from spherical electron distributions are termed *diamagnetic* (*dia* meaning opposite) because these currents oppose the applied H_o. A perfectly spherical electron distribution will produce a larger local shielding than one that is distorted from spherical symmetry. Shieldings arising from distortions of the spherical electronic distributions are termed *paramagnetic* because they act to produce induced local fields in the same direction as H_o. The result is that the distorted electron distributions appear to have fewer electrons surrounding the nucleus than the corresponding spherical distributions.

Measurement of Chemical Shifts

The chemical shift for a given isotope is measured as a dimensionless value defined by

$$\delta = 10^6 \left[(\nu_{\text{sample}} - \nu_{\text{reference}})/\nu_{\text{reference}} \right] \quad (6.16)$$

where $\nu_{\text{reference}}$ is the resonance frequency of the same isotope in a reference compound. Because chemical shifts are small ($\sim 10^{-3} - 10^{-6}$), they are usually quoted in parts per million (ppm). For protons, the accepted internal standard is tetramethylsilane (TMS).

Because chemical shifts in high-resolution solution NMR spectroscopy are very small and depend on H_o, the magnetic field must be stabilized in order to make accurate measurements of the chemical shifts. This stabilizing is done with a lock system that uses the resonance of a nucleus (usually deuterium from a deuterated solvent) to continuously adjust the magnetic field so that the resonance of the lock substance remains constant. A number of different factors can modify the values of the frequencies used to measure chemical shifts, including the magnetic susceptibility (χ), solvent effects, and temperature effects. The applied magnetic field inside the sample depends on its bulk magnetic susceptibility according to the following equation:

$$H_{\text{eff}} = (1 - k\chi)H_o \quad (6.17)$$

where k is a shape factor. This correction is usually small, that is, 1 ppm or less. Chemical shifts also depend on the nature of the solvent used, and therefore a knowledge of the solvent effect on the reference material is necessary for accuracy. The temperature of the measurement also influences the value of the chemical shift, although this effect is also quite small. For protons, the temperature effect is approximately 10^{-3} ppm K^{-1}.

Interpretation of Chemical Shifts

The ^{13}C nucleus has a large range of chemical shifts (250 ppm), which offers an excellent opportunity for chemical characterization of the different carbons. Carbons that differ only by a substituent that is four or five bonds away can be distinguished from each other (*24*). Carbons with sp^3 tetrahedral hydridization are highly shielded and exhibit chemical shifts in the range of 0 – 80 ppm. Carbons with sp^2 trigonal hybridization are less shielded, and their chemical shifts fall in the range of 100 – 200 ppm; and carbons with *sp* hybridization have intermediate shielding and have chemical shifts in the range of 70 – 130 ppm. Electronegative substituents on a carbon tend to deshield that particular carbon. Extensive tables that correlate the chemical shifts with the chemical nature of the carbons have been published (*24*). A useful handbook with 58,000 ^{13}C NMR reference spectra has been published (*25*).

The ^{13}C NMR spectra of organic compounds can be interpreted by using two different approaches. In the first, the spectrum of the unknown substance is compared with those of familiar compounds, and the degree of correlation is assessed. The second approach is to use additivity rules for structure determination. The magnitude of the chemical shift for a carbon in a given molecular structure can usually be estimated (*24*). Simple additive substitution rules have been empirically generated to estimate the structural dependence of the chemical shifts (*26, 27*). For example, in hydrocarbons, each carbon substituent in the α or β position deshields the observed carbon to produce a downfield shift of approximately +10 ppm relative to an unsubstituted carbon. Carbon substituents in the γ positions shield the observed carbon, and the result is upfield shifts of approximately −2 to −3 ppm.

In order to be useful for structure analysis, the observed resonances must be assigned to chemical structures. For copolymer analysis, the enormity of the problem of assigning the observed resonances can be recognized by the following simple considerations. For the 21 most common vinyl and vinylidene monomers, there are a total of 210 possible binary copolymers and 1330 ternary copolymers (*28*). These combinations can be alternating, random, or block so that the total number of possible copolymers is quite large. Experimentally, only about 30 of these copolymers have been studied. This extremely complex situation suggests that computer simulation of the spectra of the copolymers would be helpful in the interpretation (*29*).

One approach to ^{13}C NMR simulation involves the construction and use of linear models to relate numerically encoded structural features to the observed ^{13}C NMR chemical shifts (*30*). These models have the general form:

$$S_{cs} = b(0) + b(1)X(1) + b(2)X(2) + \cdots b(p)X(p) \qquad (6.18)$$

where S_{cs} is the predicted chemical shift of a given carbon, and the $X(i)$ terms encode structural features of the chemical environment of the atom and indicate the presence or absence of substituents at certain positions (position parameter). The $X(i)$ terms can assume values of 0, 1, or 2, depending on the number of substituents present. The $b(i)$ terms are coefficients representing shift increments, and they are determined from a multiple-linear-regression analysis of a set of unambiguously assigned chemical shifts. The term p denotes the number of descriptors in the model and is seldom greater than 5 because the shielding is nearly zero at this distance from the specified carbon.

Linear and branched alkanes (*26, 27*), as well as heterocyclics and unsaturated molecules (*31, 32*), have been examined by using this approach. More recently, computer techniques have been developed to calculate the chemical shifts of complex molecules (*33–38*), including copolymers (*28*).

Structural Applications of Chemical Shifts

Only a few selected examples of the application of high-resolution NMR spectroscopy will be given in this chapter. Several excellent monographs on the application of NMR spectroscopy to polymers are available, and one cannot do better than to read Bovey (*9*), but for pedagogical reasons and completeness, some results will be given here.

Determination of Branching in Polyethylene. By using the chemical shielding factors, chemical shifts can be determined for a number of structural entities found in polyethylene (PE) (*19*), including the products of oxidation processes (*39*), and the types of branching (*40*) and cross-linking (*19*).

Polyethylenes prepared conventionally are polymers that are structurally complex and that have many versatile properties. These unusual features result from the type and concentration of the branch groups associated with a given polymer. Thus a quantitative analysis of the branching characteristics is useful. The thermodynamic, morphological, and physical properties should depend on the kind, distribution, and concentration of the branches. The ^{13}C NMR spectra can be used to determine the nature of the branches in PE. From an analysis of the NMR spectra, the concentrations of seven types of branches can be evaluated. The types of branches include ethyl, butyl, pentyl, hexyl, 2-ethylhexyl, 1,3 paired ethyl, and long six-carbon branches. Unique spectral fingerprint resonances are observed for each branch length. The chemical shifts, which can be predicted by using the appropriate shielding parameters, are shown in Table 6.2. The nomenclature used to designate the polymer backbone and side-chain carbons discriminated by ^{13}C NMR spectroscopy is also shown in the table. The distinguishable backbone carbons are designated by Greek symbols, and the side-chain carbons are numbered consecutively starting with the methyl group and ending with the methylene carbon bonded to the polymer backbone. The resonances are different and can be used to identify branch lengths up to six carbons, but a six-carbon branch produces the same ^{13}C pattern as any subsequent branch of greater length. Therefore, NMR spectroscopy is not useful for the detection of long-chain branches. The ^{13}C NMR spectrum at 25.2 MHz of a low-density PE produced by a high-pressure process is shown in Figure 6.11 (*41*). The resonance assignments of the branches are given in the figure.

Determination of Thermal Oxidation in Polyethylene. NMR spectroscopy is very useful for the study of polymer degradation. NMR spectroscopy can be used to distinguish between the various degradation products at an early stage of degradation and to produce detailed chemical identification and quantification of the oxidized species (*39*). The chemical shifts of the CH_2 groups that are α and β to carbons bearing oxygen can be used to identify the oxidation sites. The types of spectral differences are shown in Figure 6.12. The NMR results indicate the following distribution of oxidation products for the polyethylene sample that initially had 17.7 branches per 1000

Table 6.2. Polyethylene Backbone and Side-Chain ^{13}C Chemical Shifts as a Function of Branch Length

Branch Length	*Methine*	α	β	*1*	2	3	4	5	6
1	33.3	37.6	27.5	20.0					
2	39.7	34.1	27.3	11.2	26.7				
3	37.8	34.4	27.3	14.6	20.3	36.8			
4	38.2	34.6	27.3	14.1	23.4	–	34.2		
5	38.2	34.6	27.3	14.1	22.8	32.8	26.9	34.6	
6	38.2	34.6	27.3	14.1	22.8	32.2	30.4	27.3	34.6

Note: All values are given in parts per million (±0.1) downfield from TMS. The solvent was 1,2,4-trichlorobenzene, and the temperature was 125 °C.

Source: Reproduced from reference 40. Copyright 1980 American Chemical Society.

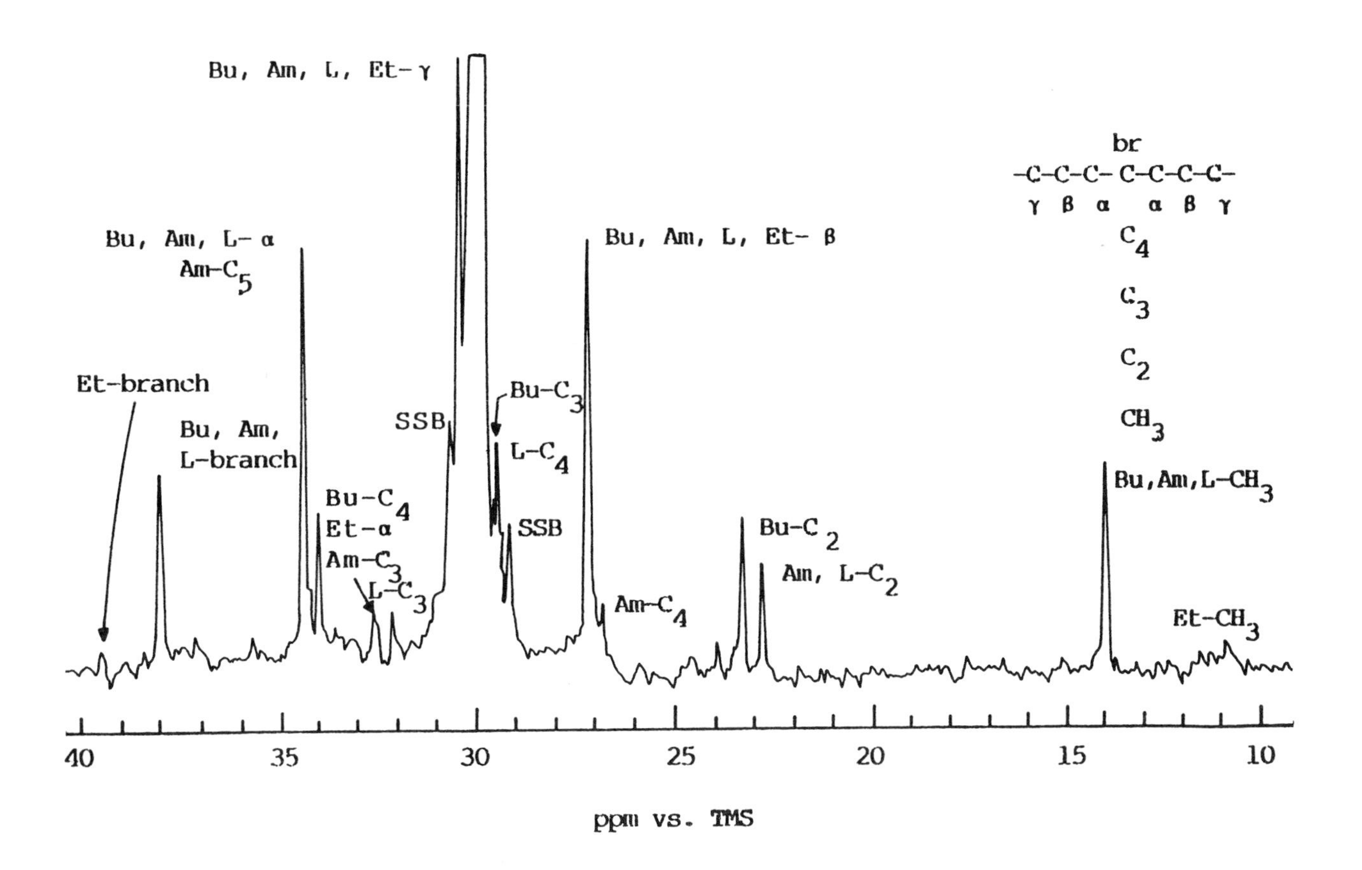

Figure 6.11. The ^{13}C NMR spectrum at 25.2 MHz of a low-density PE (LDPE) produced by a high-pressure process. The LDPE (M_w, 15,600; M_n, 13,950) was a 20% solution in trichlorobenzene, and the spectrum was obtained at 110 °C with 9500 scans. The structure on the upper right-hand side of the figure illustrates the carbon positions. The assignments are as follows: Am, amyl; br, branch; Bu, butyl; Et, ethyl; L, long; and SSB, spinning side band. (Reproduced with permission from reference 45. Copyright 1974 Steinkopff Verlag Darmstadt.)

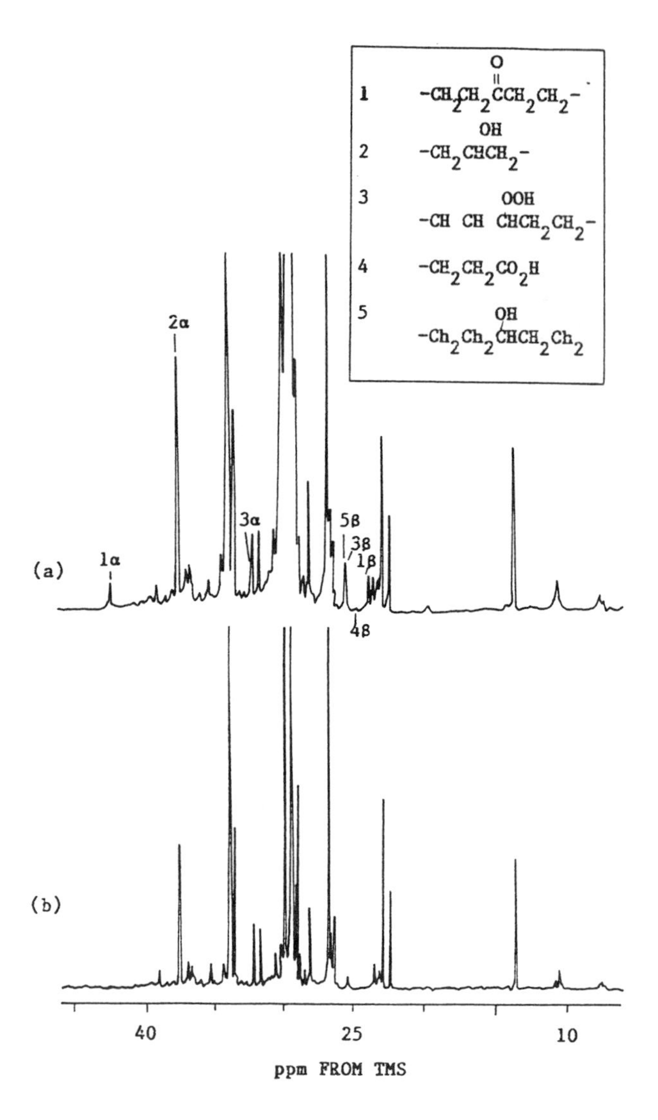

Figure 6.12. ^{13}C NMR spectra of the CH_2 region of an oxidized PE sample (a) and a control sample (b). The peak assignments are shown in the insert. (Reproduced from reference 19. Copyright 1984 American Chemical Society.)

CH_2 groups: ketone, 18%; secondary alcohol, 40%; tertiary alcohol, 7%; carboxylic acids, 4%; and secondary hydroperoxides, 32%. The total amount of oxygen was 0.6%. Some conversion of hydroperoxide to secondary alcohol occurs during the process of dissolution and recording of the NMR spectra.

Determination of Stereoregularity of Polymers. The stereoregularity of polymers has been determined by high-resolution ^1H and ^{13}C NMR spectroscopy (*9*). *Racemic* (r) or *syndiotactic* dyads are pairs of adjacent asymmetric centers that have opposite optical configuration (dl). *Meso* (m) or *isotactic* dyads have the same optical configurations (dd or ll). Triads, tetrads, pentads, etc., are denoted by a succession of dyads. In ^1H NMR spectroscopy, molecules having CM_2 groups (where M is an observable nucleus) are termed *geminally heterosteric* when the M groups can be differentiated because of differences in their average environments and are termed *geminally homosteric* when the M groups are equivalent. For a syndiotactic placement of a vinyl monomer, the protons in the methylene groups are racemic and have the same average environment and, in the absence of vicinal coupling, are not distinguishable. Consequently, a single line resulting from the syndiotactic placement or racemic dyad (r) is observed. For the isotactic placement, or meso dyad (m), the protons of each methylene group do not experience the same average environment and have different chemical shifts. The methylene protons are heterosteric, and the AB quartet of lines is observed. On this basis, ^1H NMR spectroscopy can be used to distinguish on an absolute basis the isotactic and syndiotactic dyads of some vinyl polymers. Poly(methyl methacrylate) (PMMA) was the first polymer studied in this manner, and the spectra of an isotactically rich sample and a syndiotactically rich sample are shown in Figure 6.13 (*9, 42*). It is evident from this figure that the methylene proton spectrum of the anionically initiated polymer exhibits an AB quartet (J_{gem} = 14.9 Hz) centered at 8.14 on the now-outmoded τ scale (1.86 ppm from TMS), and therefore this polymer is predominately isotactic. In the spectrum of the free-radical polymer, the methylene-proton resonance is a broad singlet, and so this polymer is predominately syndiotactic.

Actually, neither polymer is perfectly stereoregular, so both tactic features appear in the spectra. There are some obvious additional resonances observed as fine structure on the major resonances. These additional resonances result from longer sequences and will be discussed shortly.

Additionally, three resonances are observed for the α-methyl resonances centered at approximately 9τ. These three species are assigned to the triad sequences. Those monomer units that are flanked on

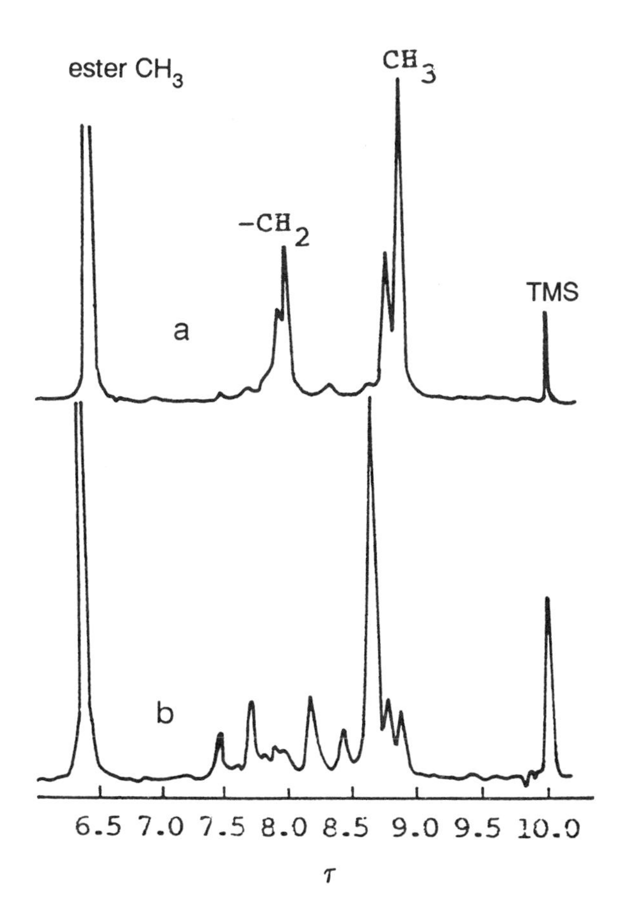

Figure 6.13. The 60-MHz spectra of 15% (w/v) solutions of PMMA in chlorobenzene prepared with a free-radical initiator (a) and an anionic initiator (PhMgBr) (b). The ester methyl resonance appears near 6.5τ, the β-methylene protons appear near 8.0τ, and the β-methyl protons give three peaks between 8.5 and 9.0τ. (Reproduced from reference 41. Copyright 1980 American Chemical Society.)

both sides by units of the same configuration are *isotactic triads* (i), those monomer units that have units of opposite configuration on both sides are *syndiotactic triads* (s), and those monomer units that have a unit of the same configuration on one side and a unit of opposite configuration on the other side are *heterotactic triads* (h). As shown in Figure 6.13, these three triad α-methyl resonances have the same chemical shifts but have very different intensities in each spectrum. The relative intensities provide a measure of the triad probabilities.

As the magnetic strength is increased, longer configurational sequences can be observed. For the β-methylene groups, the tetrad (and perhaps even the hexad) resonances appear as fine structure on the m and r dyad resonances. For the α-methyl groups, pentad and heptad sequences may be observable at high resolution. The relative concentrations of the longer configurational sequences can be calculated by using probability theory and the appropriate polymerization model. The manner in which the relative proportions of the various sequences can be calculated for the Bernoullian or terminal model of the stereochemical polymerization is shown in Table 6.3. Similar calculations can be performed for more complex mechanisms such as the penultimate and pen-penultimate models. The proton-resonance spectra of the β-CH_2 units of the two samples of PMMA at a high-field strength of 220 MHz are shown in Figure 6.14. The high-field resonance spectra of the α-CH_3 groups in PMMA are shown in Figure 6.15. The assignments are superimposed on the spectra (*9*). The spectra at higher field strengths indicate the influence

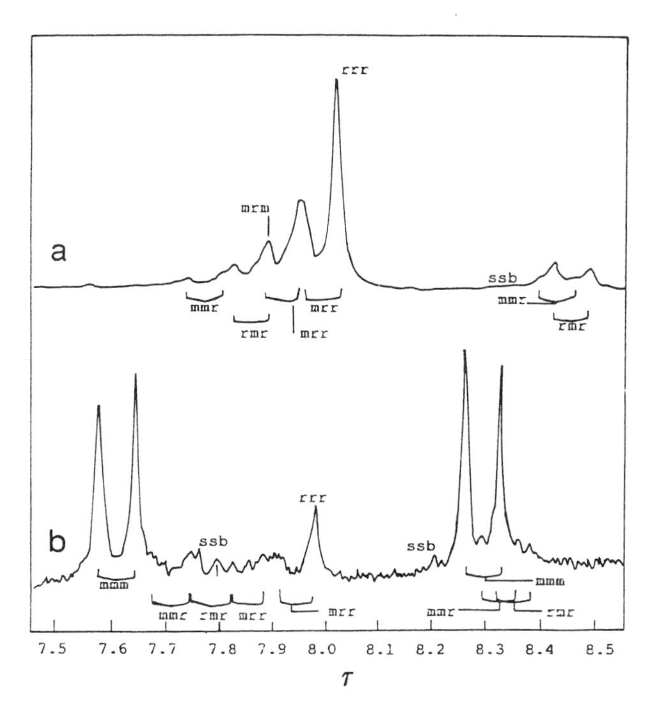

Figure 6.14. The proton resonance spectra of the β-CH_2 units of syndiotactically rich (a) and isotactically rich (b) samples of PMMA at a high-field strength of 220 MHz. (Reproduced from reference 39. Copyright 1984 American Chemical Society.)

Table 6.3. Stereochemical Sequence Designation and Bernoullian Probabilities

α-Substituent

Designation		Projection	Bernoullian probability
Triad	Isotactic, *mm* (*i*)		P_m^2 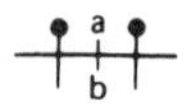
	Heterotactic, *mr* (*h*)		 $2P_m(1-P_m)$
	Syndiotactic, *rr* (*s*)		$(1-P_m)^2$ 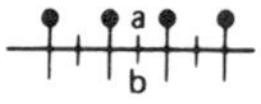
Pentad	*mmmm* (isotactic)		P_m^4 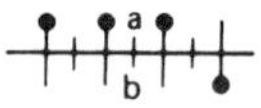
	mmmr		$2P_m^3(1-P_m)$ 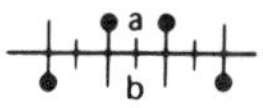
	rmmr		$P_m^2(1-P_m)^2$ 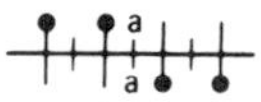
	mmrm	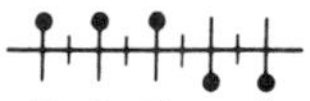	$2P_m^3(1-P_m)$ 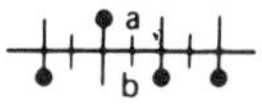
	mmrr		$2P_m^2(1-P_m)^2$ 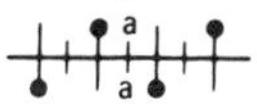
	rmrm (heterotactic)		$2P_m^2(1-P_m)^2$
	rmrr		$2P_m(1-P_m)^3$
	mrrm		$P_m^2(1-P_m)^2$
	rrrm		$2P_m(1-P_m)^3$
	rrrr (syndiotactic)		$(1-P_m)^4$

β-CM_2

Designaton		Projection	Bernoullian probability
Dyad	meso, *m*		P_m
	racemic, *r*		$(1-P_m)$
Tetrad	*mmm*		P_m^3
	mmr		$2P_m^2(1-P_m)$
	rmr		$P_m(1-P_m)^2$
	mrm		$P_m^2(1-P_m)$
	rrm		$2P_m(1-P_m)^2$
	rrr		$(1-P_m)^3$

of tetrads. The m peak shows additional resolved resonances resulting from mmm, mmr, and rmr placements. Substitution of an outside m with an r results in a downfield shift. When adding placements on either side, the change in chemical shift on substituting an m for an r is just the opposite.

The differences in stereoconfiguration between odd and even numbers of units in vinyl polymers like PVC are reflected in the splitting of the CH and CH_2 carbon resonances. The number of signals from CH and CH_2 groups are shown in Table 6.4 for the various sequences. As the magnetic field strength increases, the resolution increases, and the amount of available information improves.

A particularly good example is atactic polypropylene (PP) for which 20 heptad resonances out of a possible 36 have been clearly resolved. The 90-MHz ^{13}C NMR spectrum of atactic polypropylene is shown in Figure 6.16 (*43*). The lines on the figure represent the positions of the CH_3, CH_2, and CH sequences that were calculated by using the γ-gauche effect, which will be discussed later.

When the resonance assignments to the stereoconfigurational sequences have been made, the stereoconfigurational propagation mechanism can be determined. Accurate peak areas must be obtained. The unconditional probability of occurrence of any particular sequence S is simply its peak area, $A(S)$, divided by the total resonance area of the carbon under observation.

The relationships between the probability of meso addition, P_m, and the various stereosequences are shown in Table 6.3. Triad data are necessary and sufficient to test for conformance to Bernoullian statistics, but tetrad data are required to test first-order Markov statistics if the former treatment is inappropriate. For Bernoullian statistics, P_m can be calculated directly from the ratio of the areas, $A(rr)/A(rm)$, which is equal to $(1 - P_m)/P_m$.

The differences between the activation enthalpies

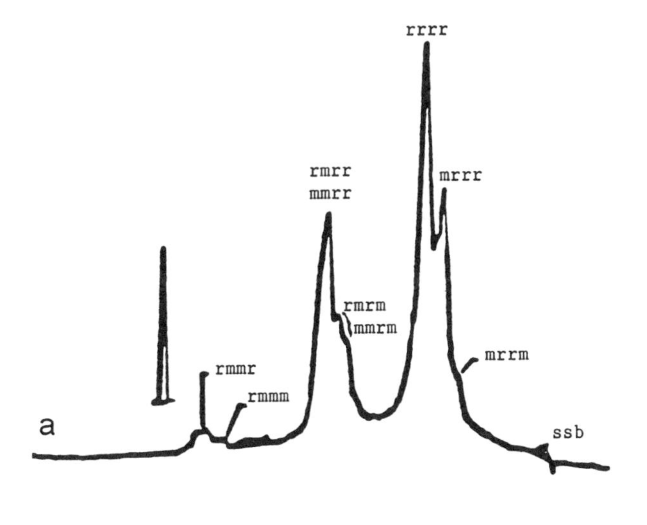

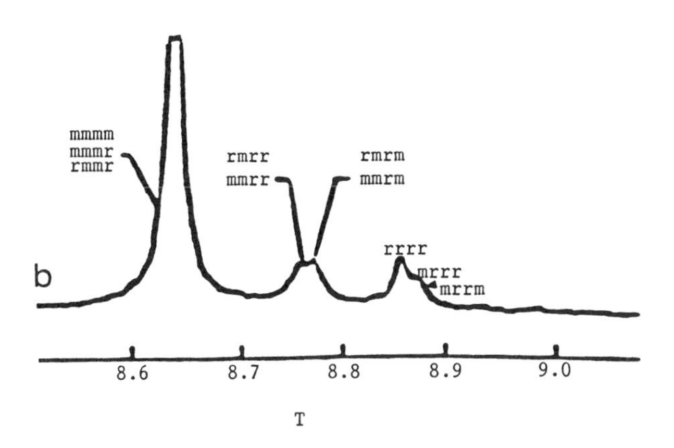

Figure 6.15. The high-field resonance spectra of the α-CH_3 groups of syndiotactically rich (a) and isotactically rich (b) samples of PMMA at a high-field strength of 220 MHz. (Reproduced with permission from reference 9. Copyright 1988 Academic Press.)

Table 6.4. Numbers of Signals of CH and CH_2 Groups for Monosubstituted Vinyl Systems in ^{13}C NMR Spectra

CH Group (Triad)	*Pentad*	*Heptad*
rr	3	10
rm	4	16
mm	3	10
CH_2 Group (Tetrad)	*Hexad*	*Octad*
rrr	3	10
rmr	3	10
rrm	4	16
mrm	3	10
mmr	4	16
mmm	3	10

and entropies for meso and racemic placement can be obtained from a plot of ln (P_m/P_r) against reciprocal temperature. Thus,

$$\ln (P_m/P_r) = \ln (k_m/k_r) = (\Delta S^m/R) - (\Delta H^m/RT) \qquad (6.19)$$

where k_m and k_r are the rates of meso and racemic addition, respectively.

In general, polymers prepared by using free-radical initiators are predominantly syndiotactic, and the tendency to form syndiotactic sequences increases as the polymerization temperature is lowered. The propagation reaction can be regarded as being subdivided into two fundamental reactions, that is, isotactic and syndiotactic propagations. In this way, the placement of a given unit (isotactic or syndiotactic) is not determined until after the next unit has been added to the end of the growing radical. This is because the end unit itself, being a free radical, is unable to maintain asymmetry and is either planar or oscillating rapidly between the two possible tetrahedral configurations. Therefore, there is not a permanent spatial position for the terminal carbon radical. However, the α-substituted atom of the penultimate unit does take up a single tetrahedral configuration and must be either isotactic or syndiotactic with respect to the previous asymmetric carbon. Theoretically, syndiotactic propagation should be slightly favored energetically (for steric reasons) over isotactic propagation in the free species.

Determination of Directional Isomerism in Polymers. There is a directional form of structural isomerism in polymers, namely *regioisomerism*, that is isomerism arising from head–tail, head–head, and tail–tail additions of an asymmetrical monomer unit. The simplest example is the addition of a vinylidene monomer, M, such as poly(vinylidene fluoride), which is directional by virtue of an asymmetrical arrangement of substituents about the double bond (i.e., protons at one end and X atoms at the other). The head will be designated as the substituted end (= CX_2), and the other end will be designated as the tail, that is, the CH_2 portion of the unit. Thus a sequence of M units may have head–head (CX_2-CX_2), head–tail (CX_2-CH_2), or tail–tail (CH_2-CH_2) junctions. A given structure is designated *isoregic*, *syndioregic*, and *aregic* for se-

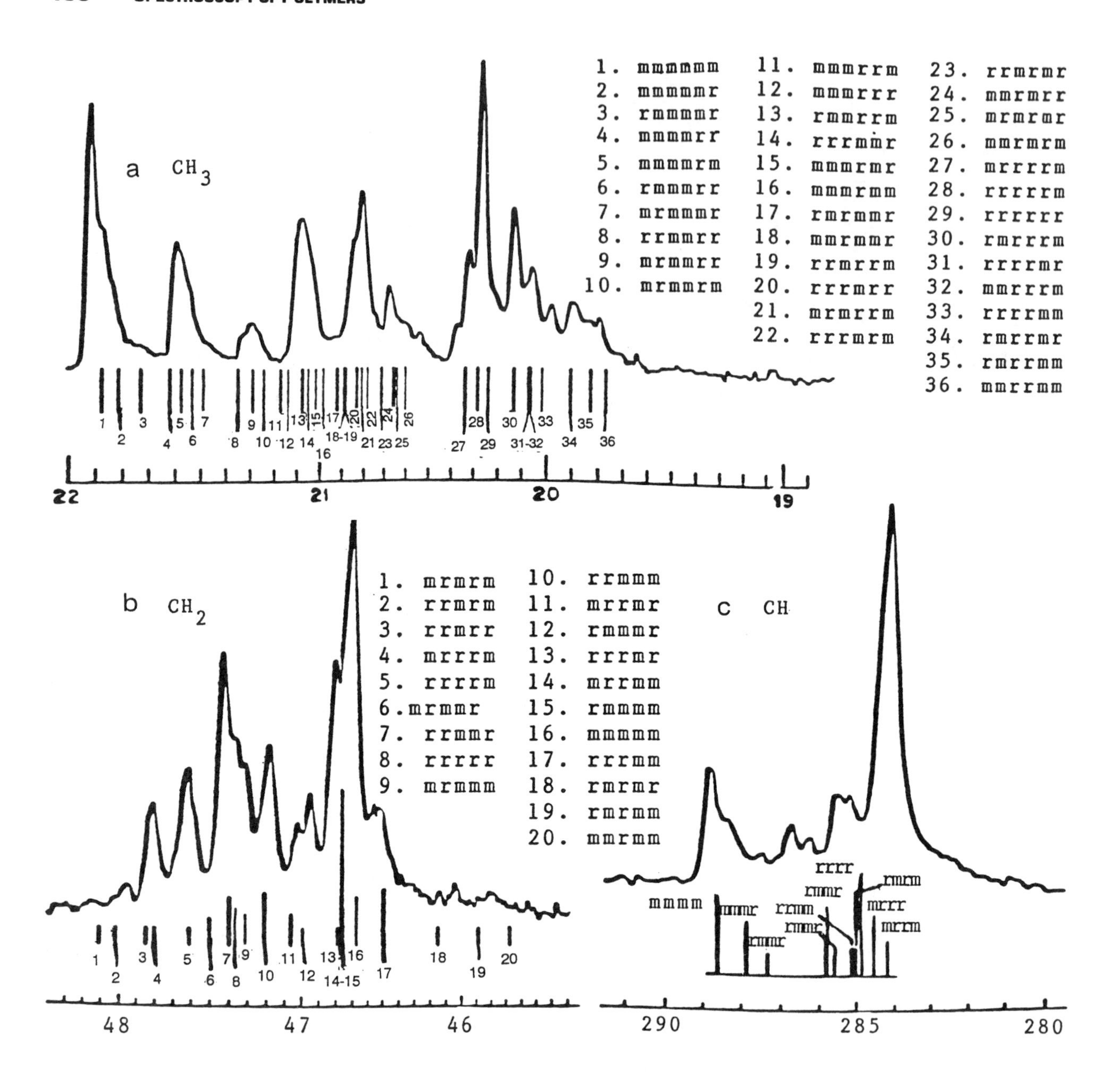

Figure 6.16. The 90-MHz ^{13}C NMR spectra of atactic polypropylene in a 20% (w/v) solution of heptane at 67 °C; (a) CH_3 , (b) CH_2 , and (c) CH. Line spectra appearing below the experimental spectra correspond to theoretically calculated resonance positions for (a) heptad, (b) hexad, and (c) pentad configurational sequences. (Reproduced with permission from reference 9. Copyright 1988 Academic Press.)

quences in which the directional sense of successive monomer units is the same, alternating, or random, respectively (*44*).

Poly(vinylidene fluoride) (PVF_2) is a simple example of regio-irregular sequences in homopolymers, because PVF_2 does not have the stereoconfigurational irregularity that complicates the spectra of other polymers. The 188.22-MHz ^{19}F NMR spectrum of PVF_2 dissolved in dimethylformamide is shown in Figure 6.17. The resonances can be assigned to heptad sequences, and for convenience, the head of M will be represented by 1 and the tail will be represented by 0.

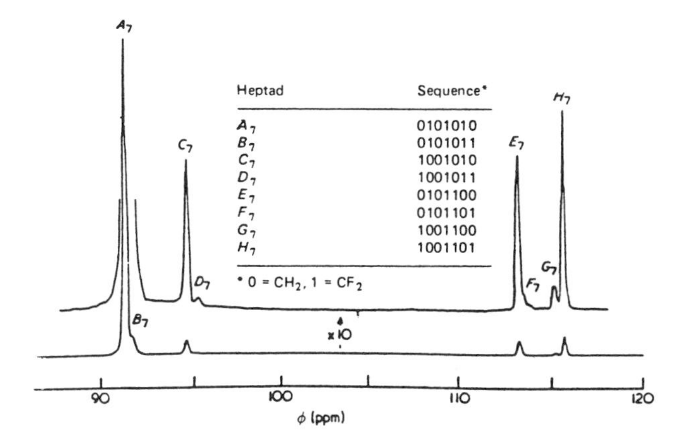

Figure 6.17. The ^{19}F NMR spectrum of PVF_2 dissolved in dimethylformamide is shown with the heptad assignments (insert). (Reproduced with permission from reference 9. Copyright 1972 Academic Press.)

Table 6.5. Assignments of Heptad Sequences for the ^{19}F NMR Spectrum of PVF_2

Heptad	*Sequence*	*Chemical Shift (±0.05)*
A_7	0101010	91.31
B_7	0101011	91.79
C_7	1001010	94.43
D_7	1001011	95.37
E_7	0101100	113.33
F_7	0101101	113.62
G_7	1001100	115.34
H_7	1001101	115.76

The resonance assignments are listed in Table 6.5.

The relative intensities of the resonances can be used to determine the nature of the mechanism for regiospecific addition. In the case of PVF_2, the data are highly suggestive of a first-order Markov-enchainment mechanism rather than a Bernoullian mechanism (*44*).

Determination of Copolymer Structure. As a technique for the analysis of sequence distribution in copolymers, high-resolution NMR spectroscopy is particularly useful when the spectral resolution is sufficient to resolve the resonances of the specific sequences. A number of copolymer structural problems can be elucidated by using NMR spectroscopy. The composition of the copolymer can be quantitatively determined. The detection of compositional dyads can be used to determine the distribution of composition, that is, whether the sample is a mixture of homopolymers, a block copolymer, an alternating copolymer, or a random copolymer. If resonances are resolved because of the triad sequences of the copolymer, sequence distributions can be determined, and the mechanism of the copolymerization can be tested in terms of Bernoullian, first-order Markov, second-order Markov, or non-Markovian statistics. In rare circumstances, the tactic nature of the copolymer can be determined if distinguishable syndio- and isotactic resonances are resolved. Such an analysis has been carried out for copolymers of methyl methacrylate – methacrylic acid, for which the α-CH_3 resonances of all 20 triads have been assigned and have been used to determine the cotacticity of the copolymer (*45*).

A number of examples may be found in the literature. A simple example is the NMR analysis of vinyl chloride (VC) – vinyl acetate (VA) copolymers (*46*). The 300-MHz ^{1}H NMR spectra of VC – VA copolymers are shown in Figure 6.18. The resonances are assigned on the basis of analogy with the assignments for PVC and PVA and are listed in Table 6.6. The methylene-proton resonances centered at 1.96, 2.14, and 2.34 ppm result from the three compositional

Table 6.6. Spectral Assignments for VC – VA Copolymers as Measured by 300-MHz ^{1}H NMR Spectroscopy

Chemical Shift[a] (ppm)	*Protons*	*Dyads or Triads*
2.01	methyl	(VA, VA, VA), (VA, VA, VC), (VC, VA, VC)
1.96	methylene	(VA, VA)
2.14	methylene	(VA, VC)
2.34	methylene	(VC, VC)
4.03	methine	(VA, VC, VA)
4.23	methine	(VC, VC, VA)
4.45	methine	(VC, VC, VC)
4.86	methine	(VA, VA, VA)
5.07	methine	(VA, VA, VC)
5.27	methine	(VC, VA, VC)

[a]Chemical shift is given with respect to internal $(CH_3)_2SO$-d_5H (2.50 ppm).

SOURCE: Reproduced from reference 46. Copyright 1983 American Chemical Society.

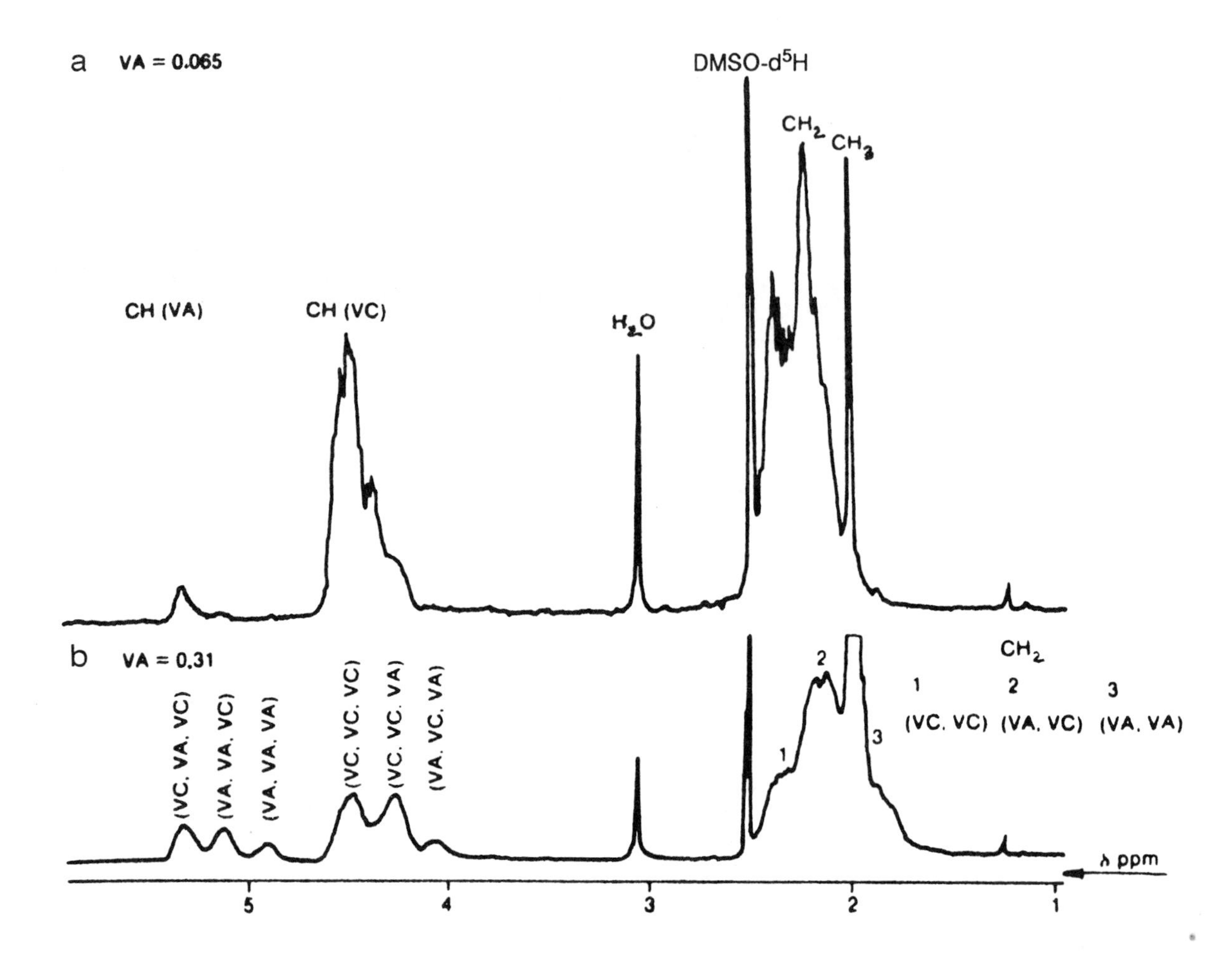

Figure 6.18. 300-MHz ¹H NMR spectra of VC – VA copolymers containing 0.065% VA (spectrum a) and 0.31% VA (spectrum b) in $(CH_3)_2SO$-d_6 at 85 °C. (Reproduced from reference 43. Copyright 1984 American Chemical Society.)

dyads. The lines are broad because of a combination of spin – spin coupling and configurational splitting. There are six methine resonances in two separate groups of three lines each. All six of these methine resonances are assigned to the six compositional triads as shown in Table 6.6. The composition and the number-average sequence length of the acetate and chloride sequences can be calculated by using the integrated areas of the triads. The results for composition, dyad, triad, and number-average sequences are shown in Table 6.7 for the VA – VC sample with VA = 0.31%. The Bernoullian model and the first-order and second-order Markov models of the copolymerization mechanism can be tested by using the observed distribution of the triads. The results for the three models are shown in Table 6.8 for the sample with VA = 0.31%. The conclusions were stated as follows:

> *It is evident that for polymer B (VA = 0.31), the Bernoullian model can be abandoned and that a significant amelioration of the results is achieved by going from first-order Markov to second-order Markov analysis. Thus the propagation statistics of polymer B is adequately described by a second-order Markov process.*
>
> – G. van der Velden (*46*)

This copolymer system demonstrates the nature of the information obtained from high-resolution NMR spectra of copolymer systems. Unfortunately, many of the copolymers do not have well-resolved resonances that can be easily integrated and used for quantitative analysis. However, as the applied mag-

Table 6.7. Composition and Mole Fraction of Dyads and Triads as Measured and Calculated by ^{1}H NMR Spectroscopy of a VC – VA Copolymer (VA = 0.31)

Dyads and Triads	*^{1}H NMR*[a]	*^{13}C NMR*[b]
(VC, VC, VC)	0.329	
(VC, VC, VA)	0.302	
(VA, VC, VA)	0.055	
(VC, VA, VC)	0.139	
(VA, VA, VC)	0.116	
(VA, VA, VA)	0.055	
(VC, VC)		0.47
(VA, VC)		0.40
(VA, VA)		0.13
n_0^{VA}	1.57	1.65
n_{2+}^{VA}	2.95	
n_0^{VC}	3.34	3.35
n_{2+}^{VC}	4.18	
(VA)	0.31	0.32

[a]In $(CH_3)_2SO$-d_6.
[b]In C_6D_6.
SOURCE: Reproduced from reference 46.

netic fields increase, the resolution will also increase, and more copolymer systems will become tractable. An alternate approach that is proving to be helpful in this regard is the two-dimensional NMR technique (the subject of Chapter 9).

References

1. Shoolery, J. N. *Ind. Res. Dev.* **1983,** 91.
2. Pople, J. A.; Schneider, W. G.; Bernstein, H. J. *High Resolution Nuclear Magnetic Resonance;* McGraw-Hill: New York, 1959.
3. Abragam, A. *The Principles of Magnetic Resonance;* Oxford: New York, 1961.
4. Zukav, G. *The Dancing Wu Li Masters;* Morrow: New York, 1979.
5. Harris, R. K.; Mann, B. E. *NMR and the Periodic Table;* Academic: London, 1978.
6. Farrar, T. C.; Becker, E. D. *Pulse and Fourier Transform NMR;* Academic: New York, 1971.
7. Shaw, D. *Fourier-Transform NMR Spectroscopy;* Elsevier: Amsterdam, 1976.
8. Andrew, E. R. *Philos. Trans. R. Soc. London, A: Mathematical and Physical Sciences* **1981,** *299,* 505.
9. Bovey, F. A. *Nuclear Magnetic Resonance Spectroscopy;* 2nd ed.; Academic: San Diego, CA, 1988.
10. Arnold, J. T.; Dharmatti, S. S.; Packard, M. E. *J. Chem. Phys.* **1951,** *19,* 507.

Table 6.8. Calculated and Observed Triad Comonomer Distributions for VC – VA Copolymers

Triad	*Distribution*	*Observed Bernoullian*[a]	*First-order Markov*[b]	*Second-order Markov*[c]
VCVCVC[d]	0.814	0.817		
VCVCVA	0.121	0.114		
VAVCVA	0.000	0.004		
VCVAVC	0.055	0.057		
VAVAVC	0.010	0.008		
VAVAVA	0.000	0.000		
VCVCVC[e]	0.329	0.328	0.316	0.332
VCVCVA	0.302	0.295	0.284	0.299
VAVCVA	0.055	0.066	0.064	0.053
VCVAVC	0.139	0.148	0.125	0.143
VAVAVC	0.116	0.133	0.160	0.117
VAVAVA	0.055	0.030	0.051	0.056

[a]$\alpha = 0.935$ (0.065% VA), $\alpha = 0.69$ (0.31% VA).
[b]$\alpha = \gamma = 0.69$, $\beta = \delta = 0.61$.
[c]$\alpha = 0.69$, $\beta = 0.71$, $\gamma = 0.74$, and $\delta = 0.51$.
[d]The copolymer contained 0.065% VA.
[e]The copolymer contained 0.31% VA.
SOURCES: Reproduced from reference 46.

11. Stothers, J. B. *Carbon-13 NMR Spectroscopy;* Academic: New York, 1972.
12. Wehrli, F. W.; Wirthlin, T. *Interpretation of Carbon-13 NMR Spectra;* Heyden: Philadelphia, 1978.
13. Levy, G. C.; Lichter, R. L.; Nelson, G. L. *Carbon-13 Nuclear Magnetic Resonance Spectroscopy;* Wiley: New York, 1980.
14. Silverstein, R. M.; Bassler, G. C.; Morrill, T. C. *Spectrometric Identification of Organic Compounds;* Wiley: New York, 1981.
15. Martin, G. E. *Org. Magn. Reson.* **1975,** *7,* 2.
16. Atta-ur-Rahman *Nuclear Magnetic Resonance: Basic Principles;* Springer-Verlag: New York, 1986.
17. Ando, I.; Webb, G. A. *Theoretical Aspects of NMR Parameters;* Academic: London, 1983.
18. Wuthrich, K. *NMR of Proteins and Nucleic Acids;* Wiley: New York, 1986.
19. Randall, J. C.; Zoepfl, F. J.; Silverman, J. In *NMR and Macromolecules;* Randall, J. C., Ed.; ACS Symposium Series 247; American Chemical Society: Washington, DC, 1984; p 245.
20. Schaefer, J.; Stejskal, E. O. *J. Am. Chem. Soc.* **1976,** *98,* 1031.
21. *High Resolution NMR of Synthetic Polymers in Bulk;* Komoroski, R. A., Ed.; VCH Publishers: Deerfield Beach, 1986.
22. Bax, A. *Two-dimensional Nuclear Magnetic Resonance in Liquids;* Reidel: Boston, 1982.
23. Ernst, R. R.; Bodenhausen, G.; Wokaun, A. *Principles of Nuclear Magnetic Resonance in One and Two Dimensions;* Oxford: New York, 1987.
24. Bremser, W.; Franke, B.; Wagner, H. *Chemical Shift Ranges in Carbon-13 NMR Spectroscopy;* Verlag Chemie: Weinheim, 1982.
25. Bremser, W.; Ernst, L.; Franke, B.; Gerhards, R.; Hardt, A. *Carbon-13 NMR Spectral Data;* Verlag Chemie: Weinheim, 1981.
26. Grant, D. M; Paul, E. G. *J. Am. Chem. Soc.* **1964,** *86,* 2984.
27. Lindeman, L. P.; Adams, J. G. *Anal. Chem.* **1971,** *43,* 1245.
28. Cheng, H. N.; Bennet, M. A. *Anal. Chem.* **1984,** *56,* 2320.
29. Cheng, H. N. *Transition Metal Catalyzed Polymerizations. Alkenes and Dienes;* MMI Press, Harwood Academic: New York, 1983; p 617.
30. Jurs, P. C.; Sutton, G. P.; M. L. Ranc *Anal. Chem.* **1989,** *61,* 1115A.
31. Lindeman, L. P.; Adams, J. G. *Anal. Chem.* **1971,** *43,* 1245.
32. Dorman, D. E.; Jautelat, M.; Roberts, J. D. *J. Org. Chem.* **1971,** *36,* 2757.
33. Small, G. W.; Jurs, P. C. *Anal. Chem.* **1983,** *55,* 1121.
34. Small, G. W.; Jurs, P. C. *Anal. Chem.* **1983,** *55,* 1128.
35. Small, G. W.; Jurs, P. C. *Anal. Chem.* **1984,** *56,* 1314.
36. Gray, N. A. B.; Crandell, C. W.; Nourse, J. G.; Smith, D. H.; Gageforde, M. L.; Djerassi, C. *J. Org. Chem.* **1981,** *46,* 703.
37. Shelley, C. A.; Munck, M. E. *Anal. Chem.* **1982,** *54,* 516.
38. Kalchhauser, H.; Robien, W. *J. Chem. Inf. Comput. Sci.* **1985,** *25,* 103.
39. Jelinski, L. W.; Dumais, J. J.; Luongo, J. P.; Cholli, A. L. *Macromolecules* **1984,** *17,* 1650.
40. Randall, J. C. In *Polymerization Characterization by ESR and NMR;* Woodward, A. E.; Bovey, F. A, Eds.; ACS Symposium Series 142; American Chemical Society: Washington, DC, 1980; p 100.
41. Bovey, F. A.; Schilling, F. C.; McCrackin, F. L.; Wagner, H. L. *Macromolecules* **1976,** *9,* 76.
42. Bovey, F. A.; Tiers, G. V. D. *J. Polym. Sci.* **1960,** *46,* 303.
43. Bovey, F. A. In *NMR and Macromolecules;* Randall, J. C., Ed.; ACS Symposium Series 247; American Chemical Society: Washington, DC, 1984; p 8.
44. Cais, R. E.; Sloane, N. J. A. *Polymer* **1983,** *24,* 179.
45. Klesper, E. In *Polymer Spectroscopy*; Hummel, D. O., Ed.; Steinkopff Verlag: Darmstadt, 1976; p 197.
46. van der Velden, G. *Macromolecules* **1983,** *16,* 1336.

7

Special Editing Techniques for High-Resolution NMR Spectroscopy of Polymers

NMR structure elucidation is very similar to synthetic organic chemistry. By using a number of experimental techniques to extract spectral information, we can construct the picture of a molecular structure by establishing the connectivities of the atoms. There are now many different 1-D and 2-D NMR techniques that can be used in conjunction with one another to obtain the information needed. As in organic synthesis, the best NMR technique or combination of techniques depends on the circumstances: the amount of material available for characterization, the nature of the information desired, the presence of certain functional groups that might result in resonances that overlap those of interest, the physical properties of the unknown, and the field strength and the capability of the NMR instrument being used.

– Peter L. Rinaldi (*1*)

For large molecules like polymers, the NMR spectra are rich with weak overlapping resonances because these molecules have a variety of structural components and their corresponding resonances. There are two challenges with the NMR spectroscopy of polymers. First, methods to increase the resolution are sought so that the resonances can be isolated. The resolution can be enhanced by increasing the strength of the magnetic fields, but this option is not always available. However, multipulse methods can be used to enhance the resolution. These methods will be described in this chapter. The second, and perhaps more demanding, challenge is the recognition and assignment of the large number of observed resonances. Resonance assignments can be made by using various means, including a comparison of the observed chemical-shift values with those calculated by empirical additivity rules, a comparison of the polymer spectrum with the spectra of model compounds, a consideration of the internal consistency of intensities, and specific chemical labeling.

Spectral editing involves determining the responses of the various resonances to different experimental conditions or pulse sequences. Differences in the responses of the resonances can be used as a basis for their assignments. With the use of proper experiments, resonances can be assigned to methyl, methylene, methine, and quaternary carbon sites on the polymer chain. This assignment can be accomplished by generating subspectra associated with the different protonated carbon species.

A powerful new approach has been the use of two-dimensional (2-D) spectroscopic techniques for correlation of spectral effects. These valuable 2-D methods have found widespread use in organic and biochemical studies but have been limited for synthetic polymers because of sensitivity problems. However, with modern high-field NMR instrumentation, some of these difficulties have been overcome, and these techniques can now be used rather routinely for the analysis of polymer systems. The techniques do require some degree of sophistication on the part of the spectroscopist, but the barrier is not a difficult one to overcome.

1904—4/92/0163$09.25/1

Spin-Manipulation Pulse Sequences

> *Many of the substantial improvements in NMR are the result of the spin gymnastics that can be orchestrated by the spectroscopist.*
> —Thomas C. Farrar (*2*)

The term *pulse sequence* has come to symbolize the recipe for preparing a nuclear spin system in a desired fashion (*3, 4*). Special pulse sequences have been designed for particular tasks, including resolution enhancement, selective pulse excitation, selective suppression, sensitivity enhancement, multiplicity selection, and connectivity (*5, 6*). These special pulse sequences utilize the influence of a variety of interactions on the spin systems. They are designed to differentiate between resonances on the basis of position, line width, multiplicity, magnitude of spin–spin coupling, magnitude of dipolar coupling, exchange rates, or relaxation times (*5*). The resultant spectra yield information concerning the spectral origin of the resonances lines.

Pulse sequences are made up of different radiofrequency (rf) pulses and delay times prior to the read pulse and acquisition of the free induction decay (FID). The total magnetic moment of the nuclear sample is denoted by the vector **M**. At thermal equilibrium, **M** is parallel to the applied field (Figure 7.1a). An rf coil on the x axis produces an rf pulse that results in a strong rf field, H_1, that tilts **M** away from the z axis. The duration and power of the rf pulse determine the direction of **M** after the pulse. If a $90°_x$ (where x is the axis) pulse or $\pi/2$ pulse is applied, **M** points along the positive y axis (Figure 7.1b), and the longitudinal or z magnetization is transformed into transverse magnetization. After the pulse is turned off, the vector, **M**, starts to dephase because of the different Larmor frequencies (Figure 7.1c). These components rotate in the xy plane at different rates and produce a voltage in the receiver coil that is detected as the NMR signal.

The larger the value of H_1, the shorter the pulse required to turn the magnetization through a given angle. At the same time, the shorter the pulse, the wider the distribution of frequency components in the rf pulse. A powerful short pulse, which affects all of the nuclei in the sample, is termed a *nonselective* or *hard pulse*. If only those nuclei corresponding to one resonance in the spectrum are to be disturbed, a much weaker and longer pulse is used, and this pulse is termed a *selective* or *soft pulse*.

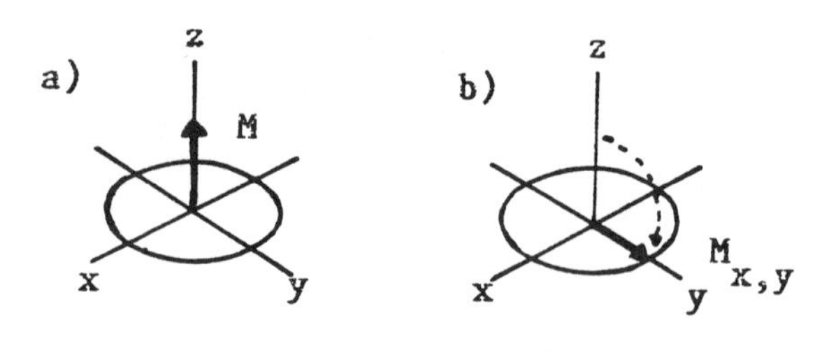

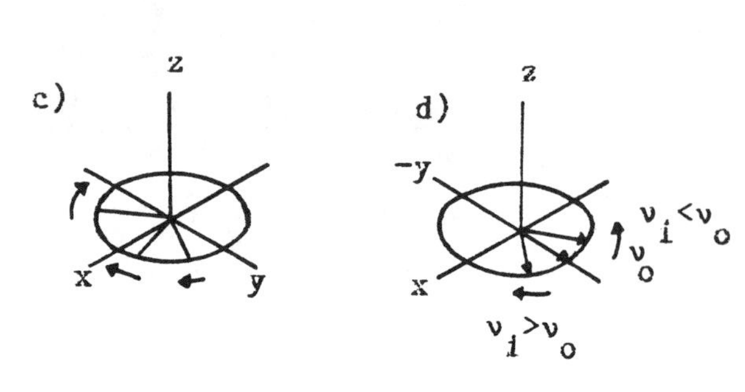

Figure 7.1. Elements of pulse sequences. (Reproduced with permission from reference 11. Copyright 1983 Steinkopff Verlag Darmstadt.)

At this point, the concept of the *rotating frame* must be reintroduced. The rotating frame is a coordinate system that rotates in the same sense and frequency as the applied rf field.

> *The concept of the rotating frame is of paramount importance in NMR spectroscopy, for almost all classical descriptions of NMR experiments are described using this frame of reference.*
> —D. E. Traficante (*7*)

Within the rotating frame, a signal in resonance with $\omega_i = \omega_o$, where ω_o is the Larmor frequency and ω_i is the frequency of the ith nucleus, is static (Figure 7.1d). Signals with frequencies $\omega_i > \omega_o$ rotate clockwise, and signals with $\omega_i < \omega_o$ rotate counterclockwise. If two nuclear species (^1H and ^{13}C) with different resonance frequencies corresponding to their magnetogyric ratios are involved, two coordinate frames rotating about the same axis (doubly rotating frame) are involved. There is no correlation between the phases of the doubly rotating frames unless they are spin coupled.

Each component of the magnetization vector is characterized by a Larmor frequency, an orientation in the rotating frame, and the component's lifetime. The chemical shift determines the Larmor frequency and thus the position of the signal in the spectrum. The orientation at the beginning of the data accumulation determines the phase, and the lifetime is determined by the effective transverse relaxation time, T_2^*

$$\omega_{1/2} = \frac{1}{\pi T_2^*} \tag{7.1}$$

where $\omega_{1/2}$ is the half width of the NMR signal.

Special multiple-pulse sequences are designed to enhance or suppress certain interactions by using spin manipulation. The most important common feature of these special pulse sequences is the addition of an evolution time, τ_e, during which the spin system evolves under the selected interaction.

pulse τ_e 90°
preparation | evolution | detection
time →

The preparation time is designed to prepare the spins in a specified state, which is most often a relaxation to thermal equilibrium. The evolution time allows for the phase modulation of the spins by J coupling or other interactions. The evolution time is selected to optimize the information content of the resultant spectra and generally corresponds to the magnitude of the selected interaction, that is, $\tau_e = 1/J$ or some multiple depending on the nature of the experiment. The read pulse terminates the sequence and is followed by detection of the FID.

Spin-Echo Techniques

The Spin-Echo Pulse

> *Hahn made the remarkable discovery that if he applied a second $\pi/2$ pulse a time* t *after the first pulse, miraculously there appeared another free induction signal at a time* 2t *after the initial pulse. He named the signal the* spin echo.
>
> —C. P. Slichter (*8*)

The introduction of multiple pulses resulted in the observation of the spin echo as first discovered by Hahn (*9*). The Hahn spin-echo pulse sequence is written

$$(90°_x - \tau_e - 90°_x - 2\tau_e - \text{Acq})$$

and is one of the most important pulse sequences in NMR spectroscopy (*9*). (Acq is acquisition.)

This spin-echo experiment causes cancellation of all effects that result from different Larmor frequencies, including those of chemical shifts and those produced by the nonuniformity of the magnetic field across the sample. The range of Larmor frequencies of these different isochromats increases the rate of decay of the observed transverse magnetization. When this dephasing process is reversed during the evolution time, spin echoes are observed.

The Carr–Purcell Echo Pulse Sequence

The Carr–Purcell pulse sequence (*10*) also generates echos and is considerably easier to visualize than the Hahn sequence. The pulse sequence is written

$$(90°_x - \tau_e - 180°_x - 2\tau_e - \text{Acq})$$

Visualization of the consequences of this pulse sequence can be made by using vector diagrams, and the vector diagram for the Carr–Purcell pulse sequence is shown in Figure 7.2 (*11*).

> A more complete diagram is shown in Figure 6.5 and this diagram should be consulted as needed.

The vectors represent the magnetizations or isochromats of the system and their responses to various pulses at the indicated times. The initial $90°_x$ pulse produces the coherence, and the dephasing of the isochromats occurs for a time, τ_e, as shown by the spread in the vectors in the diagram (Figure 7.2a). The magnetizations are then inverted by the application of a $180°_x$ pulse, and in the diagram (Figure 7.2b), the directions of all the vectors are reversed. The rotational motion of the magnetization continues in the opposite direction to reform the coherence and produce an echo at 2τ when all of the isochromats are back in phase (Figure 7.2c). The spin echo thus consists of two back-to-back FIDs. With the application of the 180° pulse, the amplitude of the spin echo is independent of field inhomogeneities and chemical shifts because these factors are refocused. The exponential decay of the amplitude of the echo arises from the spin–spin relaxation process.

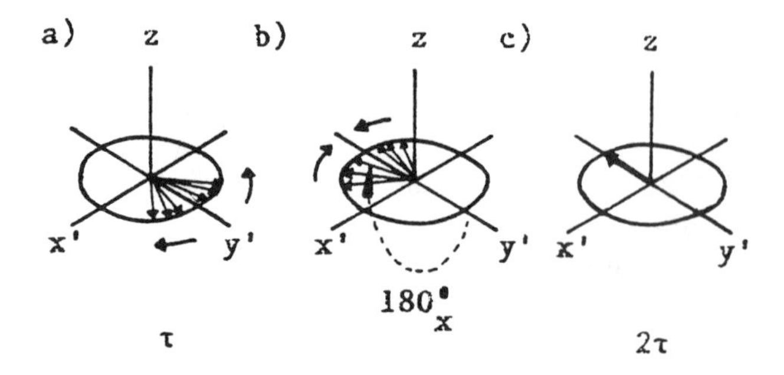

Figure 7.2. The vector diagram for the Carr – Purcell pulse sequence. In the spin-echo experiment, a is the state of the transverse magnetization at time τ after the 90°x excitation pulse, b is the effect of the 180°x pulse, and c is the state of the transverse magnetization after time 2τ. (Reproduced with permission from reference 11. Copyright 1983 Steinkopff Verlag Darmstadt.)

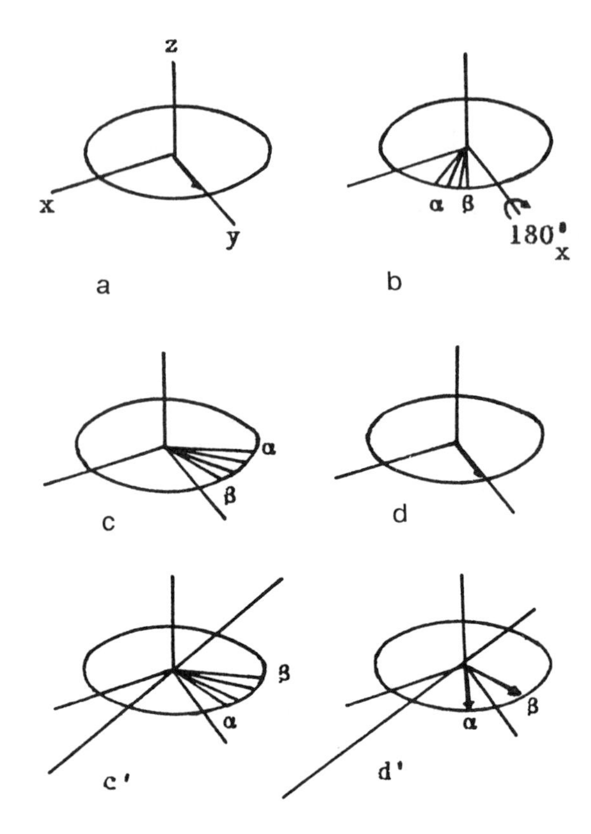

Figure 7.3. Vector diagrams illustrating the modulation of the amplitude of a Carr – Purcell echo. (a) Initial 90°x pulse at time zero. (b) Spin – spin coupling causes the magnetization to break up into fast (α) and slow (β) components, each exhibiting a spread of magnetization as a result of the inhomogeneity in the static field. The chemical shift causes a net precession of the mean frequency. (c) At time τ, a 180°x pulse rotates all the vectors into new positions that are reflections in the yz *plane. (d) If the coupling is to a nonresonance nucleus, a conventional spin echo is formed at time 2τ, and its amplitude is independent of the coupling constant, magnetic-field inhomogeneity, and chemical shift. (c′) If the coupling is to a resonant nucleus, then the 180° pulse at time τ inverts the spin states of that nucleus, interchanging the fast and slow components of the observed nucleus. (d′) At time 2τ, chemical shift and inhomogeneity effects are refocused, but spin coupling causes a continued difference between the fast and slow components, leading to a modulation of the echo amplitude according to cos (πJτ). (Reproduced with permission from reference 12. Copyright 1984 Progress in NMR Spectroscopy.)*

Separation of Spin – Spin Coupling from Chemical Shift Using the Carr – Purcell Echo Pulse Sequence

One of the most useful features of spin echoes is that they provide a method for the separation of the effects of spin – spin coupling from those of the chemical shift. Consider the case of two coupled homonuclear nuclei (an AX system) subjected to the Carr – Purcell spin-echo pulse sequence (Figure 7.3) (*12*). After a $90°_x$ pulse, the transverse magnetization from the A nucleus moves away from the *y* axis of the rotating frame because this nucleus is slightly off resonance from the resonating X proton. But the transverse magnetization will also be split into two components because of spin coupling to the X nucleus. One-half of the doublet will be farther from the transmitter frequency than the other, and will therefore move faster in the *xy* plane (Figure 7.3b). A 180° pulse is then applied, in this case about the *y* axis. The 180° pulse has two effects. It flips all isochromats into a mirror-image configuration, and it also interchanges the spin states of the X nucleus, which interchanges the slow and fast components of the A doublet. As a result of the latter reversal, the slow and fast components continue to diverge after the 180° pulse. Figures 7.3c′ and 7.3d′ show the vector diagram for this spin-coupling process. Although the chemical and inhomogeneity effects are refocused at time 2τ, two echo components occur, each dephased from the *y* axis. The phase angle between the two magnetizations becomes larger with time.

For a simple, first-order AX system, the modulation rate between the two components of the doublet is J Hz. The phase of each component of the doublet deviates from that of a singlet by an angle of $\theta\pi J\tau$ radians. Thus a first-order spectrum will refocus when $2\tau = 1/J$, and a doublet will be 180° out of phase with respect to the phase of a singlet. A similar description can be given for more complicated coupled systems. If the A and X nuclei are separate nuclear species (i.e., with heteronuclear coupling), then normal refocusing occurs.

Consequently, the echo sequence allows differentiation between chemical-shift effects (which are refocused) and homonuclear spin coupling (which modulates the echoes). This observation is the basis of some of the 1-D editing pulse sequences to be discussed, as well as of some 2-D experiments.

Proton Multiplicity of Carbons Using Heteronuclear $^1H-{}^{13}C$ Echo Modulation

Consider the addition of broad-band coupling of the protons after the 180° ^{13}C pulse.

	$90°_x$		$180°_x$		
^{13}C	——— |	$1/J$ |		$1/J$ |	FID
^1H	decoupler off		| decoupler on		

The ^{13}C NMR frequencies are shown as vectors at different times during the applied pulse sequence in Figure 7.4 (*13*). Each of the C–H types of carbons is illustrated separately. After the $90°_x$ pulse in the ^{13}C channel, all frequencies in the spectrum are in phase. The subsequent evolution of the frequency vectors is different for each type of C–H group. The signal of the quaternary carbon atom has no interaction with any hydrogen, so the signal is not affected by the fact that no ^1H decoupling is applied during the evolution period. In the rotating frame, the vector representing this signal is stationary (Figure 7.4a).

A doublet is observed for the methine C–H group in the nondecoupled spectrum. The frequency of one line of the doublet is $+J/2$ Hz from the multiplet center, and the other is shifted $-J/2$ Hz. Consequently, one vector moves faster than the rotating frame itself, and the other moves slower. In the evolution period, therefore, one frequency vector rotates clockwise, and the other rotates counterclockwise. Because both frequencies are shifted by the same amount from the center of the multiplet, the positions of the corresponding vectors are symmetric to the y axis at any moment. After a time evolution of $1/J$ seconds, both vectors have completed exactly one-half of a circle and are pointing in opposite directions. With the 180° pulse in the ^{13}C channel, a spin echo is induced to refocus all effects resulting from inhomogeneities either in the magnetic field or from chemical shifts. As the C–H coupling continues, both vectors continue to rotate independently and then combine again at exactly the same position where the quaternary vector is found.

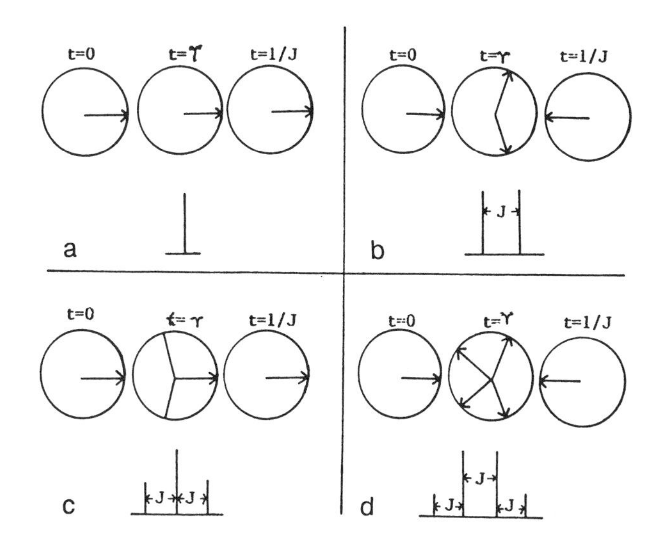

Figure 7.4 Vector modulation of ^{13}C spin evolution during the Carr–Purcell echo sequence: (a) quaternary carbon, (b) methine carbon, (c) methylene carbon, and (d) methyl carbon. (Reproduced with permission from reference 13. Copyright 1984 Magnetic Resonance.)

However, if broad-band decoupling is added, heteronuclear C–H coupling is no longer possible, and the dephasing of the two frequency vectors for the C–H stops (*14*). The C–H group is then represented by one single frequency that by convention is the frequency of the rotating frame. The superimposed vectors in this frame can no longer dephase but rather remain opposite to the quaternary carbon vector during the refocusing period. This condition results in the resonances of the C–H and quaternary carbon vectors having opposite phases in the recorded spectra. The differentiation between signals produced during the evolution period is preserved for the rest of the sequence. Similar considerations apply for the CH_2 and CH_3 groups. The CH_3 resonances will have inverted amplitudes similar to the C – H signals.

> *In general, new NMR techniques are only published when the scientist comes up with a sufficiently clever acronym for the technique.*
> –James W. Cooper and Robert D. Johnson (*15*)

The experiment just described is commonly termed the SEFT (spin-echo Fourier transform) pulse sequence (*14*) with gated-proton decoupling. The sequence relies on the modulation of the transverse magnetization through spin – spin couplings to distinguish between signals of quaternary, methine, methylene, and methyl carbons. By using appropriate delay times, singlets that differ in phase by 180° can be obtained. This result aids in the assignment process. When the delay time is equal to $1/J_{CH}$, peaks with positive amplitudes correspond to quaternary and methylene carbons, and those with negative amplitudes correspond to the methine and methyl carbons. Figure 7.5 shows the ^{13}C-NMR spectrum at 62.5 MHz of low-density polyethylene (LDPE, NBS 1476) obtained by using a SEFT pulse sequence with a delay time of τ = 8 ms. As stated previously, the positive peaks correspond to quaternary and methylene carbons, and the methine and methyl carbons have negative peaks. For LDPE, a small, positive peak at 44.47 ppm is assigned to a quaternary carbon and indicates the presence of a small concentration of tetrafunctional branch sites (*16*).

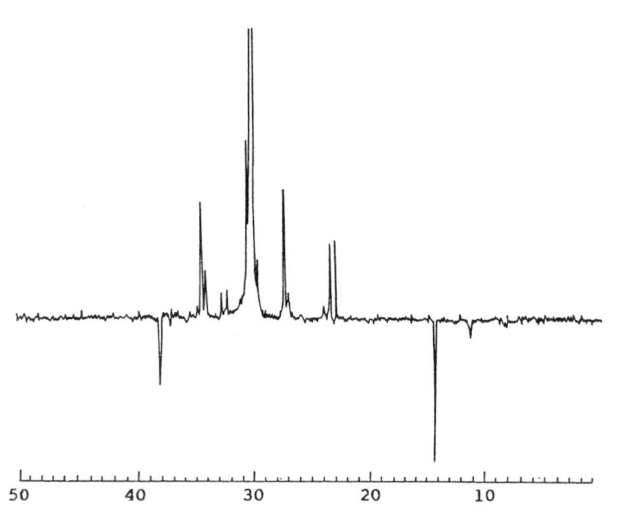

Figure 7.5. ^{13}C NMR spectrum at 62.5 MHz of LDPE (NBS 1476) obtained with the SEFT pulse sequence. (Reproduced with permission from reference 16. Copyright 1987 Pergamon Press, Inc.)

Spectral Simplification Using Decoupling Techniques

> *The most common problem in NMR spectroscopy is not a lack, but an overabundance, of information. For very complex molecules with many nuclei in subtly different environments, the spectra may consist of hundreds of lines, the assignments of which are exceedingly difficult. If there were some way to simplify the spectrum so that it depended only on the chemical shift, then it would be possible to assign the gross features of the spectrum, after which the couplings among the nuclei could be specified, with the known chemical shifts of the various nuclei as a guide. This is one of the goals of double-resonance experiments.*
> – M. D. Bruch et al. (*17*)

Double-resonance (decoupling) techniques generally result in a simplification of the NMR spectrum and yield information relative to the relationships between the resonances. This latter information reveals the internal chemical bonding of the sample under investigation.

To demonstrate these decoupling techniques, butyl rubber will be used as an example. Butyl rubber is a generic name for a family of isobutylene – isoprene copolymers that usually contain less than 3% isoprenyl groups. Butyl rubber exhibits a number of structural facets, including directional isomerism of the butyl and isoprenyl units and chemical isomerism of the isoprenyl groups, that is, 1-4, 1-2, and 3-4 structures and geometric isomerism (*Z* or *E*, which are *cis* and *trans*, respectively) (*18*). The 250-MHz ^{1}H NMR spectrum of butyl rubber in $CDCl_3$ is shown in Figure 7.6. The resonances at 1.11 and 1.41 ppm are assigned to the methyl and methylene protons, respectively, of the polyisobutylene units of butyl rubber. The isoprenyl units have resonances at 1.65, 1.94, and 5.05 ppm. A comparison of this spectrum with that of the model compound 2,2,4,8,8-pentamethyl-4-nonene shows that the isoprenyl units of butyl rubber are in a single isomeric form because the corresponding regions of the spectrum of the model compound (which is known to be a mixture of isomers) contains a pair of signals. The single triplet

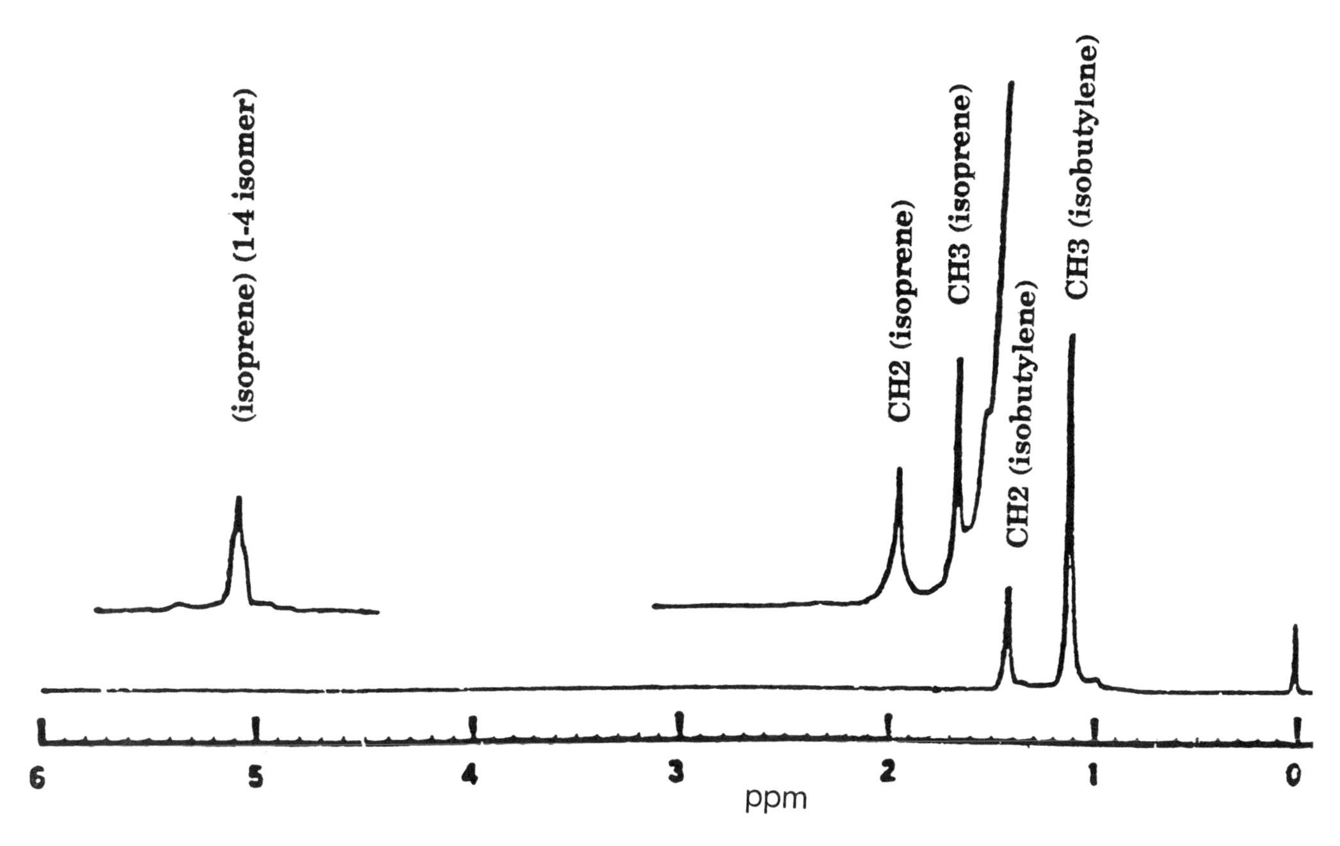

Figure 7.6. ¹H NMR spectrum of butyl rubber. (Reproduced from reference 18. Copyright 1985 American Chemical Society.)

at 5.05 ppm suggests the incorporation of the isoprenyl units in the 1-4 mode. If 1-2 or 3-4 incorporation occurred, additional signals would appear. When butyl rubber is prepared with deuterated isoprene, the signal at 1.65 ppm is absent, and the signal at 1.94 ppm is reduced to half of its original size. On the basis of these observations, the signals at 1.65 and 1.94 ppm are assigned to the methyl and methylene protons of the isoprene units, respectively. The isoprenyl groups can be in either *Z* (*cis*) or *E* (*trans*) form. The ¹H NMR spectroscopic results suggest that whereas the model compound has two sets of signals for most of the protons of isoprenyl residues, only one set is present in butyl rubber. The close agreement between the chemical shifts of the butyl rubber with the *E* isomer of the model compound suggests that the predominant geometry of the isoprenyl groups of the polymer is *E*. However, the *Z* isomer can be observed by using spectral subtraction as shown in Figure 7.7. The weak line at 1.70 ppm suggests the presence of the *Z* isomer at a level of approximately 10% (*18*).

Selective Scalar-Spin Decoupling

For the assignment of resonances, a *single* nucleus or spectral line can sometimes be selectively decoupled without perturbing the rest of the spectrum. Selective positioning of the proton frequency is used to remove a particular scalar-bond interaction. By using selective irradiation of specific protons, the coupling effects between the protons and carbon atoms to which they are attached can be eliminated. When the assignments of the proton spectrum are known, selective decoupling allows the assignment of a signal to a particular carbon atom in the ¹³C NMR spectrum. The spectral response to selective decoupling is generally quite specific and is invaluable for making assignments. The selectivity of a pulse increases with its length. In order to achieve a selectivity of 1 Hz, a pulse length of about 1 s must be used.

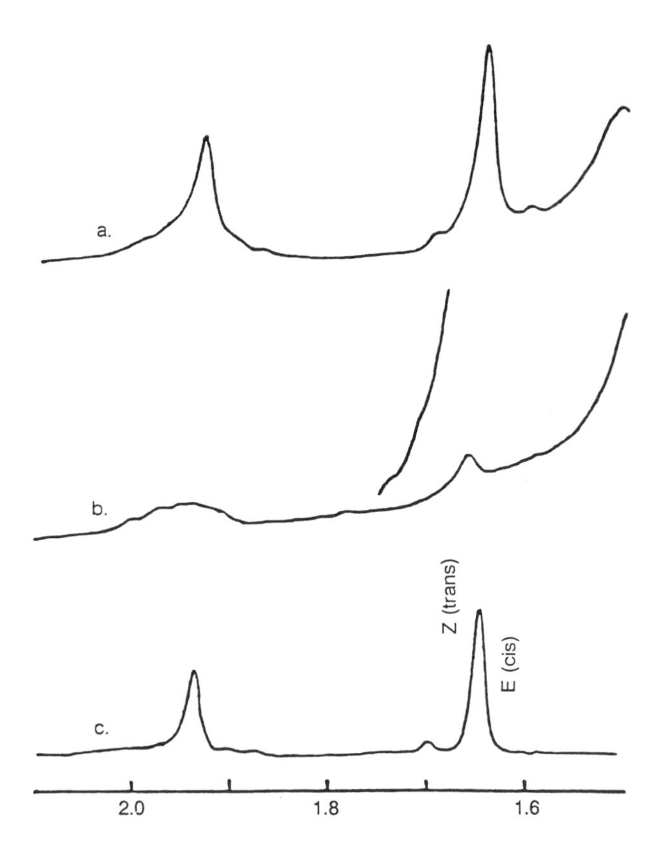

Figure 7.7. High-field region of the ¹H NMR spectrum of butyl rubber (at 250 MHz in $CDCl_3$). Spectrum a is of commercial butyl rubber, spectrum b is of specifically deuterated butyl rubber, and c is the a − b difference spectrum. The signal at 1.65 results from the E *isomer, and the weak signal at 1.70 results from the* Z *isomer. (Reproduced from reference 18. Copyright 1985 American Chemical Society.)*

For samples with well-resolved lines, the selective-decoupling method involves placing the rf field at the exact resonance frequency of the chosen resonance. Low rf power is used under these conditions so that only the selected resonance is affected. All other lines are far enough from resonance that their excitation is negligible. Because each line must be excited individually, the total time required for the selective-decoupling experiment can be long if there are a large number of lines in the spectrum.

One of the problems with selective decoupling is that the rf field actually irradiates a range of frequencies. When other resonances are within this frequency range, complete selectivity is not possible. In many cases, one spin cannot be irradiated independently of the others. Another limitation is that relaxation occurs during the pulse and limits the signal.

The spectrum of butyl rubber can also be used to illustrate selective decoupling. The assignment of the signal at 1.65 ppm to the methyl group of the isoprene residue in butyl rubber was confirmed by a selective spin-decoupling experiment. When the olefinic proton at 5.05 ppm is irradiated, the signal at 1.65 ppm becomes considerably sharper because of the removal of long-range coupling effects.

The off-resonance spectrum in the 0 – 60-ppm region of the ^{13}C NMR spectrum of butyl rubber is shown in Figure 7.8a (*18*). Selective decoupling by irradiation of the isoprenyl methylene proton resonance at 1.94 ppm causes the two triplets at 23.37 and 57.8 ppm to collapse to two singlets (Figure 7.8b). This observation indicates that these two signals result from the methylene carbons of the isoprenyl units of butyl rubber. On the basis of differences in chemicals shifts, these signals are assigned to C-4 and C-1, respectively. By using the methylene proton signals, which resonate at 1.41 ppm, of the isobutylene blocks selective spin decoupling collapses the resonances at 55.75 and 57.34 ppm into singlets (Figure 7.8c). Hence, these two resonances can be assigned to the methylene carbons from the neighboring isobutylene groups. Selective decoupling of the methyl group resonance of the isoprenyl group at 1.65 ppm collapses the methyl carbon signal at 19.21 ppm from a quartet to a doublet with residual splitting (Figure 7.8d). This residual coupling suggests the presence of long-range coupling. The measured coupling constant of 7.45 Hz for this resonance agrees with that measured for polyisoprene (J = 7.44 Hz).

Sometimes, one or more multiplets can be observed in a group of overlapping multiplets. The most direct method for selectively observing one of the multiplet patterns makes use of gated decoupling to selectively null the other multiplets. This nulling is accomplished by gating the decoupler during acquisition at the chemical shift of those protons that cause the splitting of the multiplet, which causes the splitting to be nulled. For example, by gating the decoupler on during acquisition, a doublet will be converted to a singlet with an intensity that is equal to the sum of the intensities in the rotating-frame projection (*19*).

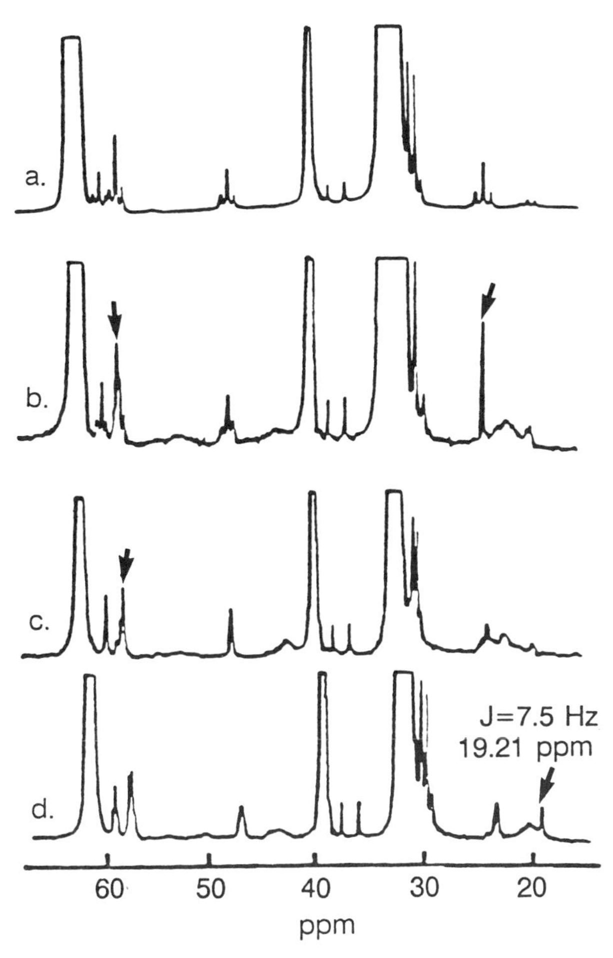

Figure 7.8. Amplified expanded ¹³C NMR spectra of butyl rubber (at 62.8 MHz in $CDCl_3$). The expanded off-resonance decoupled spectrum is shown in a. Spectra b, c, and d are the same as a except that the irradiation for the three spectra was 1.94, 1.41, and 1.65 ppm, respectively. The arrows indicate peaks that have collapsed. (Reproduced from reference 18. Copyright 1985 American Chemical Society.)

Off-Resonance Decoupling

Off-resonance decoupling is a more general and rapid procedure than selective decoupling because the entire spectrum is displayed. Off-resonance decoupling produces a coupled ¹³C NMR spectrum having collapsed multiplets and showing only the couplings to attached protons. Long-range C–H couplings are suppressed. Off-resonance decoupling reduces the J coupling to a fraction of its actual value. The reduced splitting, J_r, is related to the frequency offset of the decoupler from resonance, $\Delta\omega$, and inversely proportional to the strength of the decoupling power,

$$J_r = J_{CH}\Delta\omega\left(\frac{\gamma_H H_1}{2\pi}\right)^{-1} \qquad (7.2)$$

The signal multiplicities in the off-resonance spectrum can be used to discriminate between methyl, methylene, methine, and quaternary carbons. When the J_{CH} and the ¹H NMR chemical shifts are known, specific assignments can be made on the basis of the magnitude and multiplicity of the residual splittings.

Several disadvantages must be taken into consideration when using the off-resonance experiment. It is not completely successful in those cases where the signals appear as ill-defined, second-order multiplets. Also, severe overlap of resonance lines and slight changes in the ¹³C shieldings resulting from temperature effects can limit the value of the off-resonance experiments, especially when the spectral region contains a number of resonances. Finally, coherent off-resonance decoupling suffers from an effective sensitivity that is more than an order of magnitude lower than ordinary proton-decoupled ¹³C NMR spectroscopy.

An example of the use of off-resonance decoupling is shown in Figure 7.9 for the decoupled ¹³C NMR spectrum of butyl rubber. The signal at 31.28 ppm is associated with the methyl group carbon, the signal at 38.26 ppm is associated with the quaternary carbon bearing two methyl groups, and the signal at 59.28 ppm is associated with the carbon atom of the methylene groups.

The off-resonance experiment allows the assignment of the olefinic carbon atoms of the isoprenyl units in butyl rubber. The signal at 129.98 ppm appears as a doublet and is assigned to the olefinic (C-3) carbon. In a similar manner, the signal at 132.36 ppm, which remains a singlet in the off-resonance decoupling spectrum, is assigned to the quaternary carbon (C-2). Additionally, the off-resonance experiment allows the assignments of the methyl and methylene groups. The resonance at 19.21 ppm results from the carbon atom of a methyl group. In the same manner, resonances at 23.37, 45.63, and 55.81 ppm may be assigned to the methylene groups of isoprenyl groups and the isobutylene group attached to them (*18*).

The problem with coherent off-resonance decoupling is that the signals often appear with overlap or as ill-defined second-order multiplets, especially

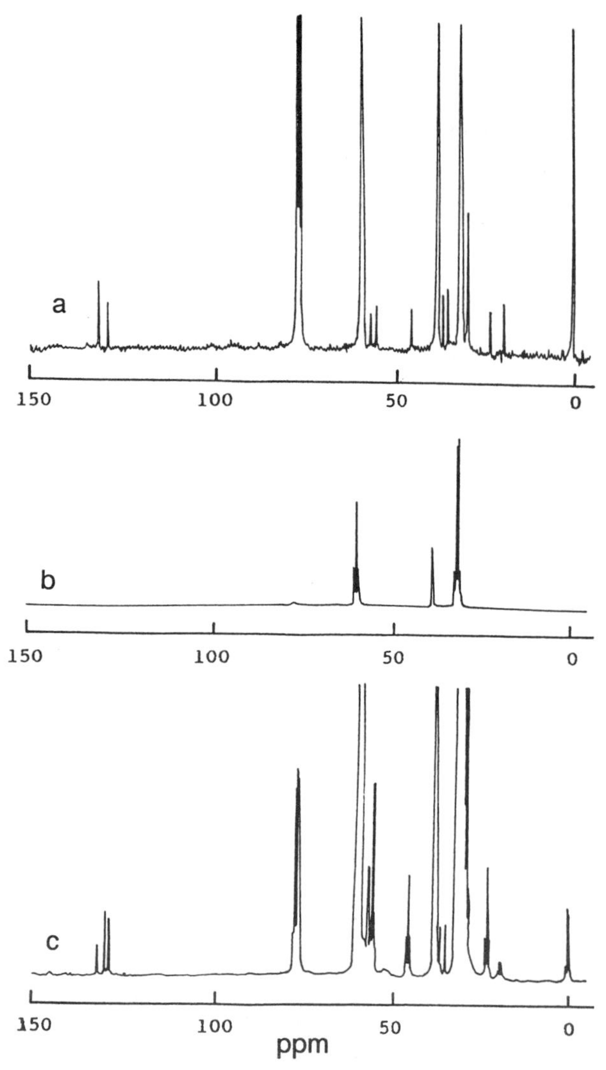

Figure 7.9. Spectra of butyl rubber at 62.8 MHz in $CDCL_3$: (a) amplified broad-band decoupled spectra; (b) off-resonance decoupled spectra of the isobutylene segment; and (c) amplified off-resonance spectra of butyl rubber. The signal at 31.28 ppm is associated with the methyl group carbon; the signal at 38.26 ppm is due to the quaternary carbon atom bearing two methyl groups, and the signal at 59.28 ppm is due to carbon atoms of the methylene groups. Reproduced from reference 18. Copyright 1985 American Chemical Society.)

when the spectral region of interest contains a number of resonances.

Selective Excitation of ^{13}C Resonances

Selective excitation of individual ^{13}C resonances is difficult because of the broad spectral window. Selective excitation can be achieved by using a tailored excitation procedure that involves the application of a *pulse train* (20 – 60 pulses). The pulse train is a series of m pulses with very small pulse angles, α, and small pulse delays, t_r, such that $m\alpha = 90°$. By using the proper values for α and t_r, ^{13}C resonances can be selectively excited. The delays are set to the inverse of the separation of the chosen resonance line from the carrier frequency position. The chosen line precesses 360° during each of the delays. The pulses therefore have a cumulative effect and turn this magnetization component down to align the y axis of the rotating frame. The remaining resonances do not have this exact synchronization. These sequences are called DANTE (delays alternating with nutations for tailored excitation) sequences and give the individual fully proton-coupled multiplets as separate spectra (*20, 21*). However, in complex spectra where extensive overlap of the lines occurs, it is difficult to distinguish the multiplets from each other.

Recently, this technique has been extended to measure $^{13}C - ^{13}C$ connectivity in solids (*22*). Another sequence, which has been termed SELDOM (selectivity by destruction of magnetization), allows selective excitation in different portions of the chemical-shift anisotropic spectrum of polymers (*23*).

> *NMR spectroscopists have an affinity for acronyms rivaled only by that of educators and generals.*
> –W. R. Croasmun and R. M. K. Carlson (*24*)

Selective Multiplet Acquisition

A suggested alternative to DANTE is based on selective acquisition of the desired signals rather than on selective excitation (*25*). A single hard excitation pulse is applied at the chemical-shift frequency of the carbon resonance whose multiplet is to be selectively observed. After a short delay time, τ, the FID is then acquired. Broad-band proton decoupling is applied during the pre-excitation period and also during excitation, but is turned off during the data acquisition. The delay time, τ, is incremented regularly, and as a result, the phase of the signal located at the pulse is unaltered. However, the off-resonance signals are altered and consequently decrease. Good selectivity is achieved if more than 25 increments are used.

Signal Enhancement Using Polarization-Transfer Pulse Sequences

Signal enhancement is often carried out by using polarization-transfer methods. These methods rely on the existence of a resolvable J coupling between two nuclei, with the polarization-rich proton serving as the polarization source and the ^{13}C nucleus serving as the receiver. Polarization is normally transferred via one-bond scalar couplings.

Selective-Frequency Polarization Transfer

Consider the mechanism of the selective polarization-transfer (SPT) experiment using a $^1H-^{13}C$ spin system (*26*). Such a weakly coupled spin system has four energy levels: one doublet each for the ^1H and the ^{13}C. The low-frequency line, H_α, in the proton spectrum arises from those molecules in which the proton is bonded to a ^{13}C in the α-spin state (lower energy). The high-frequency proton line, H_β , arises from those molecules in which the proton is bonded to a ^{13}C in the β-spin state (higher energy). A similar situation applies for the ^{13}C NMR spectral lines.

At thermal equilibrium, the polarization depends on the differences in the populations of the various levels. Because γ_H is approximately equal to $4\gamma_C$, the population differences for the carbon and proton transitions may be taken as 1 and 4 arbitrary units, respectively. Coupling of the nuclei implies that the state of the nucleus depends on the state of the coupled nucleus. When a $90°_x$ ^1H pulse is applied to the sample, the proton magnetization is rotated about the 90° axis. For the H_α resonance line, the magnetization will precess slowly in the rotating frame, and the H_β resonance line will precess more rapidly (further from the carrier frequency). These coupled spin states can be mixed by the action of an rf pulse. A 180° frequency-selective pulse can be applied to the H_α proton transition without affecting the other transition. This soft or selective 180° pulse simply interchanges the populations of the α and β levels. Inverting the H_α proton line causes the proton populations of the energy levels associated with the proton H_α-transition to be interchanged. This interchange affects the intensities of the carbon transitions associated with these same energy levels. A measurement of the ^{13}C NMR spectrum shows intensities corresponding to population differences across the carbon transitions. The intensity of the C_α and C_β signals is not a 1: 1 doublet. Instead, it is a doublet with intensities of -3 and $+5$ units, respectively. The polarization transfer shows a fourfold enhancement of the normal ^{13}C signal. In addition to this fourfold enhancement, further signal enhancement occurs because the repetition rate for signal averaging depends on the shorter ^1H T_1 rather than on the longer ^{13}C T_1.

Experimentally, the population inversion can be achieved by using a selective $180°_x$ pulse applied at the frequency of *any* selected ^1H line. The SPT experiment is carried out with the ^1H decoupler applied immediately after the ^{13}C pulse at the frequency of one of the ^{13}C satellite lines. The limitation of the SPT experiment is that only one line can be enhanced at a time.

Unfortunately, there are a number of difficulties with this experiment. The precise positions of the ^{13}C satellite lines (i.e., lines arising from the 1% of protons that are spin coupled to ^{13}C nuclei) in the ^1H NMR spectrum must be known. A soft pulse that is exactly centered on the chosen satellite line must be used. However, the principle of polarization transfer is used in a number of NMR experiments.

The SPT experiment also provides evidence of a connected proton transition. Therefore, this technique can be used to pick out the resonant frequencies of proton transitions while observing the connected ^{13}C nuclei.

General Polarization-Transfer Methods

There is a pulse sequence that results in general polarization enhancement rather than the selective enhancement of the SPT experiment. The basic INEPT (insensitive nuclei enhancement by polarization transfer) pulse sequence for nonselective polarization transfer is as follows (*27*):

^1H $90°_x$ - - τ - - - 180° - - τ - - $90°_y$ $\tau = 1/(4J_{CH})$

^{13}C ______ 180° ______ 90° acquire FID

This special pulse sequence uses strong nonselective pulses and gives general enhancement rather than specific sensitivity enhancement, as in SPT. The pulse sequence has a basic polarization-transfer portion that produces a net inversion of one of the proton spin states. Proton transverse magnetization is created by the initial proton 90° pulse and precesses for a period τ. The magnetization is then refocused as a spin echo at a time 2τ by the action of a 180° proton pulse. Application of a ^{13}C 180° pulse at the midpoint of the 2τ delay ensures that the echo is modulated by the scalar coupling, J_{CH}. If τ is chosen to be equal to $1/(4J_{CH})$, then at time 2τ, the two proton magnetizations are completely out of phase as required for polarization transfer. The second proton $90°_y$ pulse, which is phase shifted to give a rotation about the y axis, then rotates one magnetization to lie along the $+z$ and $-z$ axes. Because of the relatively small range of values for J_{CH}, a single experiment produces enhanced signals from all protonated ^{13}C sites in a molecule. Following a simultaneous ^{13}C 90° pulse, there exists enhanced magnetization in the ^{13}C multiplet. The ^{13}C doublet has equal but opposite intensity components with values of 1 and −1 to give a characteristic "up−down" pattern, or an intensity ratio of 1: −1. This inverse character can be removed by addition of an extra delay, 2Δ, after the 90° ^{13}C pulse, which allows the ^{13}C to realign to give a net signal.

The repetition rate of the experiment is determined by the T_1 of the proton, and more signal per unit time can be obtained than with nuclear Overhauser effect (NOE) enhancement, which depends on the T_1 of the carbon nucleus. Therefore, the sensitivity gain is approximately $(C_{T_1}/H_{T_1})^{1/2}(\gamma_H/\gamma_C)$.

Determination of Proton Multiplicity of Carbons

Many of the new NMR methods rely on the fact that it is possible to modify the Hamiltonian almost at will to extract desired information···. The Hamiltonian can be manipulated to such an extent that some experiments border on sorcery.

– R. R. Ernst et al. (*5*)

The INEPT Experiment

The INEPT sequence for polarization transfer just described can also be used with slight modification to determine the multiplicity of the bonding of the carbons (*28*). If a delay, Δ, is introduced prior to simultaneous decoupling and data acquisition, then selective nulling or inversion can be accomplished depending upon the number of protons coupled to the nucleus being studied. Thus, if $\Delta = (2J)^{-1}$, where J is the heteronuclear coupling constant between ^{13}C and ^1H, then only signals from carbons bearing a single proton are observed in the ^{13}C INEPT spectrum. In contrast, if $\Delta = 3(4J)^{-1}$, then the ^{13}C signals for CH_3 and CH groups appear normal, and the CH_2 carbon resonances are inverted.

For example, the INEPT technique has been applied to low-density polyethylene and ethylene−propylene copolymers (*29*). The results of the ^{13}C-INEPT experiment at 125.77 MHz for the methine and methylene regions are shown in Figure 7.10. The delay time for the INEPT experiment was $3/4J$, so methyl and methine carbon resonances appear positive and methylene carbon resonances appear negative. The observed 15 different methine carbons suggest that at least 15 different branch types exist. In addition, ethylene−1-hexene copolymers were studied with the INEPT method, and similar results were obtained (*30*). Chemical-shift assignments of regioirregular polypropylene have been verified by using the INEPT experiment (*31*). Weak lines corresponding to carbons located in the inverted portion of the chain can be assigned by using this technique.

The INEPT experiment often provides too much information. When a molecule has a large number of carbons, the spectrum can be rather complex because of all the proton-coupled fine structure. A major disadvantage of the INEPT technique is that it can lead to severe distortions of the intensities of the components of a coupled multiplet. There is also a strong dependence on J, and as a result, signals that should be null may not be, a result that leads to extraneous lines. Because variable delay times are

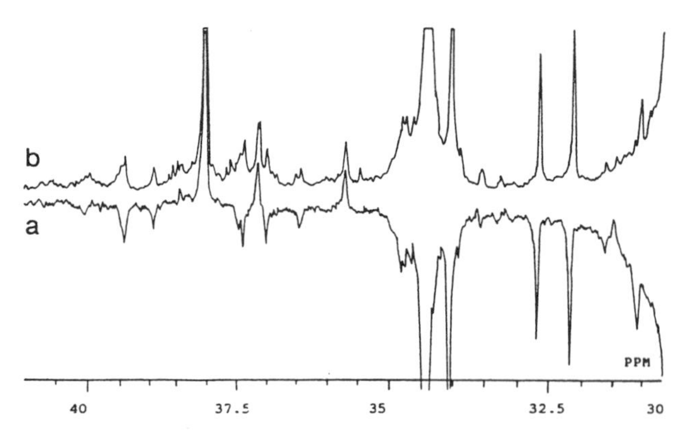

Figure 7.10. ^{13}C INEPT spectrum (a), and the proton-decoupled spectrum at 125.77 MHz (b) of LDPE. (Reproduced from reference 29. Copyright 1984 American Chemical Society.)

used for multiplicity selection, variations in T_2 can cause additional problems (*30*).

The DEPT Experiment

The DEPT (distortionless enhancement by polarization transfer) experiment was designed to eliminate many of the difficulties encountered in the INEPT experiment (*32, 33*). DEPT requires fewer pulses than INEPT and therefore is less sensitive to pulse missettings and inhomogeneities. DEPT is more sensitive to the spin-relaxation rate than the INEPT experiment because the relevant delay in DEPT is of the magnitude $3/(2J)$, and corresponding periods in INEPT typically range from $1/J$ to $5/(4J)$.

The pulse sequence is shown in Figure 7.11. By choosing the proton nutation angles of $\Theta_1 = 45°$, $\Theta_2 = 90°$, and $\Theta_3 = 135°$ in separate spectra, CH, CH_2, and CH_3 subspectra can be obtained from proper combinations of the spectra (*34*).

$$CH = \Theta_2 - c(\Theta_1 + a\Theta_3) \quad (7.3)$$

$$CH_2 = \Theta_1 - a\Theta_3 \quad (7.4)$$

$$CH_3 = \Theta_1 + a\Theta_3 - b\Theta_2 \quad (7.5)$$

where a, b, and c have theoretical values of 1.0, 0.707, and 0, respectively. Slightly different values may be required because of experimental factors such as the probe and spectrometer setups.

DEPT was used to generate subspectra of copolymers of styrene (S) with maleic anhydride (MAn) (*35*). The normal spectrum of an S – MAn copolymer with a composition of 52% styrene consists of three broad resonances that display fine structure (Figure 7.12). The three subspectra show good separation (Figure 7.12). In the CH spectrum, the multiplet at 29 ppm is a solvent impurity. The peaks at 44 and 52 ppm result from methine carbons. From the CH_2 subspectrum, the signal at 35 ppm is associated with the methylene carbons. Methylene subspectra were obtained from four copolymers with different compositions to allow the comonomer sequence assignments to be made for the observed resonances.

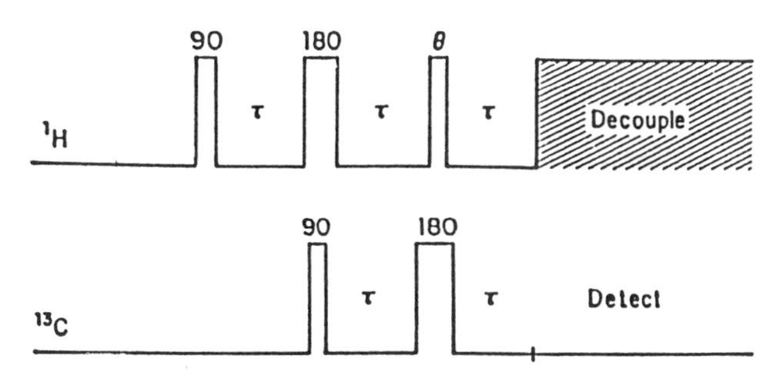

Figure 7.11. Pulse sequence for DEPT spectroscopy. The third ¹H pulse is variable with Θ = 45°, 90°, and 135°. The delay time τ is set to $(2J_{CH})^{-1}$. (Reproduced from reference 34. Copyright 1986 American Chemical Society.)

DEPT is a very simple and useful sequence for the determination of the proton multiplicity of carbons and is of considerable value for the study of polymer systems in solution and in the solid state (*34, 36*).

Determination of Carbon Connectivity–The INADEQUATE Experiment

NMR spectroscopy can be used to determine the connectivity between adjacent carbon atoms on a polymer chain. This information is usually obtained from $^{13}C - ^{13}C$ spin – spin couplings that are ordinarily lost in the NMR spectrum because the signals due to $^{13}C - ^{13}C$ couplings are 1/200th the intensity of the signal from a single ^{13}C. However, a pulse sequence that has been developed suppresses the normal ^{13}C

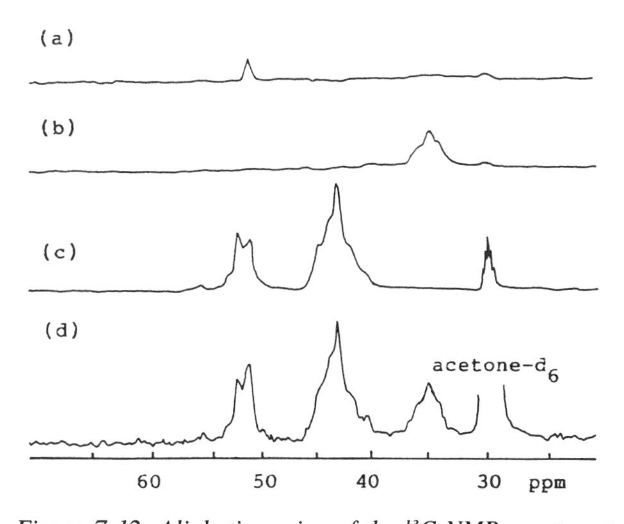

Figure 7.12. Aliphatic region of the ^{13}C NMR spectra at 75.46 MHz of S – MAn. The subspectra generated by the DEPT sequence are CH_3 (a), CH_2 (b), and CH (c), and d is the normal spectrum obtained with gated decoupling to remove nuclear Overhauser enhancement. (Reproduced from reference 35. Copyright 1984 American Chemical Society.)

NMR signal compared to the coupled $^{13}C-^{13}C$ signals on the basis of the presence or absence of the homonuclear spin – spin coupling, J_{C-C}. This experiment is termed INADEQUATE, which stands for incredible natural abundance double quantum transfer experiment (*37*).

> I wonder which was more difficult: coming up with the acronym or developing the experiment. I also wonder which came first.

This experiment takes advantage of the phase differences between single- and double-quantum coherences. The unwanted signals are removed through phase cycling. The pulse sequence for INADEQUATE is given as

$$90°_x - \tau - 180°_y - \tau - 90°_x$$

Discrimination between single- and double-quantum coherence is accomplished by setting τ equal to $1/4J_{C-C}$.

Two-Dimensional NMR Experiments

> *Probably the greatest recent change in the practice of NMR has been the explosive growth in techniques and applications of 2-D NMR.*
>
> – G. Gray (*30*)

> *Two-dimensional spectroscopy should not be dismissed as merely an alternative (and quite complicated) display mode, for it has some inherent operating advantages. The first is the possibility of disentangling spectra that are complicated by extensive and multiple overlapping of spin-multiplet structure–the problem that eventually makes very large molecules intractable for NMR.*
>
> – G. Bodenhausen et al. (*38*)

As was established in Chapter 6, high-resolution NMR spectroscopy is a highly sensitive tool for the study of polymers. But NMR spectroscopy is often limited because of the considerable overlap arising from the multiplicity and relative broadness of resonances. The usual methods of off-resonance and selective decoupling to determine the multiplicity of carbon bonding are not effective because of the inherent limitation of selectivity in the presence of the large number of overlapping resonances. Consequently, structure assignments to resonances are often difficult and arbitrary.

One method that can be used to overcome these limitations is the 2-D NMR experiment (*38*). To make a 2-D experiment out of a 1-D experiment, the evolution time must be stepped incrementally through a range of values, and a series of FIDs is stored, each with a different evolution time. After the first Fourier transformation, the resulting first-domain spectra are aligned parallel to one another to form a matrix of data points. Transposition of the matrix gives a new series of data points that can be subjected to a second Fourier transformation. This process is done for each data point of the original first-domain spectrum. When the resulting spectra from the second Fourier transformation are again arrayed in a matrix, the chemical shifts can be aligned along one axis, and the frequency, which is dependent on the phase or amplitude modulation that developed during the evolution period, can be aligned on the other axis. Thus the scale could be spin – spin coupling. The resolution in the 2-D spectrum is better than in either the ^{13}C or ^{1}H spectrum alone, but the sensitivity is lower. In 2-D NMR experiments, the resolution increases as the square of the field strength. At the higher static field strengths that are available on modern NMR instruments, the advantages of 2-D NMR experiments further increase.

In 2-D NMR spectroscopy, the raw data are a function of two time periods, an evolution time (t_e), which may also include a mixing period (t_m), and a detection time (t_d), both of which are preceded by a preparation time as shown in the following diagram.

	t_e	t_m	t_d
preparation	evolution	mixing	detection

The preparation period involves preparing the reservoir of nuclei in a known state by using T_1 processes or saturation pulses. The evolution period, t_e, (including the mixing period for some 2-D experiments) is the time in which the spins are subjected to a preselected interaction as a result of designed pulse sequences. The evolution time is systematically incremented throughout the course of the experiment to accommodate the desired spin interactions. During the detection period, t_d, a second selected interaction or combination of interactions occurs. Double Fourier transformation yields a 2-D NMR spectrum with two frequency axes

$$S(t_e, t_d) = FT^2 = S(\omega_e, \omega_d) \qquad (7.6)$$

In 2-D NMR spectroscopy, the experimental conditions (i.e., the pulse sequences) can be arranged such that only one interaction is operative during the evolution time and a second or combined interactions are predominant during the detection period. 2-D FT-NMR spectroscopy is possible if a systematic variation of the evolution period results in a modulation of the spin system, such as through spin – spin interactions, chemical-shift interactions or dipolar interactions.

In principle, the three classes of 2-D spectra are *coupling-resolved*, *coupling-correlated*, and *exchange* 2-D spectra. A coupling-resolved spectrum is characterized by one frequency axis (F_1) containing the coupling information and the other axis (F_2) containing the chemical shifts. In the coupling-correlated type of spectrum, both frequency axes contain chemical shifts. The connection between the F_1 and F_2 axes is established through homonuclear or heteronuclear scalar coupling or through dipolar coupling. In exchange 2-D spectroscopy, the interaction is the magnetization exchange resulting from either chemical exchange, conformational or motional effects, or the Overhauser effect arising from nonbonded protons.

Unfortunately, the sensitivity in 2-D experiments is a factor of 3 less than in a regular dipolar-decoupled ^{13}C experiment (*38, 39*). Because of the large volume of data required, the acquisition times are quite long. Furthermore, large data matrices are required for the 2-D experiment, and the processing time is tedious. Another concern is "t_1 noise", which generates phantom peaks in the 2-D spectra. These phantom peaks occur because of spectrometer instabilities resulting from the extended acquisition times needed for these measurements (7 – 14 h). Further limitations can arise when extensive overlap makes assignments difficult in the 2-D maps (*38, 39*).

2-D experiments utilize a particular class of pulse sequences in which the selected interaction modulates the magnitude of the resonance. Under this circumstance, when the height of any peak is plotted as a function of τ, this height will vary *sinusoidally* rather than exponentially. That is, the peak height will *oscillate* (and decay) as a function of τ. This oscillatory behavior can be analyzed to determine the frequency or frequencies contributing to the oscillations. The frequencies are determined by taking a second Fourier transform as a function of τ.

Consider the example termed 2-D *J*-resolved spectroscopy. Spin echoes, which are generated by the pulse sequence $90° - \tau - 180° - \tau - 180° - 90°$, are modulated by spin – spin coupling. If the peak heights are extracted from each of the spectra and plotted as a function of t, an interferogram that is the superposition of the contributing frequencies is obtained. When analyzed by Fourier transformation, this interferogram determines the contributing frequency or frequencies of oscillation. Double transformation generates a *J* spectrum for each chemically shifted resonance and displays this multiplet structure in the second dimension–"an operation similar to the opening of a Venetian blind" (*38*).

In the actual 2-D procedure, the first interferogram is formed from the first data point from each of the original spectra, and the second interferogram is formed from the second data point from each of the original spectra. Mathematically, this process is a simple matrix transposition. In fact, if these spectra are thought of as the rows of the data matrix $S(t_e, F_2)$ then the columns represent the variation of the intensity of a given frequency component F_2 as a function of τ. The result of this complete process is a 2-D frequency map, on which the data are plotted in the form of a series of spectra (stacked or "white-washed" plot) or in the form of a contour plot in which each successive contour represents a higher intensity.

Because the process begins with a series of "normal" 1-D spectra, the normal spectral axis will be on one axis of a 2-D plot. This axis represents the signals actually detected in the receiver of the spectrometer during a time conventionally labeled t_2, and hence this axis becomes the F_2 axis. The second axis (or *domain*), known as the F_1 axis, is the axis along which the frequencies of oscillation of the original peaks will appear following the second Fourier transformation.

An overwhelming number of 2-D experiments have been described, often in humorous acronyms that further contribute to the confusion in this area. In this chapter, the information available from selected basic 2-D experiments will be summarized; I will neglect, for the moment, methods of dynamic chemical and conformational exchange. Jelinski has compiled a useful table that also helps in this regard (*40*).

For macromolecules, which typically have broad resonances and appreciable spectral overlap, 2-D experiments are sometimes difficult. Additionally, macromolecules have long correlation times, and dipolar $^1H-^{13}C$ interactions make considerable contributions to both ^1H and ^{13}C line widths. Consequently, the sensitivity of the 2-D experiment is low for polymer systems. A 2-D experiment may be impossible to interpret because of the rapid ^1H relaxation during the evolution period, or impractical because of the insensitivity of ^{13}C detection. The types of information that can be obtained from the various 2-D experiments described in this chapter are summarized in Table 7.1.

2-D Correlation via Heteronuclear Chemical Shifts

A 2-D experiment has been designed to correlate the chemical shifts of protons with the corresponding directly bonded carbons (*41*). As a correlation experiment, the ^1H spin serves as a spy nucleus for its attached ^{13}C nucleus. The information obtained from these experiments is the resonance frequency of the proton(s) coupled directly to the various ^{13}C nuclei in the molecule (i.e., which carbon resonance is correlated with which proton resonance) (*42, 43*). Because the ^{13}C chemical shifts are generally better known with respect to their structural origin, this 2-D method permits the interpretation of the ^1H NMR spectra. With this knowledge, the highly sensitive ^1H NMR spectra can be used to determine the polymer microstructure. The main limitation of this 2-D correlation experiment is that it functions only for carbons that have directly bonded protons.

The description of the 2-D $^1H-^{13}C$ correlated spectroscopy experiment is given by Bax (*44*). Consider the case of a single $^1H-^{13}C$ system whose spectrum consists of two doublets, one in the ^{13}C spectrum and one in the ^1H spectrum. A differential proton-spin inversion is first accomplished by using two 90° proton pulses applied along the x axis of the rotating frame and separated by a time t_e. The proton signal is allowed to precess after a 90° rf pulse for an evolution time t_e, with the extent of this precession being proportional to the extent to which the peak of interest is off resonance from the ^1H transmitter frequency. This step makes the phase of the ^{13}C signal dependent on the ^1H chemical shift. Intensity is transferred from the protons to the observed nucleus by simultaneous 90° pulses.

The longitudinal proton magnetizations are proportional to $-\cos(\Omega_H + \pi J)t_e$ and $-\cos(\Omega_H - \pi J)t_e$, where Ω_H is the proton frequency offset resulting from the chemical shift, and J is the C–H coupling constant. The changes in longitudinal ^{13}C magnetizations, ΔI_1 and ΔI_2, are

$$\Delta I_1 = \cos(\Omega_H + \pi J)t_e - \cos(\Omega_H - \pi)t_e \quad (7.7)$$

$$\Delta I_2 = \cos(\Omega_H - \pi J)t_e - \cos(\Omega_H + \pi)t_e \quad (7.8)$$

for the two ^{13}C doublet components, respectively. As a result, the phases of the individual ^{13}C signals are coded with the ^1H chemical shift. These changes in longitudinal ^{13}C magnetization are observed by using the application of a 90° ^{13}C pulse after the second

Table 7.1. Information Content of the Various 2-D NMR Experiments

Experiment	F_1	F_2	*Information*
Heteronuclear *J*-resolved	J_{CH}	δ_C	Heteronuclear coupling constants
Homonuclear *J*-resolved	J_{HH}	δ_H	Homonuclear J and δ
Heteronuclear Chemical Shift	δ_H	δ_C	Correlation of δ_H and δ_C
Correlation via homonuclear scalar coupling (COSY)	δ_H	δ_C	Correlation of all scalar coupling interactions
NOE (NOSEY)	δ_H, J_{HH}	δ_H, J_{HH}	Spatial proximity of nonbonded protons
INADEQUATE	$\delta_A + \delta_X$	δ_X	Heteronuclear connectivities

proton pulse. The amplitudes of the two ^{13}C doublet components are modulated in amplitude with the proton frequencies but in an opposite manner.

By using 2-D $^{1}H-^{13}C$ correlated spectroscopy, a 2-D data matrix with the carbon chemical shifts along one axis (F_2) and the proton chemical shifts along the other (F_1) is obtained. The projections onto the two axes give the corresponding full signal, except that resonances having no corresponding shift for the other nucleus (e.g., nonprotonated carbons or hydroxyl protons) do not appear. Connectivities are established by noting the frequencies at which the peaks appear along both the carbon and the proton axes. The intensities of the (^{1}H, ^{13}C) cross peaks are proportional to the number of protons attached to the specific carbon atom. Correlations via the C – H bond are most efficient when the evolution period t_e is equal to $1/2J_{CH}$ (125 – 160 Hz), where J_{CH} is the one-bond C – H coupling constant. The value of t_d is set to $1/(3J_{CH})$ when proton-decoupled spectra are recorded. When long-range (vicinal) couplings are being studied, longer delays are required.

The 2-D heteronuclear chemical-shift correlation experiment just described is most easily understood in the simplified-doublet form that was presented. The multiplet structure complicates the results, and the protons must be decoupled from the ^{13}C nucleus by application of a 180° ^{13}C pulse at the midpoint of the evolution period. Proton decoupling can be applied during the detection period as well.

Bax (*44*) gives a detailed discussion of the experimental aspects of this 2-D heteronuclear chemical-shift correlation experiment and suggests methods for the suppression of the axial peaks as well as of errors resulting from quadrature ^{1}H detection.

For the application of this heteronuclear 2-D technique to polymers, there is the usual problem of low sensitivity resulting from limited solubility and excess viscosity of the resulting solutions. For polymers, additional problems arise because of the low ^{13}C sensitivity, the rapid relaxation of the ^{1}H nucleus during the evolution time, and large line widths that limit the resolution. The major consequence of long polymer-correlation times is that dipolar $^{1}H-^{13}C$ interactions may make considerable contributions to both ^{1}H and ^{13}C line widths. Thus, even in the 2-D experiments with polymers, considerable overlap can occur.

Because of the short relaxation times, the signal may decay to an unacceptable level before acquisition has occurred. Solutions to this problem are to use high static magnetic fields and ^{13}C enrichment. Alternatively, a large amount of sample can be used or the data collection can be limited to a small portion of the ^{1}H spectrum in order to decrease the total experimental time while maintaining the digital resolution in the second dimension (*45*).

For studies of the tacticity of polymers, the heteronuclear 2-D technique can be used to make absolute configurational-sequence assignments for the ^{1}H spectra based on the connectivity with the carbon chemical shifts. The theoretical basis of the assignment procedure for methylene systems arises from the patterns that result from the presence of homosteric and heterosteric methylene protons. Figure 7.13 shows the pattern resulting from $^{1}H-^{13}C$ correlations of methylene dyads (*46*). A single carbon shift correlated to a single proton shift indicates a sequence with equivalent or homosteric methylene protons. A single carbon shift correlated to two proton shifts indicates a sequence in which the methylene protons are nonequivalent or heterosteric. This property can be used to distinguish certain stereosequences (*40*).

An excellent demonstration of the power of 2-D heteronuclear chemical-shift correlations to determine tacticity has been given for polyvinyl chloride (PVC) (*47*). The 500-MHz ^{1}H and 125-MHz ^{13}C spectra of PVC at 65 °C are shown in Figure 7.14. The methylene region is not resolved in the 500-MHz ^{1}H spectrum, but in the $^{1}H-^{13}C$ 2-D correlation spectrum, separate proton resonances are observed for each stereosequence (Figure 7.15). For PVC, the rmr, rrm, and mmr methylene sequences are correlated with more than one proton resonance. Thus the proton methylene stereosequences can be correlated by using a knowledge of the ^{13}C assignments and the correlation with one or two proton shifts of the corresponding ^{13}C sequence. In Figure 7.15, the methine and methylene assignments for both the carbons and the protons are indicated. In the methylene region, there is no simple relationship in the ordering between the chemical shifts of the stereosequences in the ^{13}C and ^{1}H spectra. However, with a knowledge of the ^{13}C triad assignments, as is the case with PVC, the proton assignments can be made. In the methine region, the rr, mr, and mm triads have the same

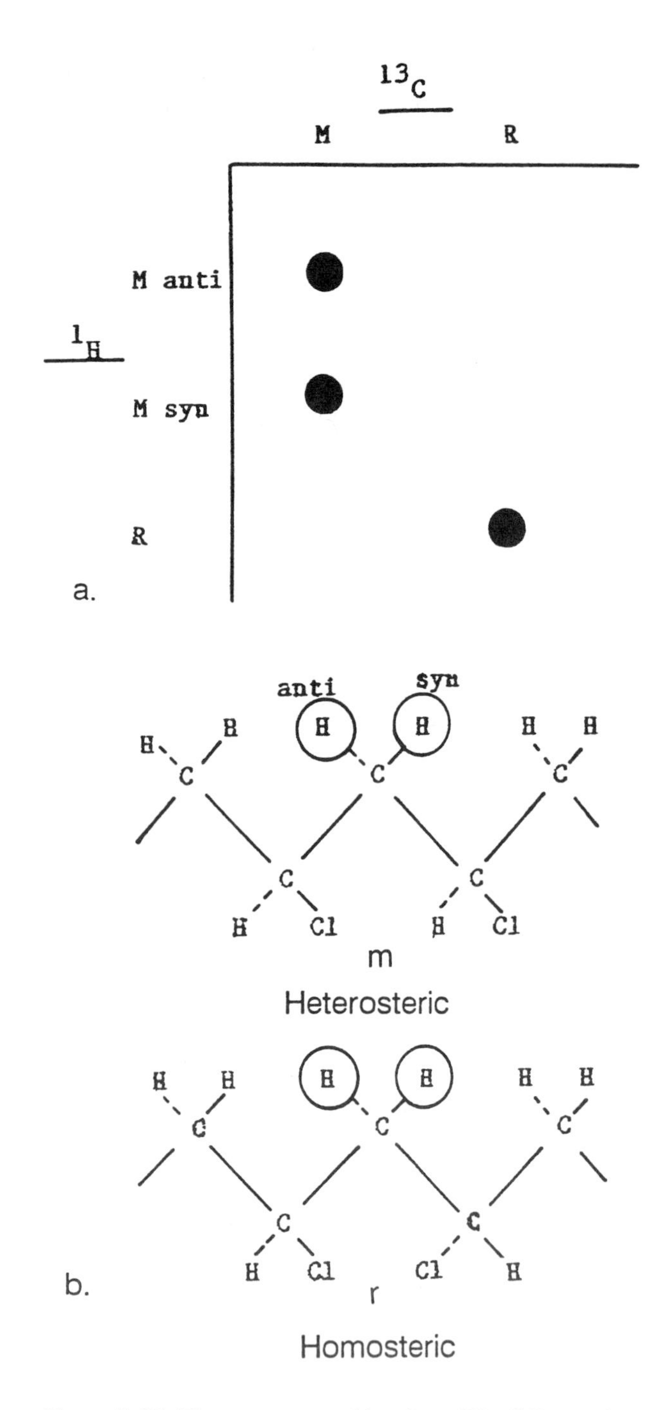

Figure 7.13. The pattern resulting from $^1H - ^{13}C$ correlations of methylene dyads (a) and the heterosteric and homosteric configurations (b). (Reproduced from reference 46. Copyright 1986 American Chemical Society.)

relative ordering along both the ^{13}C and ^1H axes. These differences in the ordering of the chemical shifts for the ^{13}C and the ^1H spectra indicate that different factors influence the ^{13}C and ^1H chemical

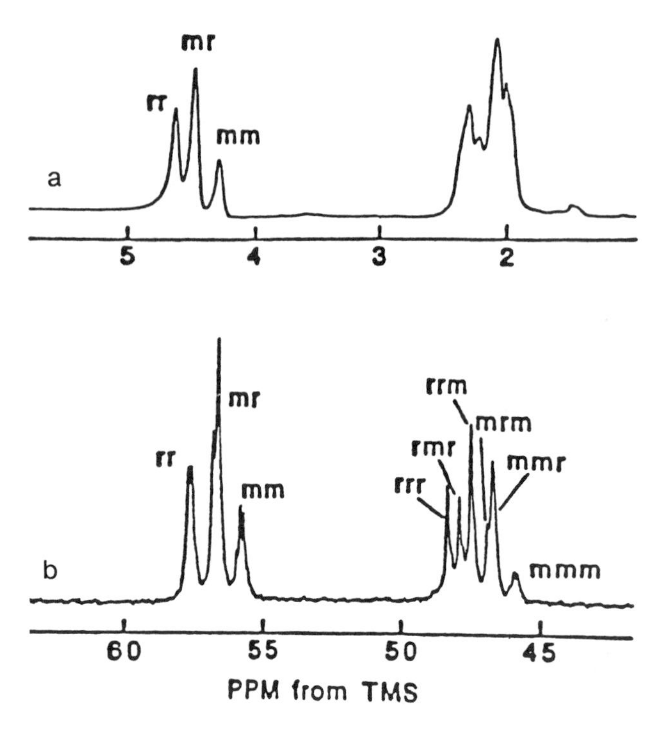

Figure 7.14. The 125-MHz ^{13}C (a) and 500-MHz ^1H (b) NMR spectra of PVC at 65 °C. (Reproduced from reference 47. Copyright 1986 American Chemical Society.)

shifts. Mirau and Bovey (*47*) suggest that "The ^1H chemical shifts are determined by the carbon type (methyl, methine, or methylene) and the inductive effects due to substituents and ionizable groups. ^{13}C chemical shifts are sensitive to these factors but also to the rotational isomeric state of the polymer."

The $^1H - ^{13}C$ heteronuclear-shift correlation has been used to assign the tactic sequences observed in the ^1H NMR spectrum of polypropylene (PP) (*48*). The $^1H - ^{13}C$ heteronuclear-shift correlation map is shown in Figure 7.16 for isotactic PP (*49*). The correlated assignments for both the ^{13}C and ^1H spectra of isotactic PP are given in Table 7.2. The $^1H - ^{13}C$ heteronuclear-shift correlation map for atactic PP is shown in Figure 7.17. This correlation map is considerably more complicated. The methine and methyl portions of the $^1H - ^{13}C$ heteronuclear-shift correlation map are given in Figure 7.18. For the methine region, even the 2-D technique does not resolve all the resonances, and individual assignments cannot be made. However, the chemical shifts for the pentads of the methyl groups occur in the same order in the ^{13}C and ^1H spectra. Therefore, the

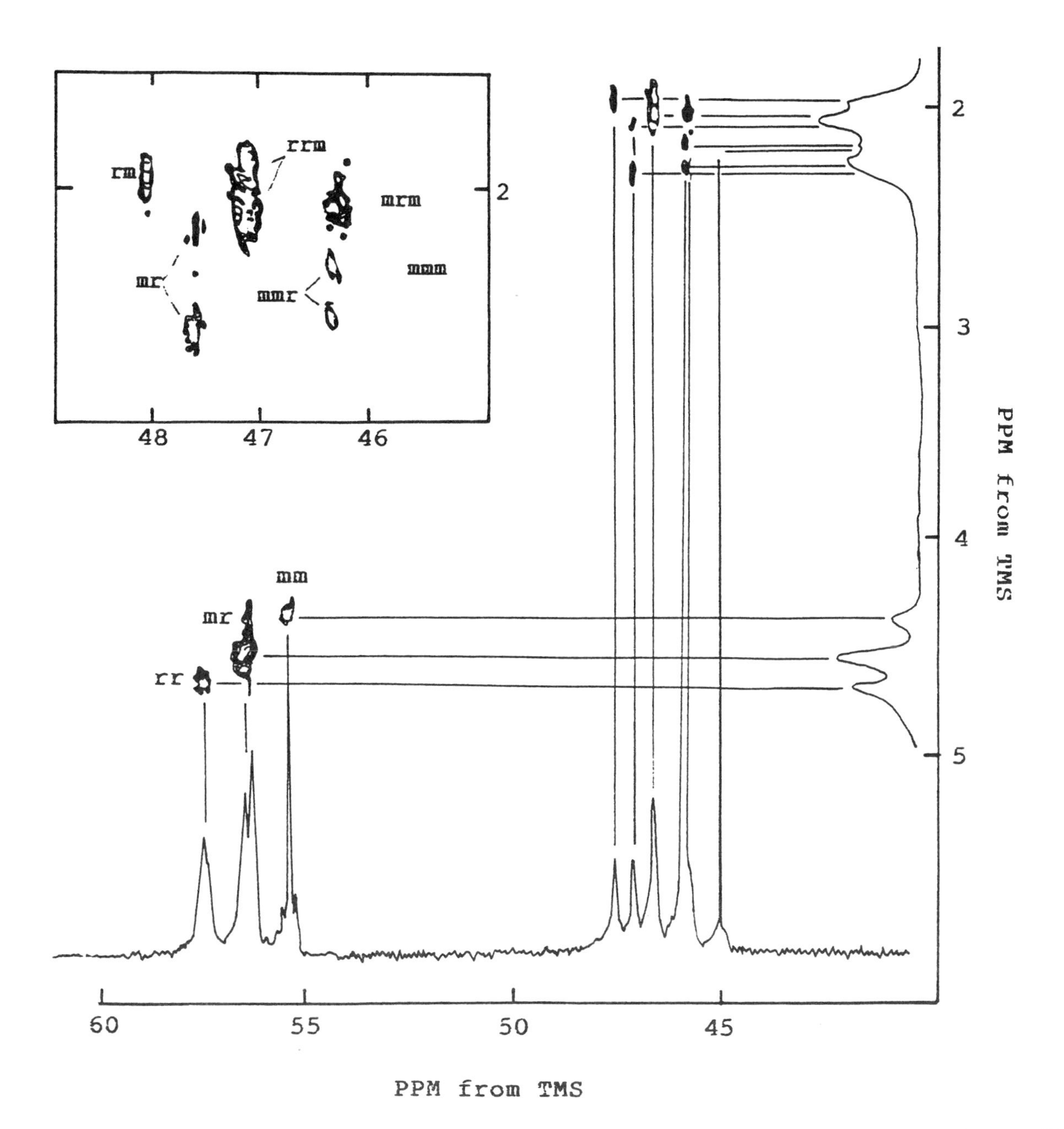

Figure 7.15. ${}^{1}H-{}^{13}C$ *heteronuclear-shift 2-D correlation NMR spectrum of PVC at 65 °C. Insert is an expansion of the methylene region. (Reproduced from reference 47. Copyright 1986 American Chemical Society.)*

correlations and assignments can be made, and these are given in Table 7.3.

The methylene region of the contour plot of stereo-irregular PP is very complex, and correlations of the ${}^{13}C$ hexads with the ${}^{1}H$ tetrads have been made (*48*). This 2-D technique was also used for ethylene – propylene copolymers (*49, 50*).

One of the ways to overcome the sensitivity problem of polymer molecules in 2-D NMR spectroscopy is to use ${}^{13}C$-enriched systems. Not only is the signal of the enriched carbon higher relative to the other naturally abundant resonances, but a very narrow line width (corresponding to the resonance being observed) can also be recorded to give high data-point

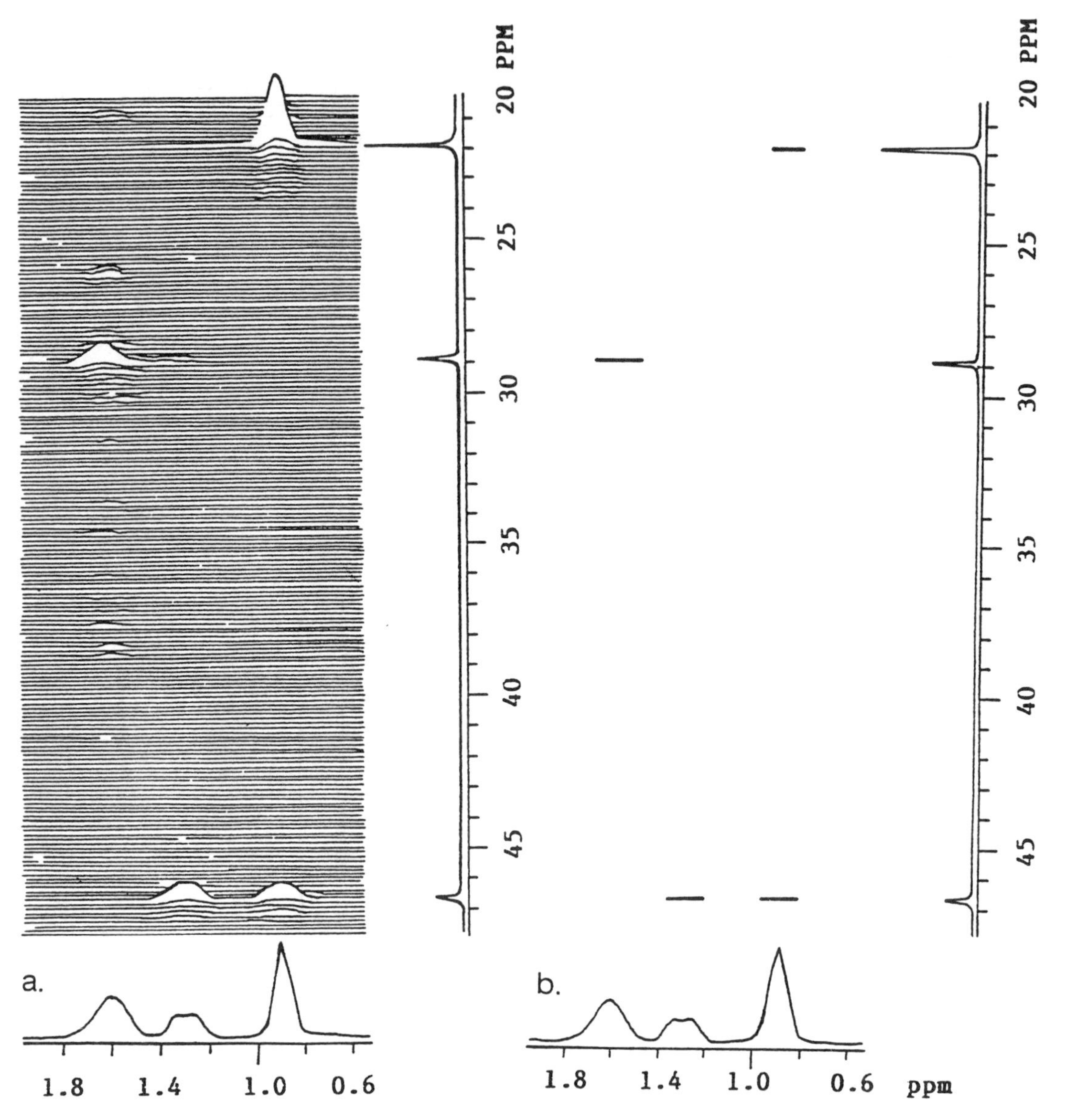

Figure 7.16. The $^1H-^{13}C$ heteronuclear-shift correlation map for isotactic PP: (a) correlation map, (b) contour plot. (Reproduced with permission from reference 49. Copyright 1985 Springer-Verlag.)

Table 7.2. NMR Spectral Assignments for Isotactic Polypropylene

Group	^{13}C Shifts (ppm)	^1H Shifts (ppm)	^1H Coupling Constants (Hz)
CH_3	21.8	0.87	$J_3(CH_3 - CH) = 6.56$
CH	28.7	1.60	$J_3(CH - CH_2, s) = 6.9$
			$J_3(CH - CH_2, a) = 6.1$
CH_2	46.6	0.89 (syn)	$J_2(CH_2, gem) = -13.6$

SOURCE: Reproduced with permission from reference 48. Copyright 1985.

resolution. This approach has been used for the $^1H-^{13}C$ heteronuclear-shift correlation to assign the tactic sequences observed in the ^1H NMR spectrum of poly(methyl methacrylate) (PMMA) by using a $^{13}C=O$-enriched sample (*51*). Figure 7.19 shows cross sections through the 2-D NMR spectrum. These cross sections are parallel to the ^1H axis and pass through the most intense ^{13}C signals assigned to the $^{13}C=O$ pentads. Because of the presence of long-range coupling, correlations occur between the $^{13}C=O$ pentads and the methylene tetrads. The mrr

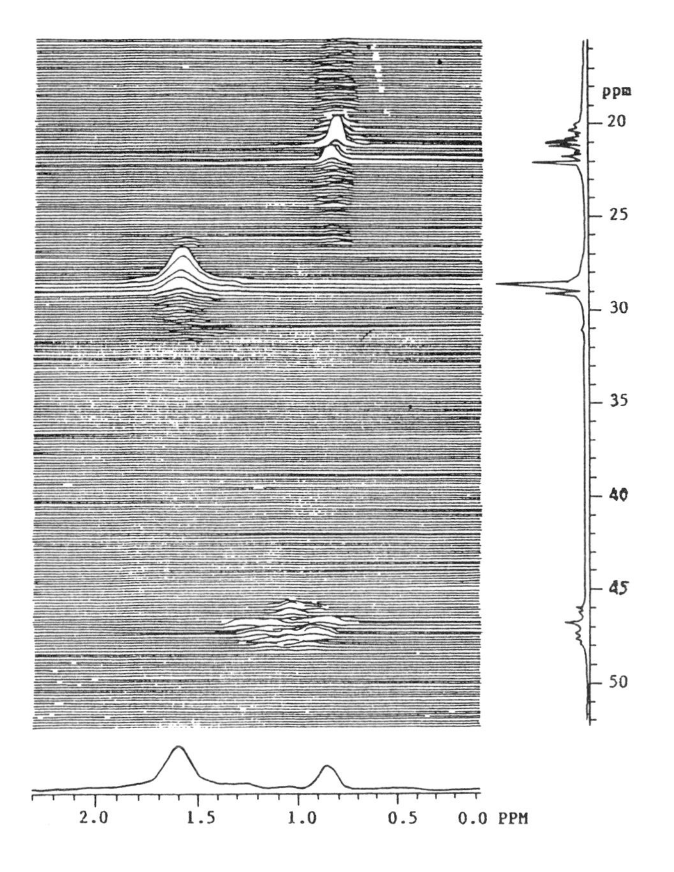

Figure 7.17. The $^{1}H-{}^{13}C$ heteronuclear-shift correlation map for atactic PP. (Reproduced with permission from reference 49. Copyright 1985 Springer-Verlag.)

Table 7.3. Methyl Assignments for Stereoirregular Polypropylene

Pentad	*^{13}C Shifts*	*^{1}H Shifts*
mmmm	21.8	0.871
mmmr	21.6	0.867
rmmr	21.4	0.859
mmrr	21.0	0.856
mrmm	20.8	0.852
rmrr	20.8	0.852
mrmr	20.6	0.846
rrrr	20.3	0.845
rrrm	20.2	0.837
mrrm	19.9	0.835

NOTE: All values are given in parts per million.

SOURCE: Reproduced with permission from reference 48. Copyright 1985.

and mrm tetrads give rise to two ^{1}H-proton signals as a result of the two types of methylene hydrogens, and the rrr and mrm tetrads give only one signal because all four methylene hydrogens are equivalent. The proton stereosequence assignments are shown in Figure 7.19.

For poly(vinyl butyral) (PVB), the structural problem is even more complex because the system is actually a copolymer of butyraldehyde (BA) rings and residual hydroxy groups from the poly(vinyl alcohol) (PVA). Thus PVB has both stereosequences and comonomer sequences in the chain. This structural complexity is reflected in the ^{1}H and ^{13}C NMR spectra. However, rather detailed ^{13}C assignments have been made for PVA, and the BA rings can be assigned as well. The 75-MHz ^{13}C spectrum of a 5% solution of PVB in $(CH_3)_2SO$-d_6 is shown in Figure 7.20. The numberings for the structures in PVB are also shown in this figure. The ^{1}H NMR spectrum is complex and badly overlapped, as shown in Figure 7.21. The line assignments for the ^{13}C NMR spectrum are transferred to the proton spectrum by 2-D $^{1}H-{}^{13}C$ correlated spectroscopy, as shown in Figure 7.22. In the case of PVA, the 2-D $^{1}H-{}^{13}C$ correlated results do not reflect the hydroxyl resonances because the hydroxy protons have no directly bonded carbons (*52*). The chemical shifts and homonuclear coupling constants for PVB at 100 °C in $(CH_3)_2SO$-d_6 are shown in Table 7.4.

2-D Correlation via Homonuclear Scalar Coupling (COSY)

The first 2-D NMR experiment was based on phase modulation by the spin–spin coupling (*53*). The standard sequence is $90° - t_1 - 90° - t_2$. After the preparation period, a 90° pulse causes the protons to precess and acquire phases that differ according to their differences in chemical shifts and the homonuclear *J* interactions. The final 90° pulse samples the orientations of the various magnetization components and records them by the spectrum in t_2. Because each proton in the spectrum has its own chemical shift as a modulation in t_e, the resulting doubly transformed data has peaks with identical ω_1 and ω_2 values, which are further modulated in ω_1 by the homonuclear *J* coupling. The correlated spectroscopy (COSY) spectrum shows all the proton resonances correlated with themselves (i.e., autocorrelated) along the diagonal.

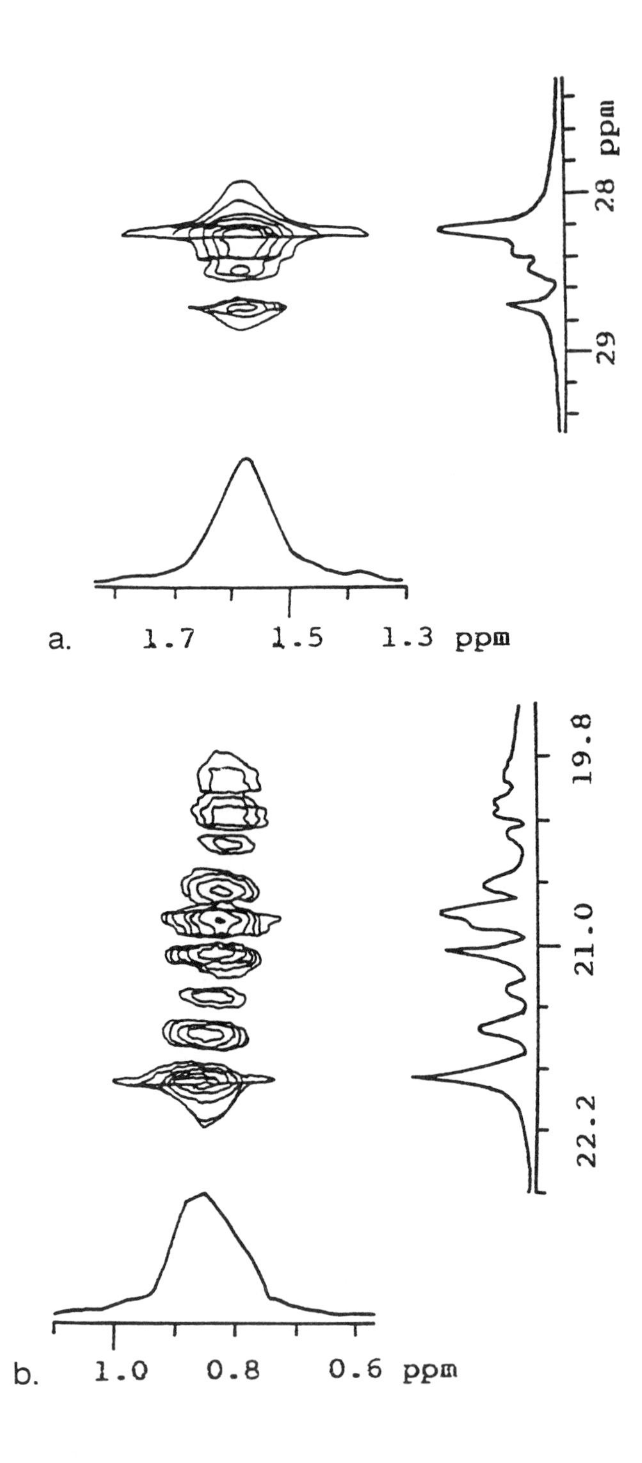

Figure 7.19. The ¹H NMR spectrum of 10%-enriched poly(methyl[carbonyl-¹³C] methacrylate) in chlorobenzene-d_5 at 108 °C (a). Cross sections are shown through the 2-D NMR spectrum in b. (Reproduced from reference 51. Copyright 1986 American Chemical Society.)

Figure 7.18. The contour plots of the methine portion (a) and the methyl portion (b) of stereoirregular PP. (Reproduced with permission from reference 49. Copyright 1985 Springer-Verlag.)

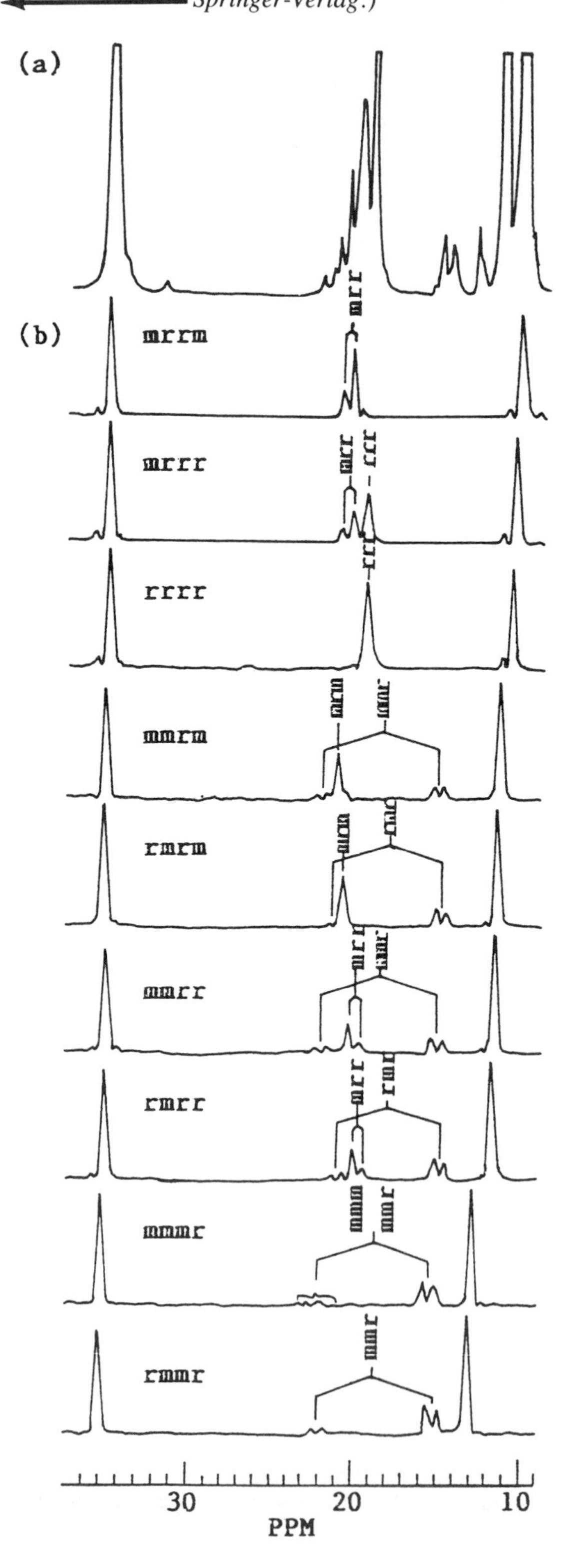

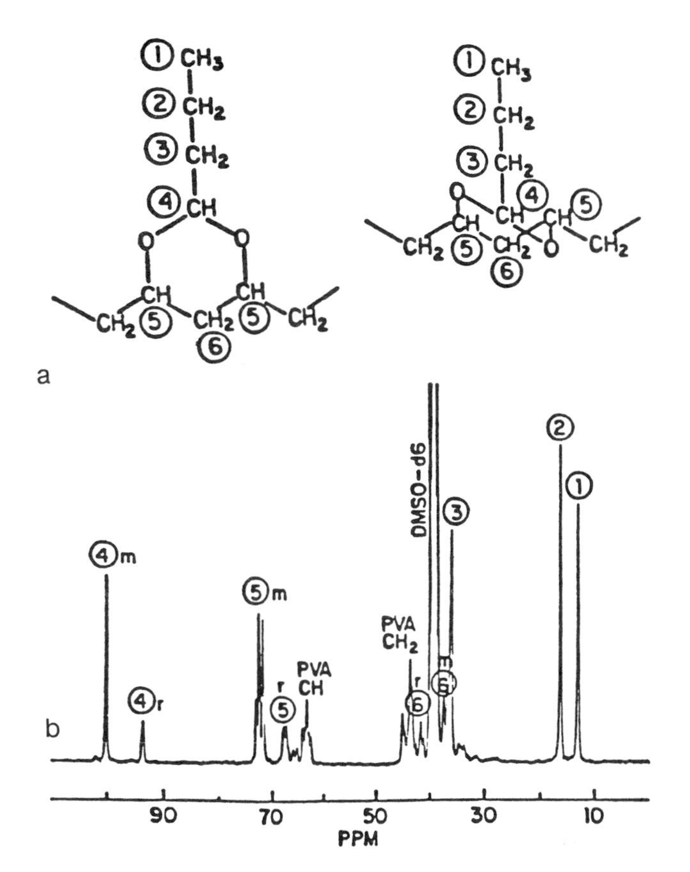

Figure 7.20. *(a) Structures for various stereosequences in PVB, meso (m) and racemic (r) butyraldehyde rings. The circled numbers are used to identify carbons and protons in the rings. (b) The 75-MHz* ^{13}C *NMR spectrum of a 5% solution of PVB in* $(CH_3)_2SO$-d_6. *(Reproduced from reference 52. Copyright 1986 American Chemical Society.)*

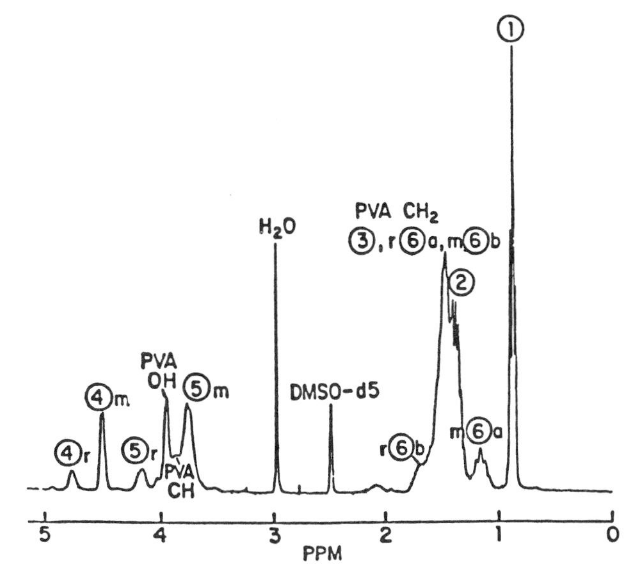

Figure 7.21. The 300-MHz ^1H *NMR spectrum of a 2% solution of PVB in* $(CH_3)_2SO$-d_6 *at a temperature of 100 °C. The labeling is the same as that in Figure 7.20. (Reproduced from reference 52. Copyright 1986 American Chemical Society.)*

The off-diagonal peaks or cross-correlated peaks show which protons are spin coupled to one another. In this fashion, it is possible to trace the vicinal and geminal coupling interactions between protons. The pattern of the cross-peak network is unique for the connectivity of the protons. This technique provides a map of the *J*-coupling network within the polymer.

Vinyl polymers have a highly coupled proton system, and these proton–proton interactions can be used to correlate both the nearest and next-nearest neighbors. The complexity can be illustrated by using the ^1H spectra of poly(vinyl alcohol) (PVA) (*54*). An ^{1}HCOH hydrogen has three-bond scalar couplings to four different CH_2 hydrogens, and the HCOH resonances consist of 16 different frequencies. By using the methine–methylene three-bond

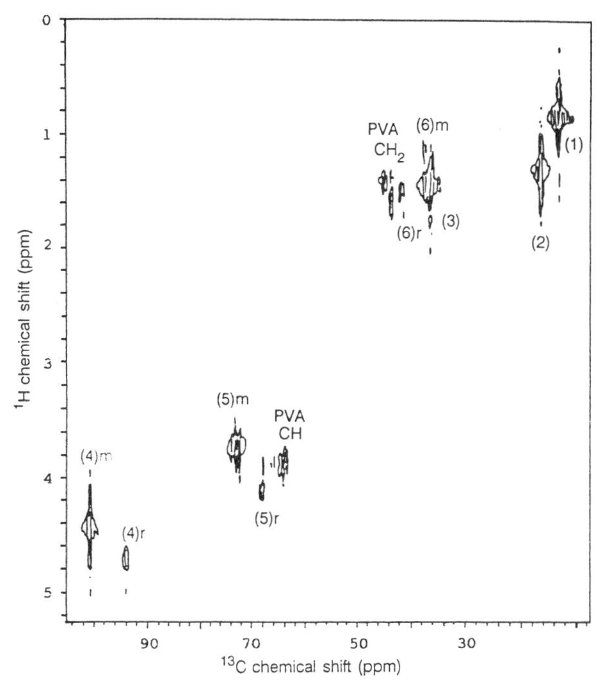

Figure 7.22. The 2-D $^1H-^{13}C$ *correlated NMR spectrum of PVB at 100 °C. The* ^{13}C *shift of each peak is indicated on the horizontal axis, and the corresponding* ^1H *chemical shifts of all directly attached protons are indicated on the vertical axis. The labeling is the same as that in Figure 7.20. (Reproduced from reference 52. Copyright 1986 American Chemical Society.)*

Table 7.4. Chemical Shifts and Homonuclear Coupling Constants for PVB in $(CH_3)_2SO$-d_6 at 100 °C

Assignment	*^{13}C Chemical Shift (ppm)*	*^{1}H Chemical Shift (ppm)*	*Homonuclear Coupling Constant (Hz)*
meso BA CH_4	100.2	4.51, 4.49, 4.47	5.0
rac BA CH_4	93.5	4.79, 4.77, 4.75	4.9
meso CH_5	73.1, 72.5, 71.8	3.77	*e*[a]
rac CH_5	68.1 – 67.3	4.15	*e*
PVA CH	65.5 – 62.2	3.91	*e*
PVA CH_2	45.4 – 42.8	1.44	*e*
rac CH_2 6: a	42.0, 41.4	1.50	*e*
rac CH_2 6: b	42.0, 41.4	1.71	*e*
meso CH_2 6: a	37.7, 36.9	1.16	*e*
meso CH_2 6: b	37.7, 36.9	1.48	*e*
meso BA CH_2 3	36.2	1.48	*e*
rac BA CH_2 3	36.2	1.41	*e*
BA CH_2 2	16.4	1.38	*e*
BA CH_3 1	13.2	0.88	7.4
mm PVA OH		4.05	6.1
mr PVA OH		3.99	5.8
rr PVA OH		3.95	5.5

[a] The letter *e* designates those assignments for which a homonuclear coupling constant was not observed.

SOURCE: Reproduced from reference 52.

coupling, the COSY experiment can correlate both the nearest and next-nearest neighbors. This correlation is illustrated in Figure 7.23a. In a vinyl homopolymer, the two hydrogens of the central CH_2 group of an m dyad are homosteric; that is, they have different environments (H_b and H_c), but the two hydrogens of the central CH_2 of an r dyad have identical chemical shifts (H_a). The COSY experiment for a polymer with dyad – triad resolution in the ^1H spectrum should show the cross peaks between dyad and triad hydrogens in the pattern shown in Figure 7.23b when the two m dyad hydrogens have different chemical shifts. The addition of more units on the sequences has similar effects overall and allows the lower stereosequences to be correlated with higher ones through the COSY experiment.

An example is the COSY spectra of the HCOH hydrogens and CH_2 hydrogens of PVA, with some scaling of the frequency axis for clarification of presentation, shown in Figure 7.24 (*54*). Twelve different multiplet cross peaks can be observed. The required compositional relationships result in a complete assignment of the resonances in PVA, and these assignments are listed in Table 7.5.

The 500-MHz COSY spectrum of PVC at 65 °C is shown in Figure 7.25 (*47*). The solid lines show the correlation between triad and tetrad resonances. When the methylene protons are equivalent, two correlations are expected for rr (rrr and mrr) and mm (mmm and mmr), and four correlations are expected for mr (mrr, mrm, rmr, and mmr). Some assignments can be made from the correlation spectrum. For example, the resonance at the highest field triad has a single cross peak in the methylene region and must result from the mm – mmr interaction. The largest number of cross peaks is in the middle triad sequence, so this resonance must be assigned to mr. The lowest field methine peak results from rr and is expected to show cross peaks to mrr and rrr. However, because of the missing cross peaks and the

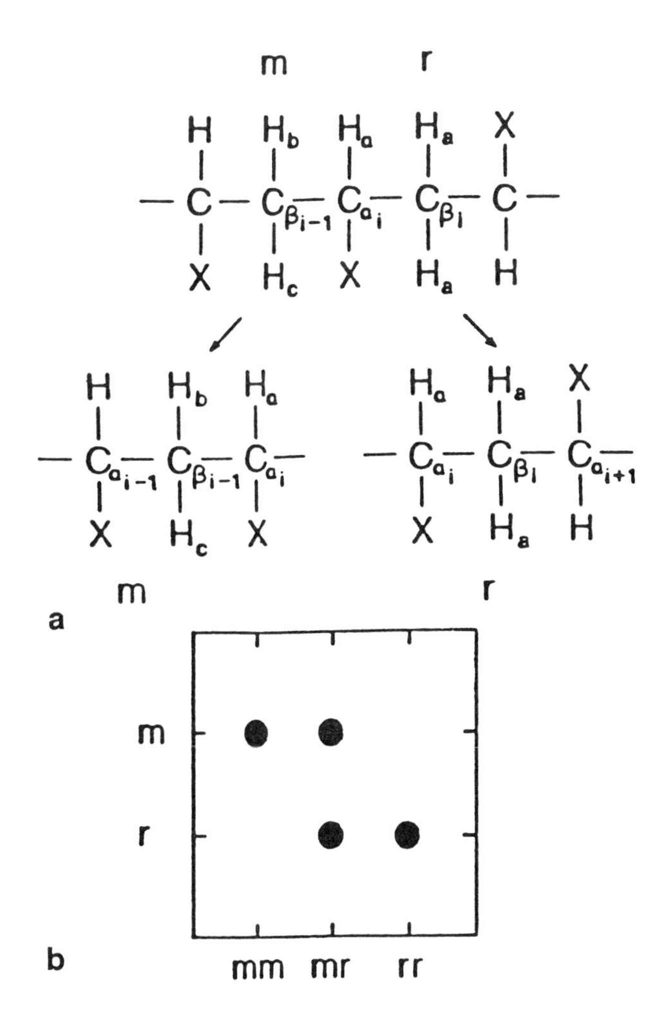

Figure 7.23. The structure of a meso-racemic (mr) triad of a vinyl polymer (a), showing the constituent meso (m) and racemic (r) dyads. The compositional pattern that would be observed for a COSY experiment is shown in b. (Reproduced with permission from reference 54. Copyright 1984 Springer-Verlag.)

Table 7.5. Assignment of Resonances in PVA

Configuration	*δ (ppm)*
HCOH	
rr	4.062
mr	4.037
mm	3.985
CH_2	
mmm	1.769, 1.675
mmr	1.719
rmr	1.754, 1.670
rrr	1.647
mrr	1.696, 1.610
mrm	1.663

SOURCE: Reproduced from reference 52. Copyright 1986 American Chemical Society.

extensive overlap, the proton spectrum of PVC cannot be completely assigned by using the COSY spectrum alone.

The 300-MHz proton spectrum and the COSY spectrum of a terpolymer of ethylene, methyl acrylate, and carbon monoxide (E – MA – CO) are shown in Figure 7.26 (*55*). By combining a knowledge of the chemical shifts and the nature of the coupling pattern with an analysis of the COSY spectrum, the proton line assignments can be made. The most-downfield resonance (labeled 1) in the proton spectrum has a chemical shift of 3.7 ppm and is not coupled to any other resonance in the COSY spectrum. This resonance is therefore assigned to the $-OCH_3$ protons of MA. Resonances 2 to 4 in the 2.2 – 3.0-ppm range correspond to protons that are α to carbonyl groups. In the sequence CEC, all four methylene protons are equivalent, and no J coupling is observed. Therefore, no cross peaks are expected in the COSY spectrum for this CEC sequence. Based on the COSY results, resonance 3 is assigned to the CEC sequence. This logical process is continued with consideration of the chemical shifts and coupling patterns. The proton line assignments for the E – MA – CO terpolymer are given in Table 7.6. These line assignments were confirmed by studies of the corresponding copolymer series.

Similar considerations apply for ^{19}F spectra. The COSY technique has been used to make assignments of the microstructure in the ^{19}F spectrum of poly(vinyl fluoride) (*56*). No $^{19}F - ^{19}F$ scalar coupling can be detected from resonance splitting in the normal spectrum because of the large ^{19}F line widths (approximately 14 Hz). Model compounds indicate that the four-bond coupling constant is approximately 7 Hz. Although these coupling constants are too small to appear as resonance splitting, they are large enough to show off-diagonal peaks in the COSY spectrum. These off-diagonal peaks arise from the coupling between the central pair of fluorines in pentad sequences that share a common hexad. For the hexad sequence

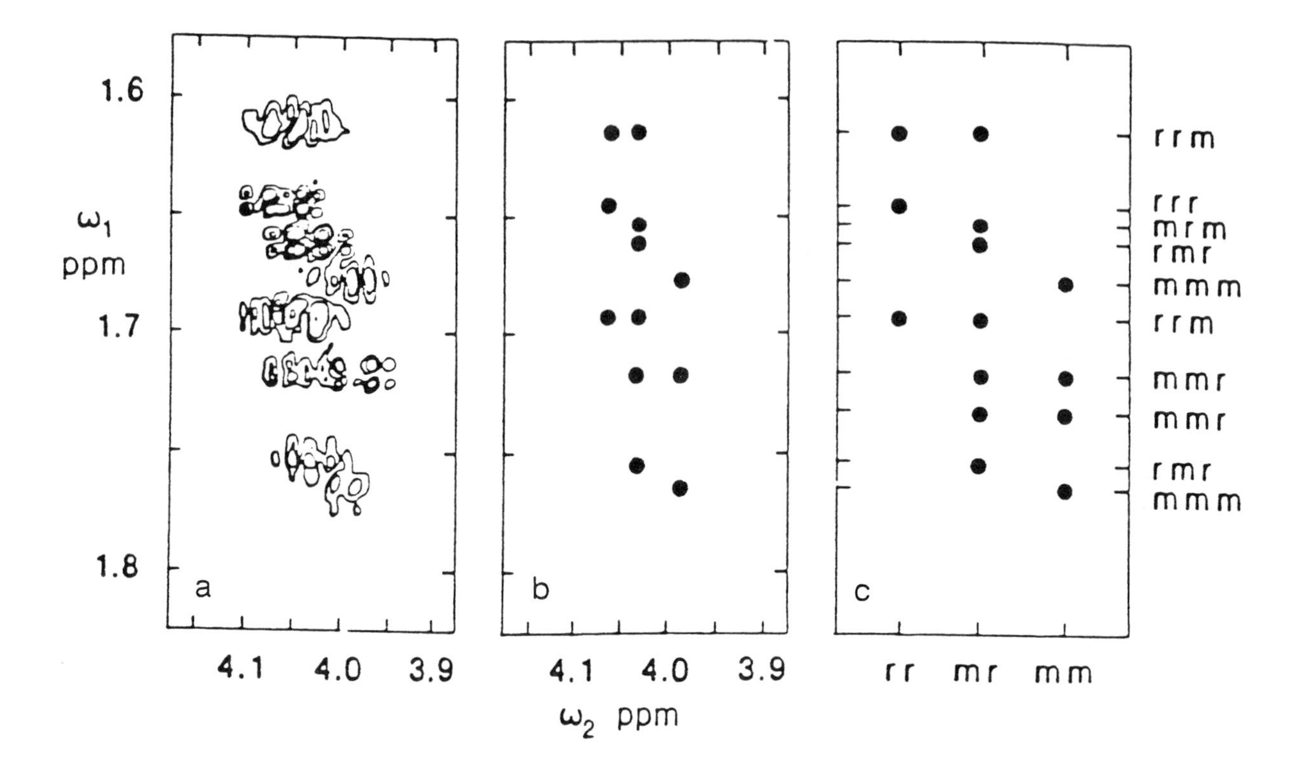

Figure 7.24. Experimental data and compositional relationships used to assign the ¹H spectrum of PVA. (a) Absolute-value contour plot of the COSY experiment with frequency scaling showing the HCOH – CH_2 cross peaks. Chemical-shift frequencies are expanded 1.7 times, and scalar couplings are reduced to 0.3 of the natural value. (b) Schematic representation of the data showing the chemical shifts of the multiplet resonances along ω_1 and ω_2. (c) Schematic representation of the necessary compositional relationships at the triad – tetrad level assuming that the heterosteric tetrads mmm, mmr, rmr, and rrm show two different ¹H resonances. (Reproduced with permission from reference 54. Copyright 1984 Springer-Verlag.)

```
  H         F         F*        F*        H         H
  |         |         |         |         |         |
- C --CH2-- C --CH2-- C --CH2-- C --CH2-- C --CH2-- C -
  |         |         |         |         |         |
  F    r    H    m    H    m    H    r    F    m    F
```

four-bond coupling between the indicated fluorines (*) will result in cross peaks connecting the rmmr and mmrm resonances in the COSY spectrum. All pentad assignments can be made unambiguously from the correlations observed in the COSY spectrum (*56*).

2-D Homonuclear *J*-Resolved Spectroscopy

Coupling constants that cannot be measured from the conventional ¹H NMR spectrum of a polymer can often be measured from the 2-D *J*-resolved spectrum. This technique separates the effects of the chemical shift and *J* coupling along two frequency axes in the 2-D spectrum and allows the coupling constants to be measured, even in the case of severe spectral overlap. The power of homonuclear 2-D J-resolved spectroscopy is in the separation of overlapping resonances in the proton spectrum of a polymer in order to determine the sequence-length sensitivity of the spectrum.

The basis of the method can be illustrated by the following simple example. In proton NMR spectroscopy, the lines in the individual patterns are phase modulated by the spin – spin coupling during the spin-echo experiment by using variations in t_e. For

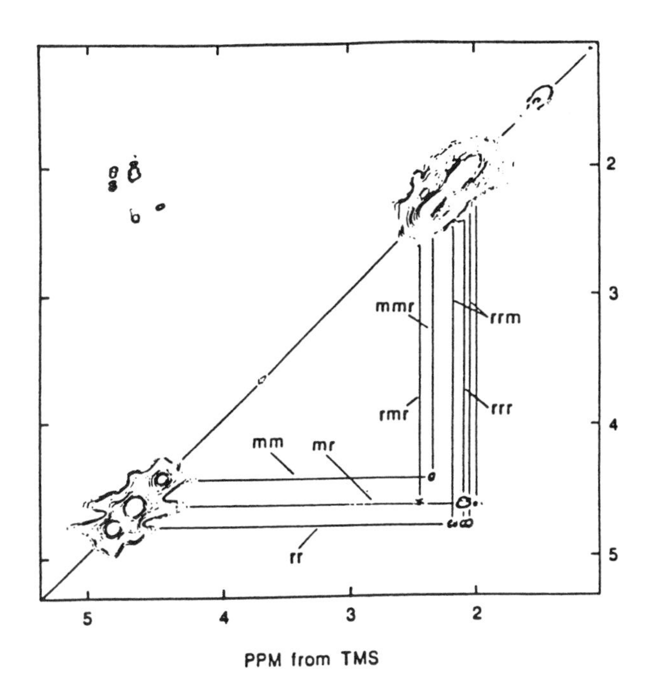

Figure 7.25. The 500-MHz COSY spectrum of PVC at 65 °C. (Reproduced from reference 47. Copyright 1986 American Chemical Society.)

example, the two components in a first-order doublet of a two-spin system are phase modulated at a frequency of $J/2$. The phase modulation of the resonance intensity results from the fact that the spin states of the protons producing the multiplet pattern are different. If a methyl proton is interacting with a methine proton, a doublet is observed because there are two possible energy orientations of the methine proton relative to H_o.

The preparation period is a relaxation delay to ensure that the spin system is in equilibrium. Following a 90° pulse, the magnetizations from each of the lines of the doublet are colinear. If the carrier frequency is equal to the frequency midway between the two lines, then during the delay period immediately following the 90° pulse, the higher-frequency component of the doublet will move ahead of the carrier frequency, and the lower-frequency component will fall behind. At time t_e, a broad-band 180° pulse is applied to reverse the orientation of the methyl protons. Because it is a broad-band pulse, however, it simultaneously reverses the orientation of the methine proton. Thus, those methyl protons that were

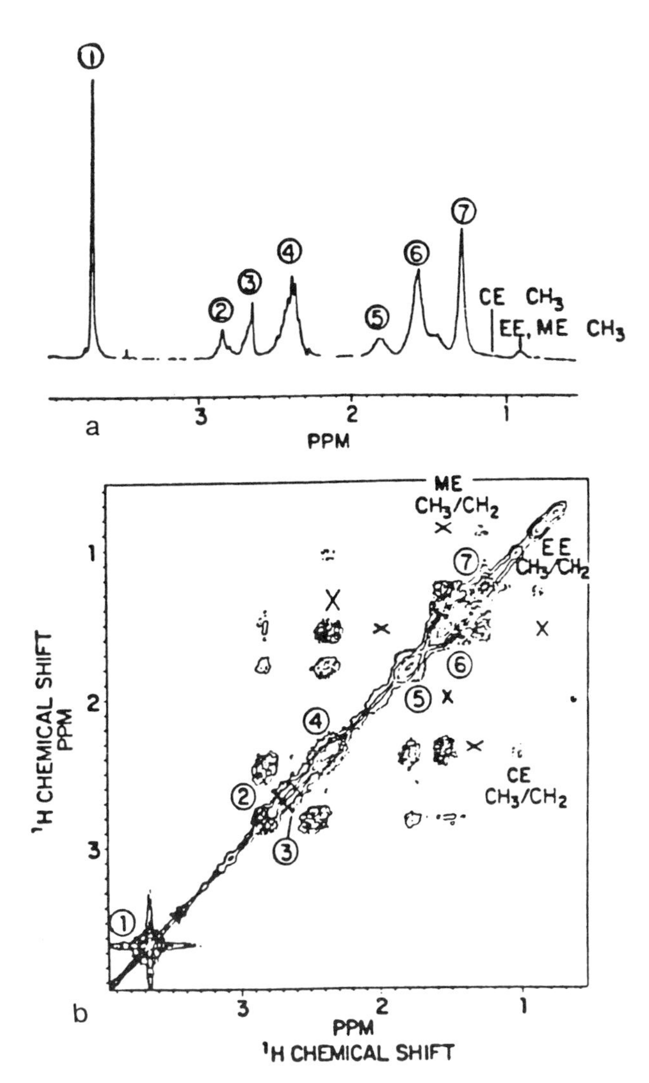

Figure 7.26. (a) Conventional 300-MHz ¹H NMR spectrum of a 5% solution of E–MA–CO terpolymer in tetrachloroethane-d₂ at 100 °C. (b) Contour plot of the 300-MHz proton 2-D COSY spectrum of E–MA–CO terpolymer. An × represents weak peaks visible at lower contour levels. The circled numbers correspond to the resonance numbers listed in Table 7.6. (Reproduced from reference 55. Copyright 1986 American Chemical Society.)

originally coupled to a methine proton of the original orientation (which gives rise to the faster rotating component of the doublet) are now coupled to a proton of the opposite orientation. Therefore the magnetization from this component precesses more slowly during the second delay period, and conse-

Table 7.6. Proton Line Assignments for E – MA – CO Terpolymer

Resonance	*Chemical Shift*[a] *(ppm)*	*Sequences*	*Description*
1	3.66	all	MA methyls in all sequences
2	2.84	CME, CMM	central methine
2, 4	2.83, 2.43	CME, CMM	nonequivalent methylenes α to carbon monoxide
3	2.65	CEC	all four ethylene protons
4	2.34	EMM, MMM[b]	central methine
4	2.31	EME	central methine
4	2.41	MEC	equivalent methylenes α to carbon monoxide
4	2.37	EEC	equivalent methylenes α to carbon monoxide
5	1.79	MM	equivalent methylenes in racemic dyads
6	1.54	CEE, CEM	equivalent β methylenes
6	1.62, 1.38	EM	nonequivalent methylenes
7	1.28	EEE, EEC	central ethylene methylenes and methylenes γ to carbon monoxide
7	1.26	MEE, MEM	central equivalent methylenes γ to the carbonyl
	1.05	CE	terminal methyls
	0.90	EE	terminal methyls
	0.87	ME	terminal methyls

[a] The chemical-shift values were obtained relative to TMS.
[b] Tentative
SOURCE: Reproduced from reference 55.

quently does not refocus at time $2t_e$. Likewise, the slower component will be out of phase with the carrier frequency at time $2t_e$, giving rise to the phase modulation. For example, if J = 6.25 Hz for the methyl doublet, the component that was precessing more rapidly immediately following the 90° pulse will be 90° out of phase when $2t_e$ = 0.08 s (t_e = 0.04 s). The other component will be −90° out of phase. When $2t_e$ is 0.16 s, the two components will be 180° and −180° out of phase, respectively. When $2t_e$ is 0.32 s, the two components will be 360° and −360° out of phase. As t_e is increased, this phase modulation is repeated at a frequency of J. Thus, in general, a doublet will be inverted when t_e = $1/2J$, $3/2J$, and $5/2J$. The doublet will be positive when t_e = $1/J$, $2/J$, and $3/J$. In 2-D spin-echo FT-NMR experiments, FIDs are obtained at a series of closely spaced t_e values. Each FID is Fourier transformed to give spectra that are arranged as a function of t_e at particular frequencies over the chemical-shift region of interest. The amplitude in the *echo interferogram* at a particular chemical shift is modulated at frequencies that are determined by the nature of the resonance patterns at that particular chemical shift. Thus, Fourier transformation of the echo interferogram separates overlapping resonances in the J dimension according to their modulation frequency.

The results of 2-D J-resolved spectroscopy can be plotted in two dimensions, with F_1 representing the J dimension and F_2 representing the chemical-shift dimension. Projection of the individual spectra on the F_2 axis gives the completely proton–proton decoupled spectrum of the sample. Hence, the projection can be used to determine the number of distinct spin species in a molecule and to measure the chemical shift of each of these species. Projection onto the F_1 axis gives the J spectra from which the homonuclear J couplings can be measured. The individual components of a given multiplet appear at frequencies in the F_1 and F_2 dimension such that the individual lines of a given multiplet overlap for a projection at an angle of 45° relative to F_1 and F_2. The important result is that single lines are obtained for the chemically shifted protons rather than the multiplet patterns of normal spectra.

The intensities of the cross peaks depend on a number of features, including the coupling constants, the complexity of the coupling pattern, the relaxation rates, and the relative abundance of the coupled moieties. A number of variations of the 2-D *J*-resolved experiment have been suggested, including one designed particularly for macromolecular systems to yield increased sensitivity and resolution (*57*).

Poly(propylene oxide) has been studied by using 2-D *J*-resolved NMR spectroscopy (*58*). The scalar coupling is observed only between protons of the same monomer unit because the oxygen atom separates the monomer units in the chain. The 500-MHz proton NMR spectrum of a commercial atactic poly(propylene oxide) is shown in Figure 7.27. Extensive overlap resulting from the presence of stereosequences is apparent. In Figure 7.28, an expansion is shown of the 500-MHz 2-D *J*-resolved spectrum of the methyl region of poly(propylene oxide). Five main doublets are apparent in this region, and at least five very weak doublets appear at higher shielding. The five major doublets correspond to different stereosequences, and the weak doublets correspond to inverted units (*58*). Interestingly, although all of the sequences have different chemical shifts, the same coupling constant of 6.5 Hz is observed for all sequences. By using these results and by assuming the occurrence of a completely random polymer, the spectrum of the methyl region was simulated, as shown in the inset of Figure 7.27. The agreement is excellent.

The homonuclear 2-D *J*-resolved experiment was also used as an aid in assigning the copolymer sequences in poly(ethylene-*co*-vinyl alcohol) (*59*). The problem is that the coupling constants are of the same order of magnitude as the chemical-shift differences between the coupled methylenes. The EEE (ethylene) sequences give rise to a singlet because all neighboring methylenes are equivalent, and these sequences can be assigned from the 2-D spectrum.

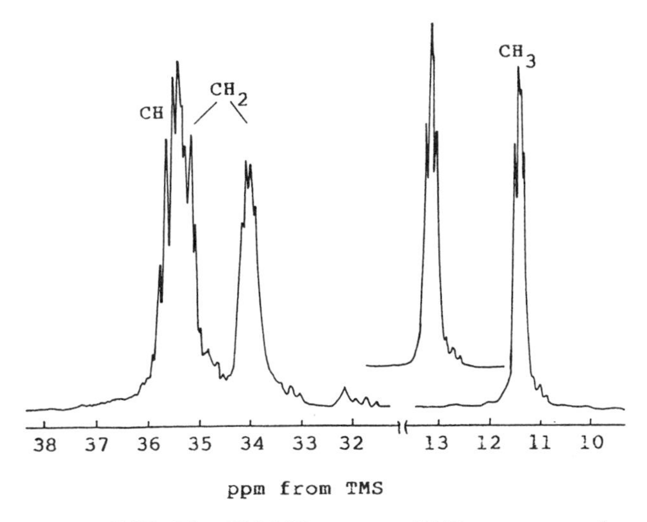

Figure 7.27. The 500-MHz proton NMR spectrum of a commercial atactic poly(propylene oxide) in $CDCl_3$ at 25 °C showing extensive overlap resulting from the presence of stereosequences. The simulated spectrum (inset) was calculated by using the values of the J *couplings that were obtained from the 2-D spectrum. (Reproduced from reference 58. Copyright 1985 American Chemical Society.)*

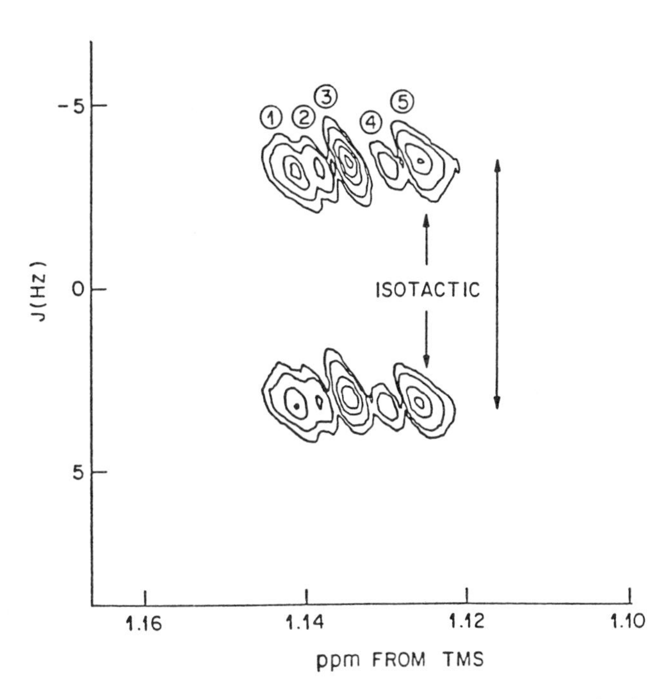

Figure 7.28. Expansion of the 500-MHz 2-D J*-resolved spectrum of the methyl region of atactic poly(propylene oxide). The circled numbers correspond to the five main doublets. The splittings due to the* J *coupling can be observed in the upper and lower portions of the figure. The measured $J_{CH_3} - {}_{CH}$ value of 6.5 Hz corresponds to the expected value for this system. (Reproduced from reference 58. Copyright 1985 American Chemical Society.)*

Methylenes in other sequences are coupled to neighboring methylenes that have similar, yet different chemical shifts. The CH – OH coupling constants are sequence dependent. The coupling constant is large (5.8 Hz) when the neighboring units are either both ethylene or both racemic vinyl alcohol. The coupling constant drops to 3.9 Hz for any sequence containing at least one meso VV (vinyl alcohol) dyad, which indicates that the average dihedral angle between the methine and hydroxyl protons is different for these sequences.

The homonuclear 2-D *J*-resolved experiment was also used as an aid in assigning resonances in the epoxy resin diglycidyl ether of bisphenol A (*60*) and the composite of bis[*N*,*N*-bis(2,3-epoxypropyl)-4-aminophenyl]methane and the curing agent bis(4-aminophenyl)sulfone (*61*). The complex proton spectrum resulting from the glycidylamine moiety can be completely elucidated in this manner.

2-D Heteronuclear *J*-Resolved Spectroscopy

Heteronuclear *J*-resolved 2-D spectroscopy is used when the normal $^1H - ^{13}C$ coupled spectra are too complicated to interpret because of the overlap of multiplets. This technique separates the chemical shift of the ^{13}C along one axis and the heteronuclear *J*-coupling along the other. Thus the heteronuclear *J*-coupling information can be separated in two dimensions from chemical-shift information. The gated-decoupling experiment has the following pulse sequence:

^{13}C: $90°_x$ - $t_e/2$ - $180°_y$ - $t_e/2$ - FID($t2$)
^1H: BB - - - - - off BB

The conventional spin-echo pulse sequence refocuses the chemical shift effects as usual. The trick is to turn off the broad-band (BB) decoupling during one $t_e/2$ period. When the decoupling is turned off following the 180° pulse, different $^1H - ^{13}C$ multiplet components can precess away or diverge from the vector associated with the chemical shift. These dephasing components do not refocus after the second $t_e/2$ period. Acquisition of the FID signal with the BB decoupling turned on again means that only the vector sum of all multiplet components is observed, and the signal amplitude is thus *J* modulated by the associated *J* couplings as t_e is varied. Because the *J* couplings are present during only one $t_e/2$ period, the F_1 transform gives all *J*s at half of their actual value.

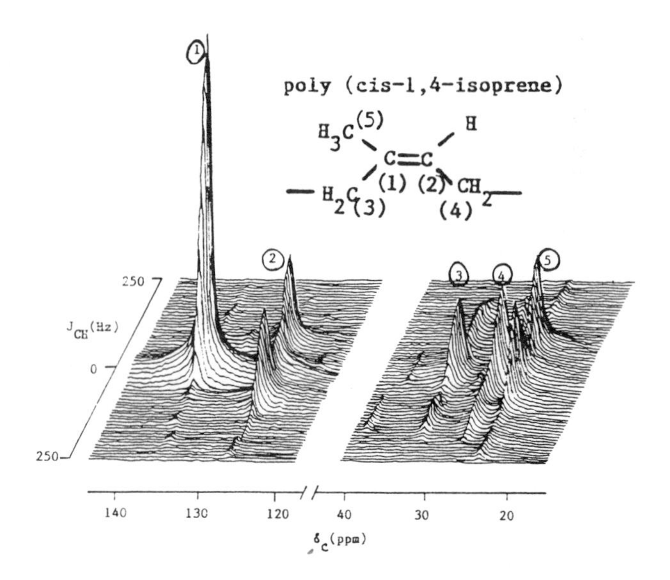

Figure 7.29. The heteronuclear 2-D J-resolved spectrum of cured, carbon-black-filled natural rubber at room temperature. The circled numbers labeling the peaks correspond to the numbered carbons in the inset structure. (Reproduced from reference 62. Copyright 1987 American Chemical Society.)

In Figure 7.29, the heteronuclear 2-D *J*-resolved spectrum of cured, carbon-black-filled natural rubber at room temperature is shown (*62*). In the F_1 dimension, the full unscaled scalar $^1H - ^{13}C$ couplings are equal to 150 Hz for the C-2 carbon and 127 Hz for carbons C-3, C-4, and C-5. These values are identical with those obtained for natural rubber in a chloroform solution.

2-D NOE-Correlated Spectroscopy

The existence of intermolecular interactions between two polymer chains can be measured by using the 2-D NOE-correlated spectra. The nuclear Overhauser and exchange spectroscopy (NOESY) sequence generates a 2-D spectrum that contains the 1-D spectrum on the diagonal and nondiagonal peaks that reflect through-space dipole – dipole couplings. If two protons are relatively close together, a cross peak that reflects the dipole – dipole interactions between the protons will appear, even if the protons are not chemically bonded. Because the NOE is proportional to the inverse sixth power of the internuclear distance, it is assumed that NOE-correlated cross peaks do not appear beyond a distance of about 4 Å between participating protons (*63*).

A 2-D NOE-correlated spectrum was used to demonstrate the presence of dipolar through-space coupling between the methoxy protons of poly(MMA-*co*-4VP) and the aromatic protons of poly(styrene-*co*-styrenesulfonic acid) [poly(S-*co*-S)] (*64*). This 2-D NOE-correlated spectrum shows that the main cross peaks appear from the ortho-aromatic coupling and the $CH-CH_2$ coupling of the main chain. These protons are separated from each other by three chemical bonds and represent scalar couplings. Their close proximity generates the cross peaks in the 2-D NOE-correlated spectrum. One cross peak correlates the two polymer chains, and this cross peak appears to belong to the outer part of the poly(MMA-*co*-4VP) chain and to the external part of the poly(S-*co*-S) chain. By using different mixing times, it was established that the NOE correlation was in fact between these two specific types of protons.

The 2-D NOE spectrum of an alternating styrene – methyl methacrylate copolymer shows a number of cross peaks connecting both resonances within a monomer unit and also connecting neighboring units. The resonance intensities vary as a function of mixing time and depend on the molecular motions and the inverse sixth power of the proton – proton separation (*65*). For this system, four different proton – proton distances can be measured. These four proton – proton distances place restraints on the possible solution conformations of the copolymer. The data are best satisfied by a solution structure that is predominately tt (*65*). The best fit was obtained with an average solution conformation of 58% tt, 24% tg^-, and 19% g^+t.

The special editing techniques described simplify or correlate the complex and multiline NMR spectra measured. The resulting spectra are considerably easier to use to obtain the structural information. Progress in this field has been rapid as the NMR instrumentation and software has improved. In the future, we will see considerably more use of the 2-D methods (and 3-D and 4-D, as well) in the study of synthetic polymer systems.

References

1. Rinaldi, P. L. In *Two Dimensional NMR Spectroscopy;* Croasmun, W. R.; Carlson, R. M. K., Eds.; VCH Publishers: Deerfield Beach, 1987; p 425.
2. Farrar, T. C. Source unknown.
3. Slichter, C. P. *Principles of Magnetic Resonance;* Harper & Row: New York, 1980.
4. Atta-ur-Rahman *Nuclear Magnetic Resonance: Basic Principles;* Springer-Verlag: New York, 1986.
5. Ernst, R. R.; Bodenhausen, G.; Wokaun, A. *Principles of Nuclear Magnetic Resonance in One and Two Dimensions;* Oxford: New York, 1987.
6. Bovey, F. A. *Nuclear Magnetic Resonance Spectroscopy;* 2nd ed.; Academic: New York, 1988.
7. Traficante, D. E. In *Two Dimensional NMR Spectroscopy;* Croasmun, W. R.; Carlson, R. M. K., Eds.; VCH Publishers: Deerfield Beach, 1987; p 92.
8. Slichter, C. P. *Principles of Magnetic Resonance;* p 39.
9. Hahn, E. L. *Phys. Rev.* **1950,** *80,* 580.
10. Carr, H. Y.; Purcell, E. M. *Phys. Rev.* **1954,** *94,* 630.
11. Benn, R.; Gunther, H. *Angew. Chem. Int. Ed., Engl.* **1983,** *22,* 350.
12. Turner, C. J. *Prog. NMR Spectrosc.* **1984,** *16,* 311.
13. Gerhards, R. In *Magnetic Resonance, Introduction, Advanced Topics and Applications to Fossil Energy;* Petraks, L.; Fraissard, J. P., Eds.; 1984, p 377.
14. Brown, D. W.; Nakashima, T. T. *J. Magn. Reson.* **1981,** *45,* 302.
15. Cooper, J. W.; Johnson, R. D. *FT NMR Techniques for Organic Chemists;* IBM Instruments: 1986.
16. Bugada, D. C.; Rudin, A. *Eur. Polym. J.* **1987,** *23,* 809.
17. Bruch, M. D.; Dybowski, C.; Lichter, R. L. In *NMR Spectroscopy Techniques;* Dybowski, C.; Lichter, R. L., Eds.; Marcel Dekker: New York, 1987; p 170.
18. Chu, C. Y.; Vukov, R. *Macromolecules* **1985,** *18,* 1423.
19. Rabenstein, D. L.; Nakashima, T. T. *Anal. Chem.* **1979,** *51,* 1465A.

20. Bodenhausen, G.; Freeman, R.; Morris, G. *J. Magn. Reson.* **1976,** *23,* 171.

21. Morris, G.; Freeman, R. *J. Magn. Reson.* **1978,** *29,* 433.

22. Bork, V.; Schaefer, J. *J. Magn. Reson.* **1988,** *78,* 348.

23. Tekely, P.; Brondeau, J.; Elbayed, K.; Retournard, A.; Canet, D. *J. Magn. Reson.* **1988,** *80,* 509.

24. *Two Dimensional NMR Spectroscopy;* Croasmun, W. R.; Carlson, R. M. K., Eds.; VCH Publishers: Deerfield Beach, 1987; p 497.

25. Sadler, I. H. *J. Chem. Soc., Chem. Commun.* **1987,** 321.

26. Sorensen, S.; Hansen, R. S.; Jakobsen, H. J. *J. Magn. Reson.* **1974,** *14,* 243.

27. Morris, G. A.; Freeman, F. *J. Am. Chem. Soc.* **1979,** *101,* 760.

28. Doddrell, D. M.; Pegg, D. T. *J. Am. Chem. Soc.* **1980,** *102,* 6388.

29. Hikichi, K.; Hiraoki, T.; Takemura, S.; Ohuchi, M.; Nishioka, A. In *NMR and Macromolecules;* ACS Symposium Series 247; American Chemical Society: Washington, DC, 1984; p 119.

30. Gray, G. A. In *NMR and Macromolecules, Sequence, Dynamic and Domain Structure;* Randall, J. C., Jr., Ed.; ACS Symposium 247; American Chemical Society: Washington, DC, 1984.

31. Asakura, T.; Nishiyama, Y.; Doi, Y. *Macromolecules* **1987,** *20,* 616.

32. Bendall, M. R.; Pegg, D. T.; Doddrell, D. M.; Field, J. *J. Am. Chem. Soc.* **1981,** *103,* 936.

33. Dodrell, D. M.; Pegg, D. T.; Bendall, M. R. *J. Magn. Reson.* **1982,** *48,* 323.

34. Bruch, M. D.; Bonesteel, J. K. *Macromolecules* **1986,** *19,* 1622.

35. Barron, P. F.; Hill, D. J. T.; O'Donnell, J. H.; O'Sullivan, P. W. *Macromolecules* **1984,** *17,* 1967.

36. Komoroski, R. A.; Shockcor, J. P.; Gregg, E. C.; Savoca, J. L. *Rubber Chem. Technol.* **1985,** *59,* 328.

37. Bax, A.; Freeman, R.; Kempsell, S. P. *J. Am. Chem. Soc.* **1980,** *102,* 4849.

38. Bodenhausen, G.; Freeman, R.; Niedermeyer, R.; Turner, D. L. *J. Magn. Reson.* **1977,** *26,* 133.

39. Bax, A.; Morris, G. A. *J. Magn. Reson.* **1981,** *42,* 501.

40. Jelinski, L. W. "Modern NMR Spectroscopy", *Chem. Eng. News* **1984,** *11(5),* p 34.

41. Sarkar, S. K.; Bax, A. *J. Magn. Reson.* **1985,** *62,* 109.

42. Maudsley, A. A.; Ernst, R. R. *Chem. Phys. Lett.* **1977,** *50,* 368.

43. Bodenhausen, G.; Freeman, R. *J. Magn. Reson.* **1977,** *28,* 471.

44. Bax, A. In *Topics in Carbon-13 NMR Spectroscopy;* Levy, G. C., Ed.; Wiley: New York, 1984; Vol. 4.

45. Westler, W. M.; Ortiz-Polo, G.; Markley, J. L. *J. Magn. Reson.* **1984,** *58,* 354.

46. Crowther, M. W.; Szeverenyi, N. M.; Levy, G. C. *Macromolecules* **1986,** *19,* 1333.

47. Mirau, P. A.; Bovey, F. A. *Macromolecules* **1986,** *19,* 210.

48. Cheng, H. N.; Lee, G. H., *Polym. Bull.* **1985,** *13,* 549.

49. Cheng, H. N.; Lee, G. H., *Polym. Bull.* **1984,** *12,* 463.

50. Cheng, H. N.; Lee, G. H. *J. Polym. Sci., Part B: Polym. Phys.* **1987,** *25,* 2355.

51. Moad, G.; Rizzardo, E.; Solomon, D.; Johns, S. R.; Willing, R. I. *Macromolecules* **1986,** *19,* 2496.

52. Bruch, M. D.; Bonesteel, J. *Macromolecules* **1986,** *19,* 1622.

53. Jeener, J.; Meier, B.; Bachmann, P.; Ernst, R. *J. Chem. Phys.* **1979,** *71,* 4546.

54. Gippert, G. P.; Brown, L. R. *Polym. Bull.* **1984,** *11,* 585.

55. Bruch, M. D.; Payne, W. G. *Macromolecules* **1986,** *19,* 2712.

56. Bruch, M. D.; Bovey, F. A.; Cais, R. E. *Macromolecules* **1984,** *17,* 2547.

57. Macura, S.; Brown, L. R. *J. Magn. Reson.* **1983,** *53,* 529.

58. Bruch, M. D.; Bovey, F. A.; Cais, R. E.; Noggie, J. H. *Macromolecules* **1985,** *18,* 1253.

59. Bruch, M. D. *Macromolecules* **1988,** *21,* 2707.

60. Jagannathan, N. R.; Herring, F. G. *J. Polym. Sci., Polym.Chem.* **1988,** *26,* 1.

61. Herring, F. G.; Jagannathan, N. R.; Luoma, G. *J. Polym. Sci., Polym. Chem.* **1985,** *23,* 1649.

62. Kentgens, A. P. M.; Veeman, W. S.; van Bree, J. *Macromolecules* **1987,** *20,* 1234.

63. Schilling, F. C., Bovey, F. A.; Bruch, M. D.; Kozolowski, S. A., *Macromolecules* **1985,** *18,* 1418.

64. Natansoh, A.; Eisenberg, A. *Macromolecules* **1987,** *20,* 323.

65. Mirau, P. A.; Bovey, F. A.; Tonelli, A. E.; Heffner, S. A., *Macromolecules* **1987,** *20,* 1701.

Suggested Reading

Bax, A. *Two-Dimensional Nuclear Magnetic Resonance in Liquids;* Reidel: Boston, 1982.

Bovey, F. A.; Mirau, P. A. *Acc. Chem. Res.,* **1988,** *21,* 37.

Ernst, R. R.; Bodenhausen, G.; Wokaun, A. *Principles of Nuclear Magnetic Resonance in One and Two Dimensions;* Oxford University Press: New York, 1987.

Hatada, K.; Ute, K.; Tanaka, K.; Imanari, M.; Fujii, N. *Polymer* **1987,** *19,* 425.

Kessler, H.; Gehrke, M.; Griesinger, C. *Angew. Chem. Int. Ed., Engl.,* **1988,** *27,* 490.

8

High-Resolution NMR Spectroscopy of Solid Polymers

The history of high-resolution n.m.r. in solids has been, inter alia, a quest for narrow spectral lines.
– A. N. Garroway et al. (*1*)

In polymer solutions, sharp NMR lines are obtained because the local fields are averaged to zero by the rapid isotropic motions of the nuclei. Anisotropic interactions, such as dipolar and quadrupolar interactions, are averaged to zero by the molecular motions to effectively remove them from contributing to the spectra. Generally, in solids there is not sufficient motion to average the anisotropic interactions. Because the incoherent averaging (molecular motion) may not narrow the NMR lines, coherent-averaging techniques such as dipolar decoupling and magic-angle spinning must be used to produce narrow line widths. For solids, the low sensitivity of the ^{13}C nucleus is improved by transferring polarization from the magnetization-rich protons to the ^{13}C nuclei by using cross-polarization. By combining high-power decoupling (DD), magic-angle spinning (MAS), and cross-polarization (CP) experiments into one grand experiment (*2*), narrow lines and enhanced sensitivity can be obtained for polymers in the solid state. As a consequence, high-resolution NMR spectroscopy has become an important tool in the structural investigation of polymers in the solid state (*3* – *8*).

The Dipolar-Decoupling Experiment (DD)

Proton dipolar broadening in ^{13}C spectra of solids can be removed by a high-power version of the decoupling technique used in solution NMR spectroscopy.
– C. Y. Yannoni (*9*)

The major contributions to the line widths in ^{13}C NMR spectra of organic solids are the interactions arising from heteronuclear dipolar broadening by protons. Consider the general model of a rigid isolated pair of nuclei, designated μ_1 and μ_2 with spin ½, interacting through their dipoles. The dipole resulting from μ_1 precesses at the Larmor frequency about the applied magnetic field, H_o, to create a static component along the field direction and a rotating component in the plane perpendicular to the static-field direction. The static component of μ_1 produces a small static field at the site of the dipole μ_2. The magnitude of this local field, H_{loc}, depends on the relative positions of the spins and their orientations with respect to the applied magnetic field. If a sample containing this isolated pair of nuclei is placed in a static magnetic field, H_o, each nucleus then experiences an effective magnetic field (H_{eff}):

$$\begin{aligned} H_{eff} &= H_o \pm H_{loc} \\ &= H_o \pm \left(\frac{\mu}{r_{ij}^3}\right)(3 \cos^2 \Theta_{ij} - 1) \end{aligned} \qquad (8.1)$$

where Θ_{ij} is the angle between the internuclear vector, r_{ij} and the direction of the external magnetic field and μ is the magnetic moment (Figure 8.1). For a large H_o, only the components parallel or antiparallel to H_o cause significant changes in the net static field. The resulting line width should be approximately equal to the magnitude of the local static field, thereby contributing to substantial line broadening.

For unlike spins, the rotating field is off resonance, and thus the interaction is less. The ^{13}C spin experi-

1904—4/92/0197$08.75/1

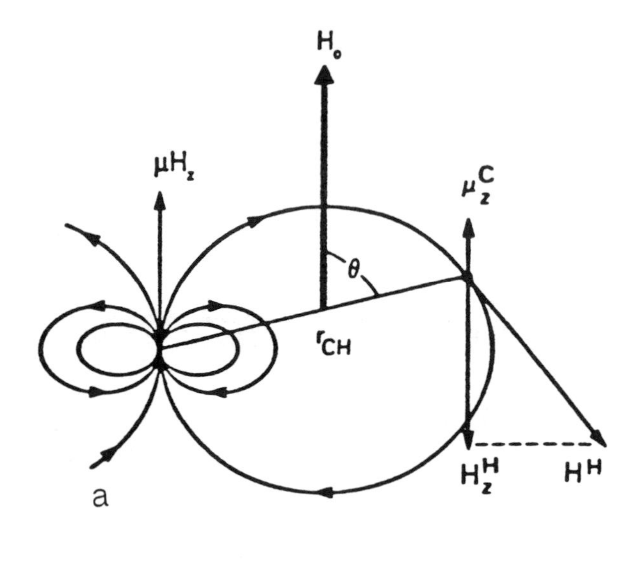

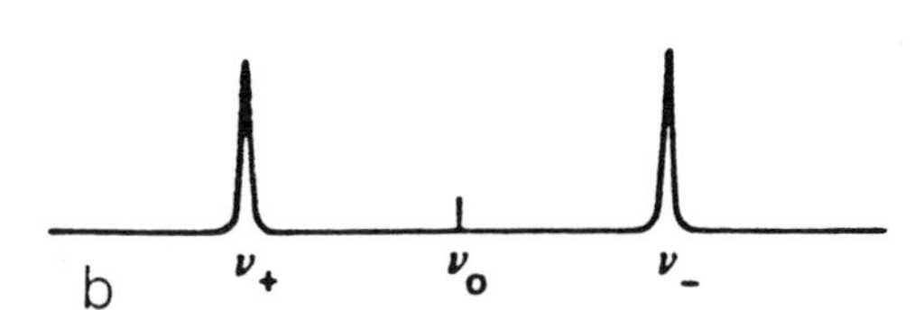

Figure 8.1. Proton – carbon dipolar coupling in an isolated C – H bond. (a) Lines of force from the proton magnetic moment (shown here in the + state, parallel to the external field H_o*) generate a static local field* H^H *at the* ^{13}C *nucleus. Because* $H^H << H_o$*, the* ^{13}C *experiences only* H_z^H*, the component of* H^H *that is antiparallel to* H_o*. (b) The resulting* ^{13}C *spectrum for a sample of isolated C – H fragments with a single orientation, θ, is shown. (Reproduced with permission from Douglas P. Burum. Copyright 1983 Bruker Reports.)*

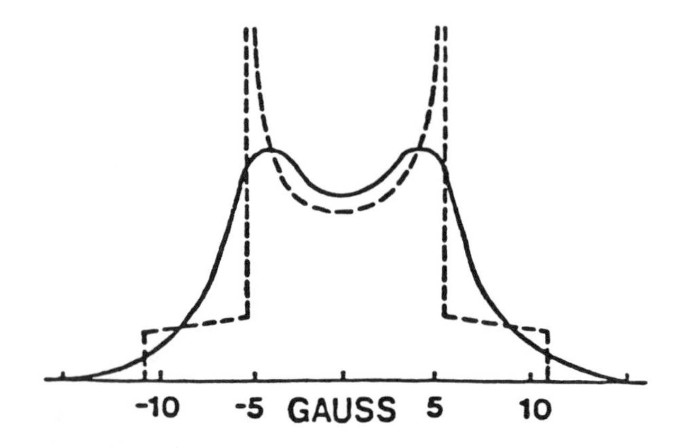

Figure 8.2. The absorption curve of a two-spin system for dipolar interactions. The dotted curve gives the absorption for an isolated pair of nuclei. The solid curve shows the effect of neighboring nuclei on the isolated system. In the NMR spectrum, a doublet will be observed, with a spacing determined by the length and orientation of the internuclear vector. (Reproduced with permission from reference 10. Copyright 1948 American Institute of Physics.)

ences a force due to the z component of the magnetic field, H_z^1, generated by the ^1H spins. This component may add or subtract from the applied external field and results in the appearance of a doublet. In the NMR spectrum, a doublet will be centered around the ^{13}C Larmor frequency ($\Delta\nu_{CH}$) with a splitting in hertz (Figure 8.2) equal to (*10*)

$$\Delta\nu_{CH} = \left(\frac{\gamma_C}{\pi}\right)H_z^1 \quad (8.2)$$

with

$$H_z^1 = \left(\frac{\mu_z^H}{r_{CH}^3}\right) < 1 - 3\cos^2\Theta > \quad (8.3)$$

where r_{CH} is the internuclear carbon-to-hydrogen distance and μ_z^H is the z component of the dipole of the ^1H nucleus.

The angular brackets denote an average that depends on the molecular motion. For dipolar $^1H - {}^{13}C$ splittings, values as high as 40 kHz are possible. However, a doublet is not observed for semicrystalline polymers. The observed result is generally a broad Gaussian-shaped resonance. The dipolar spectrum is featureless because of the many interactions of all of the spins with each other. Because of the distance dependence of the dipolar interactions, only spins within a radius of ≤20 Å contribute, but for most polymer systems there are still 10 – 100 dipolar interactions.

When a strong decoupling field is applied in the solid state, there are rapid transitions or mutual spin flips that occur at a rate approaching that of the inverse of the line width of the proton resonance. The time constant for this process is T_2 (the spin – spin relaxation time), which ranges from 10 to 100 μs for rigid solids.

There is a distinction between scalar coupling and dipolar effects. Both are eliminated by decoupling, but the sources of the effects are different. Scalar coupling arises from the coupling of the C – H energy levels, and dipole – dipole effects arise from changes in the local magnetic field due to the dipoles present. Scalar coupling is eliminated by relatively low-power decoupling, and dipole – dipole effects require high-power decoupling.

Dipolar decoupling (DD) with high-powered proton decoupling can be used to coherently average the heteronuclear dipolar interactions to zero (*10*). The dipolar decoupling forces the proton spins to change energy states at a fast rate compared to the frequency of the dipolar $^1H-{}^{13}C$ interactions. Under these circumstances, the local dipolar fields at the ^{13}C nucleus are reduced to zero (*11*). To decouple protons from carbons in solids, the magnitude of the decoupling field must be capable of exciting all proton transitions within a bandwidth of 40 – 50 kHz, which is large compared to the proton – proton dipolar coupling.

The DD experiment can be applied either continuously or gated (*12*). In the latter experiment, often termed *inverse-gated decoupling*, broad-band irradiation is applied only during acquisition. In this manner, the total power consumption is smaller, and this condition leads to less heating of the sample.

The required decoupling field can be calculated. Consider a system with chemically shifted resonances that occur in a frequency range of Δ Hz, following the method of Lyerla (*13*). The effective field in the rotating frame is given by

$$H_{\text{eff}} = (H_o - \omega/\gamma)\,z + H_1 x \qquad (8.4)$$

or in terms of interest here

$$H_{\text{eff}} = (1/\gamma)[(\omega_i - \omega)^2 + (\gamma H_1)^2]^{\frac{1}{2}} \qquad (8.5)$$

This is the relationship between the precession frequency ω_i and the effective field when H_1 is applied at ω. If H_1 is large enough to meet the condition $\gamma H_1 >> \pi\Delta'$, where Δ' is the total range of frequencies from the rf, then the first term in eq 8.4 can be neglected, and

$$H_{\text{eff}} = H_1 \qquad (8.6)$$

The magnetizations of all the nuclei in the frequency range Δ' precess about H_1 or all $\mathbf{M}_i$ are rotated through the same angle Θ. By using short, intense pulses, the entire range of resonance frequencies for the protons can be simultaneously decoupled.

The required decoupling power also depends on the sample size. The ^1H power in a sample coil is given approximately by

$$^1H = 3\left(\frac{PQ}{\omega_o V}\right)^{\frac{1}{2}} \qquad (8.7)$$

where P is the transmitter power in watts, Q is the quality factor of the probe circuit, ω_o is the Larmor frequency (MHz), and V is the sample volume (cm^3). Thus, doubling the sample volume requires a corresponding doubling of the rf power level to keep a constant ^1H field or 90° pulse width.

The effect of high-powered proton decoupling is shown in Figure 8.3 for the ^{13}C spectra at – 180 °C of the central carbon of the *tert*-butyl cation prepared from [2-^{13}C]-2-chloro-2-methylpropane and SbF_3. Spectrum a was obtained without decoupling, and spectrum b was obtained with strong (43 kHz) ^1H and

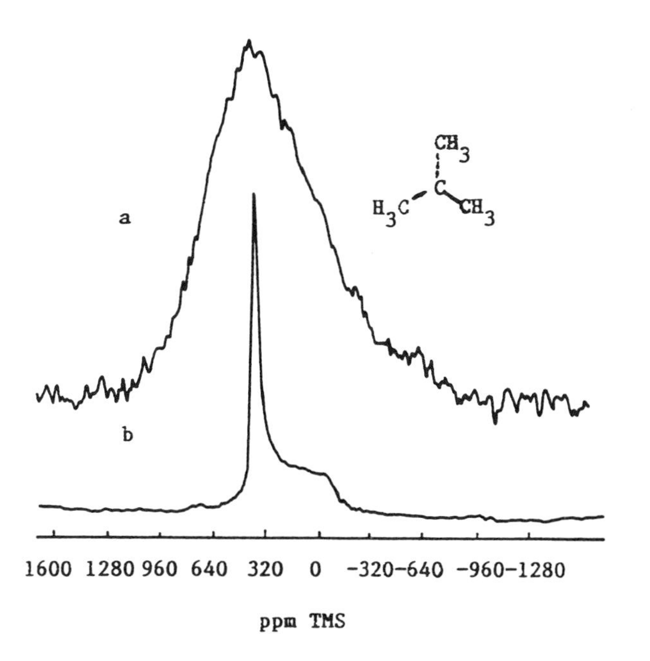

Figure 8.3. The effect of high-powered proton decoupling is shown for the ^{13}C spectra at – 180 °C of the central carbon of the tert-*butyl cation prepared from [2-^{13}C]2-chloro-2-methylpropane and SbF_3. Spectrum a was obtained without decoupling and spectrum b was obtained with strong (43 kHz) ^1H and ^{19}F decoupling. (Reproduced from reference 14. Copyright 1982 American Chemical Society.)*

^{19}F decoupling. The spectral contribution from the methyl carbons was subtracted from these spectra (*14*). The residual resonance line shape observed is due to the chemical-shift anisotropy of the central carbon in a powdered sample. The effects of chemical-shift anisotropy will be discussed later in this chapter.

Heteronuclear Decoupling Using Spin-Locking Techniques

Decoupling by spin locking has the advantage of longer relaxation times so that additional pulse sequences can be used. For spin-lock decoupling, a 90°_x pulse is applied, followed by a $\pi/2$ phase shift of the rf field, which is then left on. From the perspective of the rotating frame, this spin-locking experiment has the effect of aligning **M** along the rf field H_1 (*15*). As long as H_1 is large compared to the local dipolar field, the system cannot relax by a T_2 process because the spin-flip mechanism does not conserve Zeeman energy along H_1.

At resonance, the rotating frame is a coordinate system moving in synchronism with the precession of the proton moments around the static field at a frequency

$$\nu_{o\mathrm{H}} = \left(\frac{\gamma_\mathrm{H}}{2\pi}\right) H_o \tag{8.8}$$

In this rotating reference frame, the proton spins are stationary, but in the presence of the rf field, $H_{1\mathrm{H}}$, the proton spins precess at a frequency

$$\nu_{1\mathrm{H}} = \left(\frac{\gamma_\mathrm{H}}{2\pi}\right) H_{1\mathrm{H}} \tag{8.9}$$

While spin locked, the z components oscillate at a frequency $\nu_{1\mathrm{H}}$. The ^{13}C spins experience this field as a coherently averaged z component $< \mu_z^\mathrm{H} >$. For sinusoidal variation,

$$< \mu_z^\mathrm{H} > = 0 \tag{8.10}$$

so that the ^{1}H – ^{13}C dipolar couplings are removed. For effective decoupling, $\nu_{1\mathrm{H}}$ must be much greater than $\Delta\nu_{\mathrm{HH}}$, where $\Delta\nu_{\mathrm{HH}}$ is the homogenous proton line width. Thus, high-power heteronuclear decoupling requires spin-lock fields of 1 mT (43 kHz) or more (*13*).

Homonuclear Decoupling Using Multipulse Methods

The previous decoupling methods involved coherent spatial averaging to remove the heteronuclear dipolar coupling. The dominant line-broadening mechanism in the ^{1}H magnetic resonance of solids is normally the ^{1}H – ^{1}H homonuclear dipolar interaction, which cannot be removed by the double-resonance experiment if the ^{1}H spectrum is to be observed. In order to obtain narrow ^{1}H resonances of solid samples, the line-broadening effects of the homonuclear dipolar interactions must be eliminated while the chemical-shift interactions are retained.

The homonuclear dipolar ^{1}H – ^{1}H interactions can be removed by modulating the spin states rather than the spatial factors (*16*). This multipulse sequence reduces the line widths in rigid solids from 10^2 to 10^4 Hz. The multiple-pulse sequence used is the WAHUHA sequence (after the initials of the originators) (*16*) and is

$$\mathrm{P}_{-y} - \tau - \mathrm{P}_{-x} - 2\tau - \mathrm{P}_x - \tau - \mathrm{P}_y - 2\tau - \mathrm{P}_{-y}$$

The cycle time, $t_n = 6\tau$, can be considered as a rotation period. If the pulses are short enough and the delays τ are kept to a minimum, the "rotation rate" of the process can be made fast enough to effectively decouple homonuclear dipole – dipole interactions. During the multiple pulse sequence, the evolution of the nuclear spins in the magnetic field is not a free precession. Under multiple-pulse decoupling, the proton signal is acquired while the multiple-pulse train drives the spin states. The apparent evolution is determined by the internal interactions modulated by the rf excitation. Experimentally, the system is examined stroboscopically at such times that the detected signal is not influenced by the dipolar interactions. As a consequence of this process, the chemical shift is reduced by a factor of $3^{1/2}$ (*17*).

Unfortunately, the WAHUHA sequence does not completely eliminate dipole – dipole interactions because the applied pulses have finite widths (*18, 19*). To minimize these effects, more complex sequences have been developed.

Dipolar Decoupling Process Limitations

In the heteronuclear double-resonance dipolar decoupling process, residual line broadening will occur when the proton rf field is off resonance or when the decoupling field is not powerful enough. For solids, the proton irradiation is always off resonance because the protons have anisotropic chemical-shift tensors. It has been suggested (*20*) that the off-resonance effect (assumed to be 4 ppm) can contribute a broadening of 1.0 − 2.4 Hz at a field of 1.4T.

Line broadening can occur because of incomplete decoupling when the molecular motion has correlation frequencies that are near the frequency of the decoupling field (*21*). An example is the molecular reorientation of polymeric methyl groups (*22*). Incomplete decoupling arises when the methyl group motion about the C_3 axis is comparable to the strength of the decoupling field. Under experimental conditions, when the temperature of the polypropylene sample is decreased, the methyl resonance broadens significantly, and at − 143 °C, the methyl resonance completely disappears (*22*). Similar phenomena have been observed for methyl groups in epoxy resins (*23*) and in poly(tetrafluoroethylene) (PTFE) (CF_3 in this case) (*24*).

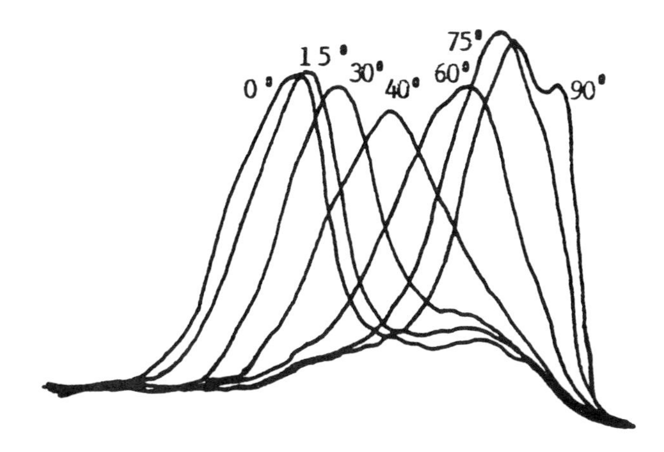

Figure 8.4. ^{19}F NMR spectra of oriented PTFE fibers at − 108 °C for various angles between the fiber direction and the magnetic field. (Reproduced from reference 26. Copyright 1980 American Chemical Society.)

Chemical-Shift Anisotropy in Solids

> *A complete chemical-shift tensor gives information about the local symmetry of the electron cloud around the nucleus and therefore presents a much more detailed picture of the chemical bonding of a certain atom than the isotropic chemical shift measured in solution.*
>
> − W. S. Veeman (*25*)

Basis of Chemical-Shift Anisotropy

The resonance frequency of a nuclear spin is determined by the shielding of the static magnetic field by the surrounding electrons. When a magnetic field is applied to the sample, a secondary magnetic field is generated by the motion of the electrons, and this secondary magnetic field partially shields the nucleus from the applied magnetic field. This shielding due to the electrons, as observed in solid-state NMR spectroscopy, is *anisotropic*.

An example of the anisotropic behavior of the chemical shift of the resonating nucleus is demonstrated in Figure 8.4. For the multiple-pulse decoupled ^{19}F NMR spectra of oriented PTFE fibers at − 108 °C, the observed chemical shift depends on the orientation angle between the fiber direction and the applied magnetic field (*26*). The PTFE fiber exhibits only one ^{19}F resonance (the CF_2), and this resonance shifts with the angle of alignment of the fiber with the magnetic field. These experimental observations, along with many others, indicate that the electron cloud surrounding the nuclei in most chemical bonds does not have spherical symmetry. That is, the shielding due to the electrons is anisotropic, and this anisotropic shielding causes a resonance shift that depends on the orientation of the chemical bond in the applied magnetic field. In solutions, the random molecular motions are fast on the NMR time scale. The anisotropic portion of the chemical shift is therefore averaged out to leave only the isotropic part of the chemical shift. In rigid solids, where molecular motion is highly restricted, the spatial dependence of this shielding determines the resonance line shape.

The chemical-shift anisotropy (CSA) is the chemical shift difference with the orientation of the bond in the static magnetic field. The ranges of chemical-shift anisotropies of several useful nuclei are listed in Table 8.1. The CSA range is much larger than the range of the isotropic chemical shift because the solution measurement reflects the average value.

The anisotropic nature of the chemical-shift tensor reflects the local symmetry of the electron cloud around the nucleus. The nuclei are shielded not only by their electrons, but also by the polarized electron

Table 8.1. Typical Chemical-Shift Anisotropy Values

Nucleus	*Chemical-Shift Anisotropy Extremes*	*Typical Line Width*	*Isotropic Chemical-Shift Range, σ*
^{1}H	100	10	20
^{13}C	425	120	250
^{19}F	1150	150	100
^{31}P	500	250	200

NOTE: The anisotropy values are given in parts per million (ppm).

clouds of the neighboring chains. Because of the existence of neighboring molecules, the electron cloud of the observed nucleus is modified through van der Waals interactions. Consequently, the CSA depends on the intermolecular distances as well as on the intramolecular environment. A summary of the available CSA data can be found in the recent literature (*17, 25, 27*).

Description of Chemical-Shift Anisotropy

The chemical shift of a nuclear spin depends on the relative spatial orientation of the external magnetic field and on the molecule to which the spin belongs. The directional nature of this chemical-shift interaction can be described in the following manner. The local field, H_{loc}, is given by

$$H_{loc} = \sigma H_o \qquad (8.11)$$

where σ is a dimensionless second-rank tensor that represents the anisotropic shift of the resonance frequency with respect to a bare nucleus. This tensor can be written as the sum of a symmetric and an antisymmetric component. The antisymmetric component affects only the NMR line positions to the second order and can therefore be neglected (*17*).

In a coordinate axis system, with x, y, and z related to the frame of the molecule, the chemical-shift tensor has nine elements but only six unique components:

$$\sigma = \begin{bmatrix} \sigma_{xx} & \sigma_{xy} & \sigma_{xz} \\ \sigma_{yx} & \sigma_{yy} & \sigma_{yz} \\ \sigma_{zx} & \sigma_{zy} & \sigma_{zz} \end{bmatrix} \qquad (8.12)$$

where $\sigma_{xy} = \sigma_{yx}$, $\sigma_{xz} = \sigma_{zx}$, and $\sigma_{yz} = \sigma_{zy}$. A special molecular-based axis system, known as the principal axis system x', y', z', can be constructed, and in this system, the chemical-shift tensor is diagonal:

$$\sigma = \begin{bmatrix} \sigma_{x'x'} & 0 & 0 \\ 0 & \sigma_{y'y'} & 0 \\ 0 & 0 & \sigma_{z'z'} \end{bmatrix} \qquad (8.13)$$

where $\sigma_{x'x'}$, $\sigma_{y'y'}$, and $\sigma_{z'z'}$ are the principal or diagonal elements of the tensor. In this molecular reference system, the chemical-shift tensor is completely determined by three principal elements and by the direction cosines (Θ_{ij}) of these three principle axes with the applied magnetic field. By convention, the principal elements are ordered according to the amount of shielding, so that σ_{11} is the least-shielded element, and σ_{33} is the most-shielded element. When the magnetic field points along the least-shielded direction, the nuclear spin resonates at the lowest field strength. Depending on the molecular symmetry, σ_{11} may be equivalent to $\sigma_{x'x'}$, $\sigma_{y'y'}$, or $\sigma_{z'z'}$, where x', y', and z' are determined by symmetry. Therefore,

$$\sigma_{zz} = \Theta_1^2\sigma_{11} + \Theta_2^2\sigma_{22} + \Theta_3^2\sigma_{33} \qquad (8.14)$$

The isotropic average of each Θ_i^2 is 1/3, so the average of σ_{zz} is the isotropic average

$$\sigma_{zz} = \sigma_i = (1/3)\,[\sigma_{11} + \sigma_{22} + \sigma_{33}\,] \qquad (8.15)$$

Effect of Chemical-Shift Anisotropy on Line Shapes

The molecules in a powder are randomly oriented. Therefore, the observed spectrum is the superposition of the signals from molecules having all possible orientations ranging from σ_{xx} to σ_{zz}. The line shape of the powder pattern $I(\sigma)$ is given by (*28*)

$$I(\sigma) = \left[\frac{K(\theta)}{\pi\Delta_1}\right]\left[\text{arc sin}\left(\frac{\Delta_2}{\Delta_1}\right)^{\frac{1}{2}}\right] \quad (8.16)$$

with

$$\Delta_1 = [(\sigma_{zz} - \sigma_{yy})(\sigma - \sigma_{xx})]^{\frac{1}{2}} \quad (8.17)$$

$$\Delta_2 = [(\sigma_{zz} - \sigma)(\sigma_{yy} - \sigma_{xx})]^{\frac{1}{2}} \quad (8.18)$$

where $K(\theta)$ is the complete elliptic integral of the first kind. This equation holds for $\sigma_{xx} < \sigma_{yy} < \sigma_{zz}$. A broad asymmetric powder pattern is observed when σ_{11} is not equal to either σ_{22} or σ_{33} (Figure 8.5a). This broad but highly characteristic band shape has break points that can be used to extract the tensor principal values of each chemically unique spin, as shown in Figure 8.5a. The line shape can usefully be discussed in terms of its width and asymmetry. The width, δ, is given as

$$\delta = \sigma_{33} - \sigma_i \quad (8.19)$$

and the asymmetry, η, is given by

$$\eta = \frac{(\sigma_{22} - \sigma_{11})}{\delta} \quad (8.20)$$

In a polycrystalline spectrum, the three principal elements can be determined, but the direction of the principal axis (which can be determined only for single crystals) cannot. For uniaxial symmetry, $\sigma_{11} = \sigma_{||}$, and $\sigma_{22} = \sigma_{33} = \sigma_{\perp}$

$$\sigma = \sigma_{||}\cos^2\phi + \sigma_{\perp}\sin^2\phi \quad (8.21)$$

The resulting polycrystalline pattern for a nucleus with an axially symmetric CSA has the shape shown in Figure 8.5b.

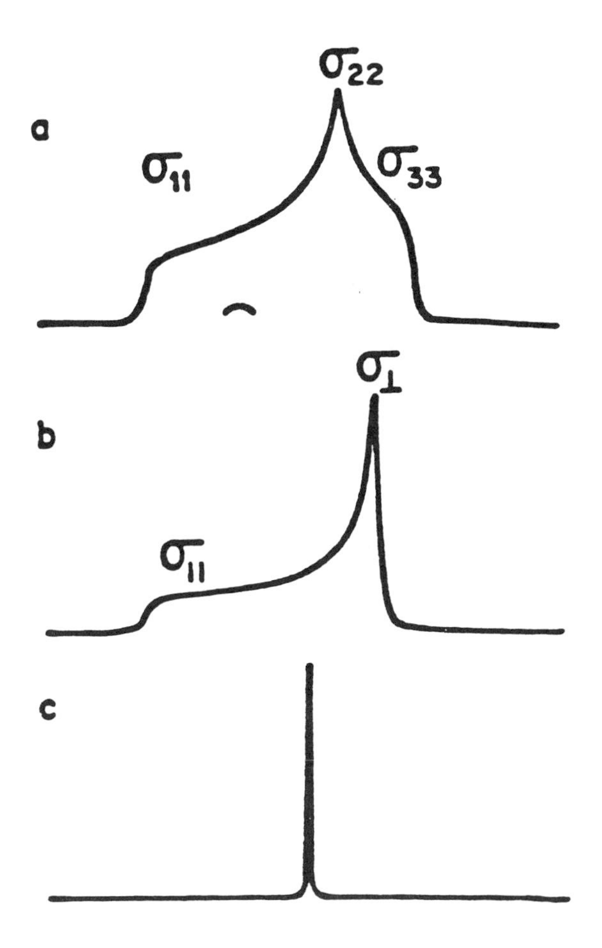

Figure 8.5. Line shapes in solid-state NMR spectroscopy: (a) axially asymmetric CSA, (b) axially symmetric CSA, and (c) solution.

A number of factors complicate the experimental measurements of CSA. These include a distribution in the values of the CSA; nonrandom orientation of the chains in the sample; inhomogeneous interactions, such as residual line broadening due to incomplete proton decoupling and carbon–carbon dipolar coupling; and molecular motion (*29*).

First, the shielding tensor, σ, of a nucleus may differ slightly from those of the other nuclei. In polymers, the CSA reflects a distribution of structural parameters such as variations in chain conformation and intermolecular distances. A range of CSA values is expected, and this spread in principal values will affect the line shapes.

Second, if the molecules are oriented, the line shape will be modified in a characteristic way that is determined by an orientation distribution function

(*30*). When there is axial symmetry in both the molecules and the sample, the spectral intensity at any given frequency is directly proportional to the population of molecules in a particular orientation (*31*). From the observed line shape, information about the distribution or orientation of the sample can be obtained (*26*).

This derivation of the theoretical line shapes assumes that the NMR lines of each spin are infinitely narrow. The introduction of a broadening function that arises from the experimental limitations makes the detection of singularities in the pattern difficult. The powder line shapes of an axially symmetric chemical-shift tensor are shown in Figure 8.6 (*25*). These powder line shapes have been convoluted with Lorentzian broadening functions of different widths of $A = [\sigma_{11} - \sigma_{22}]\omega_o$, where ω_o is the NMR frequency.

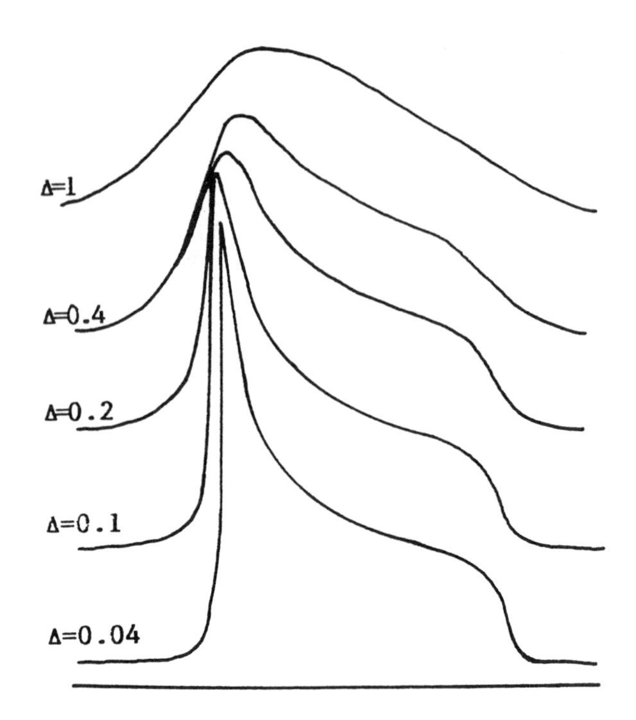

Figure 8.6. The powder line shapes of an axially symmetric chemical-shift tensor. (Reproduced with permission from reference 25. Copyright 1984 Pergamon Press, Inc.)

Experimental Determination of the Chemical-Shift Anisotropy

The NMR experiments that can be used to measure CSA are the following (*25*):

- the angular dependence of the chemical-shift shielding in single crystals
- the observed shift of molecules oriented in a liquid crystal
- the line shape of a powder
- reconstruction from sideband intensities observed with magic-angle spinning
- 2-D experiments

In most molecules, there are direct and indirect couplings between the ^{13}C and other spins such as ^{1}H. When the shift of a resonance line is the result of two tensorial interactions, the problem becomes more complicated. For polymers, the ^{13}C chemical-shift interactions can be isolated from the dipolar interactions by removal of the proton interaction with high-powered dipolar decoupling.

When the molecule contains several chemically different carbons, the powder pattern is a superposition of powder line shapes of the individual carbons. This overlap causes additional problems.

For single crystals that are sufficiently large to be precisely oriented with respect to the magnetic field, the six components of the chemical-shielding tensor can be determined (*32*). In the most general case, the orientations of the principal axes in the molecular frame must be specified. Because of the inversion symmetry of the magnetic interactions, the single-crystal spectrum is the same for two oppositely directed magnetic-field vectors. It is thus necessary to cover only half of all directions to obtain the correct result.

The chemical shift for six orientations of the single crystal with respect to the magnetic field must be measured. The change in frequency of each resonance can be determined as the sample is rotated. Analysis of rotation plots of the chemical shift yields the required information. From these measurements, the six unknowns can be determined. This method works, but it has low precision. A better approach is to rotate the crystal around three axes and to measure the shift as a function of Θ (*17*). A problem with this approach for polymers is that large single crystals are not available.

In some cases, the CSA can be determined from the line shapes of a powder spectrum. The NMR spectrum of a powder is the sum of the resonance lines resulting from all possible orientations of the nuclear sites. Such powder spectra often have characteristic band shapes with well-defined features from which the NMR parameters can be directly read. In a single isolated nucleus, the three principal values of the chemical-shift tensor can be obtained directly from the frequencies of the two shoulders and the peak in the spectrum. A least-squares curve-fitting procedure has been described (*33*) in which a set of theoretical points is produced by calculation from a set of adjustable parameters, and the theoretical points are compared to the corresponding set of experimental points.

Unfortunately, for systems with a number of different carbon nuclei, there is usually extensive overlap of the CSA signals, and the analysis of the powder pattern is difficult, if not impossible. Therefore, other techniques that partially narrow the lines must be used. Three different experiments can be used to determine chemical-shift anisotropies for powders: (1) slow magic-angle sample spinning (*34–36*), (2) off-angle magic-angle spinning (*34, 37*), and (3) rotation-synchronized rf pulse techniques (*38, 39*).

When the spinning speed of the sample is less than the CSA, rotational echoes appear as spinning sidebands that flank the averaged isotropic main peak. These echoes are spaced at the inverse of the spinning frequency. A graphical procedure has been described (*40*) for extracting the principal components of the chemical-shielding tensor from the intensities of the spinning sidebands. This method has been used for poly(butylene terephthalate) (PBT) (*41*) and dimethyl terephthalate (DMT), and the results are listed in the following table.

Sample	*Carbonyl Carbon*	*Nonprotonated Aromatic Carbon*
DMT	137 ± 4	201 ± 4
PBT[a]	127 ± 5	202 ± 5

NOTE: The values shown were obtained as $|\sigma_{33} - \sigma_{11}|$ and are in parts per million.
[a] From ref. 41.

The limitation of the slow magic-angle spinning (MAS) method is the difficulty of obtaining accurate measurements of the sideband intensities for complex systems in which there is extensive overlap. All sidebands must be resolved and properly assigned to use this method.

Rapid off-magic-angle spinning (OMAS) yields a scaled powder pattern that can be analyzed to determine the individual chemical-shielding tensors. The isotropic chemical shift must be known, and can be determined separately. The tensor components can be calculated from the OMAS-scaled components by the relation:

$$\sigma_i = \frac{(\sigma_{i'} - \sigma_{\text{ave}})}{C} + \sigma_{\text{ave}} \quad (8.22)$$

$$C = \frac{(3\cos^2\Theta - 1)}{2} \quad (8.23)$$

where Θ is the angle between the rotational axis and the static magnetic field, σ_i is one of the three tensor components, $\sigma_{i'}$ is one of the OMAS-scaled components, and σ_{ave} is the isotropic value of the three tensor components. OMAS is used to generate the tensor information at both positive and negative offsets of the magic angle because the sign of the shielding anisotropy inverts. OMAS has been used to study poly(ethylene terephthalate) (PET) (*37*) and PBT (*42*). The experimental values of the chemical-shift tensors were used to simulate the OMAS spectra. The values that were measured are listed in the following table.

Carbon Type	σ_{11}	σ_{22}	σ_{33}
Carbonyl			
$\theta = 56.3°$	165.8	165.8	165.0
$\theta = 57.6°$	167.1	167.1	158.4
Nonprotonated aromatic			
$\theta = 56.3°$	137.6	133.0	129.9
$\theta = 57.6°$	140.9	132.6	127.0
Protonated aromatic			
$\theta = 56.3°$	133.5	128.2	125.9
$\theta = 57.6°$	136.9	127.4	123.3

NOTE: All values are given in parts per million.

The CSA can also be determined by using 2-D techniques (*38, 43, 44*). The 2-D experiment results in a plot on which one axis is the anisotropic shift and the other axis corresponds to the isotropic shift. If the

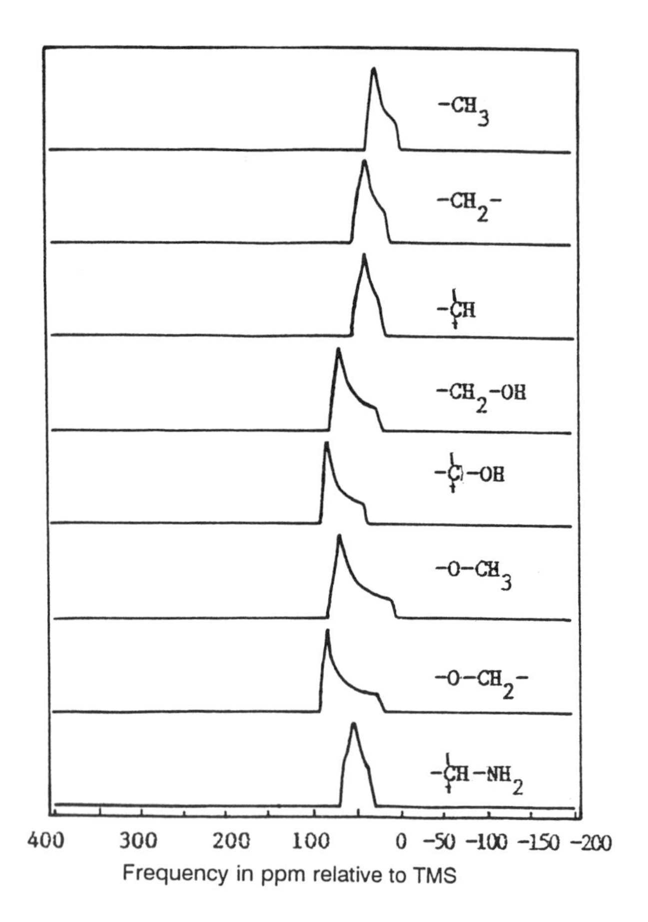

Figure 8.7. The CSA for carbons with sp³ *hybridization. (Reproduced with permission from reference 27. Copyright 1987 American Institute of Physics.)*

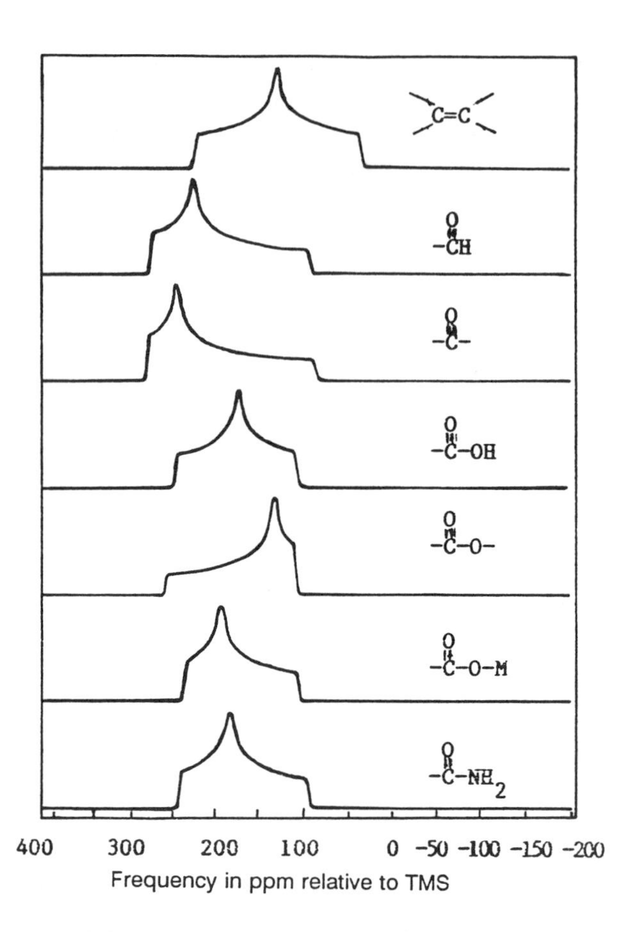

Figure 8.8. The CSA for carbons with sp² *hybridization. (Reproduced with permission from reference 27. Copyright 1987 American Institute of Physics.)*

FID is sampled at times $t_n = n(2\pi/\omega_r)$, where ω_r is the rotor frequency, the FID represents the isotropic spectrum. If the FID is observed at different times t . . . t_n, and the FIDs are Fourier transformed, the 2-D spectrum will be observed (*45*).

The aforementioned methods for determining the CSA are applicable only for samples without dipolar-modulated chemical-shift powder patterns (*46*). Under these circumstances, the analysis problem becomes quite difficult. Sample-spinning NMR techniques must be used to measure the heteronuclear dipolar–chemical shift 2-D powder patterns (*47, 48*).

CSA for Hydrogen Nuclei in Various Environments

Sufficient results have been obtained from CSA measurements to allow a comparison of the principal values of the shielding tensors for hydrogen nuclei in various environments (*17*).

CSA for Carbon Nuclei in Various Environments

The CSA patterns for the various types of carbons have been determined. The CSA for carbons with *sp*³ hybridization are shown in Figure 8.7, those for carbons with *sp*² hybridization are shown in Figure 8.8, and those for carbons with *sp* hybridization are shown in Figure 8.9 (*27*). The high sensitivity of the CSA to the nature of the carbon is reflected in these CSA patterns. The width of the CSA for a given carbon can be quite narrow, as shown for the methyl carbon of ethyl alcohol, or can be very wide, as for the aromatic carbons of benzene and toluene.

Of particular interest to polymer scientists is the CSA of methylene carbons. For the methylene car-

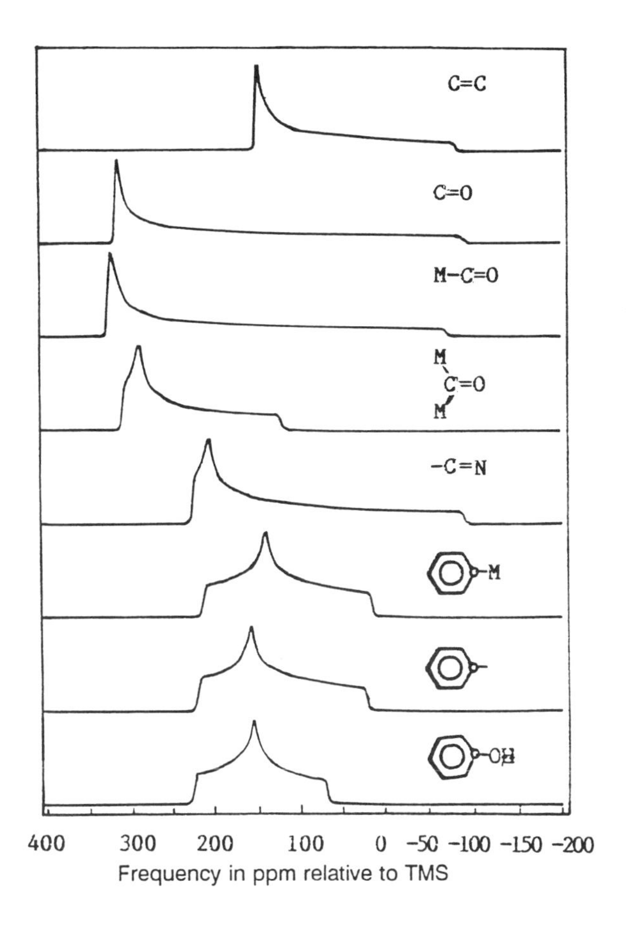

Figure 8.9. The CSA for carbons with sp *hybridization. The circle in the structures represents the carbon that is being observed. (Reproduced with permission from reference 27. Copyright 1987 American Institute of Physics.)*

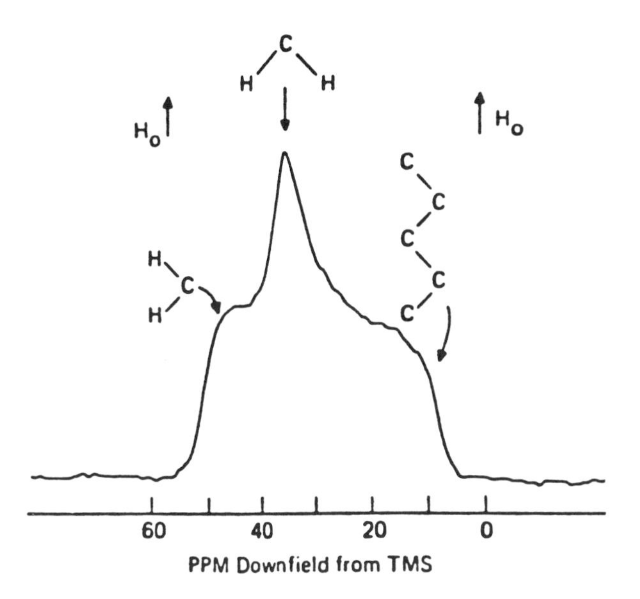

*Figure 8.10. The DD – CP ^{13}C spectrum of polycrystalline PE. The three elements of the chemical-shielding tensor and their assignments in an all-*trans *polymethylene chain are shown. (Reproduced with permission from reference 49. Copyright 1980 Academic Press.)*

bon, the values of the chemical-shift tensors are as follows: σ_{11}, which bisects the H – H bond, has a value of 50.5 ± 15 ppm; σ_{22}, which bisects the H – C – H angle, has a value of 37 ± 10 ppm; and σ_{33}, which is perpendicular to the H – C – H plane, has a value of 16 ± 17 ppm (*32*). The three elements of the chemical-shielding tensor and their assignments in an all-*trans* methylene chain are shown in Figure 8.10. For the methylenes in polyethylene (PE), σ_{11} = 49 ppm, σ_{22} = 35 ppm, and σ_{33} = 12 ppm (*49*). On the basis of these assignments, the carbons of a PE chain with its axis aligned along H_o will resonate at 12.9 ppm (*49*).

Effect of Motion on Chemical-Shift Anisotropy

Changes in the NMR line shapes can define the geometry of the dominant motions in the polymers. Motionally averaged powder patterns reflect the rates, amplitudes, and angles of the motions involved. Rapid rotation about one bond axis of a molecule causes a partial averaging of the shielding tensor, and the average shielding tensor displays axial symmetry about the rotation axis. Consider the simple case of rotation of a CF_3 group around the C – C axis (Figure 8.11). The CSA changes from an asymmetric pattern to a uniaxially symmetric pattern as a result of the averaging of σ_{11} and σ_{22} because of rotation about the σ_{33} axis, and the line shapes are characterized by only two values of the chemical-shift tensor.

The chemical line shape is influenced by the rate and type of motion that takes place in the sample. Three types of motions make important contributions to the line shapes, and these are macroscopic sample rotation, molecular reorientation, and molecular conformational changes.

Macroscopic sample rotation is performed in a variety of ways, including magic-angle spinning (MAS), which is discussed later in this chapter. When the solid sample undergoes magic-angle spinning, the average shift over one cycle of rotation

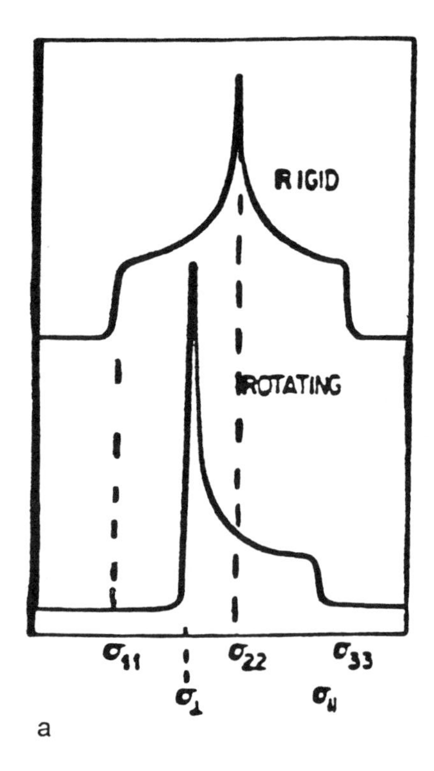

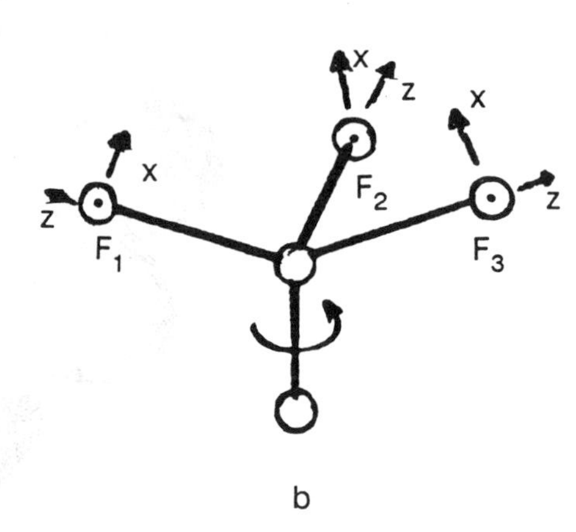

Figure 8.11. (a) The ¹⁹F chemical-shift spectra of polycrystalline CF_3COOAg for rigid and rotating $-CF_3$. (b) Diagram of rotation of the CF_3 group about the C−C axis.

becomes independent of orientation and is given by the isotropic shift (*49*). However, the sample rotation must be rapid compared to the anisotropy spread ($\sigma_{33} - \sigma_{11}$) in order for the coherent averaging to occur. This condition means that a polycrystalline sample will have only a single narrow line for each magnetically inequivalent nucleus. Chemical shifts observed in the solid state are usually within a few parts per million of those observed in solution.

When the rate of molecular motion is much faster than the spectroscopic time scale (the *fast-exchange limit*), only the angles are needed for line-shape calculations (*50*). For example, the effect of fast-diffusional rotation about an arbitrary axis is calculated by transforming the static tensor to a rotationally averaged tensor. Spiess has reviewed the method of calculating the chemical-shift line shapes in the presence of rotational motion of molecules (*51*).

Consider the example of motions of phenyl rings (*52*). Two different motional models are possible. Reorientation about the chain ring $C_\beta - C_\nu$ bond axis may occur either by 180° flips (jumps between two indistinguishable conformations) or by fast-diffusional rotational motion with small angular displacements. The principal axis system is shown in Figure 8.12. The chemical-shielding tensor for aromatic-ring carbons has σ_{zz} bisecting the C−C−C angle, σ_{yy} perpendicular to the plane of the ring, and σ_{xx} orthogonal to these two axes. With this orientation, $\sigma_{yy} > \sigma_{zz} > \sigma_{xx}$, and their sum equals zero. Consequently,

$$\sigma_{zz} = -(1/2)(1 + \delta)\sigma \tag{8.24}$$

$$\sigma_{xx} = -(1/2)(1 - \delta)\sigma \tag{8.25}$$

$$\sigma_{yy} = \sigma \tag{8.26}$$

with $\delta = (\sigma_{xx} - \sigma_{zz})/\sigma_{yy}$. The powder pattern calculated for σ = 3.87 kHz and δ = 0.73 is given in Figure 8.12. The effects of fast-diffusional rotation have been calculated by transforming the tensor from the static system to a rotating system. For comparison, the motional model that considers rapid jumps between two equivalent sites is calculated by using rotation matrices to construct an average flipping-frame tensor from the static one. The results are compared in Figure 8.12 (*52*).

NMR studies of the motions of the aromatic amino acids of the fd bacteriophage protein were made (*52*), and the results are shown in Figure 8.12. The experimental spectrum fits the calculated flip-averaged spectrum better than it fits the rotationally averaged spectrum. Therefore the rings are executing 180° flips in less than 1 ns, and probably in less than 0.01

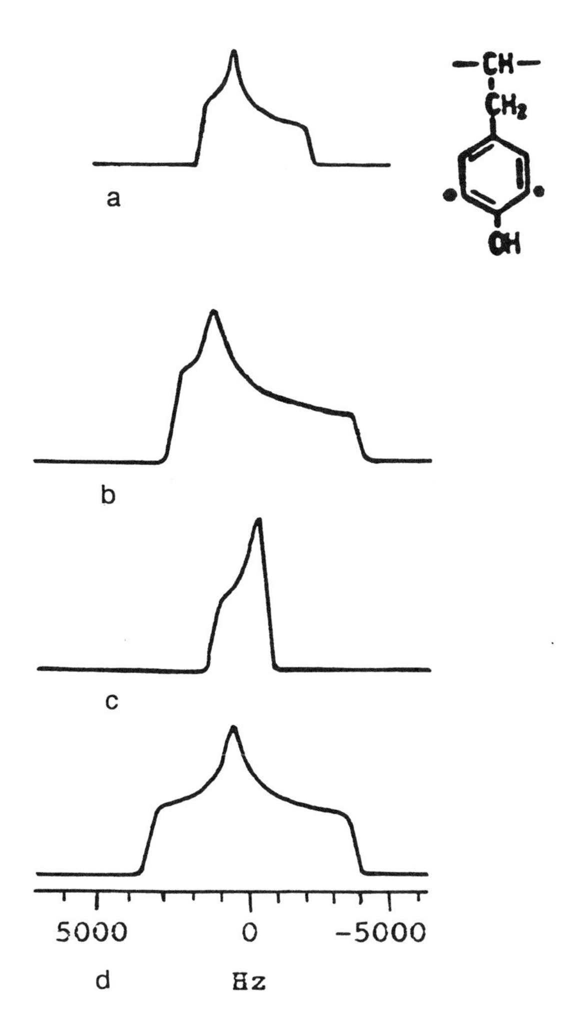

Figure 8.12. The observed spectra of the phenyl ring (a), and the simulated spectra for motion involving ring flips (b), rapid rotation (c), and a static system (d). The principal axis is the C – C – phenyl ring bond axis shown in a.

ns. Additionally, the experimental spectrum is narrower than expected from the static chemical-shift tensor, and so additional molecular motion of the main chain is also involved.

NMR line-shape experiments of molecular motion have been made for the polycarbonates of bisphenol A (BPA-PC) (*53 – 55*). Two motional processes occur: (1) restricted rotational diffusion over a limited angular amplitude around the carbon-ring bond axis, and (2) 180° flips between two potential minima around the same axis; these 180° flips constitute the primary motion. These motions correlate well with mechanical and dielectric results for this polymer.

Structural Applications of Chemical-Shift Tensors

A number of applications of CSA have been developed with the experimentally determined band shapes. Differences in the CSA band shapes are a consequence of changes in the electronic environment associated with the chemical nature of the carbon site, and these differences can be used to determine the amount of the unique carbon in the sample. In practice, powder patterns are simulated and compared with the experimental spectra to obtain structural information or the relative fractions of the carbon types (*56*).

CSA Analysis of Coal Samples. CSA line shapes were used for the analysis of carbon types in coal samples (*56*). The derived experimental line shapes were fit to the experimental coal spectra for samples with high and low oxygen contents, and the fractions of the different types of carbons in the coal were determined (*56*). Differences in the CSA band shapes for aromatic compounds are due to differences in the bond orders of the aromatic-ring carbons. Aromatic carbons can be classified into three different subgroups according to distinct chemical environments. These groups are benzenelike sp^2 hybridized protonated carbons on the periphery of the aromatic ring, substituted peripheral carbons (e.g., alkylated carbons), and condensed inner and bridgehead carbons. Protonated benzenelike carbons are characterized by a highly asymmetric powder pattern. Upon substitution of an aliphatic group, the two in-plane tensor components move closer to reduce the asymmetrical features of the band. The inner carbons in a large polycondensed aromatic system have nearly axial symmetry, and the band shape is nearly axially symmetric (*57*). These band shapes were used to analyze the different carbon types in fusinite and anthracite coals (*57*).

CSA Determination of Crystallinity. The CSA line shapes of the crystalline and amorphous portions of polymers are substantially different from each other and can be used to determine the crystallinity of polymer systems. An analysis of the NMR crystallinity of poly(tetrafluoroethylene) (PTFE) was accomplished by using ^{19}F CSA line-shape analysis with multiple-pulse techniques for the removal of dipolar effects (*26*). The ^{19}F spectral changes of PTFE as the

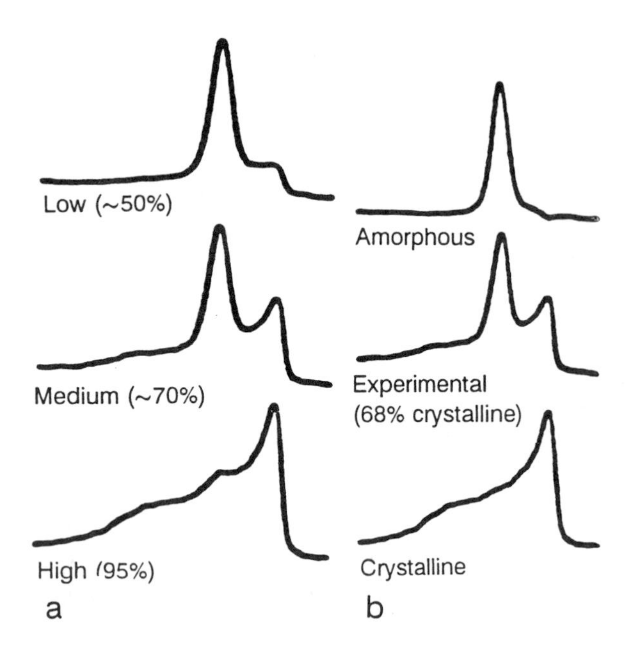

Figure 8.13. (a) The ¹⁹F REV-8 chemical shift of PTFE samples of varying crystallinity obtained at 259 °C. (b) The ¹⁹F REV-8 chemical shift of a PTFE sample with 68% crystallinity, showing the decomposed line shapes of the amorphous and crystalline fractions. (Reproduced from reference 26. Copyright 1980 American Chemical Society.)

crystalline content is varied are shown in Figure 8.13. The analysis of the ¹⁹F spectrum was accomplished by decomposing the line shapes into the amorphous and crystalline fractions.

CSA Determination of Orientation. A number of methods are available for the determination of the degree of molecular orientation in solid polymers. Techniques that selectively give information about the molecular orientations in the crystalline and amorphous regions are especially valuable. A potential method for measurement of orientation is the determination of the CSA.

This approach has been used to study the uniaxial drawing of PTFE. The ¹⁹F multiple-pulse spectrum of semicrystalline PTFE samples as a function of the angle, β, between the direction of stress and the magnetic field is shown in Figure 8.14 (*58*). By varying the moments of these spectra about the isotropic chemical shift, an approximate orientational probability distribution of the chain axes about the

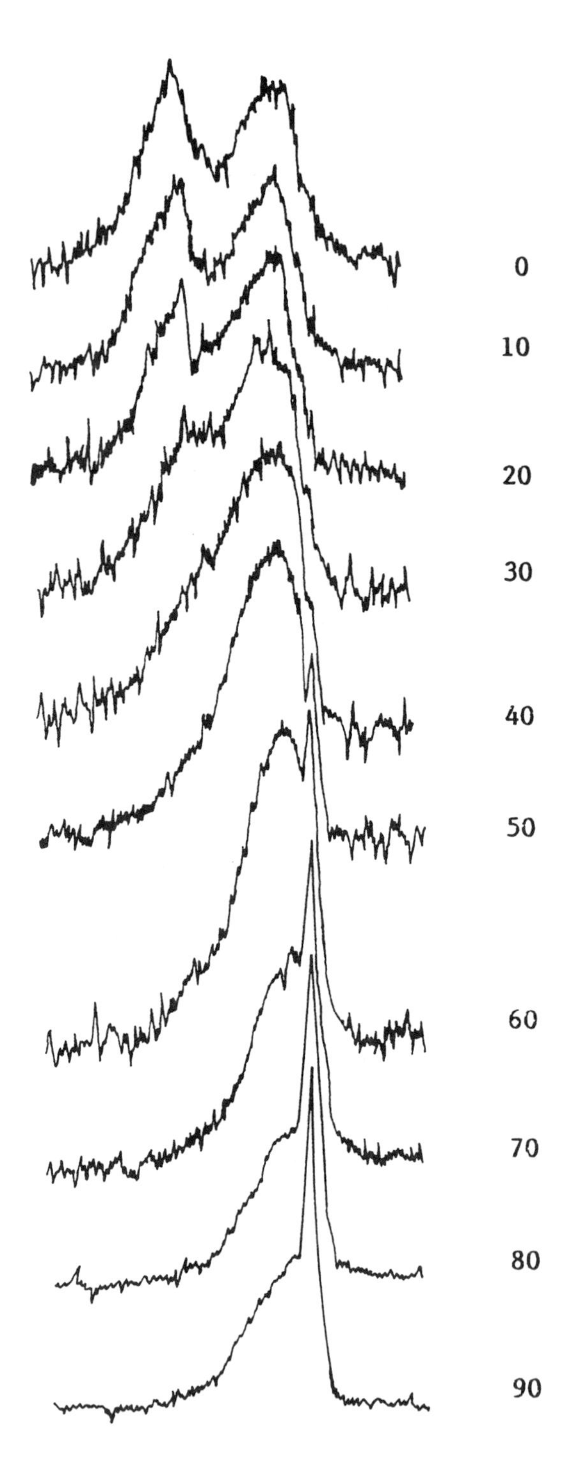

Figure 8.14. The ¹⁹F NMR chemical-shift spectra of deformed PTFE samples as a function of the orientation, β, of the stretch axis relative to the static magnetic field. (Reproduced with permission from reference 58. Copyright 1982 Butterworth & Co., Ltd.)

direction of stress can be determined. These results are shown in Figure 8.15 for samples with different draw ratios (*59*). The increase in the intercept reflects the greater alignment of the chains with the draw direction.

Biaxial films of poly(ethylene terephthalate) (PET) were studied by using the NMR CSA technique (*60*). There is a highly ordered component of the film in which the planes of the aromatic rings lie close to, but not in, the plane of the film. There is also a fast-relaxing component that is much less oriented and results from polymer chains that are relatively mobile, that is, the amorphous phase.

The Magic-Angle Spinning (MAS) Experiment

As suggested by Andrew (*11*), for solids in which the molecular motion is insufficient, "one may seek to emulate nature by imposing a motion on the nuclei". This was first accomplished by rapidly spinning the solid sample at 57.4° (the *magic angle*) (*61, 62*). Under this circumstance, each molecule experiences a continuous series of orientations with respect to the external magnetic field. The result is an isotropic average of the chemical shift for the solid similar to that observed for the same sample in solution.

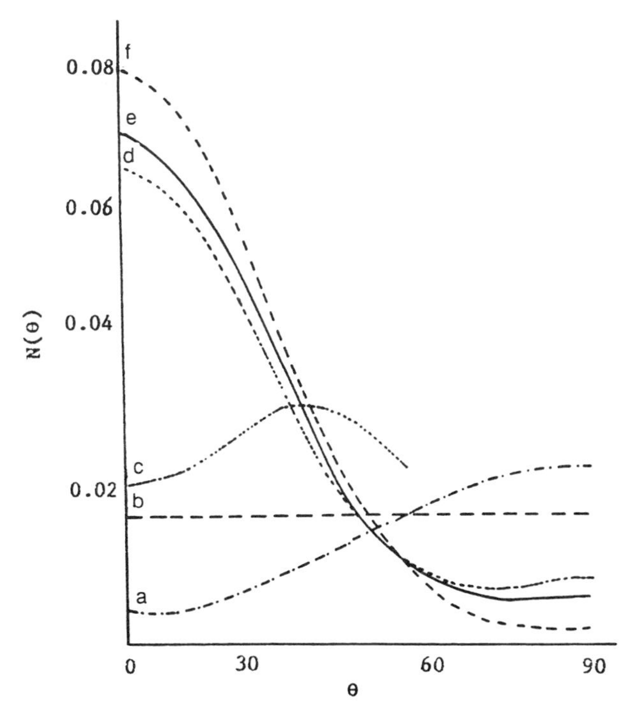

Figure 8.15. The distribution function, N(θ), *derived from the* ^{19}*F NMR spectrum of PTFE at different draw ratios. (a)* λ = *0.82, (b)* λ = *1.0 (unoriented), (c)* λ = *1.40, (d)* λ = *1.75, (e)* λ = *2.10, and (f)* λ = *2.40. (Reproduced with permission from reference 59. Copyright 1983 John Wiley & Sons, Inc.)*

Utility of MAS

MAS can potentially remove any of the magnetic interactions responding to the ($\cos^2 \beta - 1$) geometrical dependence. These interactions include dipolar interactions, CSA, and quadrupole interactions. However, there are severe limitations to the use of MAS. First, the spinning rate must exceed the magnitude of the interactions to be removed. Because of the large magnitude of some of these interactions, sufficiently high spinning rates are difficult to achieve. The magnitude of the homonuclear dipolar field for protons is approximately 20 kHz, and generally, it is not practical to spin at such high speeds. High-speed spinners are now commercially available, and in the future it may be possible to use MAS to remove the proton dipolar interactions. Currently, homonuclear dipolar interactions for protons are generally removed by using multipulse techniques. For ^{13}C nuclei, the homonuclear dipolar effects are much smaller because of the large distances between the ^{13}C nuclei resulting from the low natural abundance of the ^{13}C. MAS will therefore remove the ^{13}C homonuclear dipolar interactions.

The substantial removal of anisotropic-broadening interactions from the NMR spectra of solids by rapid rotation about the magic angle reveals the chemical shift and spin-multiplet fine structures for solids, and these are similar to those obtained in the high-resolution NMR spectra of liquids. The MAS NMR method is used when a material must be studied in the solid state in order to examine the solid-state structure, or in cases where the material will not dissolve in a suitable solvent or melt without decomposition.

Removal of Chemical-Shift Anisotropy

Because CSA has the same angular dependence as the dipolar interactions, but the magnitude of the effect (1 – 3 kHz) is much smaller, MAS is a viable technique for the removal of CSA. The chemical shift can be written as follows:

$$\sigma_{z'z} = \langle \sigma \rangle + \sigma^{a} \qquad (8.27)$$

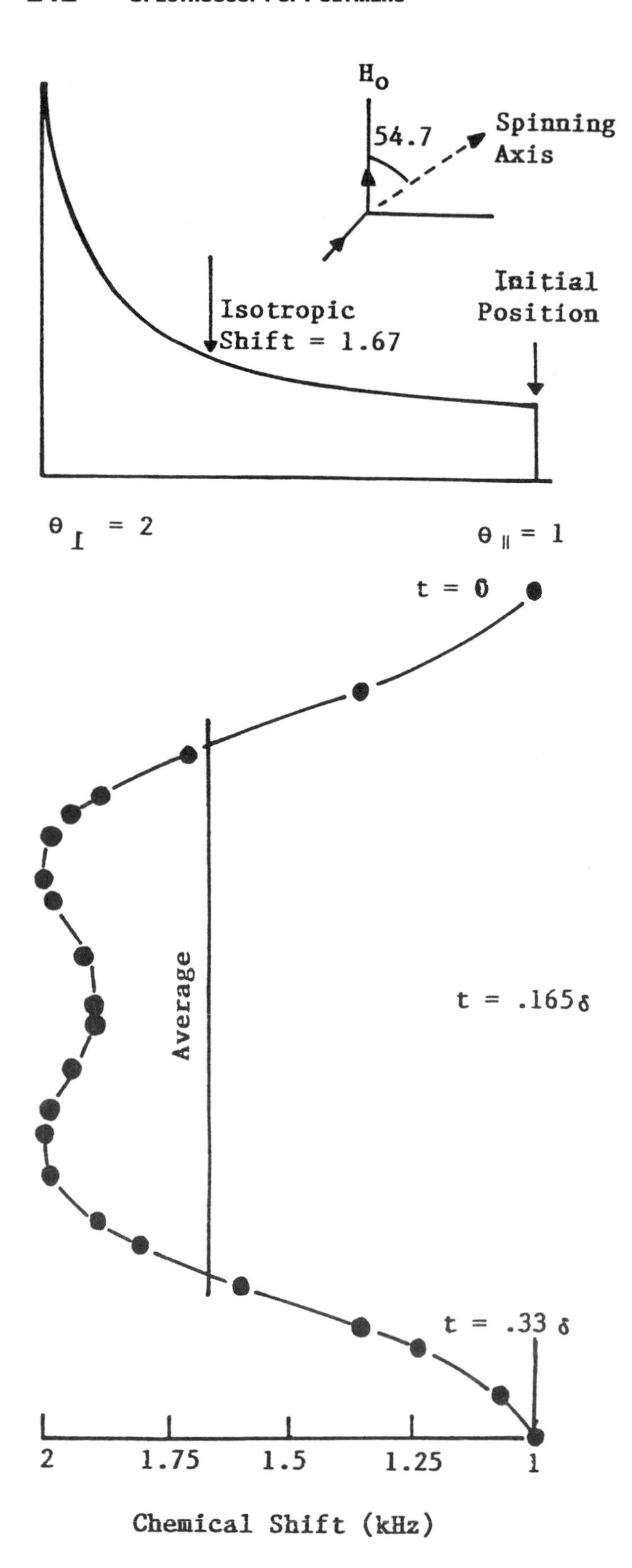

Figure 8.16. The change in chemical shift with rotation is shown for a uniaxial chemical shift. In this case, an average isotropic shift of 1.67 will be measured when the anisotropy range is 1−2. The top figure is the powder pattern for a CSA going from 2 to 1. The bottom figure shows the specific shift as a function of rotation angle. These measurements should agree very closely with the value measured for the corresponding carbon in the liquid or solution state if other solid-state effects are neglected. (Reproduced with permission from reference 89. Copyright 1982 Plenum Publishing Corporation.)

where $<\sigma>$ is the isotropic part that is equal to $(1/3)Tr(\sigma)$ and σ^a is the anisotropic part. For a uniaxially symmetric tensor, the anisotropic part can be written

$$\sigma^a = (1/3)[\sigma_1 - \sigma_2(3\cos^2\phi - 1)] \quad (8.28)$$

and when the sample is rotated at the magic angle (ϕ = 57.4°)

$$<\sigma^a> = 0 \quad (8.29)$$

Under these circumstances, the CSA is removed, and only the isotropic average is measured. This is illustrated in Figure 8.16, where the changes in the chemical shifts with rotation are shown for a uniaxially symmetric CSA. In this case, an average isotropic shift of 1.67 ppm will be measured when the anisotropy ranges from 1 to 2 ppm.

However, another complication is observed in the spectra obtained with MAS. When the sample spinning speed is less than the CSA, rotational echoes appear as spinning sidebands that flank the isotropic peak. These echoes are spaced at the inverse of the spinning frequency. The effects of spinning on the ^{31}P spectra are shown as a function of the spinning rate in Figure 8.17 (*63*). As the rate of rotation increases, the rotational spinning sidebands move further out and become weaker. At very high spinning speeds, the intensities of the sidebands become negligible, and the spectrum consists of the narrowed central line at the Larmor frequency, ω_o. The intensities of the first satellites are expected to decrease at a rate of ω_r^{-2}, thus preserving their contributions to the second moment of the entire spectrum. Indeed, the magnitude of the second moment of the spectrum should be invariant with respect to rotation. The intensities of the second and higher satellites fall even more rapidly with an increase in ω_r, as ω_r^{-2n} for the nth satellite.

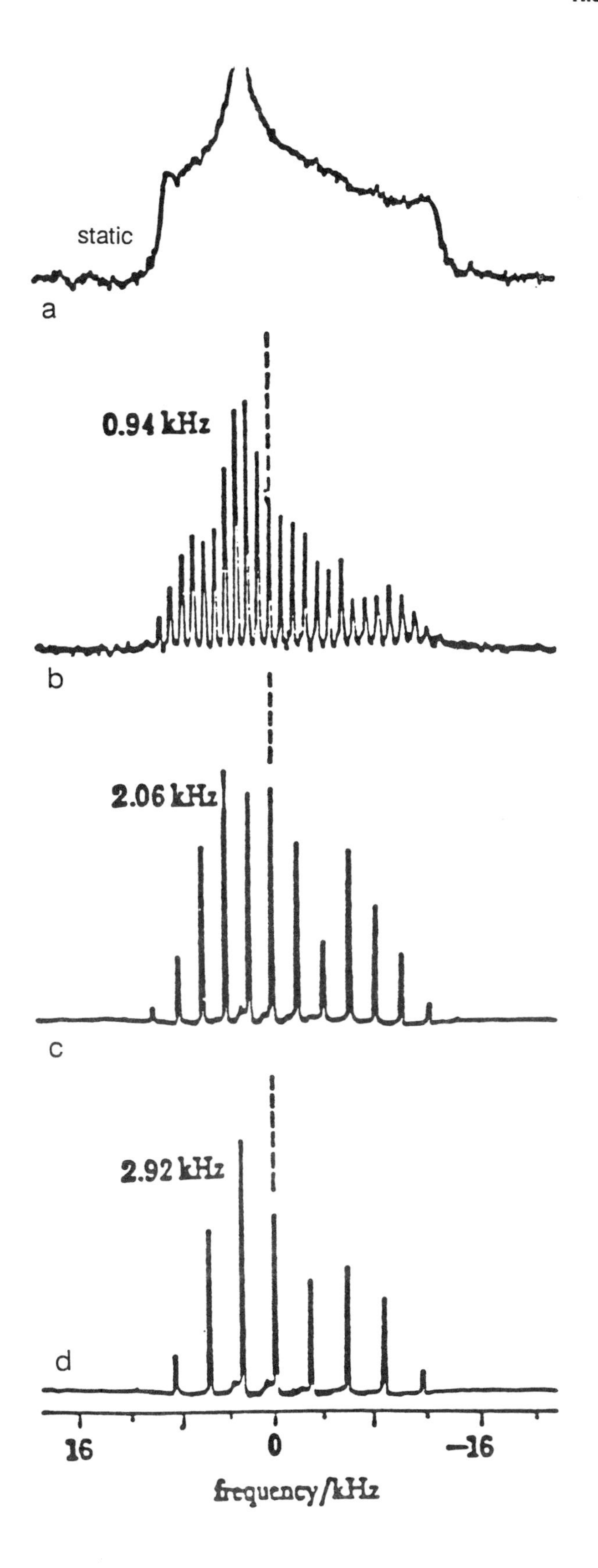

Setting the magic angle is very critical to the success of the MAS experiment. If the error in setting the magic angle is 0.1° for a shielding anisotropy of 200 ppm (not uncommon for ^{13}C in organic solids) this misalignment error results in a line broadening of 1.02 ppm (*64*). The effect on the measured spectra of misalignment of the magic angle is shown in Figure 8.18. In the case of polycrystalline sucrose, the non-spinning spectrum is shown as well as the spectra for a misadjustment of the magic angle by 0.7° (*65*). The line width increases and a loss of resolution occurs with the magic-angle misadjustment, although the lines are clearly narrowed by the spinning. A series of spectra of a single crystal of sucrose are shown in Figure 8.19; these spectra were obtained with misadjustment of the magic angle (*65*). In this case, chemical-shift differences for a single crystal are observed with the missetting of the magic angle.

MAS is most effective in narrowing the resonances of carbons in solids with a narrow CSA. For aliphatic carbons, the CSA is small, and moderate spinning speeds suffice. For carbons with multiple bondings, the CSA is much larger, and higher spinning speeds are required to coherently average the CSA. For aromatic carbons, the CSA is nearly 150 ppm. At a carbon Larmor frequency of 15 MHz, this CSA corresponds to 2.25 kHz, and the sample must be rotated at this frequency to coherently average the CSA. Because the magnitude of the CSA increases with the applied magnetic field, higher spinning speeds are required for higher applied magnetic fields.

The Cross-Polarization (CP) Experiment

The ^{13}C isotope presents a challenge for NMR spectroscopy because of its inherent low sensitivity. The low natural abundance, the low gyromagnetic ratio, and the long spin–lattice relaxation time of the ^{13}C nucleus limit the observable ^{13}C signal. The sensitiv-

Figure 8.17. ^{1}H decoupled ^{31}P MAS spectra of BDEP. The powder pattern in (a) is approximately 23 kHz wide and narrows to a spectrum exhibiting sidebands of 60 Hz full width at n_r = 2.92. *The dashed line marks the isotropic shift. (b – d) The effects of spinning on the ^{31}P spectra as a function of spinning rate. (Reproduced with permission from reference 63. Copyright 1980 The Royal Society of Chemistry.)*

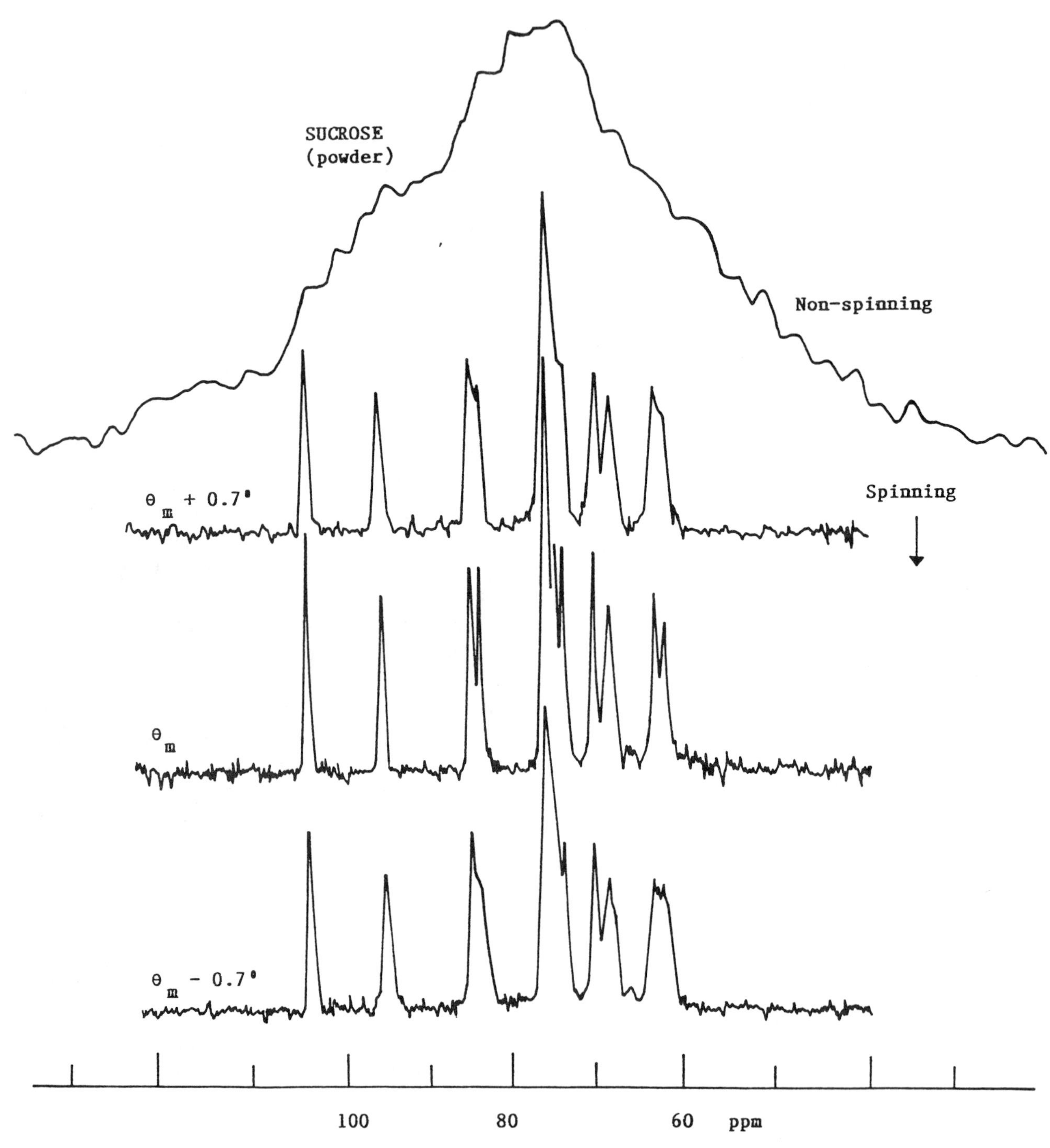

Figure 8.18. The effect on the measured spectra of misadjustment of the magic angle. (Reproduced with permission from reference 65. Copyright 1982 Academic Press.)

ity must be increased for the detection of the resonances of the ^{13}C nuclei. In response to this need, polarization-transfer methods are used to transfer the large spin-state polarization of the protons to the weakly polarized ^{13}C nuclear species. Perfect transfer of polarization from protons to ^{13}C would improve the ^{13}C signal by a factor of $\gamma_H/\gamma_C = 4$.

In NMR spectroscopy, improvements in the sig-

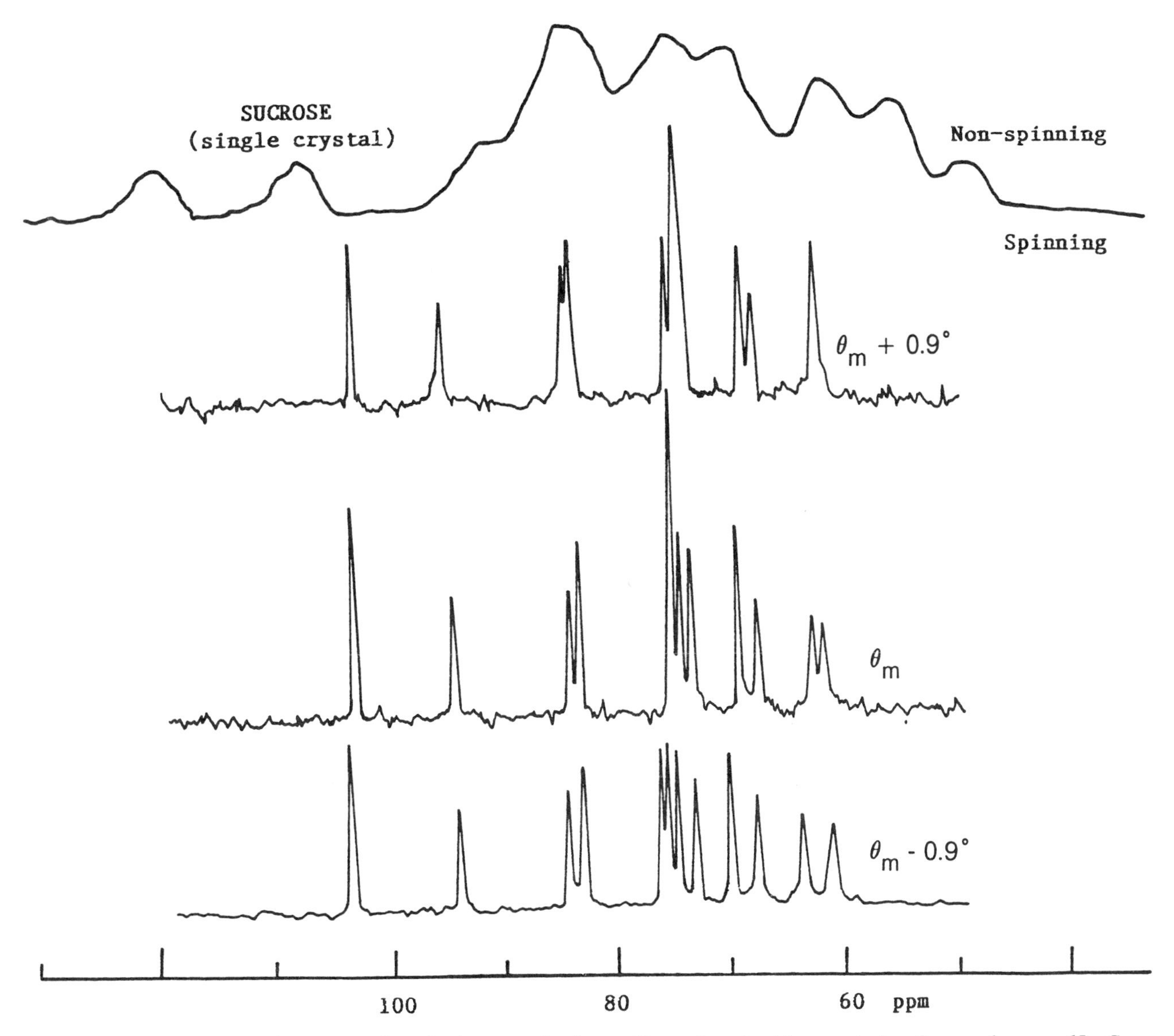

Figure 8.19. A series of spectra of a single crystal of sucrose obtained with misadjustment of the magic angle. (Reproduced with permission from reference 65. Copyright 1982 Academic Press.)

nal-to-noise ratio are often accomplished by signal averaging through the coadding of the FIDs. The delay between pulse sequences that can be coadded is limited by the time required to repolarize the nuclear spin system ($\sim 5 \times T_1$). The ${}^{C}T_1$ spin–lattice relaxation times for solids are significantly longer than the ${}^{H}T_1$ times, so the maximum number of pulses that can be used is less. However, with polarization transfer, the delay time between cross-polarization pulse sequences depends on the shorter ${}^{H}T_1$ times, and further gains in signal enhancement occur to the extent of $({}^{C}T_1/\,{}^{H}T_1)^{1/2}$.

Experimental Implementation of Cross-Polarization

Cross-polarization (CP) is a double-resonance experiment in which the energy levels of the ^{1}H and ^{13}C spins are matched to the Hartman–Hahn condition (*66*) in the rotating frame. Under this condition, energy may be exchanged between the two coupled spin systems. The result is a growth of the ^{13}C magnetization at the expense of the ^{1}H magnetization. The basic pulse sequence used for the cross-polarization experiment is shown in Figure 8.20. The method

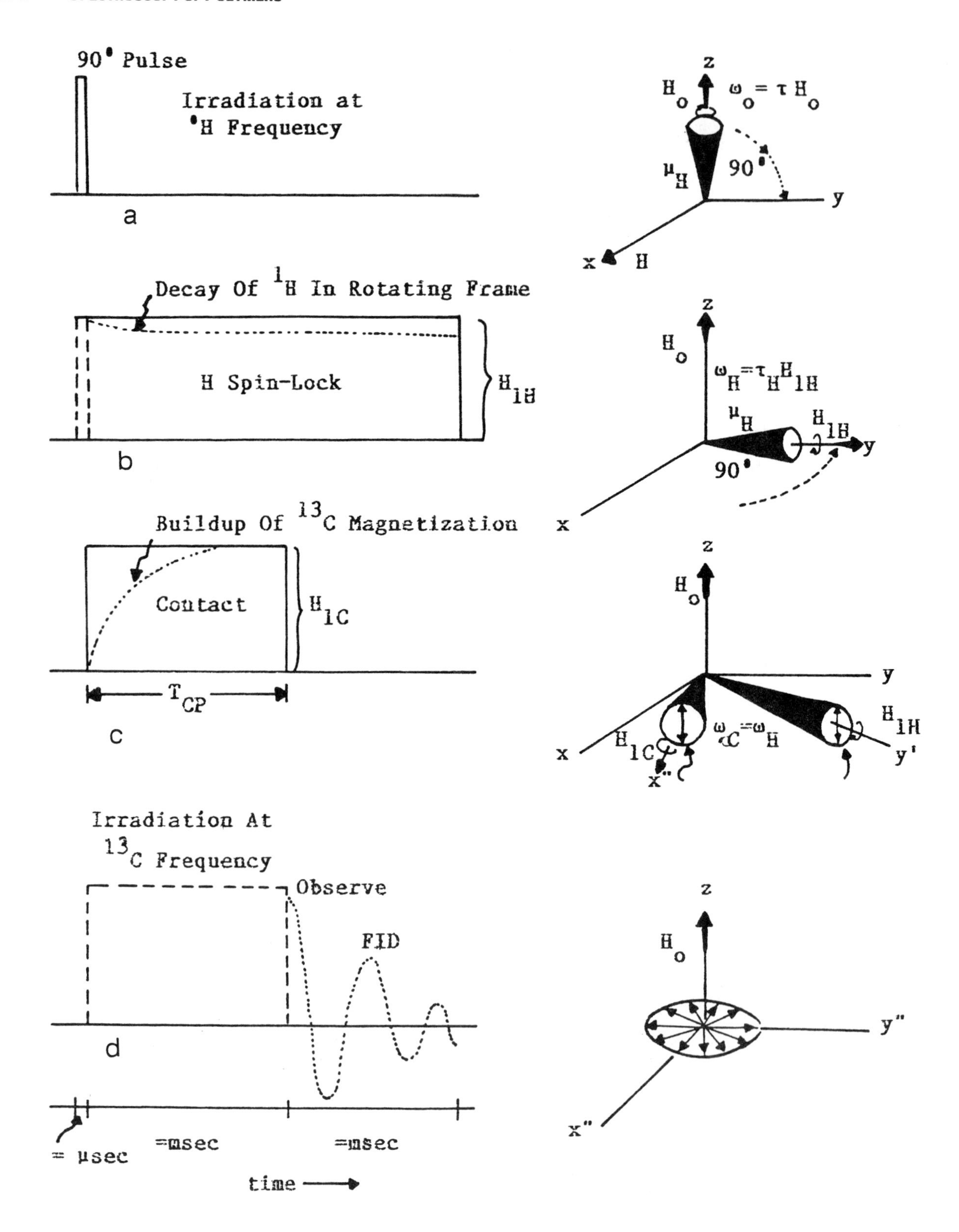

Figure 8.20. The timing sequence for the cross-polarization experiment. (Reproduced with permission from reference 67. Copyright 1979 International Scientific Communications, Inc.)

involves four steps: (a) polarization of the ^1H spin system, (b) spin-locking of the protons in the rotating frame, (c) establishment of contact between the carbons and the protons, and (d) measurement of the magnetization of the carbon nucleus.

The application of a 90° ^1H pulse rotates the ^1H longitudinal magnetization into transverse magnetization (the *xy* plane is in the rotating reference frame as shown in Figure 8.20a). The second step is the nearly immediate (within microseconds) application of a 90° phase-shifted proton pulse, which makes the effective field colinear with the proton magnetization and results in *spin locking* (Figure 8.20b). Spin locking is used because the magnetization of the protons in the spin-locked state decays by a rotating frame spin – lattice relaxation time, $^HT_{1\rho}$, at a rate that is orders of magnitude slower than the ordinary spin – spin relaxation (T_2) process. Thus, the magnetization loss of the protons resulting from relaxation effects during the contact time is small.

Simultaneously with the spin locking of the protons, in the third part of the experiment (Figure 8.20c), a pulse is applied in the ^{13}C channel, and this pulse is carefully adjusted so that the energy gap for spin flips corresponds exactly to that of the protons. This pulse is maintained for a time, t_{CP}, and is the *contact time*. This contact time allows for the exchange of energy between the abundant proton-spin reservoir and the rare carbon-spin system. This exchange, called *cross-polarization*, occurs when the Hartmann – Hahn match (*68*), which is defined as

$$\omega_H = \omega_C \tag{8.30}$$

is satisfied in the rotating frame.

The fourth part of the CP experiment (Figure 8.20d) is to terminate the ^{13}C pulse and to observe the FID while the ^1H field is used for decoupling. The resulting FID is Fourier transformed to give the frequency-domain spectrum.

In the CP experiment, the ^{13}C spin system is polarized by the ^1H spin system under the Hartmann–Hahn condition. Therefore, the longer $^CT_1$ (1 – 1000 s) is replaced by the shorter $^HT_1$ spin – lattice time (1 – 100 ms). The optimum recycle delay has been calculated, and it is 1.25 $^HT_1$ for the situation in which spin diffusion dominates the spin system (*69*). For the case in which spin diffusion is not efficient, the optimum recycle delay must be longer (4 – 5 $^HT_1$) to avoid saturation.

Dynamics of the CP Experiment

A typical increase in carbon magnetization with contact time is shown in Figure 8.21. In the CP experiment, polarization is transferred from the magnetically rich protons to the magnetically poor carbons via their static dipolar interactions. When both spins are locked in the rotating rf fields and their amplitudes are matched by using the Hartmann – Hahn condition, the time constant for the transfer of polarization under conditions of spin locking is called the T_{CH} relaxation time. This transfer is a spin – spin process, and it generally has a magnitude in the range of 100 μs. In polymers, the T_{CH} for methine and methylene carbons ranges from 15 to 50 μs and from 45 to 100 μs for protonated aromatic carbons. These T_{CH} values are essentially independent of the rate of sample rotation in the MAS experiment.

The result of CP is an initial growth of the ^{13}C magnetization at a rate that is inversely proportional to the cross-polarization-transfer rate constant (T_{CH}) (Figure 8.21). After a suitable period (typically 1 – 10 ms), a maximum ^{13}C magnetization is reached, after which the ^{13}C signal begins to decay at

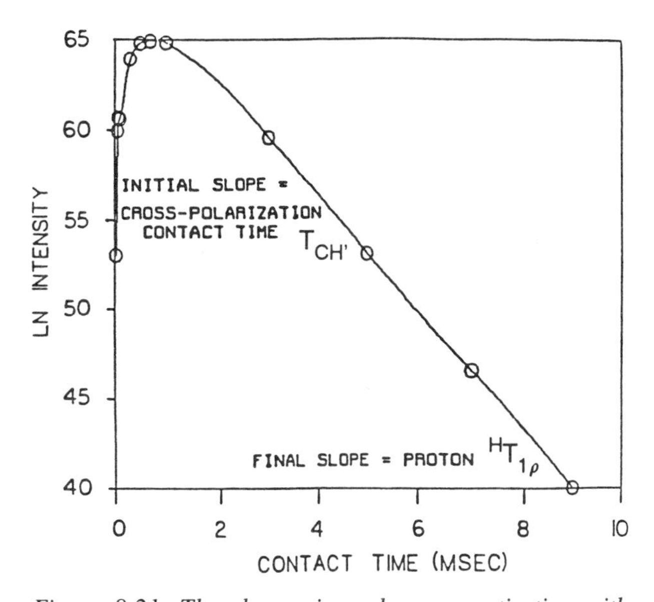

Figure 8.21. The change in carbon magnetization with contact time for the cross-polarization experiment. The initial rise is due to the cross-polarization contact time, T_{CH}, and the relaxation decrease is governed by the $^HT_{1\rho}$.

a rate proportional to the inverse of the ${}^{H}T_{1\rho}$ time. The carbon magnetization, S, is given by

$$S = \exp\left(\frac{t_m}{{}^{H}T_{1\rho}}\right)\left[1 - \frac{\exp(-t_{CP})}{T_{CH}}\right] \qquad (8.31)$$

where t_{CP} is the contact time. Thus two opposing relaxation mechanisms operate during the contact time. For short contact times, the T_{CH} process dominates, and the carbon magnetization increases exponentially. At longer contact times, the process is dominated by the ${}^{H}T_{1\rho}$ process, and the magnetization decreases exponentially because of the proton spin–lattice relaxation in the rotating frame. ${}^{H}T_{1\rho^{-1}}$ is approximately proportional to the spectral density function near the rotating-frame precession rate, $\omega_H = \gamma^{H_1}$, and provides a direct measure of the low-frequency motion of the resonant group. T_{CH} is related to the carbon–proton dipolar line-width contribution in an undecoupled conventional ${}^{13}C$ NMR spectrum by

$$T_{CH}{}^{-1} = C_{CH}\left[\frac{M_2^{CH}}{(M_2^{HH})^{\frac{1}{2}}}\right] \qquad (8.32)$$

where C_{CH} is a geometry-dependent term, M_2^{CH} is the (C, H) second moment, and M_2^{HH} is the (H, H) second moment. In general, more motion of a particular resonant group results in smaller residual dipolar interactions and a longer T_{CH}.

The cross-polarization transfer of magnetization is inversely proportional to the sixth power of the C–H internuclear distance. Provided that the carbons in a polymer repeat unit are subjected to the same motions, carbons with directly bonded protons are expected to cross-polarize more rapidly than carbons without direct interactions because the shorter interaction distance results in a larger local dipolar field. Rapid molecular motion attenuates the cross-polarization mechanism. For protonated carbons in rigid solids, most of the signal buildup occurs over about the first 100 μs. For nonprotonated carbons, most of the signal appears in the first few milliseconds.

The rate of signal decay depends on ${}^{H}T_{1\rho}$ and can differ among carbon types, although ${}^{H}T_{1\rho}$ is generally the same for all carbon groups with highly coupled protons. If ${}^{H}T_{1\rho}$ is very short, the cross-polarization process is endangered because of the rapid decay process. By varying the contact time prior to observation of the spectrum, T_{CH} and ${}^{H}T_{1\rho}$ may be determined for the protons coupled with each resolved carbon resonance.

The preceding analysis assumes that the rf fields H_{1H} and H_{1C} are much larger than their respective line widths. Also, the relaxation parameters are assumed to have the following order: ${}^{C}T_1 > {}^{H}T_1 > {}^{H}T_{1\rho} > T_{CH}$. When these conditions are not met, cross-polarization may be ineffective.

Factors Influencing Cross-Polarization

Cross-polarization is more effective for rigid systems than for mobile structures. When extensive motion occurs, the C–H dipolar interactions are reduced, and this reduction causes a lower cross-polarization rate. In some systems, mobile components can exhibit sufficient motional anisotropy to permit cross-polarization because the anisotropy can impart a residual static component to the overall motion (*69*). This appears to be the case for elastomeric systems.

Performing cross-polarization experiments on mobile samples is sometimes difficult. If the ${}^{H}T_{1\rho}$ is in the range of hundreds of microseconds, the ${}^{1}H$ rotating-frame magnetization relaxes before appreciable ${}^{13}C$ magnetization can be cross-polarized. Polyethylene oxide (PEO) is an example of a polymer that exhibits this effect. The ${}^{H}T_{1\rho}$ is 100 μs for PEO, and it is approximately 1 ms for polyethylene (PE). Cooling the PEO sample changes the ${}^{H}T_{1\rho}$ and T_{CH} sufficiently to allow effective cross-polarization.

In PE, the CP enhancement varies with the phase of the polymer. For the crystalline regions, the CP enhancement is 3.5 ± 0.2, and in the disordered regions, the enhancement factor ranges from 2.4 to 3.0 (*70*). The smaller enhancement in the disordered regions arises because molecular motion shortens ${}^{H}T_{1\rho}$ and lengthens T_{CH}.

The magnitude of $T_{CH}{}^{-1}$ is related to the strength of the static dipolar interactions between the carbons and protons. T_{CH} should be less than 0.13 ms for a rigid methylene group (*71*). In general, T_{CH} values decrease in the following order: nonprotonated carbons > methyl (rotating) carbons > methylene carbons > protonated aromatic–aliphatic methine carbons > methyl (static) carbons.

MAS has only a small influence on the cross-polarization process for rigid polymer systems. In principle, magic-angle spinning reduces the C – H dipolar decoupling, but in order for this averaging process to be efficient, the local $^1H - ^{13}C$ coupling must remain static for at least half of a revolution, or $(2\omega_r)^{-1}$, where ω_r is the spinning frequency. When the rotor speed is 2 kHz, the rotor period rate is 500 μs $(1/\nu_r)$. For a speed of 5 kHz, the period is 200 μs. Essentially all of the T_{CH} values of rigid polymers are shorter than these typical rotor periods. Consequently, MAS has little effect on the CP process for glassy polymers. Because the T_{CH} is approximately 100 μs for protonated aromatic carbons in glassy polymers, the cross-polarization is virtually completed by the time the spinner has rotated very far. For the unprotonated aromatic and quaternary carbons, relaxation times increase by about 50% when the speed is increased from 1 to 3 kHz.

MAS does have an effect on the cross-polarization efficiency for mobile systems like elastomers (*72*). For a static sample, the transfer rate is a function of ω_C and has a maximum polarization-transfer rate at the Hartman – Hahn match and a slower polarization transfer with increasing mismatch. However, when the sample is spun with MAS, there is a very slow transfer of polarization at the Hartman – Hahn match, where the transfer is expected to be most rapid. A series of maxima in the transfer rate occurs when ω_{1C} and ω_{1H} are mismatched by a multiple of the spinning frequency. Thus, mismatching the ω_{1C} and ω_{1H} produces high signals at multiples of the MAS frequency. A series of maxima in T_{CH} are observed as a result of an amplitude-modulated pattern due to C – H heteronuclear coupling (*73*).

Cross-Polarization as a Tool for Resonance Assignments

Cross-polarization can be used as a tool for the assignment of resonance lines. As the contact time is varied, the spectra change dramatically. The signals observed at the smallest contact times (0.01 ms) are assigned to the carbons that have the largest dipolar $^1H - ^{13}C$ interactions. The strong dipolar interactions experienced by these carbons imply that their motions are hindered. This conclusion is usually supported by the broad line shapes that are observed. Evidently, the molecular motions of these carbons cannot average the chemical-shift anisotropies effectively, and the resulting lines are broad. Sharper signals emerge when the contact time is in the 0.25 – 1-ms range. These signals are assigned to carbons that experience greater motional averaging of their $^1H - ^{13}C$ dipolar interactions and their ^{13}C chemical-shift anisotropies. For example, the carbonyl signals have line widths of 30 ppm compared to the 150-ppm line width observed for a rigid lattice. The sharpest signals are obtained from long contact times because of the efficient relaxation of the motionally hindered carbons.

For example, spectra of poly(p-oxybenzoyl) as a function of contact time are shown in Figure 8.22. For the very short contact times, only the resonances assigned to *ortho* ring carbons $C_{2,2'}$ and $C_{3,3'}$) are present. This result reflects the larger static dipolar

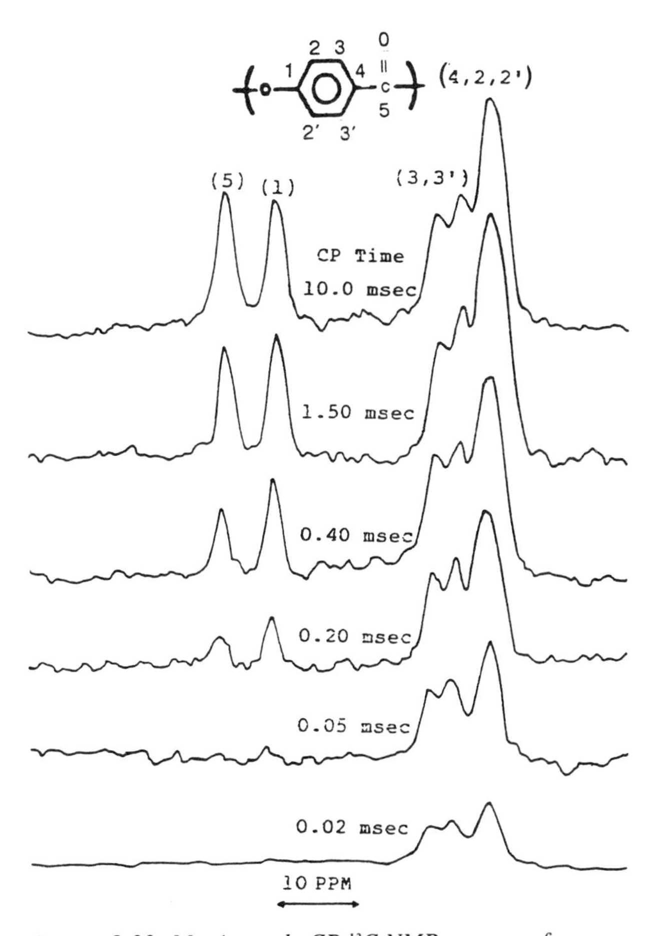

Figure 8.22. Magic-angle CP ^{13}C NMR spectra of a p-ox*ybenzoyl polymer as a function of time. The numbered peaks correspond to the numbered carbons in the structure. (Reproduced from reference 73. Copyright 1979 American Chemical Society.)*

interaction for the proton-bearing carbons. At long contact times, the line between $C_{2,2'}$ and $C_{3,3'}$ begins to grow at about the same rate as the line associated with C_4. The line assigned to the carboxyl carbon polarizes at the slowest rate.

The Interrupted-Dephasing CP Experiment

An experiment that is useful for assigning nonprotonated carbon resonances in the NMR spectrum is the interrupted-decoupling CP experiment (*74, 75*). In this experiment, the decoupling of the protons is interrupted (by turning off the proton rf field) for a predetermined time. During the interrupted-decoupling period, ^{13}C magnetizations due to protonated immobile carbons are efficiently dephased by static ^{13}C dipolar coupling. This dephasing is terminated by switching off the carbon field, restoring the proton field, and acquiring the data. The quaternary carbons with no bonded protons dephase very slowly and survive the decoupling period. This effect allows a discrimination of the quaternary carbons from the carbons with bonded protons. Unfortunately, motions that are effective in averaging dipolar couplings to reduced values can also cause signals of highly mobile protonated carbons to persist in the interrupted-decoupling experiments. The most common example is the methyl group, which rotates very rapidly even in rigid solids, and the methyl resonance can survive the interrupted-dephasing process.

An example of the use of the interrupted-decoupling experiment for suppression of the protonated carbon resonances is shown in Figure 8.23. For the diacetylene polymer, the carbons labeled B, C, H, and I survive the long interrupt cycle and are therefore quaternary carbons (*76*).

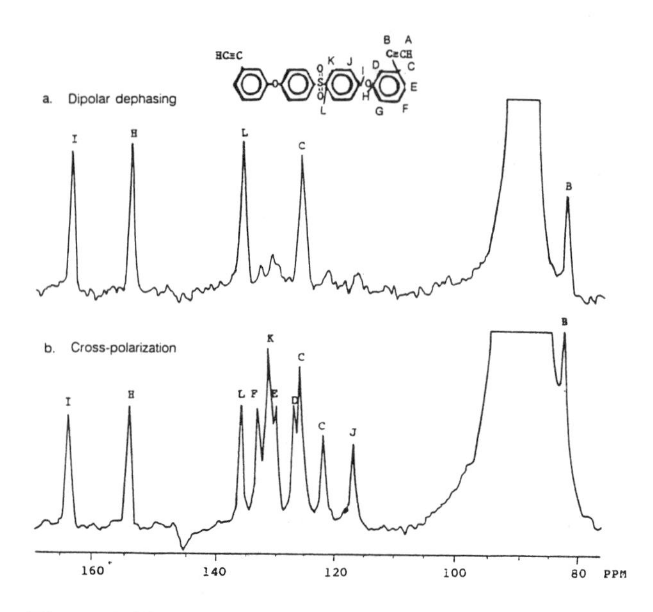

Figure 8.23. An example of the interrupted-decoupling experiment for suppression of the protonated carbon resonances is shown. The labeled peaks correspond to the labeled carbons in the structure. (Reproduced from reference 76. Copyright 1984 C. Shields.)

Quantitative Aspects of Cross-Polarization Spectra

Under proper experimental conditions, the integrated areas in a ^{13}C NMR spectrum are related to the fraction of the various carbon types in the polymer structure. One of the problems in a CP ^{13}C spectrum is the possible distortion of the peak areas because of differential cross-polarization times for the various carbons.

In fact, not all of the carbons may be observed in the CP experiment because of two different problems. Carbons in hydrogen-deficient domains will have minimal dipolar couplings to remote protons and hence will not cross-polarize efficiently. For complex samples like coals, aromatic carbons are likely to have this problem, and the aliphatic carbons will not. Second, a fraction of the carbons in the sample may be near free-radical sites, which either broaden the resonances beyond detectability or greatly reduce the $^{H}T_{1\rho}$ in the vicinity of the radical.

For CP experiments, a spin-counting procedure should be devised to quantify the fractions of detectable carbons in the sample under observation. Basically, the procedure involves measuring a ^{13}C CP–MAS spectrum of a known quantity of the sample that has been coground with a weighed quantity of a suitable quantitation standard, such as glycine. From the weight fraction of carbon and the integrated peak areas, the fraction of detectable carbons can be calculated.

By comparing known masses of polypropylene (PP) with a calibration standard of hexamethylbenzene, approximately 80% of the carbons in PP were observed (*77*). The basis for this loss of detectability of the carbons in PP is the relative values of T_{CH} and $^{H}T_{1\rho}$. When $^{H}T_{1\rho}$ is very short, equilibrium between the two spin systems cannot be obtained before the protons decay.

True relative intensities in the cross-polarization experiment are obtained only when the following three criteria are satisfied (*78*):

1. The cross-polarization times are short compared to the shortest proton rotating-frame relaxation time, ${}^{H}T_{1\rho}$, for the carbons (if there is extensive motion in the mid-kilohertz range, then ${}^{H}T_{1\rho}$ becomes approximately 10^{-4} s and does not meet this criterion).
2. The proton rf field is much larger than the local proton – proton dipolar fields (rf signals in the 50 – 60-kHz range create a variation of only a few percent in polyethylene).
3. Cross-polarization time constants, T_{CH}, are short enough to allow the ^{13}C nuclei to equilibrate with the protons during cross-polarization (if there is extensive motion, the CP process can become very inefficient in the mobile region compared to the more rigid portions) (*78*).

The recommended procedure is to run a series of different delay times using only DD and MAS to check if the cross-polarization is distorting the CP – MAS spectrum. The best method to obtain quantitative areas is to vary the cross-polarization contact time and then to construct a graph of the magnetizations and extrapolate the linear portion to zero contact time, as shown in Figure 8.24 (*79*).

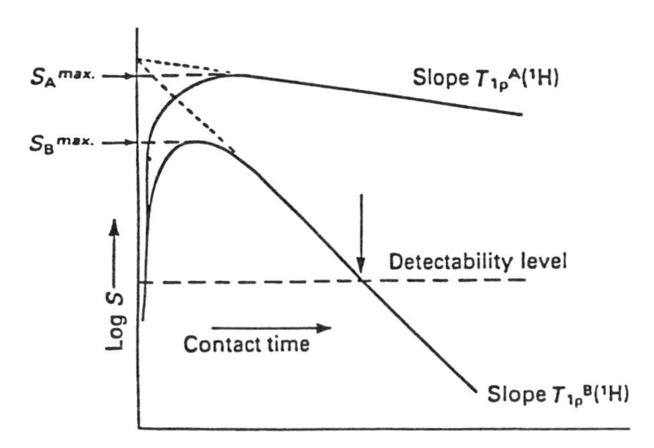

Figure 8.24. The extrapolation method for obtaining quantitative intensities is to vary the cross-polarization contact time and to construct a graph of the magnetizations and extrapolate the linear portion to zero contact time. (Reproduced with permission from reference 79. Copyright 1985 The Royal Society of Chemistry.)

Quantitative Applications of Cross-Polarization

Quantitative measurements on solid polymers have been made (*80*) by using MAS – DD – CP. A cross-polarization technique has been devised not only to determine the number of defects (end-group and branch defects) in PE but also to determine the distribution of the defects between the crystalline and noncrystalline phases (*80*). By using a large number of transients to average the signal, carbon fractions on the order of 2×10^{-4} can be detected, and resonance areas accurate to $\pm 10^{-4}$ can be detected if flat base lines are present.

The partitioning of the defects in the two phases was accomplished by generating pure crystalline and amorphous spectra of PE (*80*). Three different CP sequences were used. In one sequence, the normal CP experiment was used in which all the carbons had comparable intensities. In the second sequence, a preparation pulse that favored the crystalline phase was used. In the third sequence, the preparation pulse favored the amorphous phase. In this manner, spectra representing different mixtures of the two phases were obtained, and by taking the appropriate linear combination of these three spectra, the "pure" spectra of the two components were determined. The resulting pure crystal and pure noncrystal spectra generated from linear combinations of the spectra of the samples are shown in Figure 8.25 (*80*). The distribution of the defects was then determined with curve fitting of the pure spectra to the spectra of the samples.

Use of Cross-Polarization for Separation Based on Mobility

If a ^{13}C resonance line (without dipolar broadening) is not substantially broadened, then cross-polarization discriminates against liquidlike lines. In fact, cross-polarization can be used to distinguish mobile components from rigid components. This ability is illustrated by the study of styrene – butadiene copolymers. For a cross-polarization contact time of 10 ms, only resonances assigned to the butadiene carbons appear, whereas at a contact time of 1.5 ms,

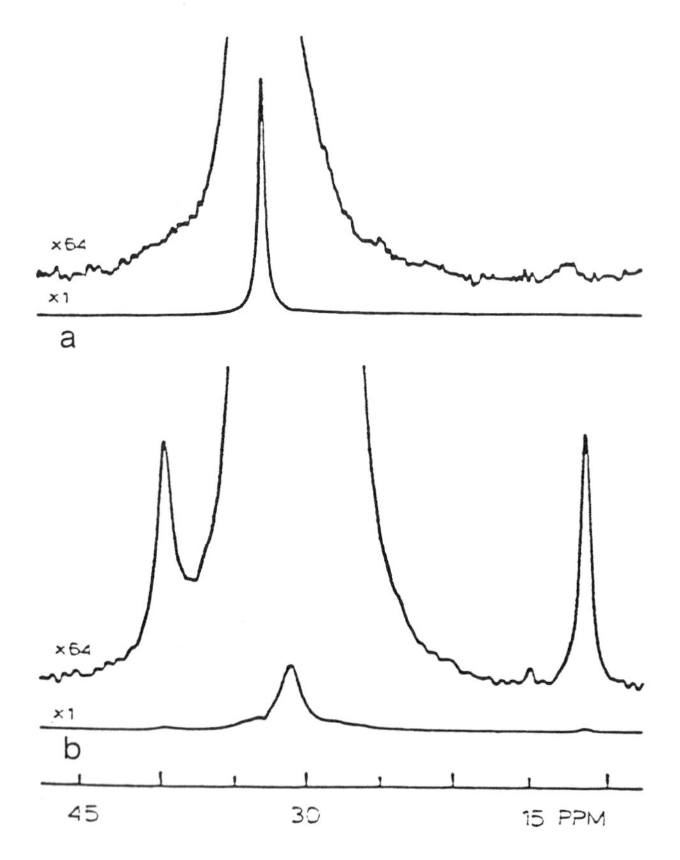

Figure 8.25. The pure crystal (a) and pure noncrystal (b) spectra of PE generated from linear combinations of the spectra obtained by using different pulse sequences. (Reproduced from reference 80. Copyright 1987 American Chemical Society.)

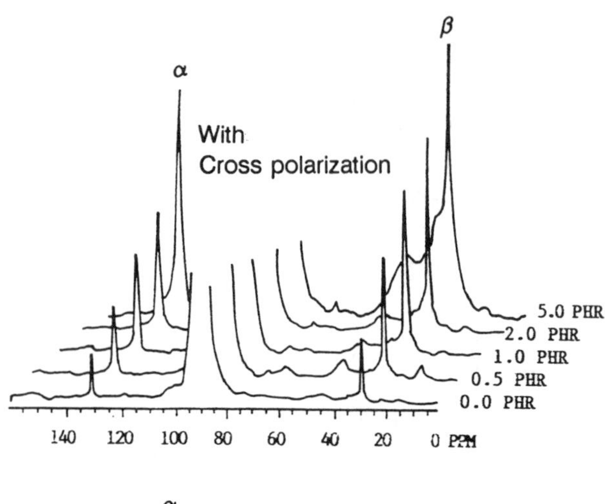

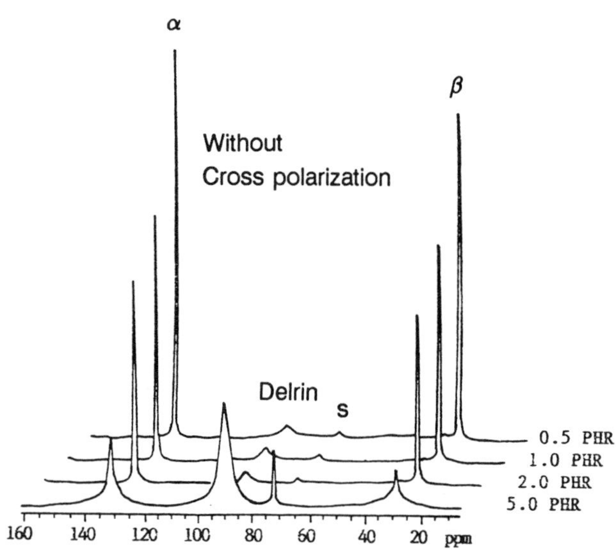

Figure 8.26. The separation of the network and nonnetwork structures in the peroxide cross-linking of polybutadiene. (Reproduced from reference 81. Copyright 1984 American Chemical Society.)

both styrene and butadiene signals are observable (*79*).

This sensitivity of the cross-polarization experiment to more rigid components has been used in the separation of the network and nonnetwork structures in the peroxide cross-linking of polybutadiene (*81*). In Figure 8.26, the gated high-powered decoupled (GHPD) ^{13}C spectrum with MAS but without cross-polarization is shown for polybutadiene networks with increasing amounts of peroxide. The initial polybutadiene lines decrease significantly as the peroxide level increases. On the other hand, Figure 8.26 also shows the cross-polarization spectrum, and in this case, the mobile samples have little intensity and the highly cured samples have high intensity, a result that reflects the rigidity of the cured networks.

The Cross-Polarization Experiment for the Study of Polymer Blends

Intermolecular interactions can be probed by using cross-polarization. In the cross-polarization experiment, only carbons that are dipolar-coupled to protons give rise to ^{13}C resonances, so deuterated carbons will not give rise to an NMR signal unless they are in close contact with protons on adjacent molecules. The appearance of deuterated-carbon resonances in a cross-polarized system indicates that the deuterated carbons have neighboring hydrogens only a few angstroms away. The resonances of the deuterated carbons may appear slightly shifted from the resonances of their protonated counterparts as a result of a small deuterium-isotope effect. If the chemical-shift differences are not sufficient, then the deuterated-carbon resonances can be distinguished from the protonated carbons by the use of interrupted decoupling delay. The protonated-carbon resonances are suppressed by the interrupted decoupling experiment, but the deuterated carbon resonances are only slightly affected. The cross-polarization experiment can be used to probe blends and mixtures and to study spatial order for mixtures of protonated and deuterated polymer blends (*82*). Similarly, the degree of mixing of a plasticizer can be studied with this deuterium approach (*83*).

Contributions to Line Broadening in Solids

Resonance line broadening arises from both homogeneous and inhomogeneous processes. *Homogeneous* processes result from a *single* structural source and are derived from relaxation effects. Among the homogenous processes are the natural line width that is usually associated with dipolar relaxation and residual static C – H dipolar broadening from incomplete motional narrowing. Processes that produce a distribution of resonance frequencies are *inhomogeneous*. Inhomogeneous processes arise from *multiple* structural sources producing different chemical shifts, each with small intrinsic widths or spin packets. Among the inhomogeneous processes are inhomogeneity in the static field, a variation in bulk magnetic susceptibility within the sample, a distribution of isotropic chemical shifts, and a residual CSA resulting from incomplete motional narrowing.

Experimentally, homogeneous and inhomogeneous processes can be differentiated in several ways. Higher MAS rotation rates are required to effectively narrow homogeneous processes, and inhomogeneous processes can be narrowed with slower rotation rates (*84*). A spin-echo experiment can be used to distinguish between these two mechanisms (*85*). The inhomogeneous processes can be refocused by 180° rf pulses, but the homogeneous processes cannot. Any decrease in the echo corresponds to a homogeneous broadening mechanism.

The most obvious source of line broadening is lack of homogeneity of the applied magnetic field. However, instrumental improvements have reduced this effect to a few milligauss out of 104 G. For solids, the homogeneity of the field can be measured by using adamantane, and nonuniformities in the magnetic field result in line broading of only 2 Hz (*20*). For studies as a function of temperature, efforts should be made to readjust the homogeneity of the magnetic field because of changes in the sample geometry with changes in temperature as well as other factors. For studies of polymer melts, a broadening contribution of 50 – 60 Hz from this source has been reported (*86*).

In solids, the anisotropy of the bulk magnetic susceptibility for a configuration of irregularly shaped particles can be a factor in the broadening of the resonance lines. Under MAS, the shape-demagnetization factors are axially averaged so that the sample appears as if it were a spheroid with a symmetry axis along the spinning axis (*87*). Additionally, macroscopic or microscopic inhomogeneities within the sample can cause differences in the bulk magnetic susceptibility. The *macroscopic inhomogeneities* can arise from fissures, bubbles, and other mechanical or structural irregularities in the sample and can be minimized by careful sample preparation. The *microscopic inhomogeneities* can arise from different phases, interfaces, and other structural variations. For example, the bulk susceptibility of a crystalline *n*-alkane is different (approximately 2% smaller) from that of the corresponding liquid (*20*).

The induced magnetic polarization from a heterogeneous sample can produce a nonuniform screening field, which leads to line broading. However, the susceptibility fields are small for organic materials, and the bulk-volume susceptibility is typically on the order of 1 ppm. The bulk magnetic susceptibility is inherently anisotropic for crystalline solids and cannot be removed by MAS (*20*). However, the magnitude of the effect is small (~ 0.7 ppm).

The T_1 processes effectively limit the lifetime of the Zeeman states, and this broadening effect amounts to $h/2\pi T_1$. Because the T_1s for carbons are long (10 – 1000 s), this effect is negligible for solids.

The broadening introduced by residual dipolar coupling can be estimated from the magnitude of the local magnetic field, H_{loc}. This local field is

$$H_{loc} = \frac{\mu}{r^3} \tag{8.33}$$

By using $r = 2$ Å and $m = 10^{-23}$ erg/G, H_{loc} is approximately 1 G (*88*). The homogeneous contribution to the line width, which is primarily attributable to motional modulation of the C–H coupling, is $(\pi {}^{C}T_{1\rho})^{-1}$.

Two types of motion can influence the dipolar coupling and broaden the line widths. In one case, the nuclei undergo random isotropic motion at a rate such that all angles are not averaged within the time interval T_2. Alternatively, the motions may be quite rapid, but the accessible angular motion is severely restricted. The effect of molecular motion on the line widths behaves the same as chemical exchange, which will be discussed in the next section.

Line Shapes: Chemical-Exchange Effects

If a nucleus exchanges between two sites B <=> B′, with resonance frequencies ω_B and $\omega_{B'}$ and lifetimes τ_B and $\tau_{B'}$ (corresponding to exchange rates of $\tau_B{}^{-1}$ and $\tau_{B'}{}^{-1}$), the effects of chemical exchange on the line shape may be summarized under three different conditions:

1. *Slow exchange*, where $[\omega_B - \omega_{B'}] >> \tau_B^{-1}\tau_{B'}^{-1}$. The resonances at ω_B and $\omega_{B'}$ merely acquire additional widths given by $(\pi\tau_B)^{-1}$ and $(\pi\tau_{B'})^{-1}$ (in units of hertz), respectively.
2. *Fast exchange*, where $[\omega_B - \omega_{B'}] << \tau_B^{-1}\tau_{B'}^{-1}$. The two resonances merge into a single resonance with a frequency given by the average of these quantities in the absence of exchange weighted by the fractional concentration of the two species. In the absence of exchange, the averaging of the line widths requires $(\omega_B - \omega_{B'})^2\ \tau_B$ and $(\omega_B - \omega_{B'})^2\ \tau_{B'}$ to be negligible compared to the line widths of the B and B′ resonances, respectively.
3. *Intermediate exchange*, where $[\omega_B - \omega_{B'}] = \tau_B^{-1}$ and $\tau_{B'}^{-1}$. Both frequency shifts and line-width changes occur to yield a complex line shape.

All of these spectral features can be described quantitatively by means of a simple theory based on the Bloch equations. Computer simulations of the effects of a two-site exchange with no multiplicity from spin–spin effects are shown in Figure 8.27 (*88*).

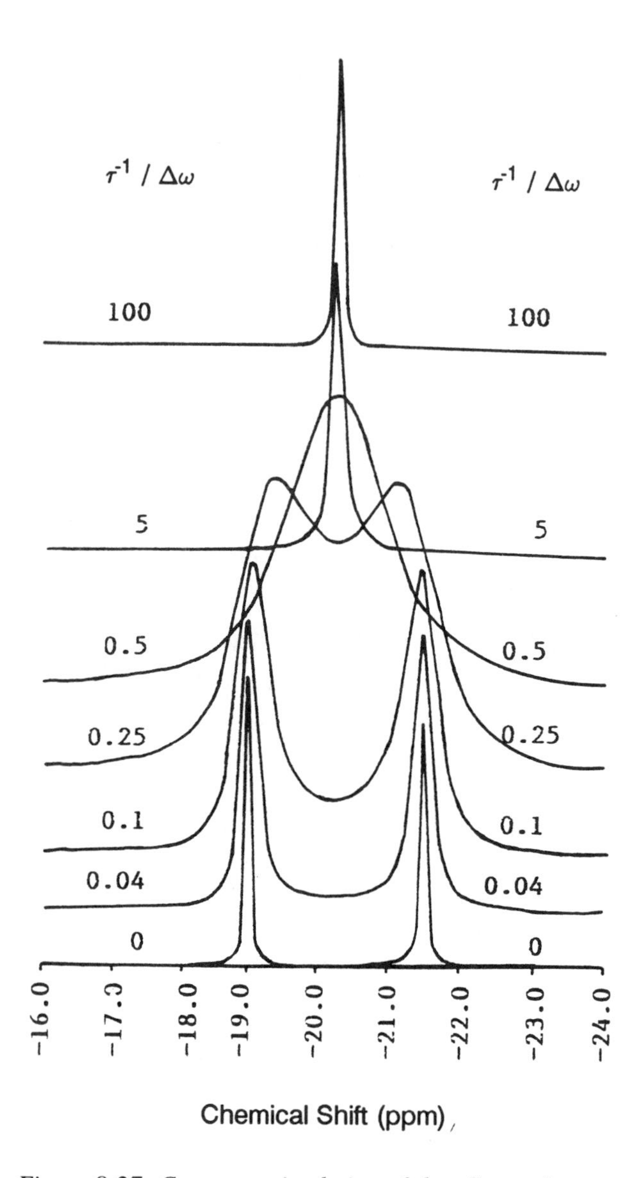

Figure 8.27. Computer simulation of the effects of a two-site exchange with no multiplicity from spin–spin effects. The exchange rate is expressed as a fraction of the chemical shift (in units of radians per second) at 81 MHz. The concentrations of the two species are taken to be equal. (Reproduced with permission from reference 88. Copyright 1984 Academic Press.)

The Grand Experiment (MAS–DD–CP)

By combining the three previously described experiments into one grand experiment, high-resolution ^{13}C NMR spectra of solids can be obtained (*2*). The benefits are demonstrated by comparing the ^{13}C solid-state NMR spectra of a sample of poly(methyl methacrylate) (PMMA) obtained under different conditions (Figure 8.28). Spectrum a of the stationary PMMA sample was obtained with scalar decoupling but without CP and shows no observable resonances. When the stationary sample is examined by using DD and CP (spectrum b), broad resonances are observed, and these resonances reflect the effects of chemical-shift anisotropy. With the addition of MAS to the DD experiment and elimination of CP, the lines are narrowed considerably, but the signal-to-noise ratio is poor (spectrum c). Finally, with the grand experiment (MAS – DD – CP), the spectrum is resolved into the specific carbons of the PMMA molecule with good signal-to-noise ratio (spectrum d).

Such are the benefits of the grand experiment. Of course, what applies to PMMA can be used for other

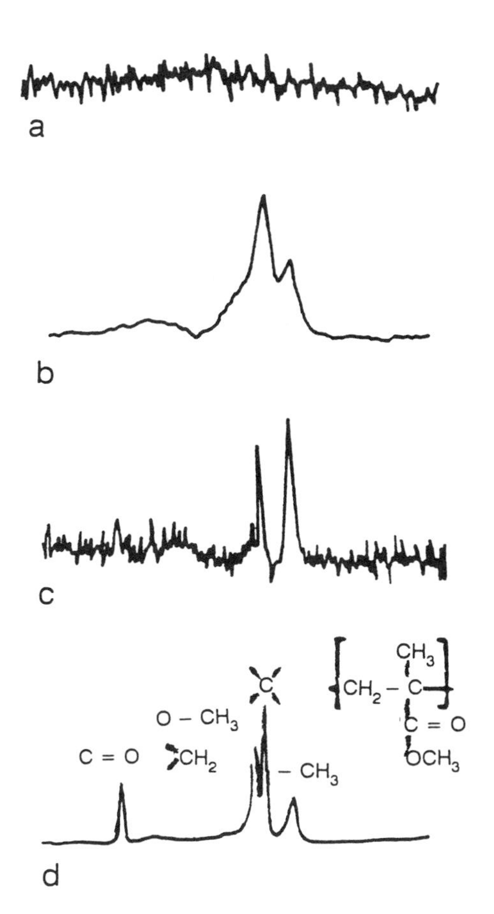

Figure 8.28. The ^{13}C solid-state NMR spectra of a sample of PMMA obtained under different conditions. (Reproduced from reference 82. Copyright 1981 American Chemical Society.)

samples as well with proper manipulation of the various experimental factors. With modern spectrometers, the grand experiment can easily be accomplished in a routine manner after a preliminary determination of the relevant parameters.

Applications of the grand experiment to polymers will be discussed in Chapter 9.

References

1. Garroway, A. N.; VanderHart, D. L.; Eare, W. L. *Philos. Trans. R. Soc. London, A:* **1981,** *299,* 609.
2. Schaefer, J.; Stejskal, E. O. *J. Am. Chem. Soc.* **1976,** *98,* 1031.
3. Fyfe, C. A. *Solid State NMR for Chemists;* C. F. C. Press: Guelph, 1984.
4. Axelson, D. E. *Solid State Nuclear Magnetic Resonance of Fossil Fuels: An Experimental Approach;* Multiscience: 1983.
5. Axelson, D. E.; Russell, K. E. *Prog. Polym. Sci.* **1985,** *11,* 221.
6. Havens, J. R.; Koenig, J. L. *Appl. Spectrosc.* **1983,** *37(3),* 226-249.
7. *High Resolution NMR of Synthetic Polymers in Bulk;* Komoroski, R. A., Ed.; VCH Publishers: Deerfield Beach, 1986.
8. *Solid State NMR of Polymers;* Mathias, L., Ed.; Plenum Publishers: New York, 1989.
9. Yannoni, C. S. *Acc. Chem. Res.* **1982,** *15,* 203.
10. Pake, G. E. *J. Chem. Phys.* **1948,** *16,* 327.
11. Andrew, E. R.; Eades, R. G. *Discuss. Faraday Soc.* **1962,** *34,* 38.
12. Freeman, R.; Hill, H. D. W.; Kaptein, R. *J. Magn. Reson.* **1972,** *7,* 327.
13. Lyerla, J. R. *Methods Exp. Phys. Part A* **1980,** *16,* 241.
14. Yannoni, C. Y. *Acc. Chem. Res.* **1982,** *15,* 201.
15. Abraham, A. *The Principles of Magnetic Resonance;* Oxford: New York, 1961; 571.
16. Waugh, J. S.; Huber, L. M.; Haeberlen, U. *Phys. Rev. Lett.* **1968,** *20,* 180.
17. Mehring, M.; *High Resolution NMR Spectroscopy in Solids;* Springer-Verlag: New York, 1983.

18. Rhim, W. K.; Elleman, D.; Vaughan, R. *J. Chem. Phys.* **1973,** *58,* 8.

19. Burum, D.; Rhim, W. K. *J. Magn. Reson.* **1979,** *34,* 241.

20. VanderHart, D. L.; Earl, W. L.; Garroway, A. N. *J. Magn. Reson.* **1981,** *44,* 361.

21. Rothwell, W. P.; Waugh, J. S. *J. Chem. Phys.* **1981,** *74,* 2721.

22. Fleming, W. W.; Lyerla, J. R.; Yannoni, C. S. *Polymer Characterization: Spectroscopic, Chromatographic and Physical Instrumental Methods;* Craver, C. D., Ed.; Advances in Chemistry Series 203; American Chemical Society: Washington, DC, 1983; Chapter 26, p 457.

23. Garroway, A. N.; Moniz, W. B.; Resing, H. A. *Faraday Discuss. Chem. Soc.* **1979,** *13,* 63.

24. Fleming, W. W.; Fyfe, C. A.; Lyerla, J. R.; Vanni, H.; Yannoni, C. S. *Macromolecules* **1980,** *13,* 460.

25. Veeman, W. S. *Prog. NMR Spectrosc.* **1984,** *16,* 193.

26. Vega, A. J.; English, A. D. *Macromolecules* **1980,** *13,* 1635.

27. Duncan, T. M. *J. Phys. Chem. Ref. Data* **1987,** *16,* 125.

28. Haeberlen, U. "High Resolution NMR in Solids-Selective Averaging"; In *Advances in Magnetic Resonance, Supplement 1;* Waugh, J. S., Ed.; Academic: New York, 1976; Chapter III.

29. Spiess, H. *NMR Basic Principles and Progress;* Diehl, P.; Fluck, E.; Kosfeld, R. Eds.; Springer Verlag: New York, 1978; Vol. 15.

30. Hempel, G.; Schneider, H. *Pure Appl. Chem.* **1982,** *54,* 635.

31. Henrichs, P. M. *Macromolecules* **1987,** *20,* 2099.

32. Veeman, W. S. *Philos. Trans. R. Soc. London, A:* **1981,** *299,* 629.

33. Alderman, D. W.; Solum, M. S.; Grant, D. M. *J. Chem. Phys.* **1986,** *84,* 3717.

34. Stejskal, E. O.; Schaefer, J.; McKay, R. A. *J. Magn. Reson.* **1977,** *25,* 569.

35. Maricq, M.; Waugh, J. S. *Chem. Phys. Lett.* **1977,** *47,* 327.

36. Waugh, J. S.; Maricq, M.; Cantor, R. *J. Magn. Reson.* **1978,** *29,* 183.

37. Murphy, P. D.; Taki, T.; Gerstein, B. C.; Henrichs, P. N.; Masa, D. J. *J. Magn. Reson.* **1982,** *49,* 99.

38. Yarim-Agrev, Y.; Tutunjian, P. N.; Waugh, J. S. *J. Magn. Reson.* **1982,** *47,* 51.

39. Bax, A. *J. Magn. Reson.* **1983,** *51,* 400.

40. Herzfeld, J.; Berger, A. E. *J. Chem. Phys.* **1980,** *73,* 6021.

41. Jelinski, L. W. *Macromolecules* **1981,** *14,* 1341.

42. Jelinski, L. W.; Dumais, J. J.; Engel, A. K. *Macromolecules* **1983,** *16,* 403.

43. Aue, P.; Ruben, D. J.; Griffith, R. G. *J. Magn. Reson.* **1981,** *43,* 472.

44. Bax, A.; Szeverenyl, N. M.; Maciel, G. E. *J. Magn. Reson.* **1983,** *52,* 147.

45. Bax, A.; Szeverenyl, N. M.; Maciel, G. E. *J. Magn. Reson.* **1983,** *51,* 400.

46. Harris, R. K.; Packer, K. J.; Thayer, A. M. *J. Magn. Reson.* **1985,** *62,* 284.

47. Naki, T.; Ashida, J.; Terao, T. *J. Chem. Phys.* **1989,** *88,* 6049.

48. Urbine, J.; Waugh, J. S. *Proc. Natl. Acad. Sci. USA* **1974,** *71,* 5062.

49. VanderHart, D. L. *J. Chem. Phys.* **1976,** *64,* 830.

50. Andrew, E. R. *Progress in NMR Spectroscopy;* Pergamon: New York, 1971; Vol. 18.

51. Spiess, H. W. *NMR* **1978,** 15.

52. Gall, C. M.; Cross, T. A.; DiVerdi, J. A.; Opello, S. J. *Proc. Natl. Acad. Sci. USA* **1982,** *79,* 101.

53. Inglefield, P. T., Amici, R. M.; O'Gara, J. F.; Hung, C.; Jones, A. A. *Macromolecules* **1983,** *16,* 1552.

54. O'Gara, J. F.; Jones, A. A.; Hung, C.; Inglefield, P. T. *Macromolecules* **1985,** *18,* 1117.

55. Roy, A. K.; Jones, A. A.; Inglefield, P. T. *Macromolecules* **1986,** *19,* 1356.

56. Wemmer, D. E.; Pines, A.; Whitehurst, D. D. *Philos. Trans. R. Soc. London, A:* **1981,** *300,* 15.

57. Sethi, N. K.; Pugmire, R. J.; Facelli, J. C.; Grant, D. M. *Anal. Chem.* **1988,** *60,* 1574.

58. Brandolini, A. J.; Apple, T. M.; Dybowski, C.; Pembleton, R. G. *Polymer* **1982,** *23,* 39.

59. Brandolini, A. J.; Dubowski, C. *J. Polym. Sci., Polym. Lett. Ed.* **1983,** *21,* 423.

60. Henrichs, P. M. *Macromolecules* **1987,** *20,* 2099.

61. Andrew, E. R.; Bradbury, A.; Eades, R. G. *Nature (London)* **1959,** *182,* 1659.

62. Lowe, I. J. *Phys. Rev. Lett.* **1959,** *2,* 285.

63. Herzfeld, J.; Roufosse, A.; Haberkorn, R. A.; Griffin, R. G.; Gilmcher, M. J. *Philos. Trans. R. Soc. London, B:* **1980,** *289,* 459.

64. Gerstein, B. C. *Anal. Chem.* **1983,** *55,* 899A.

65. Earl, W. L.; VanderHart, D. L. *J. Magn. Reson.* **1982,** *48,* 35.

66. Pines, A.; Gibby, M. G.; Waugh, J. S. *J. Chem. Phys.* **1973,** *59,* 569.

67. Miknis, F. P.; Bartuska, V. J.; Maciel, G. E. *Am. Lab.,* **1979,** *11(11),* 20.

68. Hartmann, S. R.; Hahn, E. L., *Phys. Rev. A* **1952,** *137,* 2042,

69. Sullivan, M. J.; Maciel, G. E. *Anal. Chem.* **1982,** *54,* 1615.

70. VanderHart, D. L.; Khoury, F., *Polymer* **1984,** *25,* 1589.

71. Alemany, L. B.; Grant, D. M.; Pugmire, R. J.; Alger, T. D.; Zilm, K. W. *J. Am. Chem. Soc.* **1983,** *105,* 2133.

72. Curran, S. A.; Padwa, A. R. *Macromolecules* **1987,** *20,* 625.

73. Fyfe, C.; Lyerla, J. R.; Volkson, W.; Yannoni, C. S. *Macromolecules* **1979,** *12,* 757.

74. Opella, S. J.; Frey, M. H. *J. Am. Chem. Soc.* **1979,** *101,* 1854.

75. Opella, S. J.; Frey, M. H.; Cross, T. A. *J. Am. Chem. Soc.* **1979,** *101,* 5866.

76. Shields, C. Master's Thesis, Case Western Reserve University, **1984**.

77. Cudby, M. E. A.; Harris, R. K.; Metcalfe, K.; Packer, K. J.;Smith, P. W. R. *Polymer* **1985,** *26,* 169.

78. VanderHart, D. L.; Khoury, F. *Polymer* **1984,** *25,* 1589.

79. Harris, R. K. *Analyst* **1985,** *110,* 649.

80. Perez, E.; VanderHart, D. L.; Crist, B., Jr.; Howard, P. R. *Macromolecules* **1987,** *20,* 78.

81. Patterson, D. J.; Koenig, J. L. "Peroxide Cross-linked Natural Rubber and *cis*-Polybutadiene: Characterization by High Resolution Solid State Carbon-13 NMR". In *Characterization of Highly Cross-linked Polymers;* Labana, S. S.; Dickie, R. A., Eds.; ACS Symposium Series 243; American Chemical Society: Washington, DC, 1984.

82. Schaefer, J.; Sefcik, M. D.; Stejskal, E. O.; McKay, R. A. *Macromolecules* **1981,** *14,* 188.

83. Hewitt, J. M.; Henrichs, P. M.; Scozzafava, M.; Scaringe, R. P.; Linder, M.; Sorriero, L. J. *Macromolecules* **1984,** *17,* 2566.

84. Andrew, E. R. *Philos. Trans. R. Soc. London, A:* **1981,** *299,* 505.

85. Slichter, P. *Principles of Magnetic Resonance;* Harper & Row: New York, 1980.

86. Dechter, J. J.; Komoroski, R. A.; Axelson, D. E.; Mandelkern, L. *J. Polym. Sci., Part B: Polym. Phys.* **1981,** *19,* 631.

87. Garroway, A. N. *J. Magn. Reson.* **1982,** *49,* 168.

88. Nagewara Rao, B. D. *Phosphorus-31 NMR: Principles and Applications;* Gorenstein, D. G., Ed.; Academic: New York, 1984; p 63.

89. Yannoni, C. S. In *Contemporary Topics in Polymer Science;* Shen, M. Ed.; Plenum Press: New York, 1983; Vol. 3, p 155.

9

Applications of High-Resolution Solid-State NMR Spectroscopy to Polymers

The difficulty encountered in using NMR for solids isn't the result of a dearth of chemical information as compared with solution spectra. If anything, the NMR spectrum of the typical solid contains more information on the structure and environment of the molecule than does the spectrum of the same material in solution.

– Cecil Dybowski (*1*)

Although high-resolution NMR spectroscopy of the solid state has been available for only approximately 10 years, the number of applications has grown rapidly. Solid-state NMR spectroscopy has been used for two general classes of samples: (1) samples that are insoluble, such as cross-linked or intractable systems, and (2) samples for which solid-state NMR spectroscopy provides information about the nature of the solid state, such as the conformation, crystallographic forms, and the morphological character of the solid.

Interpretation of the spectra from solid-state NMR spectroscopy is more complicated than the interpretation of solution NMR spectra. In solution NMR spectroscopy, the nature of the chemically inequivalent nuclei that give rise to resolved resonances must be determined. In the solid state, however, not only must assignments of the chemically inequivalent carbons be made, but also whether the chemically inequivalent nuclei are magnetically inequivalent as a result of their solid-state environment must be determined.

Theoretically, solid-state NMR spectra are expected to be more complicated and to have more resonance lines than solution NMR spectra, but because the lower inherent resolution of the solid spectra limits the number of resolvable lines, the spectra of solids are instead often simpler. This effect is shown in Figure 9.1 for poly(vinyl chloride) (PVC), in which solution and solid-state ^{13}C NMR spectra are compared for the same sample (*2*). These spectra suggest that the information content of a solid-state spectrum is considerably less.

In amorphous or glassy polymers, a considerable range of conformational states is accessible to the polymer chains. Theoretically, these conformational sequences have different chemical shifts, but broad resonances often occur because the spectral dispersion is not sufficient to resolve the expected lines. Therefore, considerable inhomogeneous line broadening occurs in the solid-state NMR spectra of glassy and amorphous polymers because of configurational, conformational, and sequential effects.

Conformational Analysis Using Chemical Shifts in the Solid State

By using the grand experiment, well-resolved ^{13}C NMR spectra for polymers in the solid state can be observed. One of the first questions that must be resolved concerns the magnitude of the measured

1904—4/92/0229$08.00/1

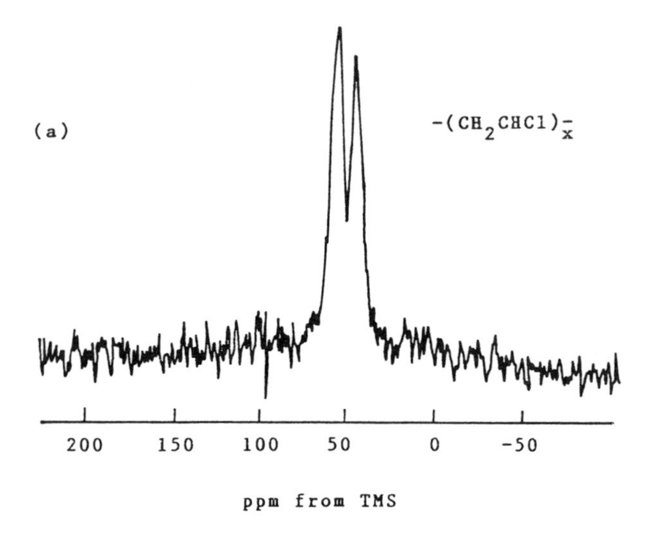

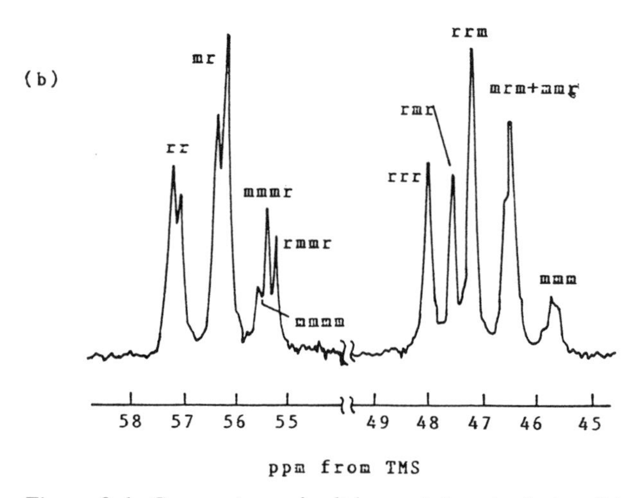

Figure 9.1. Comparison of solid-state (a) and solution (b) ^{13}C spectra of poly(vinyl chloride). Meso and racemic placement are designated by m and r, respectively. (Reproduced from reference 2. Copyright 1986 American Chemical Society.)

isotropic chemical shifts of solids compared to that of solutions. It has been concluded that:

> *The same influences which determine the dependence of carbon-13 shifts on chain microstructure in the solution NMR spectra of macromolecules appear to be operative also in the solid state.*
>
> —L. A. Belfiore et al. (*3*)

To a first approximation, there should be very little difference between the chemical shifts of the same chemical species of carbons found in the liquid or solid states, except for substances exhibiting either rigid conformational isomeric species or in substances in which intermolecular effects in the solid state arise from differences in molecular packing.

Changes in molecular packing in the solid state can lead to differences in the chemical shifts, although this effect appears to be small. For example, the ^{13}C chemical shifts of methylene carbons with the *trans* zigzag structure in the orthorhombic, triclinic, and monoclinic forms are 33, 34, and 35 ppm, respectively (*4*). It has been suggested that molecular-packing interactions beyond 1 nm do not influence the isotropic chemical shifts in amorphous systems (*5*).

Grant and Cheney developed a steric-hindrance model that predicts the effect on the chemical shifts of the additional crowding of the C–H bonds in the solid state (*6*). In this model, a C–H bond is compressed or expanded by mutual repulsions of the bonded hydrogens and nearby nonbonded hydrogens. The chemical-shift differences depend on the hydrogen–hydrogen distance, r (nm), the angle ϕ between the C–H bond, and the interhydrogen separation vector. Thus,

$$\Delta^{13}C = -1680 \cos \phi \exp(-26.71r) \quad (9.1)$$

The solid-state spectrum of diglycidyl ether of bisphenol A (DGEBA) has been compared with the liquid-state spectrum (*7*), and the chemical shifts were examined using the steric-hindrance model. This model has also given consistent results in the analysis of chemical shifts of the α and β conformers of *trans*-1,4-polyisoprene in the solid state (*8*). The steric-hindrance model predicts shifts that are the same order of magnitude as those observed.

Thus, the chemical shifts observed in solution can be used as a first approximation to ascertain the chemical nature of the carbons. However, caution must be exercised because conformational and intermolecular effects can play a role in solids.

Chemical-Shift Studies of Polymer Conformation

> *The NMR spectrum of a solid is much like a complex piece of beautiful music. Its texture is so rich that it's impossible to analyze except in a cursory way. Modern solid-state NMR spectroscopy provides the same control of information that a careful analysis of an individual piece provides in appreciation of music, the ability to separate and examine the elements and to present them coherently and conve-*

niently. In this manner, a greater understanding of the total structure is obtained, be it a symphony or a spectrum.

– Cecil Dybowski (*1*)

Solid-state NMR experiments using chemical shifts can lead to a determination of the conformational states in the solid. The restricted motion in the solid state creates unique chemical environments for nominally equivalent carbons. The chemical-shift differences generated by these environments in the solid state are seen as additional line splittings when compared to the observations in solution. This effect was first observed with poly(phenylene oxide). The protonated aromatic carbon resonance is a doublet in the solid state, whereas in solution it is a singlet (*9*). The doublet in the solid state arises because the two protonated carbons of the ring are magnetically inequivalent in the solid state due to the nonlinearity of the C–O–C bond. However, these two carbons are magnetically equivalent in solution, where free rotation occurs.

NMR evidence exists for more than one conformation of the monomer of DGEBA based on the number of resonance lines observed in the solid state (*7*). For the monomer, although there are only *meta* and *ortho* carbons next to the oxygen carbon in the two phenylene rings, the spectrum shows four and five (possibly six) lines, respectively. At low temperatures, the resonances from the carbons *ortho* and *meta* to the oxygen are partly resolved doublets. At higher temperatures, these lines become a singlet. This result is consistent with molecular motion involving a 180° flip–flop of the phenylene ring at higher temperatures.

Conformationally related chemical-shift variations in polymers are generally reflected through two effects: the γ-*gauche* effect and the vicinal *gauche* effect. In a model with three conformations for each bond, there are two magnetically distinguishable γ positions, the *trans* (or anti) position and the *gauche* position (Chart I). Replacement of a *trans* by a *gauche* position in the polymer conformation results in an upfield chemical shift. The magnitude of the shift depends on the type and number of carbons involved and the relative orientations of the substituents (*5*).

The C–C backbone in organic polymers usually adopts the three staggered rotational states of *trans* (t), *gauche*$^+$ (g^+), and *gauche*$^-$ (g^-), ($\phi = 0°, \pm 120°$) as illustrated in Chart I. A rotational barrier of 2 kcal/mol exists between these staggered states for polyethylene. In the *gauche* conformation, the observed and γ carbons are in close proximity, and they are distant in the *trans* arrangement. This spatial difference produces a shielding effect for the *gauche* isomer relative to the *trans* conformation. This effect is termed the γ-*gauche* effect and has a value of approximately −5 ppm but can exhibit angular dependence (*5*).

For the vicinal *gauche* effect, the conformation of

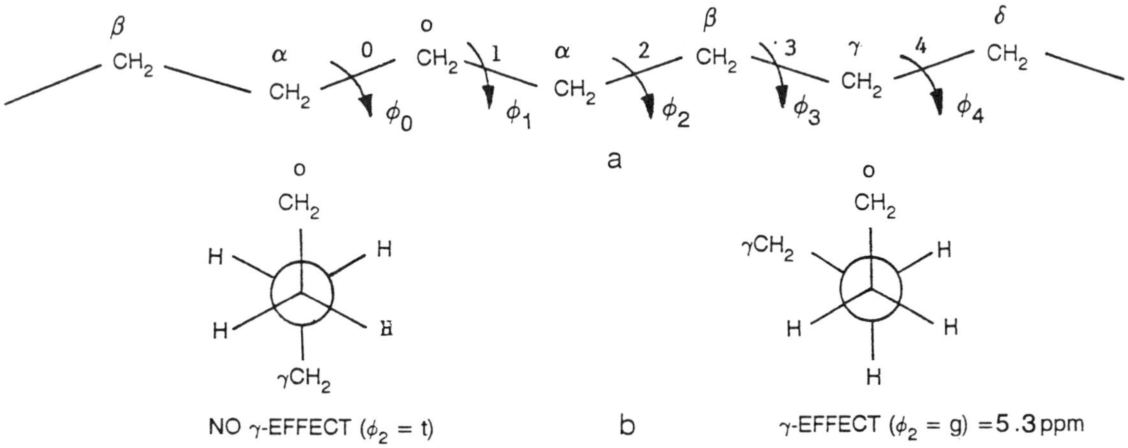

Chart I. (a) The C–C backbones in organic polymers usually adopt the three staggered rotational states t, g^+, g^- ($\phi = 0°, \pm 120°$). (b) In the gauche *($g^\pm$) conformation, the observed and γ carbons are close; they are distant in the* trans *(t) arrangement.*

the α bonds affects the chemical shift if the α and β carbons exchange positions. The vicinal *gauche* effect results in an upfield shift of about 6 ppm compared with the chemical shift of all-*trans* carbons of cyclic paraffins.

The ^{13}C NMR spectrum of solid polyethylene (PE) has a sharp peak separated from a weak shoulder by 2.36 ppm (± 0.1 ppm) (*4*). By comparing the CP–MAS spectra with the gated high-powered decoupled (GHPD) spectra with different delay times, the upfield peak can be assigned to the amorphous region, and the downfield peak can be assigned to the crystalline region (Figure 9.2). The differences in the chemical shifts are associated with conformational differences between the all-*trans* crystalline region and the equilibrium populations of *trans* and *gauche* conformations in the amorphous regions. The chemical shift of the methylene carbons in the disordered region can be approximated by the following equation:

$$\sigma = \sigma_o - 2\gamma f_g \qquad (9.2)$$

where σ_o is the chemical shift of the all-*trans* methylene chain and is taken as 33.0 ppm; γ is the γ-*gauche*

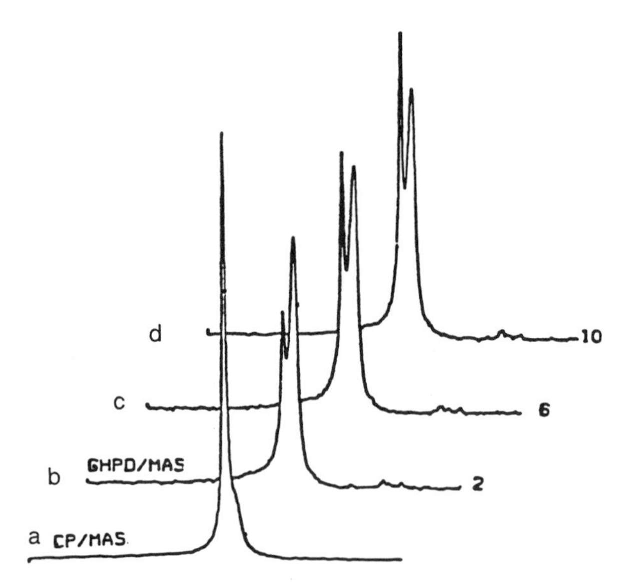

Figure 9.2. The high-resolution solid-state NMR spectra at 37.7 MHz of PE. The ^{13}C CP–MAS–DD spectrum (a) and the GHPD–MAS spectra (b–d) of the PE sample with recycle times of 2 s (b), 6 s (c), and 10 s (d). (Reproduced with permission from reference 54. Copyright 1987 Society for Applied Spectroscopy.)

effect, which is -5.3 ppm; and f_g is the equilibrium fraction of *gauche* bonds. The calculated value of f_g is 0.25, which corresponds to the known liquidlike distribution of *gauche* and *trans* bonds.

Studies of cyclic paraffins in the solid state have also been made (*10, 11*). For cyclic paraffins, the conformations are the same in solution and in the solid state up to $C_{32}H_{64}$, and the observed chemical shifts are the same (*12*).

CP–MAS ^{13}C NMR studies of folding in solution-grown crystals of PE have been reported. If the fold carbons are in a sharp-folded structure with little molecular motion, they should resonate 5–10 ppm upfield from the all-*trans* carbon resonance. However, the observed shift for single crystals of PE is only 2.3 ppm, a result suggesting that "the fold carbons are not in a sharp-folded structure but are in a mobile state with slightly less *gauche* character than truly amorphous or molten PE" (*10*).

Assuming that cycloalkanes are suitable model compounds for adjacent reentry tight folds in polyethylenes, solid-state ^{13}C NMR spectroscopy yields information about the conformation in the fold. Generally, the ggtgg fold conformation along the *b* axis is well-established for the monoclinic modification. With increasing ring size, the formation of lamellar-type crystals, in which pairs of adjacent stems are connected by tight folds, is essential for the arrangement of the molecules within a crystal lattice. The NMR spectra are consistent with two parallel all-*trans* arranged stems connected by tight ggtgg folds (*11*).

Molecular packing in the various crystalline forms can produce additional shifts. In *n*-paraffins, differences of 1 ppm are observed between the triclinic and orthorhombic crystal forms (*13*).

Chemical-Shift Effects for Helical Formation

Differences in bond angles of the various helical conformations also can produce variable chemical shifts in the solid state. Consequently, measured chemical shifts can be used to diagnose the helical conformations of polymers.

Polypropylene. Isotactic polypropylene (iPP) exhibits a stable crystalline α form consisting of chains in the 3_1 helical conformation (tgtgtg) packed in a monoclinic unit cell. Left- and right-handed helices face each other in the α form. The metastable

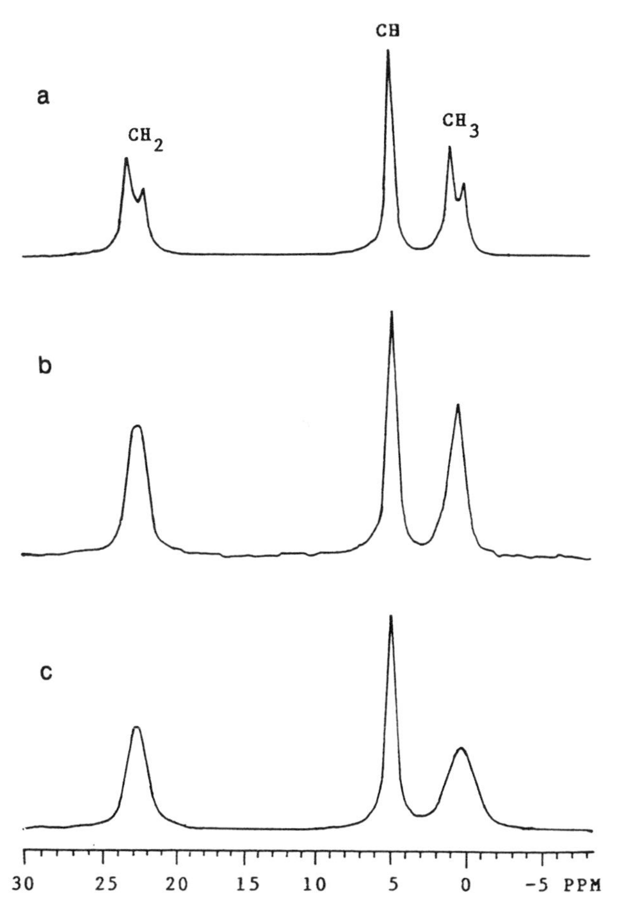

Figure 9.3. The CP – MAS ^{13}C spectra of isotactic polypropylene in the α form (a), the β form (b), and the smectic form (c). (Reproduced with permission from reference 16. Copyright 1987 Butterworth - Heinemann Ltd.)

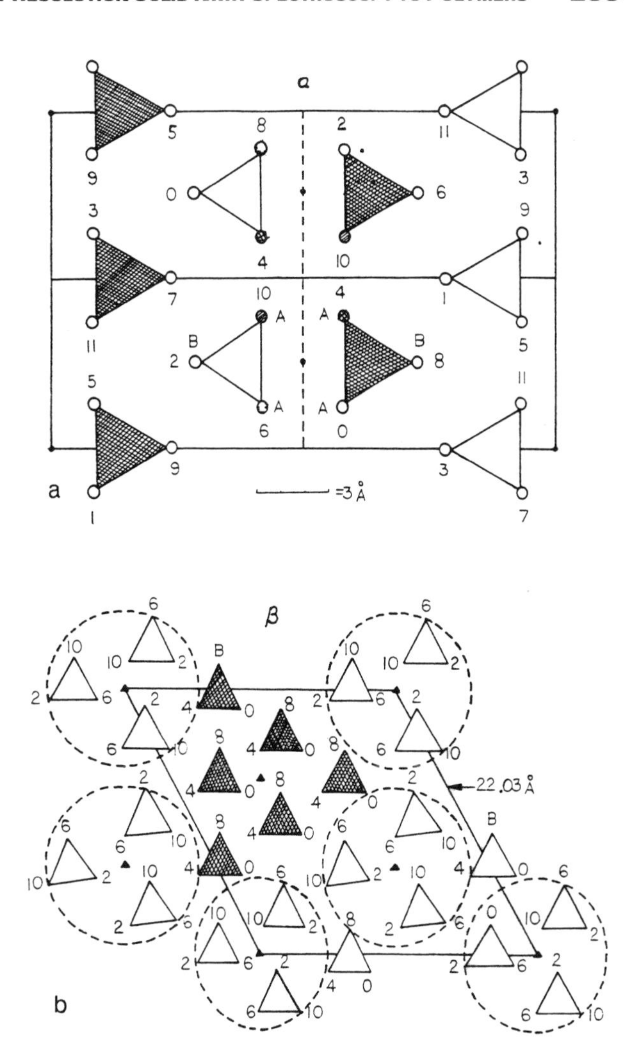

Figure 9.4. Crystal structures of the α form (a) and β form (b) of iPP. Full (RH) and open (LH) triangles indicate 3_1 helical iPP chains of different handedness. The labels A and B designate the inequivalent sites. Numerals at the triangle vertices indicate heights of methyl groups above a plane perpendicular to the c axis in twelfths of c. The circles at the triangle vertices in a correspond to methyl carbons, and the cross-hatched and stippled pairs of circles correspond to the enmeshed A-site methyls. (Reproduced with permission from reference 16. Copyright 1987 Butterworth–Heinemann, Ltd.)

β-form crystals contain hexagonally packed 3_1 helical chains arranged in groups of the same helical handedness (left or right). There is also a smectic, or partially ordered, form consisting of hexagonally packed 3_1 helical chains with disorder occurring in the intermolecular packing of the chains.

The CP – MAS ^{13}C NMR spectra of the different forms of iPP are shown in Figure 9.3 (*14 – 16*). Both the methylene and methyl carbon resonances in the α form are split by approximately 1 ppm, but the methine resonance shows only a shoulder. When the spectra of the annealed and quenched polypropylenes are compared, the ratio of the methylene and methyl resonance intensities is 2:1, with the more intense peak being the high-frequency resonance in each case. These chemical-shift differences are interpreted in terms of the unit cell having paired left- and right-handed 3_1 helices that generate distinguishable sites A and B for the methyl, methine, and methylene carbons in a 2:1 ratio through interchain interactions (Figure 9.4). Thus two interhelical distances give rise to the observed splitting. The A sites apparently correspond to a separation of 5.28 Å between helical axes, and the helices of the B sites are 6.14 Å apart. The CP–MAS ^{13}C NMR spectrum of the β form of iPP exhibits only a single resonance for each carbon, and the chemical shifts nearly correspond to those of the B sites in the α form. The packing of iPP helices in the β form is 6.36 Å and is thus closer to the more distant of the two interhelical packings (B); this closeness results in the near coincidence of the resonances. For the quenched smectic iPP, the chemical shifts correspond to those of the β form, and hence the smectic packing of the helices closely resembles the β form. It was suggested that the smectic form consists of 3_1 helices of opposite handedness but without the interlocking of the methyl groups of the A sites of the α form (*16*). Thus, the local packing arrangement in the smectic form of iPP can be characterized by interhelical distances comparable to the α-form B sites. This model of the smectic phase allows for a transformation to the α form through rotations and translations about and along the 3_1 helix axis rather than an unwinding and rewinding of the 3_1 portions of helical chains.

The CP–MAS ^{13}C NMR spectrum of syndiotactic polypropylene (sPP) is compared with the spectrum of iPP in Figure 9.5. The spectra are substantially different because of the differences in the conformations in the solid state. The sPP has a 2_1 helix with a (gg)(tt)(gg)(tt) structure. This structure involves *external* (outside the helix) and *internal* (inside the helix) methylene units that have substantially different resonances (*17*). In sPP the methylene carbon resonance is split 1:1 at 8.7 ppm. Consideration of the shielding effects of γ substituents suggests a shift difference of approximately 8 ppm between the external and internal methylene carbons.

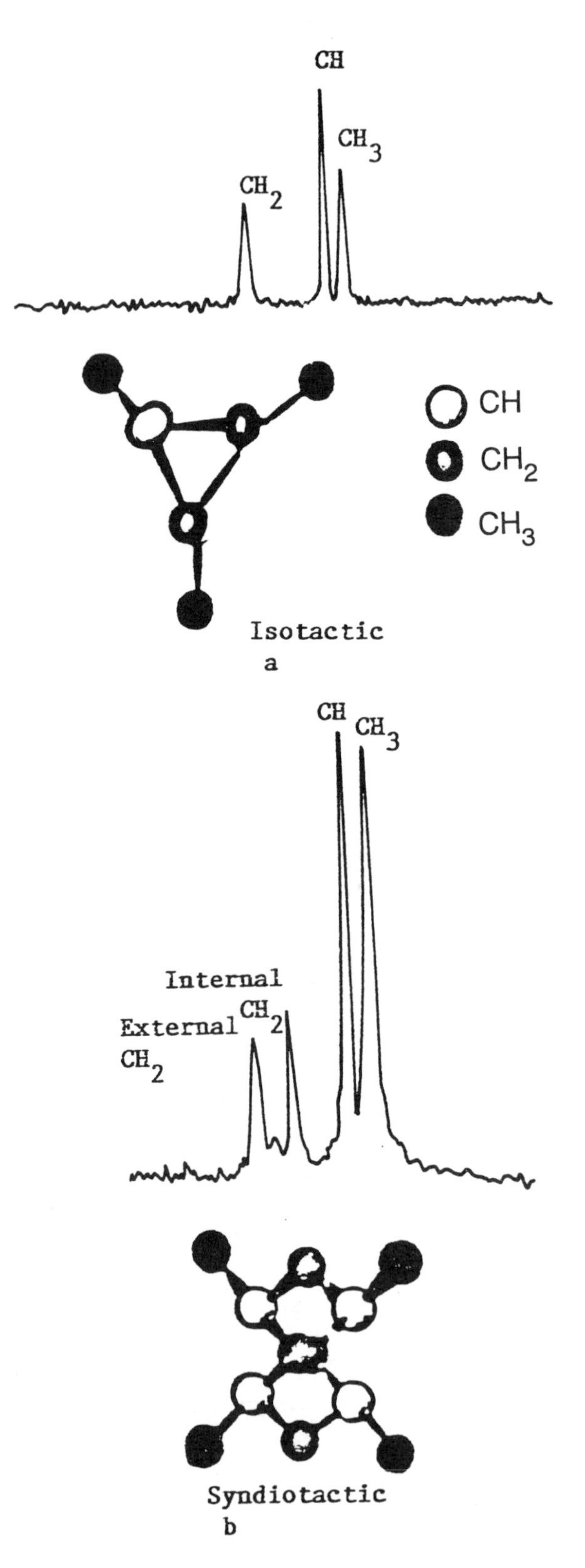

Figure 9.5. The solid-state ^{13}C spectra and helical conformations of iPP (a) and sPP (b). (Reproduced from reference 2. Copyright 1986 American Chemical Society.)

Poly(1-butene). Poly(1-butene) exists in three different helical conformations in the solid state (Figure 9.6) (*18*). The CP–MAS spectra of the three crystalline polymorphs and the amorphous polymer of poly(1-butene) (Figure 9.7) have been reported, and the spectra were interpreted by using a γ-*gauche* shielding parameter that is dependent on the dihedral angle. The γ-*gauche* shielding parameter is reduced to half its standard value when the angle is increased to 82 ± 1°.

This angular dependence of the interaction parameter can be extended by using the rotational isomeric

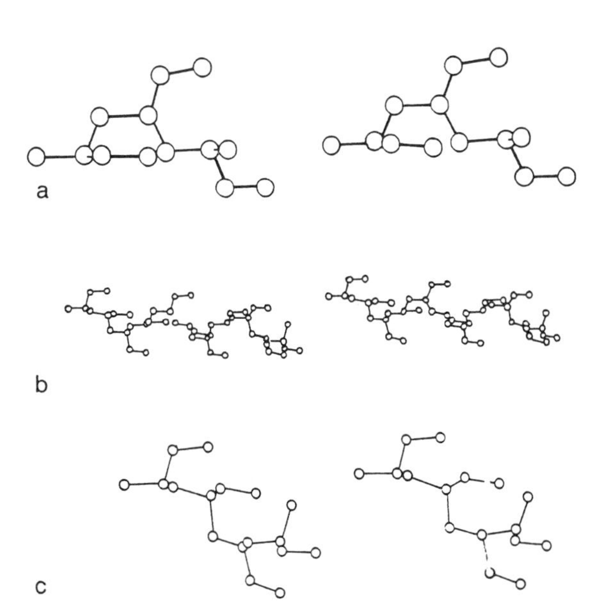

Figure 9.6. Stereoscopic drawings of the three crystalline polymorphs of isotactic poly(1-butene): (a) form I, 3_1 helix; (b) form II, 11_3 helix; and (c) form III, 4_1 helix. (Reproduced from reference 18. Copyright 1984 American Chemical Society.)

state approximation to calculate the conformer populations. This approach was used to interpret the solid-state spectrum of poly(3-methyl-1-pentene), which in the amorphous state can have either a four-fold right-handed helix or a nearly equienergetic four-fold left-handed helix. Three of the six chemical shifts of a right-handed unit differ from the corresponding resonances of a left-handed unit, and these stereochemical differences are observed (*19*).

Poly(oxymethylene). Poly(oxymethylene) (POM) exists in two crystallographic forms: a trigonal form with a 9_5 helical (all-*gauche*) conformation and an orthorhombic form with a 2_1 helical conformation. The chemical shifts for the orthorhombic form are displaced upfield by about 2 ppm compared with the trigonal form (*20*). The ^{13}C NMR spectrum of trigonal POM shows a sharp peak at 88.3 ppm, and the noncrystalline peak is displaced downfield to 89.6 ppm (*21*). The crystalline and amorphous peaks can be separated by taking advantage of the differences in mobility. This is accomplished by performing the NMR experiments with different delay times, which let the magnetization of the crys-

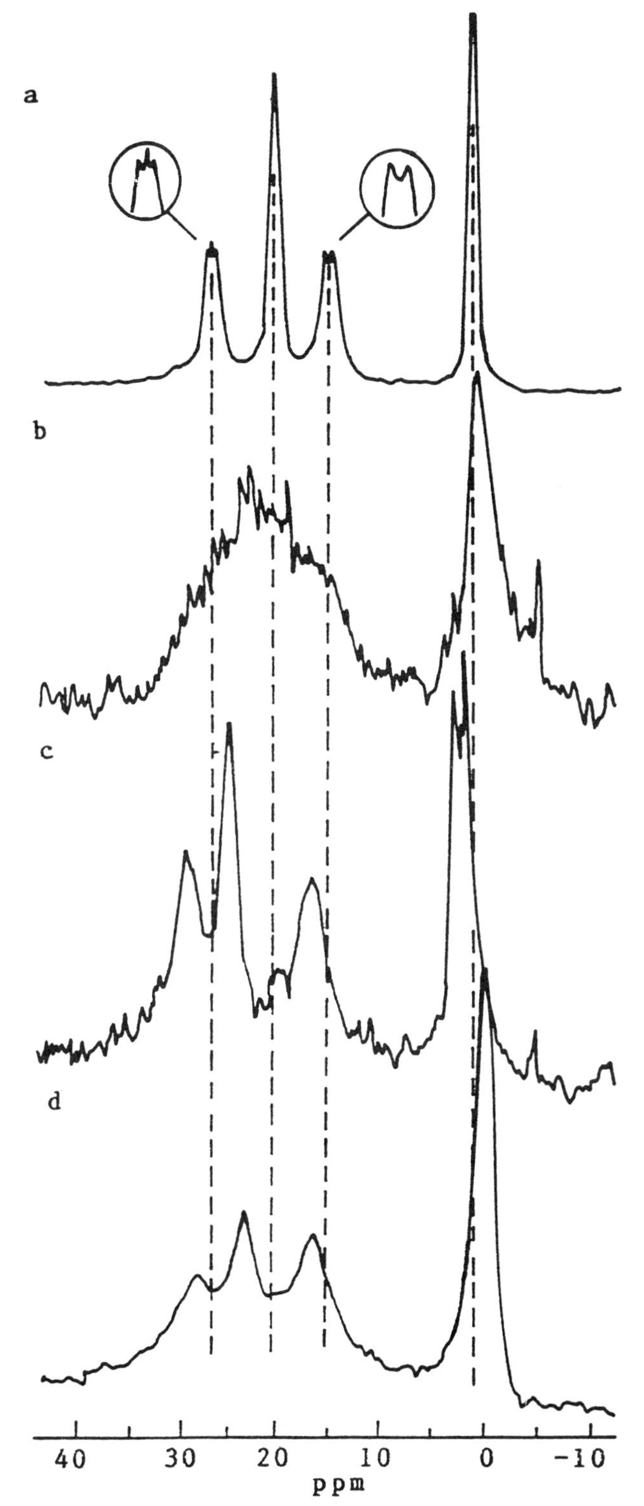

Figure 9.7. High-resolution solid-state ^{13}C NMR spectra at 50.3-MHz of poly(1-butene): (a) form I at 20 °C, (b) form II at −60 °C, (c) form III at −10 °C, and (d) the amorphous form at 43 °C. The vertical dashed lines represent the peak positions of form I. (Reproduced from reference 18. Copyright 1984 American Chemical Society.)

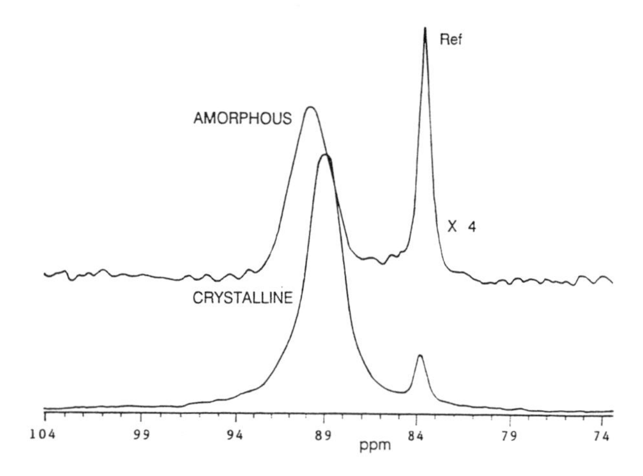

Figure 9.8. Separation of the amorphous and crystalline phases of POM by using a modified CP – MAS experiment. The amorphous spectrum has been enlarged by a factor of 4. (Reproduced with permission from reference 21. Copyright 1983 Marcel Dekker.)

talline peak decay to allow observation of the amorphous resonance (Figure 9.8).

Poly(ethylene terephthalate). The dipolar decoupling (DD) – CP – MAS ^{13}C NMR spectra of a glassy poly(ethylene terephthalate) (PET), a PET sample annealed for 76 h at 200 °C, and a solution-crystallized PET sample are compared in Figure 9.9 (*22*). The spectra show substantial differences that reflect the differences in the conformations and molecular packing of the chains. An increase in the crystallinity narrows the lines by limiting the number of conformational sequences to long sequences rather than to a series of short, disordered ones.

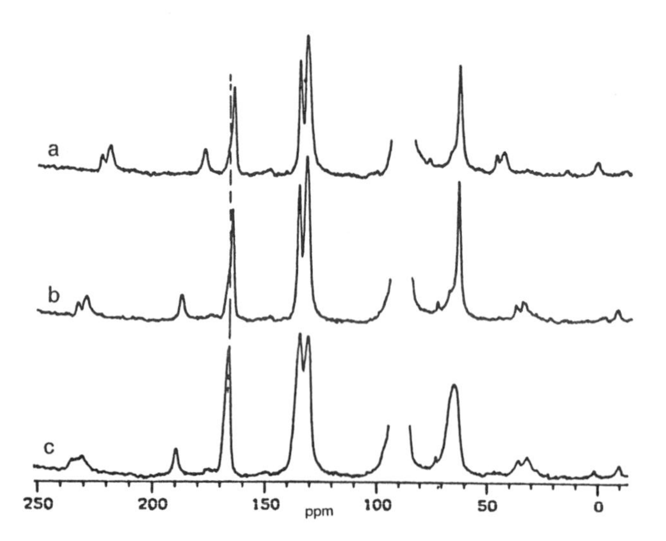

Figure 9.9. DD – MAS – CP ^{13}C NMR spectra of solution-crystallized (a), annealed 76 h at 200 °C (b), and glassy-state (c) PET. (Reproduced with permission from reference 22. Copyright 1980 Academic Press.)

Poly(butylene terephthalate). Poly(butylene terephthalate) (PBT) is an unusual material because uniaxial extension is accompanied by a reversible crystal – crystal transition. The α (relaxed) to β (strained) transition occurs at strains as low as 5%, and the transition is complete at 15% strain. When the tension is released, the polymer reverts rapidly back to the α phase with little hysteresis. Static measurements have shown that the major difference between the phases is in the conformation of the tetramethylene segment. The α phase is A – T – A (where A = non-*trans*, non-*gauche*, and T = *trans*), and the β phase is the extended T – T – T configuration (*see* Chart II).

The spectrum of the β phase is shown in Figure 9.10 (*23*). There are five peaks: the carbonyl (C = O) peak at 116.1 ppm, the nonprotonated aromatic peak at 135.2 ppm, the protonated aromatic peak at 131.5 ppm, the exterior methylenes (OCH_2) at 66.3 ppm, and the internal methylene peak at 25.8 ppm. The chemical shift of the internal methylene carbon resonance is substantially different between the two

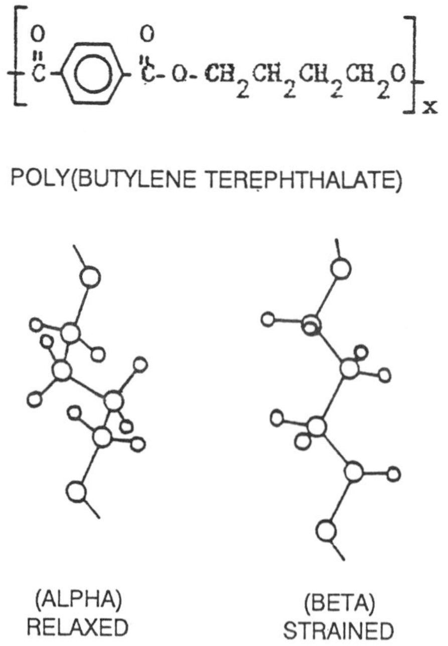

$$\left[\overset{O}{\overset{\|}{C}} - C_6H_4 - \overset{O}{\overset{\|}{C}} - O - CH_2CH_2CH_2CH_2O \right]_x$$

POLY(BUTYLENE TEREPHTHALATE)

(ALPHA) RELAXED

(BETA) STRAINED

Chart II. The repeat unit of PBT and the conformations of the α and β phases.

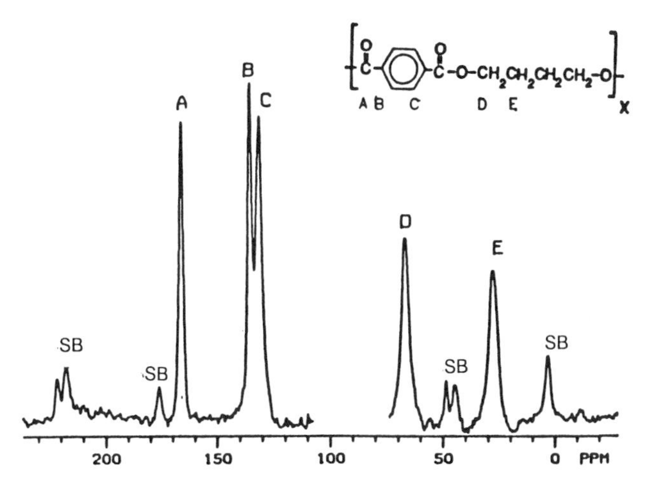

Figure 9.10. DD – MAS – CP spectrum of the stretched (β) form of PBT fibers (>50% crystallinity). The letters labeling the peaks correspond to the labeled carbons in the structure, and the peaks labeled SB are spinning side bands. (Reproduced from reference 23. Copyright 1987 American Chemical Society.)

phases; for the undrawn fiber it appears at 27.8 ppm, and for the drawn fiber it appears at 25.8 ppm. This shift of the CH_2 carbon resonance confirms past data that indicated that the main difference between the two phases involves the conformation of the tetramethylene segment. The shift is due to the spatial effects of the oxygen atom relative to the interior methylene carbons in the crumpled form, which shield the CH_2 carbon to a greater extent.

The aforementioned results are in conflict with other literature data, in which no shifts in the internal methylene resonances were reported for the phase transition in PBT (*24*). On this basis, it was concluded that no differences exist in the the conformation of the methylene carbons between the two phases. The change in length of the repeat unit (10%) during the phase transformation from the α to the β phase is, on this basis, associated with changes in the conformation of the terephthaloyl residue but not of the glycol residue.

Polypeptides. The differences in ^{13}C chemical shifts of amino acid residues in polypeptides are as large as 2 – 7 ppm, depending on the particular conformations, such as the right-handed α helix, the left-handed α helix, the ω helix, and the β sheet forms. Surprisingly, the ^{13}C chemical shifts are not affected by the neighboring units in solution unless

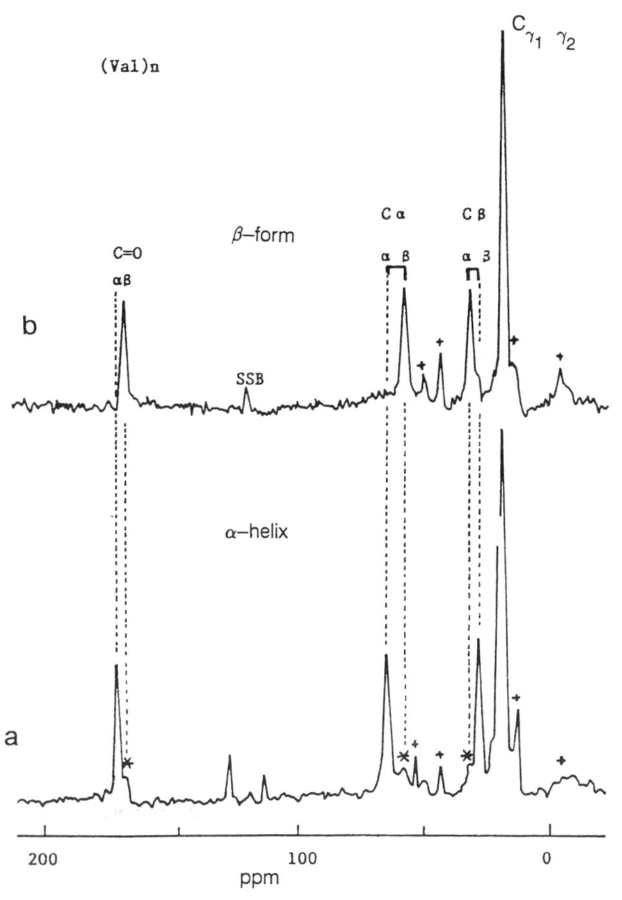

Figure 9.11. CP – MAS ^{13}C NMR spectra of solid $(Val)_n$. Spectrum a is of the α form, and spectrum b is of the β form. Peak labels are as follows: C_α, backbone carbons in valine; C_β, side chain carbons in valine; $C_{\gamma 1,\gamma 2}$, carbons in side chain of valine. (Reproduced with permission from reference 25. Copyright 1981 The Chemical Society of Japan.)

the neighbor is proline. Thus, ^{13}C chemical shifts of the individual amino acid residues can be used to probe the local conformations of the residues both in solution and in the solid state.

For polypeptides in the solid state, the resonances of the backbone carbon of the β sheet are shifted upfield by 5.4 – 7.1 ppm compared to the α helix (Figure 9.11) (*25*). The chemical shift of the random coil in solution is between the shifts of the α-helix and β-sheet forms. This result is consistent with the concept that the chemical shift observed for the random coil arises from an averaging of the ^{13}C shifts of at least two energetically favored conformations due to rapid chain isomerization (*25*).

Analysis of Cross-Linked Systems by Solid-State NMR Spectroscopy

Thermosetting Systems

Since the discovery of cured phenolic formaldehyde resins and vulcanized rubbers, interest has grown in such areas as the mechanisms of network formation, the chemical structures of cross-linked systems, and the dynamics of networks. Although the vast majority of polymers used industrially are cross-linked to a greater or lesser degree, the ability to characterize these systems is highly limited. Traditional techniques such as high-resolution NMR spectroscopy cannot be used because of limited solubility, although efforts have been made to swell the networks and examine the NMR spectra under high-resolution solution conditions. The spectra obtained by this solvent-swelling method usually have broad resonance lines, but these spectra can still provide valuable information.

The problem of limited solubility is overcome by the use of the grand NMR experiment to obtain well-resolved spectra of solid samples. However, the resolution achieved is still less than that for solution spectra, and the finer details (such as information on tacticity) are lost.

A second problem that makes analysis of cross-linked systems difficult is the low level of cross-linking required to produce substantial effects on the physical and mechanical properties. For rubbers, the level of cross-linking involves fewer than 1% of the carbons. This low level makes detection difficult because very small amounts of structure are being analyzed. Fortunately with solid-state NMR spectroscopy, the combination of available enhancement techniques and the use of a bulk sample (rather than a dilute solution) improve the sensitivity substantially.

Because of the importance of epoxy resins, it is not surprising that they received early attention from researchers using the solid-state NMR technique (*26, 27*). A number of other cross-linked systems have been investigated, including acetylene-terminated polyimide resins (*28, 29*), poly(*p*-phenylene) (*30*), 2-propenenitrile polymer with 1,3-butadiene and ethenylbenzene (ABS) resins (*31*), furfuryl alcohol resins (*32*), phenolic resins (*33*), vulcanized rubber (*34*), acrylic resins (*35, 36*), melamine – formaldehyde resins (*37*), polynuclear hydroxymethyl phenol (resol) – formaldehyde resins (*38*), and plasma-polymerized materials (*39*).

Because of the large number and the broad range of chemical systems studied, these individual systems will not be discussed here. The references listed in the preceding paragraph provide further study as well as a review that deals specifically with NMR spectroscopy of cross-linked systems (*40*). Instead, we will examine one class of systems in considerable detail in order to demonstrate the methods used for cross-linked systems.

Elastomeric Materials

Elastomeric materials are of industrial interest because of their specific dynamic properties. They are particularly suitable for solid-state NMR studies because of their extensive reorientational motion above their glass transition. Consequently, relatively weak dipolar couplings occur in elastomers compared to solid glassy polymers. Thus, the ^{13}C resonance lines are intrinsically narrow, and high-resolution spectra of this group of polymers can be obtained by applying relatively low spinning speeds and proton-power decoupling. The effects of applying the different techniques used in high-resolution solid-state NMR spectroscopy (such as magic-angle spinning, multiple-pulse, and high-power decoupling) on the spectra of an elastomer produce narrow resonances above that induced by the molecular motion. The resulting spectra are thus of high quality and contain substantial information concerning the network structure.

More than 140 years have passed since the initial discovery of sulfur vulcanization of natural rubber (NR) by Charles Goodyear and Thomas Hancock. Because of the many uses of rubber vulcanizates, the components in the curing mixture have been changed to produce finished products with good physical and mechanical properties and to allow much shorter cure times than required by the original inefficient recipes containing only sulfur and NR.

The vast range of properties that can be produced with cured NR is due to the structural variations of the cross-linked network that can occur during the curing process. These reactions produce a variety of alkyl – alkenyl types of structure, and two of these structures are shown in the following figure.

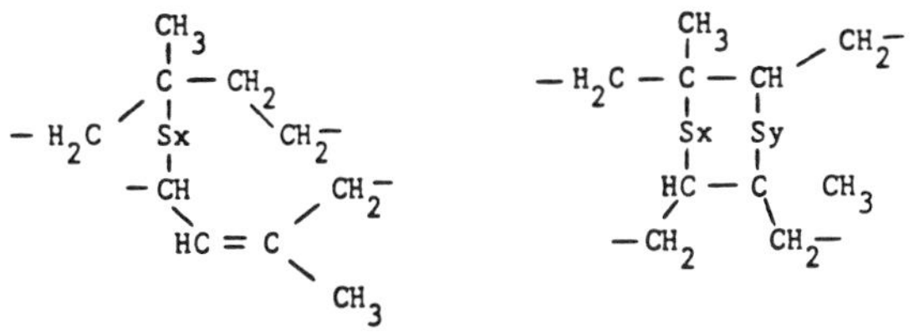

These structures include polysulfides, disulfides, monosulfides, cyclic sulfides, pendant sulfides, and conjugated diene and triene cross-links. The alkenyl and alkyl groups are designated as follows:

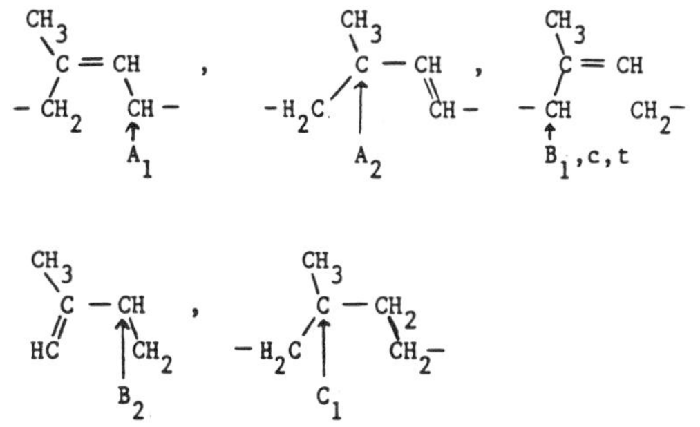

A_1, A_2, $B_{1,c,t}$, B_2, C_1

The use of sulfur to cure NR is inefficient (requiring 45 – 55 sulfur atoms per cross-link) and tends to produce a large portion of cyclic and polysulfidic structures. To overcome this problem, accelerator cure recipes that produce a higher ratio of mono- and disulfidic cross-link structures were developed.

Although these systems make more efficient use of sulfur, they suffer maturing reactions. The modulus and the cross-link density go through their maximums but then decrease with additional cure times (due to the loss of network structure by nonoxidative thermal aging). This process is known as *reversion* and can occur in addition to the maturing reactions. Reversion occurs when the desulfurization reactions are faster than the cross-linking reactions. Rubber researchers have divided these sulfur maturing reactions into two categories (*see* Scheme I). The first category is the *desulfurization* of polysulfides to di- and monosulfidic cross-links. This pathway is affected by the Zn – accelerator complex (found in accelerated sulfur vulcanization). The routes in the other category are characterized as *thermal decomposition*, in which the cross-links and the sulfuration species decompose into conjugated species, cyclic sulfides, shorter sulfur cross-links, and main-chain modifications.

Several laboratories have contributed a considerable body of information concerning the structure of the networks of vulcanized elastomers by using solid-state NMR techniques (*41 – 48*).

Polyisoprene. The ^{13}C NMR spectrum of cross-linked polyisoprene (NR) consists of two well-separated aliphatic and olefinic regions. A stack plot of the MAS – scalar decoupling (MAS – SD) spectra of NR cured with 10% sulfur for different cure times is shown in Figure 9.12 (*49*). With cure time, the spectral lines broaden because of a decrease in the molecular motion arising from the formation of cross-links, and new lines appear because of the cross-links and modifications of the main chain (*cis* to *trans* isomerism). The CP – MAS – DD ^{13}C NMR spectra at 38 MHz of the same system are shown in Figure 9.13 (*49*). The two collections of spectra are quite different, and this result reflects the different aspects of the samples being viewed in the different NMR experiments. The CP – MAS – DD experiment is sensitive to the more rigid portion (cross-linked regions) and the MAS – SD experiment is sensitive to the more mobile portion (the non-cross-linked or lightly cross-linked regions). Unlike the MAS – SD ^{13}C NMR spectra at 38 MHz, in which only one resonance (57.6 ppm) is observed between 90 and 50 ppm, the CP – MAS – DD spectra show at least four resonances in the same region. The observed chemical shifts for these new resonances are 82.7, 76.3, 67.8, and 57.9 ppm. These latter lines arise from the new chemical cross-linked structures in the vulcanized NR network.

However, structural information is hidden because the resonance lines are quite broad. One additional experimental approach to narrowing the lines is to elevate the measurement temperature, which increases the mobility of the molecules. A comparison of the GHPD – MAS ^{13}C NMR spectra at 38 MHz measured at both 80 °C and at room temperature is shown in Figure 9.14 (*41*). The resonance peaks measured at 80 °C are much sharper and better resolved (within 0.5 ppm). This enhanced resolution can also be obtained by swelling the vulcanizate with a solvent (*41*).

Another method to enhance the resolution is the use of computer simulations of solid-state ^{13}C NMR relaxation measurements. This allows decomposition of overlapped spectral absorptions and separation between signals having small differences in chemical

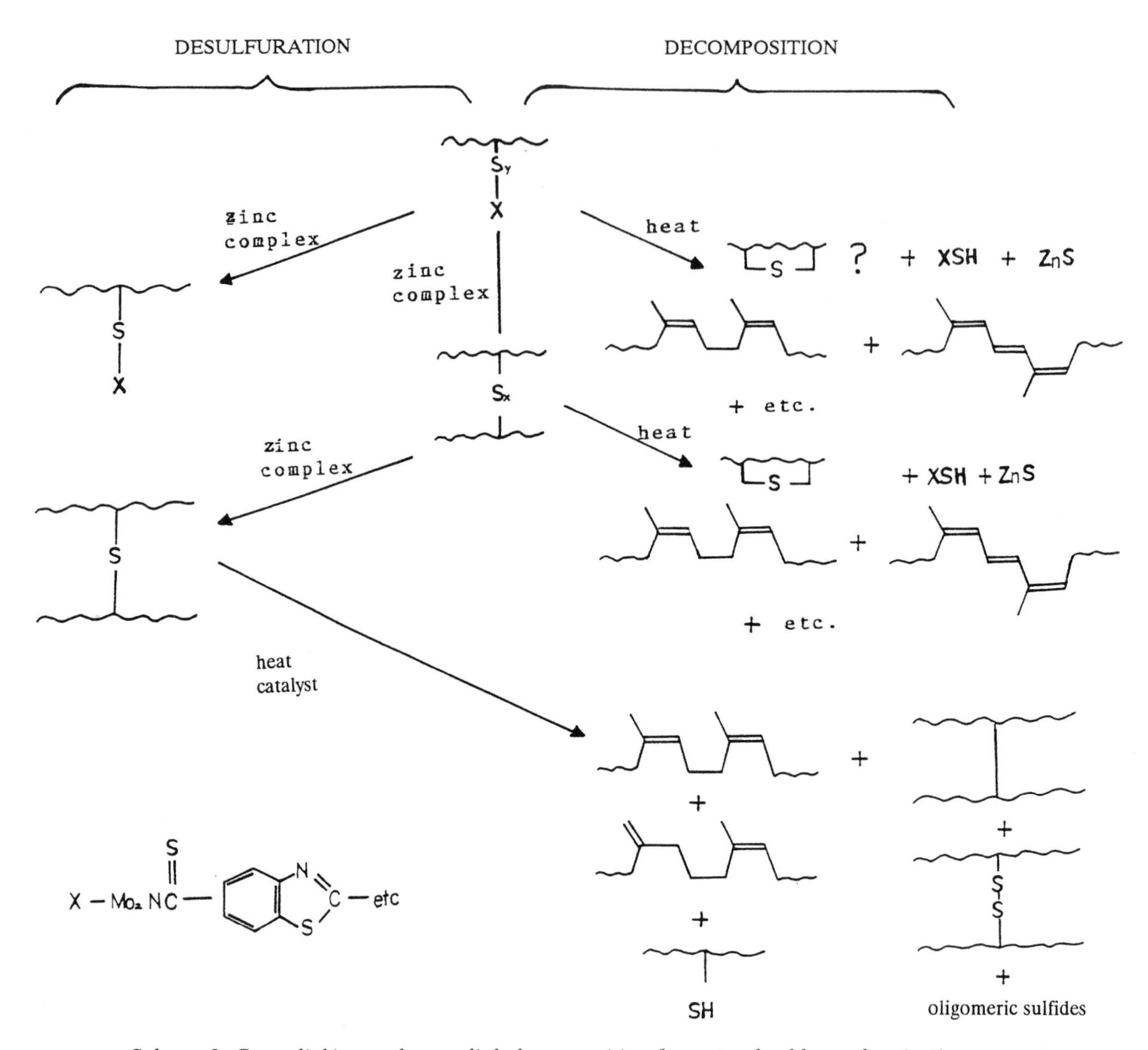

Scheme I. Cross-linking and cross-link decomposition for natural rubber vulcanizations.

shifts but different relaxation rates (*44*). The high-field region of the inversion-recovery spectra of sulfur-vulcanized NR are shown in Figure 9.15 (*44*). Curve fitting of the inversion-recovery spectra permits the decomposition of the aliphatic region into at least 18 signals. This method is particularly useful for spectra of highly cross-linked samples at lower temperatures when broad absorption lines are observed and the contributions of neighboring signals to the area of an observed resonance are significant.

In order to utilize the chemical information provided by these ^{13}C NMR spectra, chemical structures must be assigned to the new resonances. Chemical shifts for the postulated structures formed during sulfur vulcanization can be calculated by using additivity rules. The shielding parameters for the various sulfur groups are listed in Table 9.1. These shielding

Table 9.1. Aliphatic Sulfur Additivity Constants

Aliphatic Sulfur	*C-1*	*C-2*	*C-3*	*C-4*
HS—	+10.5	+11.5	−3.6	−0.2
CH_2—S	+20.4	+6.2	−2.7	+0.3
R—S—S	+25.2	+6.6	−3.4	−0.1
R—S	+17.9	+7.1	−3.0	−0.1

NOTE: All values are taken from the Sadtler C-13 NMR Index.

SOURCE: Reproduced with permission from reference 41. Copyright 1987.

values are added to the chemical-shift values of the *cis*-1,4-polyisoprene carbons to calculate the chemical shifts for the sulfurated species. The results are shown in Table 9.2 for the calculated chemical shifts of monosulfidic and polysulfidic cross-links. The data in this table suggest that the various structures can be differentiated from one another on the basis of differences in chemical shifts. No differences in chemical shifts occur between the di- and polysulfidic structures.

To confirm these initial expectations, model compound systems should be examined when they are available. For NR, model compounds based on 2-methyl-2-pentene were studied *(41)*. The available compounds and the nomenclature used to identify these structures are given in Chart III. The observed and calculated chemical shifts of the model compounds are listed in Table 9.3. The agreement be-

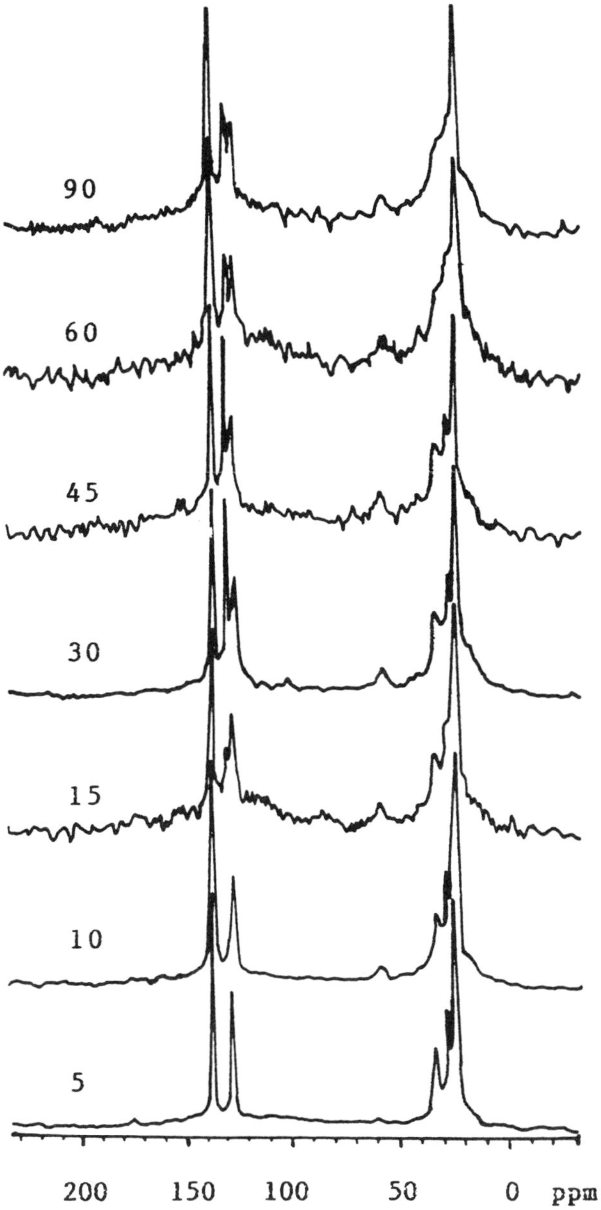

Figure 9.12. The stack plots of the MAS – SD ^{13}C NMR spectra of 10% sulfur-cured NR measured at room temperature and at 38 MHz. The different cure times are indicated on the figure. (Reproduced with permission from reference 49. Copyright 1987 Elastomers and Rubber Technology.)

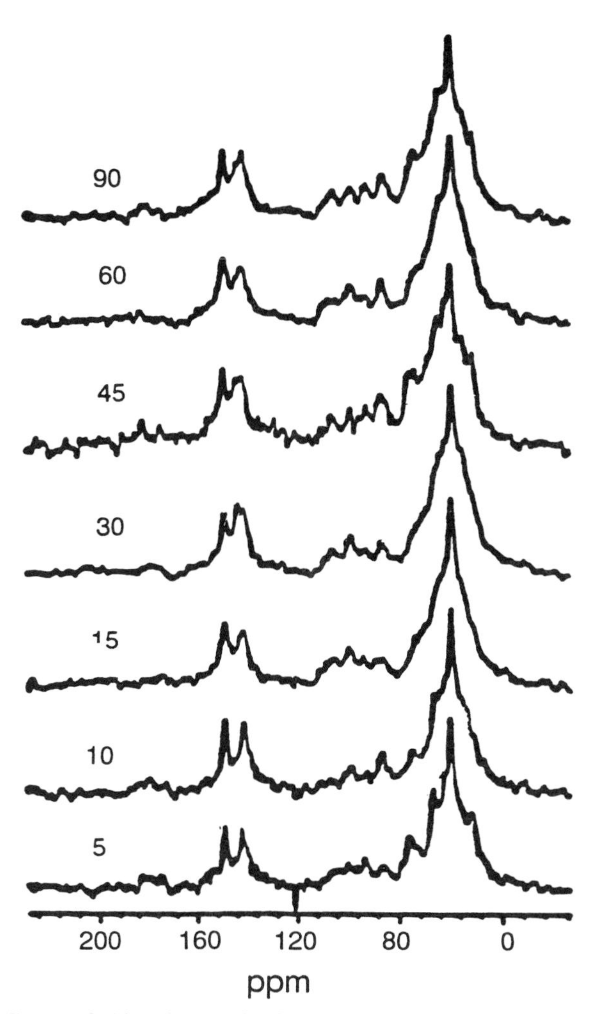

Figure 9.13. The stack plots of the CP – MAS – DD ^{13}C NMR spectra of 10% sulfur-cured NR measured at room temperature and at 38 MHz. The different cure times are indicated on the figure. (Reproduced with permission from reference 49. Copyright 1987 Elastomers and Rubber Technology.)

Table 9.2 Calculated Chemical Shifts for Monosulfidic and Polysulfidic Cross-Links, PPM

Possible structures	*Carbon*	*Mono*	*Di(poly)*
A 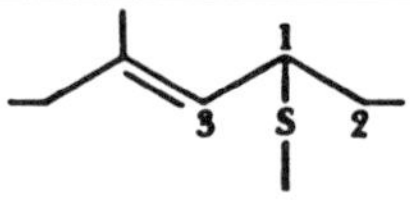	1	26.7 + 17.9 = 44.6	+ 25.2 = 51.9
	2	32.5 + 7.1 = 39.6	+ 6.6 = 39.1
	3	125.3 + 7.1 = 132.4	+ 6.6 = 131.9
B 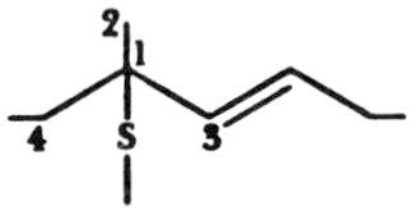	1	33.1 + 17.9 = 51.0	+ 25.2 = 58.3
	2	20.1 + 7.1 = 27.2	+ 6.6 = 26.7
	3	130 + 7.1 = 137.1	+ 6.6 = 136.6
	4	37.5 + 7.1 = 44.6	+ 6.6 = 44.1
C 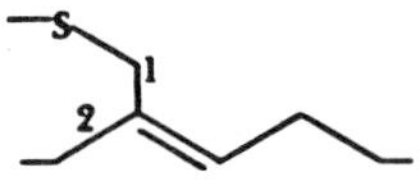	1	23.6 + 17.9 = 41.5	+ 25.2 = 48.8
	2	134.8 + 7.1 = 141.9	+ 6.6 = 141.4
D 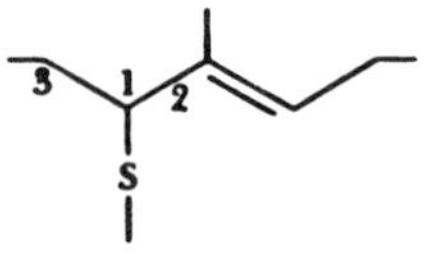	1	32.5 + 17.9 = 50.4	+ 25.2 = 57.7
	2	134.8 + 7.1 = 141.9	+ 6.6 = 141.4
	3	26.7 + 7.1 = 33.8	+ 6.6 = 33.3
E 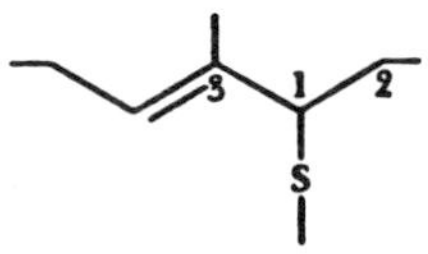	1	37.5 + 17.9 = 55.4	+ 25.2 = 62.7
	2	27.6 + 7.1 = 34.7	+ 6.6 = 34.2
	3	134.8 + 7.1 = 141.9	+ 6.6 = 141.4

SOURCE: Reproduced with permission from reference 41. Copyright 1987.

Table 9.3. Observed and Calculated Chemical Shifts of Model Compounds for Natural Rubber

Model Compound	*C – S Bond*	*Observed Chemical Shift*	*Calculated Chemical Shift*
A_1SA_1	HC—S—	36.8	21.5 + 17.9 = 39.4
A_1SA_2	HC—S—	36.8	21.5 + 17.9 = 39.4
	H_3C – C—S—	47.1	26.4 + 17.9 = 44.3
A_1SSA_1	HC – S—S—	43.6, 44.0	21.5 + 25.2 = 46.7
A_1SSSA_1	HC – S – S—S—	44.1	21.5 + 25.2 = 46.7
B_1SB_1	H_2C—S—	40.0	25.7 + 17.9 = 43.6
B_1SSB_1	H_2C – S—S—	48.9	25.7 + 25.2 = 50.9
B_1SB_2	H_2C—S—	40.0, 40.4	25.7 + 17.9 = 43.6
	HC—S—	53.2, 54.7	40.3 + 17.9 = 58.2

NOTE: All values are given in parts per million.
SOURCE: Reproduced with permission from reference 41. Copyright 1987.

tween the two methods is satisfactory, but the calculated chemical shifts could be improved with the use of better shielding parameters.

Additional evidence for the resonance band assignments can be obtained from spectral-editing methods, particularly the DEPT (distortionless enhancement by polarization transfer) experiment, which generates subspectra of the various types of carbons. The DEPT experiment can be performed on vulcanized NR in spite of the short spin – spin relaxation times (*50*). The results are shown in Figure 9.16 (*50*).

Utilizing all of the available data, the resonance assignments for the various structures observed in

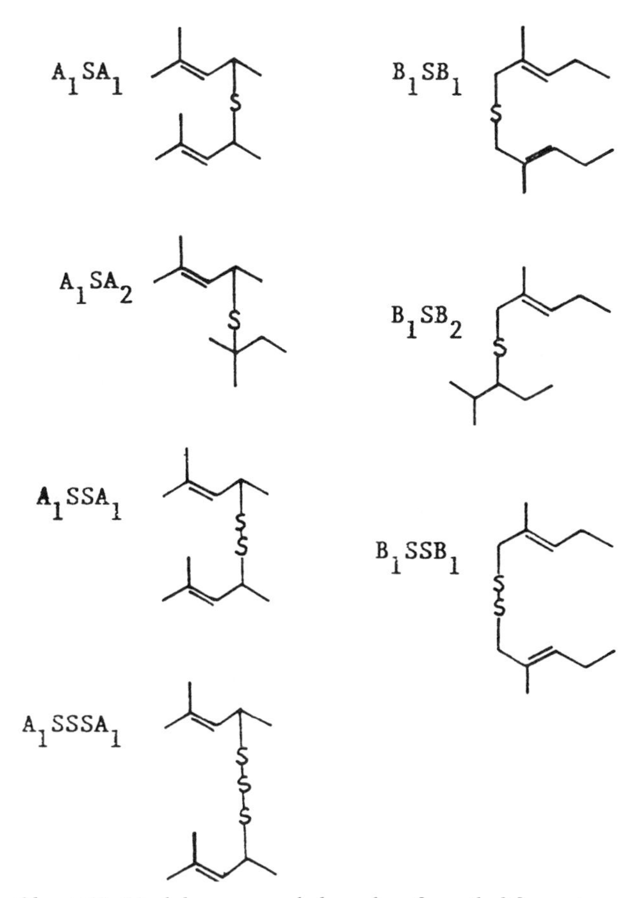

Chart III. Model compounds based on 2-methyl-2-pentene that are used as models for sulfur-vulcanized NR.

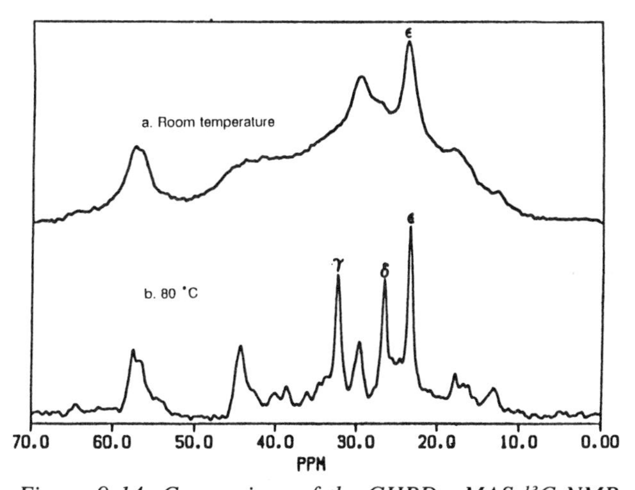

Figure 9.14. Comparison of the GHPD – MAS ^{13}C NMR spectra at 38 MHz measured at room temperature and at 80 °C of the NR cured for 60 min with 10% sulfur. (Reproduced with permission from reference 41. Copyright 1987 Rubber Chemistry and Technology.)

vulcanized NR were obtained. The structures are shown in Chart IV, and the resonance assignments are given in Table 9.4 (*44*).

cis-*1,4-Polybutadiene.* Solid-state NMR spectroscopy has been used to study the structures of polybutadiene that was vulcanized with sulfur (*43, 46*) and with an accelerated sulfur system (*43, 48*). The GHPD – MAS ^{13}C NMR spectra of uncured high *cis*-1,4-polybutadiene and *cis*-1,4-polybutadiene cured with 10 parts per hundred parts of resin (phr) of

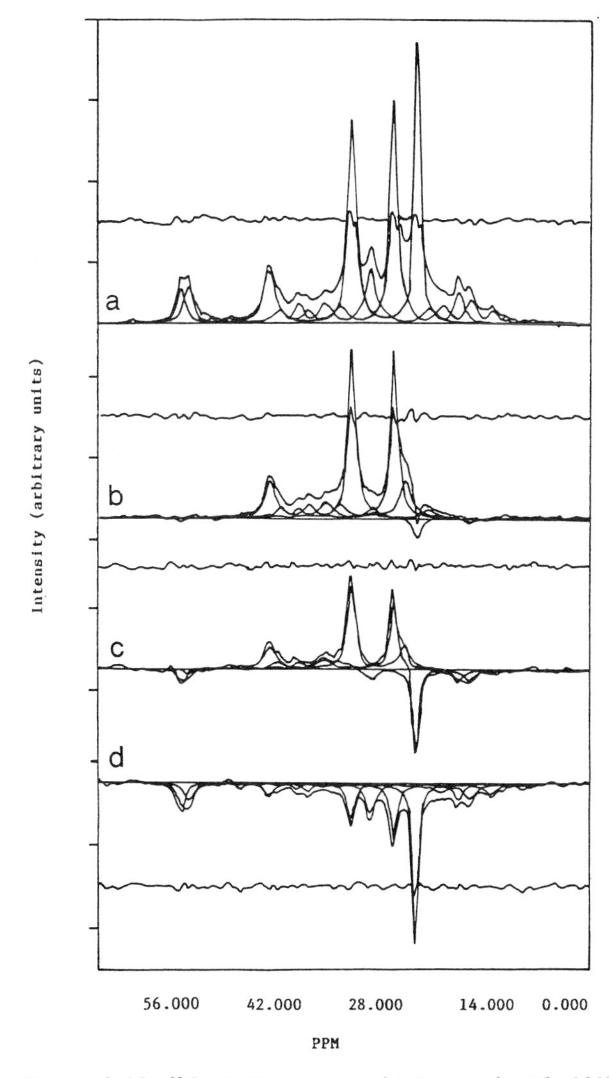

Figure 9.15. ^{13}C NMR spectra of NR cured with 10% sulfur at 150 °C for 120 min. The spectra were obtained by the inversion-recovery sequence, $180° - \tau - 90° - t_d$. The τ values are 4 s (a), 0.2 s (b), 0.1 s (c), and 0.03 s (d). (Reproduced with permission from reference 44. Copyright 1989 J. Polym. Sci. Polym. Phys.)

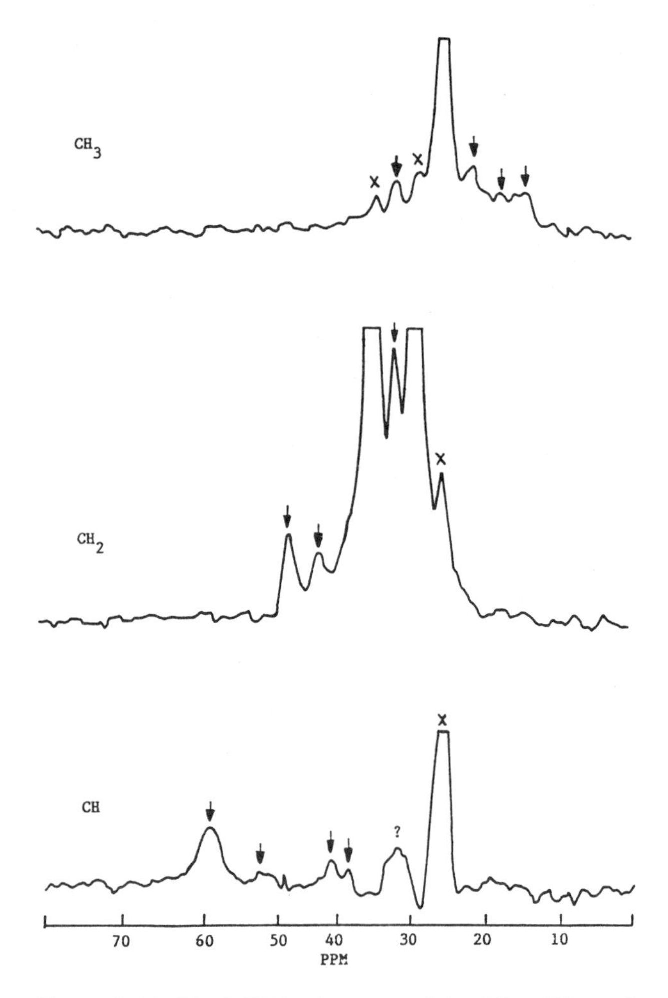

Figure 9.16. The DEPT subspectra of the CH_3, CH_2, and CH carbons of sulfur-cured NR. The label × marks residual peaks from other subspectra. Arrows indicate peaks due only to cross-links. (Reproduced with permission from reference 50. Copyright 1986 Rubber Chemistry and Technology.)

sulfur at 150 °C for various cure times are shown in Figure 9.17 (*46*). The most apparent spectral changes are the decrease in the *cis* isomer (129.6 and 27.6 ppm) and the increase in the *trans* isomer (130.1 and 32.9 ppm) with increasing cure time. The rate of change is initially rapid, then levels off after 45 min of cure. Magnifications of the spectra of the aliphatic region of the vulcanizate that was cured for 30 min are shown in Figure 9.18, and these spectra contain a wealth of information. Upon curing, many new resonances with the chemical shifts that are expected for cross-link and cyclic structures have appeared.

The results of the chemical-shift calculations for the structures in Chart V are shown in Table 9.5 (*46*). These calculations indicate that the peaks at 48 – 52 ppm represent carbons bonded to sulfur (α to sulfur) in polysulfidic cross-links and tertiary carbons in monosulfidic cyclic structures. The resonances at 34 – 40 ppm are due to methylene carbons β to sulfur in monosulfidic and polysulfidic cross-links and methylene carbons in the monosulfidic cyclics.

The chemical-shift calculations suggest that the 48 – 52-ppm region contains all methine carbon resonances and that the resonances in the 34 – 40-ppm region are all methylene carbons. To further validate these assignments, DEPT experiments were performed on the vulcanizate that was cured for 15 min. The CH_3, CH_2, and CH subspectra and the normal spectrum for the aliphatic region are shown in Figure 9.19 (*46*). The DEPT results confirm the proposed assignments. Peaks 1 – 4 at 48 – 52 ppm are methine carbons, and peak 6 at 43.7 ppm is due to a methine carbon in the 1,2 structure. Peak 7 at 40.9 ppm was identified as a methine carbon by using DEPT, and

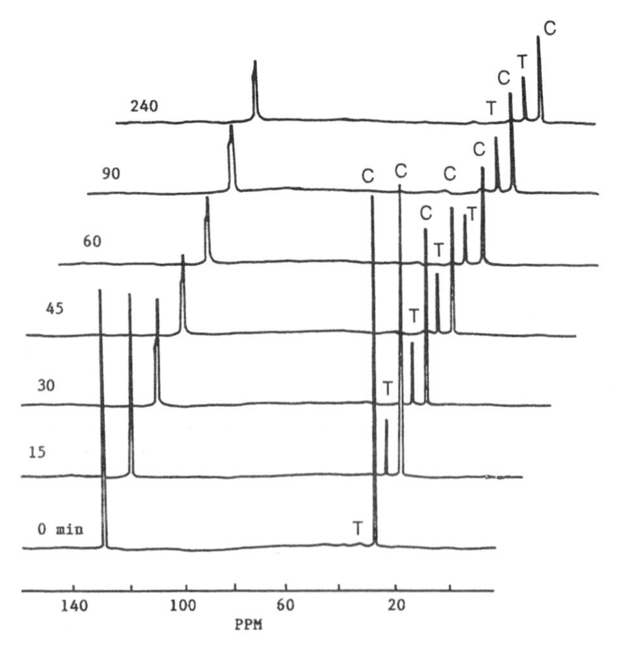

Figure 9.17. The GHPD – MAS ^{13}C NMR-spectra at 75.5 MHz of high cis*-1,4 polybutadiene cured with 10 phr of sulfur at 150 °C for the times indicated on the figure. Peak labels are as follows: C, cis; T, trans. (Reproduced with permission from reference 46. Copyright 1989 Rubber Chemistry and Technology.)*

Table 9.4 Calculated and Observed Aliphatic ^{13}C Chemical Shifts (δ) of Cross-Linked Structural Units in a Sulfur-Vulcanized Rubber

		δ (ppm)			
		Calculated			
Structure	*C atom*	*A*[a]		*B*[b]	*Observed*
I[c]	—S_n—CH(γ)<	57.7		57.5	56.8
	—CH_2(δ)—	33.3		32.7	34.2
	CH_3(ε)—	20.2		19.6	~20.0
II	—S—CH(δ)<	44.6		44.7	44.6
	—CH_2(γ)—	39.6		38.5	40.1
	CH_3(ε)—	23.5		23.5	~23.5
III	—S—CH(γ)<	57.9		58.0	57.9
	—CH_2(δ)—	37.1		36.0	36.4
	CH_3(ε)—	13.1		12.1	13.1
IV	—S—CH(δ)<		55.2[d]		56.8
	—S—CH(γ)<		51.4[d]		50.2[e]
	CH_2(δ)—		~31.9[d]		32.5[f]
	CH_3(ε)—		15.0[d]		16.1
V	—S—CH(β)	55.9		55.8	56.8
	—CH(α)<	40.2		39.1	38.8
	—CH_2(δ)—	31.8		30.7	~30.5
	CH_3(ε)—	16.8		15.8	16.1
VI	—S_n—C(α)<	58.3		58.0	57.9
	—CH_2(β)—	44.5		43.9	44.6
	—CH_2(γ)—	44.5		43.9	44.6
	CH_3(ε)—	25.5		25.5	—[g]
	—CH_2(δ)—	31.3		30.7	~31.0
VII	CH_3(ε)—	30.7		29.6	~30.0
VIII	—S—CH_2(ε)—	41.5		41.6	42.9
	—CH_2(γ)—	29.5		28.5	~30.0

[a]Calculated with the assumption of monosulfidic sulfur chemical-shift substituent effects $\alpha = +17.9$, $\beta = 7.1$, and $\gamma = -3.0$ ppm.

[b]Calculated with the assumption of modified monosulfidic sulfur chemical-shift substituent effects $\alpha = +18.0$, $\beta = +6.0$, and $\gamma = -4.0$ ppm.

[c]The disulfidic/polysufidic chemical-shift substituent effects used for calculation are as follows: $\alpha = +25.2$, $\beta = +6.6$, and $\gamma = -3.4$ ppm (A); $\alpha = +25.0$, $\beta = +6.0$, and $\gamma = -4.0$ ppm (B).

[d]The chemical shifts are estimated from the ^{13}C-NMR data of substituted 2,5-dihydrothiophenes.

[e]This resonance is not clearly resolved in all spectra. Its existence was observed previously.

[f]It is also possible that the resonance contributes to the overlapped absorption at about 31 ppm.

[g]The resonance is not resolved in the highly overlapped spectral region.

SOURCE: Reproduced with permission from reference 44. Copyright 1989.

NOTE: All values are given in parts per million.

this peak is assigned to carbons bonded to sulfur in monosulfidic cross-links. Peaks 19 – 25 are most likely methylene carbons in monosulfidic cyclic structures.

The GHPD – MAS ^{13}C NMR spectrum of high *cis*-1,4-polybutadiene cured with thiuram (tetramethylthiuram disulfide, TMTD) and ZnO is shown in Figure 9.20 along with the sulfur-vulcanized spectrum (*48*). Substantial differences occur between the two spectra, and these differences are consistent with the expectations of the chemistry involved and the differences in physical and mechanical properties of the two vulcanizates. In addition to the structures anticipated from the sulfur vulcanization, the accelerated TMTD vulcanization should exhibit pendant accelerator groups, which are the precursors to the

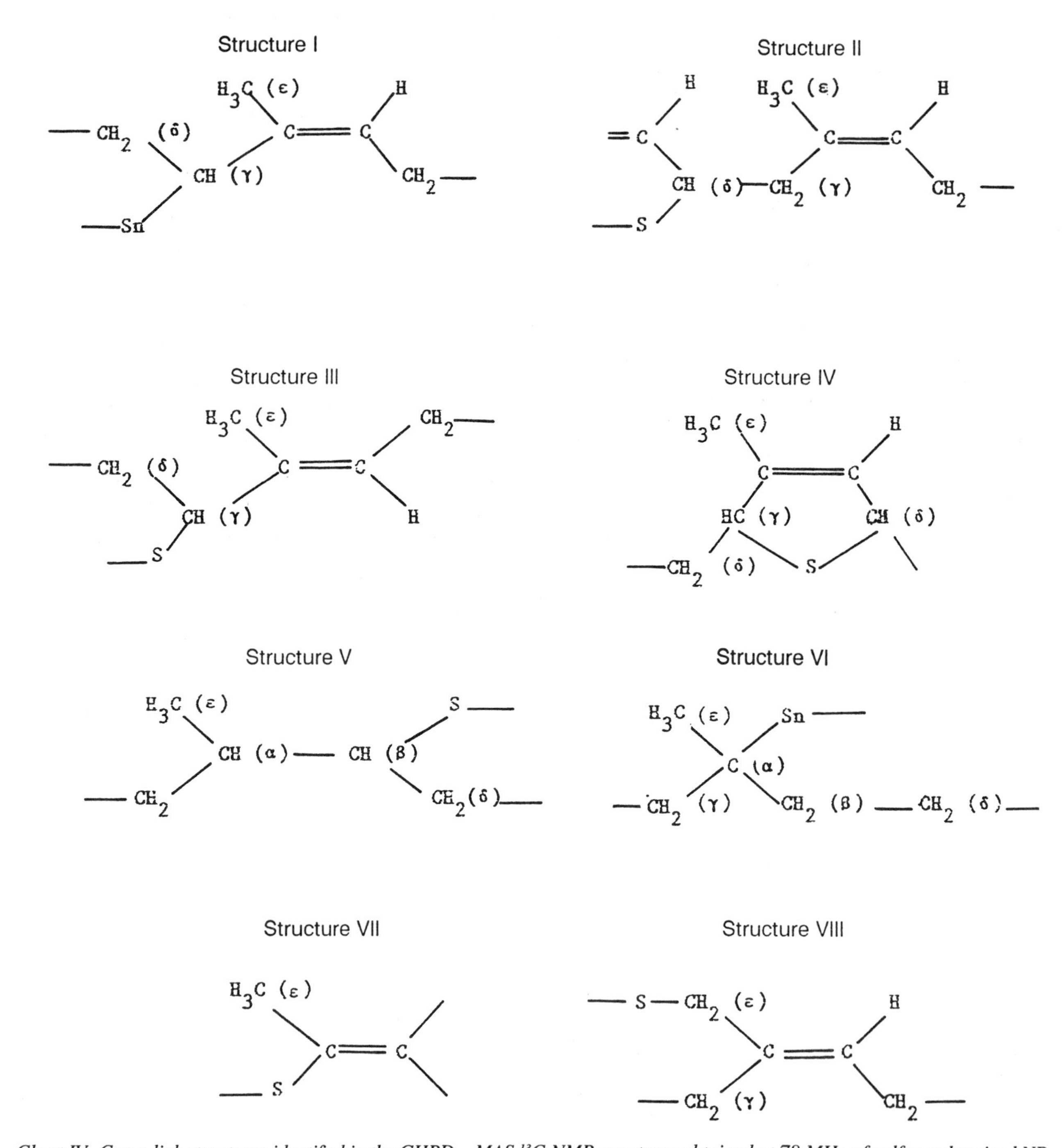

Chart IV. Cross-link structures identified in the GHPD – MAS ^{13}C NMR spectrum obtained at 78 MHz of sulfur-vulcanized NR.

cross-link structures. Model-compound studies indicate that there is very little difference between the chemical shifts due to disulfidic cross-links and those due to pendant accelerator groups. However, the TMTD moiety itself shows a chemical-shift difference when it is pendant to the rubber chain, and this chemical-shift difference can be observed (Figure 9.21). The TMTD peak at 194 ppm shifts to 195.5 ppm when the TMTD is a pendant accelerator group in the vulcanizate. A comparison of the contributions of the various structures for sulfur- and TMTD-vulcanized systems is shown in Table 9.6 (*48*).

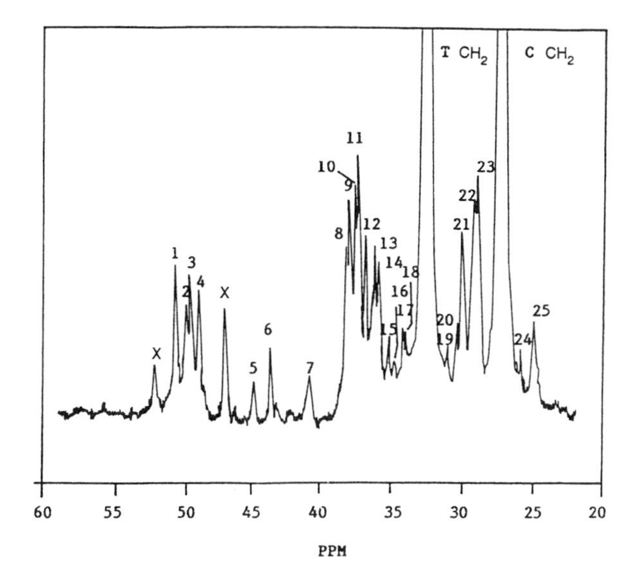

*Figure 9.18. The aliphatic region of the GHPD – MAS ^{13}C NMR spectrum at 75.5 MHz of a high-*cis *butadiene rubber vulcanizate cured with 10 phr of sulfur at 150 °C for 30 min. The × indicates a spinning sideband, and C CH_2 and T CH_2 are* cis *and* trans *CH_2, respectively. (Reproduced with permission from reference 46. Copyright 1989 Rubber Chemistry and Technology.)*

Observed Chemical Shifts of Peaks Labelled in Figures 9.18 and 9.19

Peak no.	*Chemical shift, ppm*
1	51.0
2	50.1
3	49.8
4	49.1
5	45.0
6	43.7
7	40.9
8	38.3
9	38.2
10	37.7
11	37.5
12	36.9
13	36.3
14	36.0
15	35.3
16	34.9
17	34.4
18	34.2
19	31.3
20	30.6
21	30.2
22	29.4
23	29.1
24	26.0
25	25.1

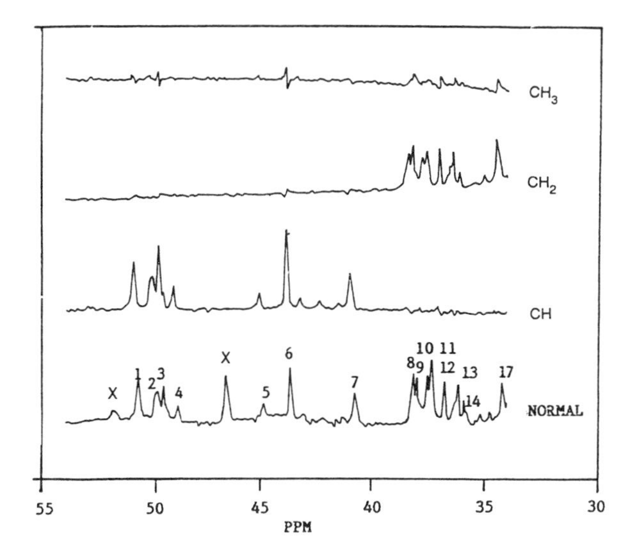

Figure 9.19. The DEPT ^{13}C NMR subspectra at 75.5 MHz of a high cis*-1,4-polybutadiene vulcanizate cured with 10 phr of sulfur for 15 min at 150 °C. (Reproduced with permission from reference 46. Copyright 1989 Rubber Chemistry and Technology.)*

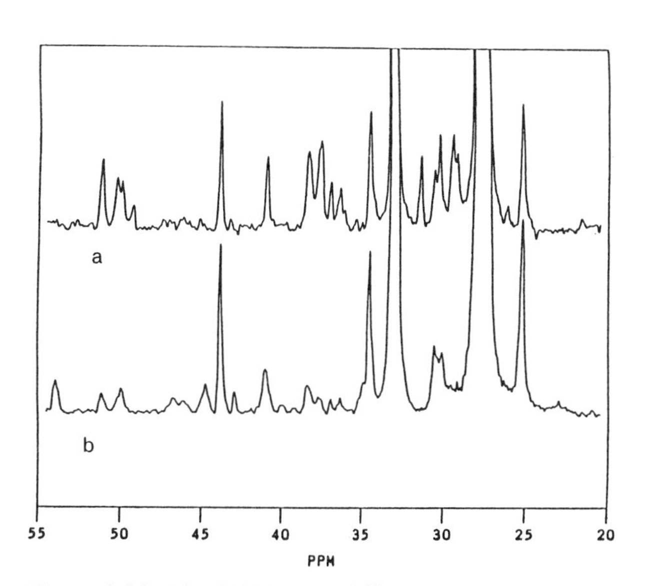

Figure 9.20. The GHPD – MAS ^{13}C NMR spectra at 75.5 MHz of high cis*-1,4-polybutadiene vulcanized with TMTD and ZnO (a) and vulcanized with 2 phr of sulfur (b). (Reproduced with permission from reference 48. Copyright 1991 Rubber Chemistry and Technology.)*

Table 9.5. Calculated Chemical Shifts of Potential Structures for Sulfur-Vulcanized Polybutadiene

Structure	*Carbon*	*Monosulfidic Cross-Links*	*Di- and Polysulfidic Cross-Links*
I	1	27.6 + 16.1 = 43.7	27.6 + 23.9 = 51.5
I	2	27.6 + 6.9 = 34.5	27.6 + 6.6 = 34.2
II	1	32.9 + 16.1 = 49.0	32.9 + 23.9 = 56.8
II	2	32.9 + 6.9 = 39.8	32.9 + 6.6 = 39.5
III	1	31.7 + 16.1 = 47.8	31.7 + 23.9 = 55.6
III	2	31.7 + 6.9 = 38.6	31.7 + 6.6 = 38.3
IV	1	27.6 + 16.1 + 6.9 = 50.6	27.6 + 23.9 + 6.6 = 58.1
		cis-*Cyclic*	trans-*Cyclic*
V	2, 6	38.3 + 8 = 46.3	33.4 + 8 = 41.4
	3, 5	36.0	34.8
	4	27.1	20.8
VI	2	35.4 + 8 = 43.4	37.1 + 8 = 45.1
	3	32.2	37.4
	4	30.6	35.5
	5	29.6 + 8 = 37.6	32.9 + 8 = 40.9
	6	33.4	36.6
VII	2, 5	44.4 + 8 = 52.4	44.3 + 8 = 52.3
	3, 4	37.9	39.5

SOURCE: Reproduced with permission from reference 46. Copyright 1989.

Elastomers can also be cross-linked by peroxides or by irradiation (*51–53*). This type of chemistry produces H-type cross-links or Y branches involving tertiary carbon moieties. With irradiation of polybutadiene, a broad resonance centered at 45.8 ppm appears for H cross-links (*52*). Another resonance is observed at 30.5 ppm and is attributed to CH_2 groups adjacent to cross-links. A similar spectrum is observed for the cumyl peroxide cure of polybutadiene (*51*).

Butyl Rubber. Changes in the microstructure of butyl rubber with cure have been studied by using high-resolution solid-state ^{13}C NMR spectroscopy (*47*). The problem with this system is that butyl rubber is a copolymer consisting of 97–98% isobu-

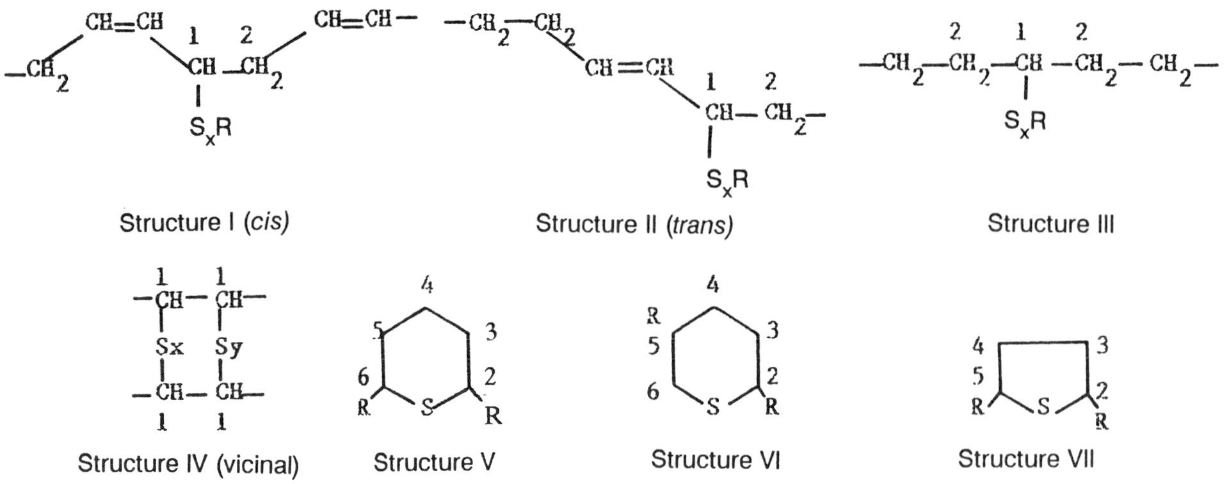

Structure I (*cis*) Structure II (*trans*) Structure III

Structure IV (vicinal) Structure V Structure VI Structure VII

Chart V. Possible cross-link and cyclic structures resulting from the vulcanization of cis-*1,4-polybutadiene with sulfur.*

Table 9.6. Comparison of Structures of TMTD- and Sulfur-Vulcanized Networks

Structure	*10 phr of TMTD* Uncured	*10 phr of TMTD* Cured 5 min	*10 phr of TMTD* Cured 60 min	*2 phr of S* Uncured	*2 phr of S* Cured 30 min
cis	96.9	89.5	88.6	98.5	90.0
trans	0.9	7.2	8.0	0.5	6.9
Vinyl	1.8	1.7	1.6	0.9	0.6
C – S bonded					
Di- and poly-	—	0.2	0.1	—	0.1
Mono-	—	0.2	0.3	—	0.2
Cyclics	—	0.4	0.6	—	1.5
Unassigned	0.4	0.5	0.5	0.1	0.4

NOTE: All values are given as percents.
SOURCE: Reproduced with permission from reference 48. Copyright 1991.

tylene and 1 – 3% isoprene, giving the polymerized elastomer only 1 – 3% unsaturation. High-resolution NMR spectroscopy has revealed considerable detail about the copolymer structure (*see* Chapter 7). The isoprene units are isolated from each other, and the isoprene double bond in the uncured elastomer is almost exclusively (90%) in the *trans*-1,4 configuration.

An accelerated-vulcanization system (accelerator: sulfur ratio of 1.75) was used to provide a relatively simple chemical microstructure. The curing system was an efficient vulcanization formulation with TMTD, zinc oxide, elemental sulfur, and stearic acid. The ^{13}C NMR spectrum at 75.5 MHz shows several distinct changes with cure time (Figure 9.22). The resonances at 24.4 and 20.3 ppm decrease with cure time, and the resonances at 26.9 and 25.2 ppm increase. New resonances appear at 23.6, 21, and 15 ppm. The *gem*-dimethyl groups do not react during the vulcanization; the reactions occur exclusively at the α carbons next to the isoprene double bond. There is no evidence of additional unsaturation, cyclization, or migration of the isoprene double bonds during the vulcanization. The initial *trans* configuration converts during vulcanization to the *cis* form. The resonance due to monosulfidic addition occurs at 21.4 ppm, and the peak at 21.1 ppm is due to polysulfidic addition. In Figure 9.23, the NMR spectrum of a butyl rubber sample cured for 15 min is shown. The three overlapping peaks arise from polysulfidic addition to the C-4 carbon (III and VI) and from the C-1 carbon (II). Examination of this system demonstrates the extremely high sensitivity of solid-state NMR spectroscopy for elastomeric systems.

Cross-Linked Polyethylene

A large number of applications of polyethylene (PE) require the polymer to be slightly cross-linked (e.g., high-voltage cable insulation). The DD – CP – MAS ^{13}C NMR spectrum has been obtained for gamma-irradiated (600 mrad) high-density polyethylene, and this spectrum is shown in Figure 9.24 (*54*). The new resonance peak at 39.7 ppm is assigned to the car-

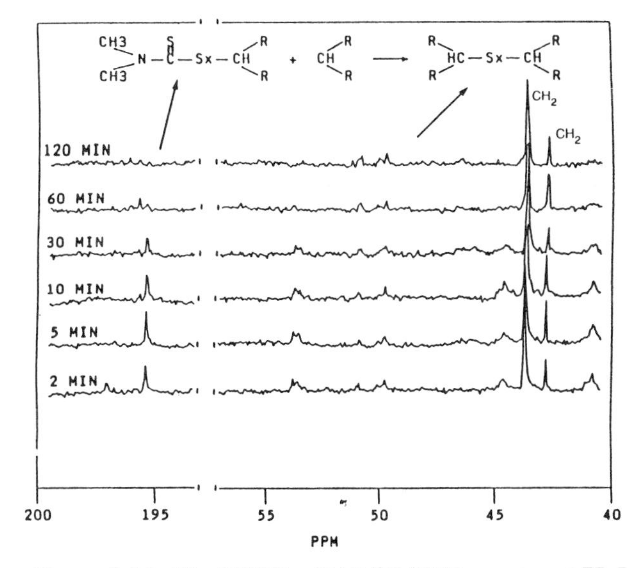

Figure 9.21. The GHPD – MAS ^{13}C NMR spectra at 75.5 MHz of high cis-*1,4-polybutadiene vulcanized with TMTD and of the pure TMTD. (Reproduced with permission from reference 48. Copyright 1991 Rubber Chemistry and Technology.)*

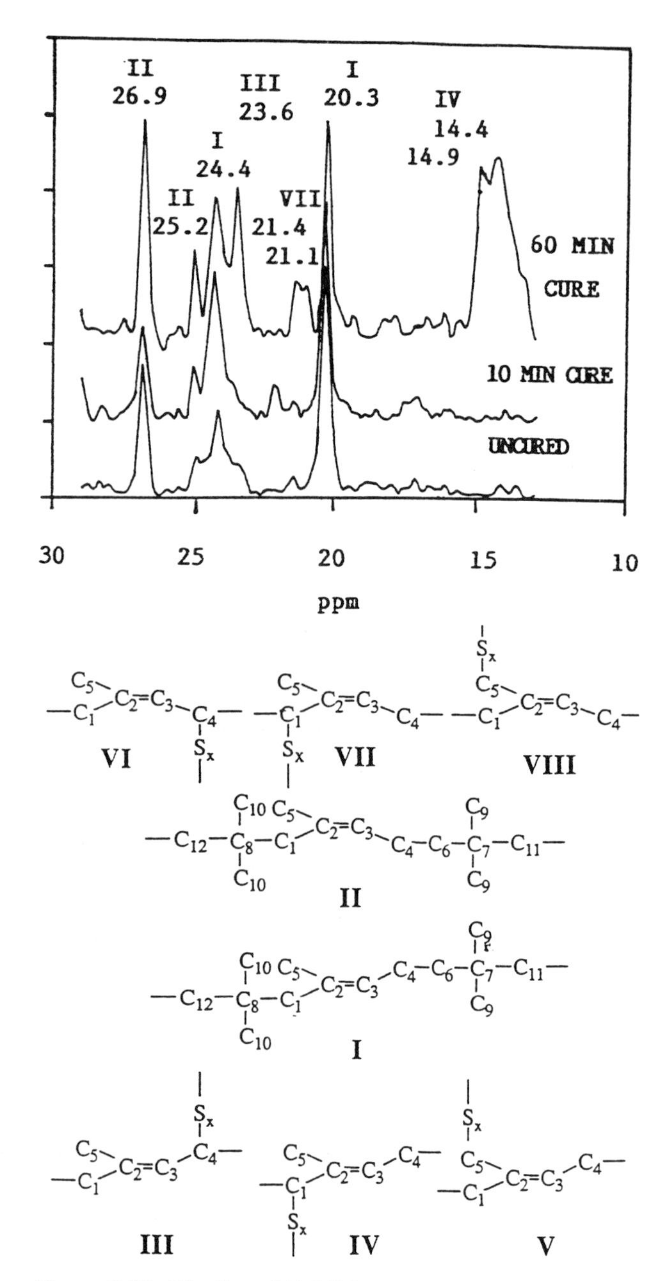

Figure 9.22. The SD – MAS ^{13}C NMR spectra at 75.5 MHz of cured (10 and 60 min) and uncured efficient-accelerated TMTD-vulcanized butyl rubber. The structural assignments are based on the structures shown in the lower portion of the figure. (Reproduced with permission from reference 47. Copyright 1991 Rubber Chemistry and Technology.)

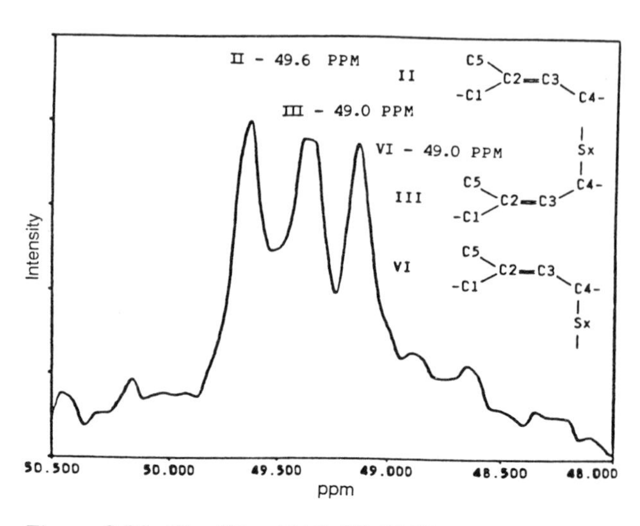

Figure 9.23. The SD – MAS ^{13}C NMR spectrum at 75.5 MHz in the 48 – 50-ppm range for a butyl rubber that was cured for 15 min. (Reproduced with permission from reference 47. Copyright 1991 Rubber Chemistry and Technology.)

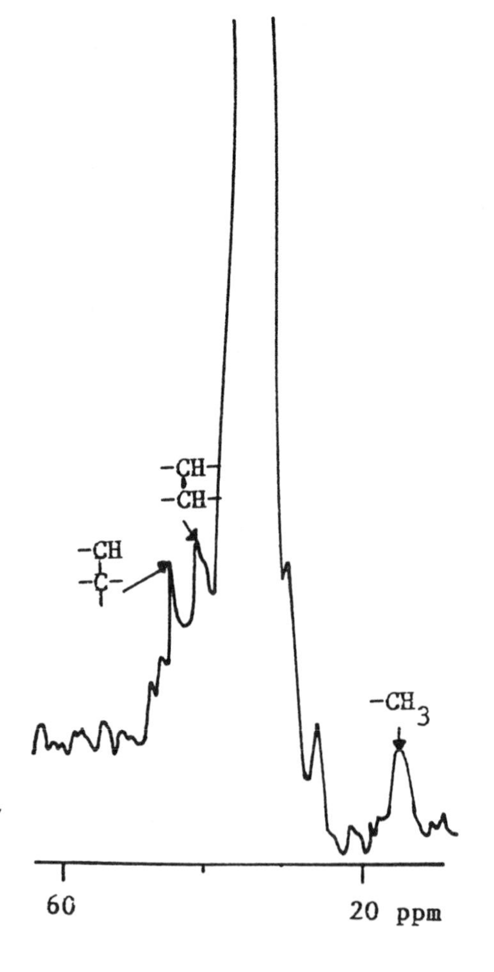

Figure 9.24. The DD – MAS – CP NMR spectrum at 37.7 MHz of gamma-irradiated high-density PE (600 mrad). (Reproduced with permission from reference 54. Copyright 1987 Society for Applied Spectroscopy.)

bons associated with carbon – carbon cross-links in PE. Model-compound work indicates that the resonance of the tertiary carbon associated with cross-links is at 39.49 ppm, and the calculated chemical shift for this structural unit is 39.5 ppm. The new resonance peak at 15.4 ppm is associated with methyl end groups. The two resonances at 30 and 27.8 ppm are associated with radiation-induced branching. The effect of gamma-radiation on each component of the crystalline and amorphous regions suggests a preferential attack on the amorphous phase. A minimal effect is observed for the crystalline phase except at very high doses of irradiation (*54*).

A more recent study (*54*) has examined the gamma-irradiation in vacuo of the molten state of linear PE, and the resulting polymer was studied with high-resolution solution ^{13}C NMR spectroscopy. H-type cross-links as well Y-type long branches were observed. The *G* values of the H-links and Y branches indicate that H-links are more effectively produced in the molten state than are the Y-type branches.

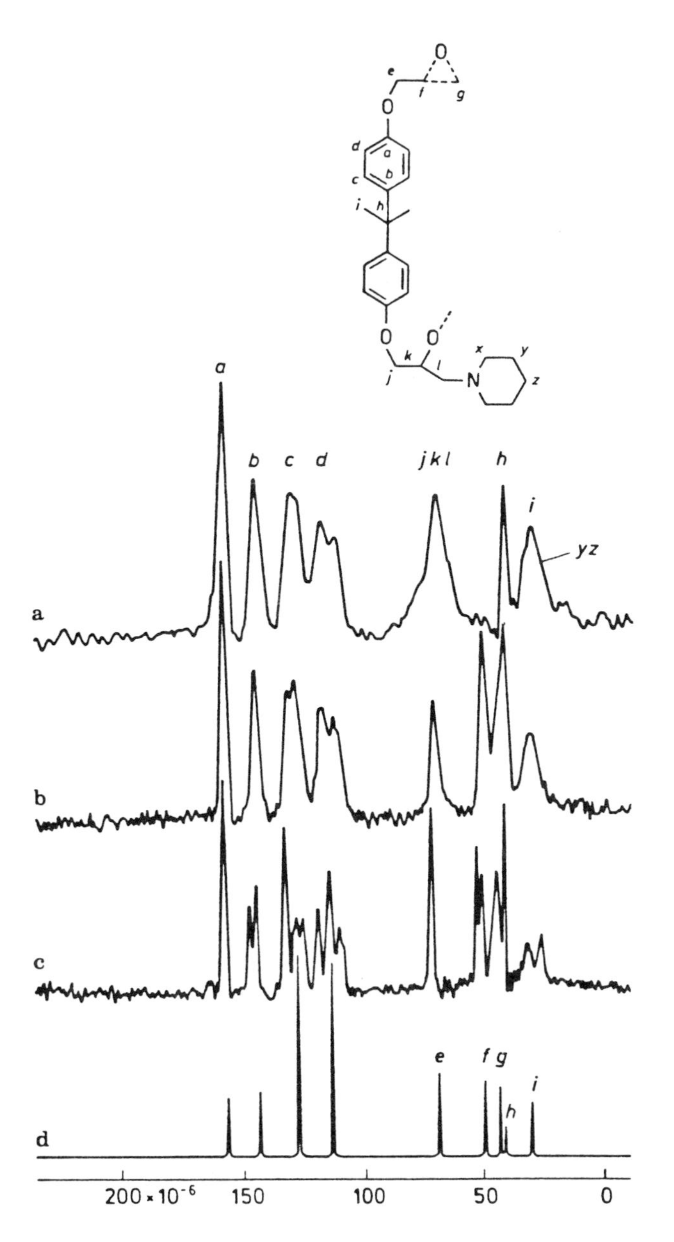

Figure 9.25. ^{13}C NMR spectra of the epoxy resin DGEBA in four different phases: (a) polymerized with 5% (by mass) piperidine, 247 K; (b) amorphous resin, 230 K; (c) polycrystalline resin, 230 K; and (d) in CCl_4, with the solvent peak deleted. (Reproduced with permission from reference 56. Copyright 1981 Academic Press.)

Epoxide Systems

Epoxy resins have been studied extensively with high-resolution solid-state ^{13}C NMR spectroscopy. The epoxy spectra are shown in Figure 9.25 of four phases of DGEBA: the resin cured with 5% piperidine, an amorphous resin, a polycrystalline resin, and the starting resin in the liquid state (*55*). In the spectrum of the polymerized epoxy resin, the opening of the epoxide ring results in the shift of the two epoxide resonances downfield. The aromatic peaks in the spectra of all phases are identical. Because the crystalline environment has high regularity, the resonance lines are narrow. The amorphous phase has a distribution of local environments and exhibits broad lines; these factors limit the detection of structural detail. The polymerized epoxy freezes in the random distribution of orientations, an effect that further decreases the resolution.

The butanediol diglycidyl ether (BDGE) cured with phthalic anhydride has also been studied with high-resolution solid-state ^{13}C NMR spectroscopy (*56*). In this system, the hydroxyl group attacks the carbonyl group of the anhydride, and this attack forces the opening of the anhydride ring to produce a carboxylic acid. Similarly, the hydroxyl group of the acid opens the epoxy ring to form a diester linkage and an alcohol. The ^{13}C NMR spectra confirm such a mechanism, but the resolution in the glassy state is not sufficient to isolate specific resonances that can be assigned.

The tetrafunctional epoxy tetraglycidyl(diaminodiphenyl)methane (TGDDM) and the tetrafunctional amine diaminodiphenyl sulfone (DDS) have also been examined with solid-state ^{13}C NMR spectroscopy (Figure 9.26) (*56*). The chemical structures resulting from the polymerization of TGDDM with DDS are complex; multiple reactions take place during cure. In the CP – MAS ^{13}C spectrum at 38 MHz, resonances due to carbons attached to hydroxy groups, intramolecular ether linkages, and intermolecular linkages are discernible. The peaks resulting from the four different carbons involved with oxygen appear between 80 and 60 ppm in all of the spectra. The carbon attached to the hydroxyl group formed by the epoxy – amine reaction (peak c) appears at 70 ppm. The peak for the two carbons involved in the intramolecular ether groups appears at 68 ppm. Both carbons in the CHOCH structure have the same environment. Therefore, only one peak is seen. The carbons involved in the intermolecular ether linkage contribute two peaks. The extent of reaction for TGDDM and the number of total epoxy junctions can be determined from the NMR experiment.

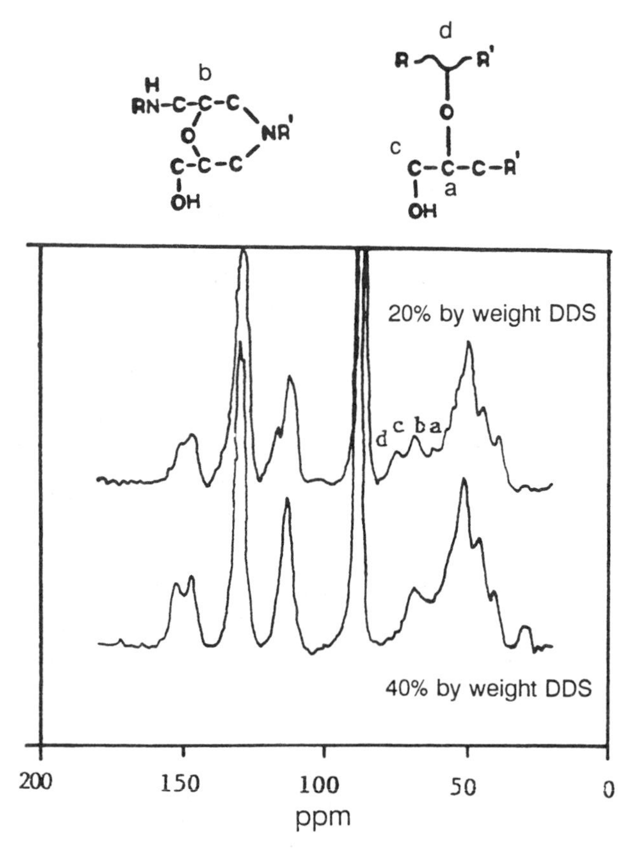

Figure 9.26. The DD – MAS – CP ^{13}C NMR spectra at 37.7 MHz of two concentrations of DDS – TGDDM. (Reproduced from reference 57. Copyright 1988 American Chemical Society.)

NMR Spectroscopy of Surface Species

Polymer composites are high-performance materials with a large range of applications. The reinforcements are typically glass fibers that are often treated with organosilane coupling agents to promote adhesion between the glass fibers and the polymer matrix. The macroscopic properties of composites depend on the nature of the glass fiber, the coupling agent – polymer interface, and the polymer matrix. These structures must be optimized in order to determine the potential of the system. Therefore, the investigation of the nature of these components on the molecular level is useful and necessary for the optimization of performance properties.

Upon the deposition of a hydrolyzed coupling agent onto a silica surface, a chemical reaction takes place, and this reaction leads to the formation of a covalent bond between the two components. Silica surface hydroxyls react with the hydrolyzed alkoxy groups of the organosilane coupling agents in a condensation reaction. In surface studies of this kind, the CP – MAS NMR experiment has the advantage of avoiding interference effects from glass, which is a problem that arises in other surface methods (*57*). Changes in the molecular structure of the material caused by sample preparation are avoided because the bulk material serves as the sample without modification.

The CP – MAS ^{13}C NMR spectra of surface-modified silicas and the corresponding 3-aminopropyltriethoxysilane (γ-APS) spectrum are shown in Figure 9.27 (*58*). When the coupling agent is bound to silica, the peaks corresponding to the carbons of the propyl chains are shifted upfield approximately 0.4 – 2.0 ppm as compared to the same resonances in the bulk polymer. Those carbons in close proximity to the binding sites (especially the α-methylene carbon) tend to show a larger upfield shift. This close proximity leads to steric hindrance of the organosilanes, which causes the chemical shift. The spectrum of condensed γ-APS shows an additional prominent peak arising at 166.0 ppm (as observed in Figure 9.27) that corresponds to the carbon resonance of a carbonyl group. If this peak is attributed to a carbonyl

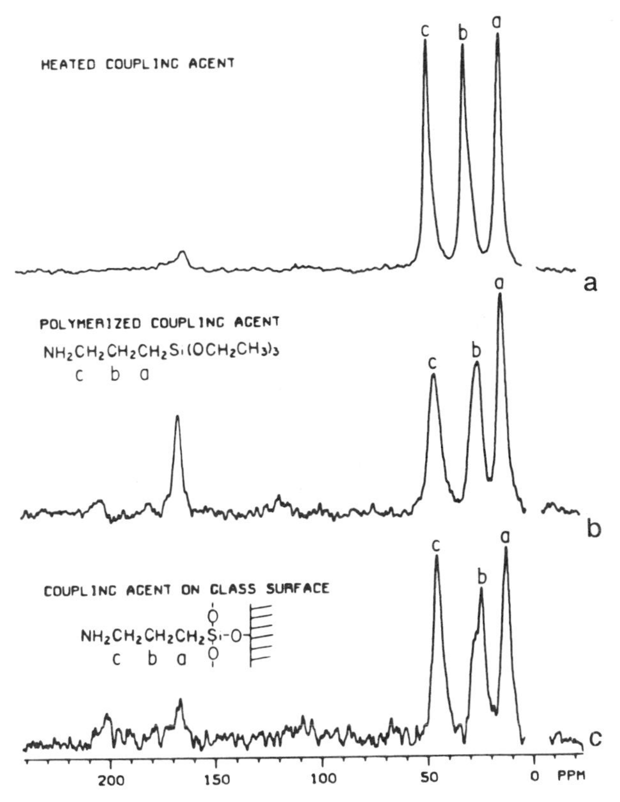

Figure 9.27. The CP – MAS NMR ^{13}C spectra at 37.7 MHz of surface-modified silicas and the corresponding γ-APS. Spectrum a is of the polymerized coupling agent subjected to heat treatment, spectrum b is of the polymerized coupling agent, and spectrum c is of the coupling agent on a silica surface. (Reproduced with permission from reference 58. Copyright 1985 Elsevier Science Publishers.)

group, then the carbonyl is most probably part of the amine – bicarbonate salt formed by a reaction of the free amine groups with carbon dioxide in the air. This reaction can occur with both bulk γ-APS and with γ-APS on the silica surface. The amine – bicarbonate salt is thermally unstable, and with heat treatment at 100 °C, the salt decomposes, and the carbon dioxide is removed from the silica surface. Removal of the carbon dioxide from the bulk polymer requires more intensive heat treatment. When the condensed amine functional coupling agent is subjected to a heat treatment of 200 °C for 12 h in a vacuum oven, the carbonyl-carbon resonance is greatly reduced in intensity as a result of the evolution of carbon dioxide (Figure 9.27a).

The problem with using CP – MAS ^{13}C NMR spectroscopy in the study of interfaces is that the structure of the silane molecule cannot be directly observed, nor can the mechanism by which the silane molecule is bonded to the glass be determined. An obvious approach is to use ^{29}Si NMR spectroscopy because this method is superior to ^{13}C NMR spectroscopy for the study of silanes because it is sensitive to the siloxane network structures. ^{29}Si CP – MAS has been used to study silane coupling agents and their surface reactions (*59, 60*). In one ^{29}Si NMR study, polyamide-6 (PA6) was used as the matrix, γ-APS was used as the coupling agent, and glass microspheres were used as the filler material (*59*). With CP for ^{29}Si, only nuclei that are near protons will be observed, and therefore silicon nuclei in the bulk glass will not be detected. The solid-state CP – MAS ^{29}Si NMR spectrum of pure glass microspheres is shown in Figure 9.28 (*59*). One intense central line at – 98.6 ppm (Q^3) has two side peaks at – 89 ppm (Q^2) and at – 109.6 ppm (Q^4). After treatment with γ-APS, a line denoted S_3 is observed at – 67.7 ppm and is assigned to ^{29}Si bonded to three other Si atoms (either on the surface or in a poly(γ-APS) polymer via oxygen bridges) with a fourth bond to the aminopropyl substituent. The – 59.5-ppm peak, denoted S_2, is due to Si bonded to two other Si atoms via oxygens, and to one hydroxyl group and an aminopropyl group. The high intensity of the S_3 peak indicates that the polysiloxane is highly cross-linked.

^{29}Si NMR spectra of bulk hydrolyzed γ-APS and silica treated with different amounts of γ-APS are shown in Figure 9.29a (*60*). The resonances are assigned as shown in Figure 9.29b. In addition, the resonances at – 100 and – 110 ppm are due to $(\equiv Si-O)_3Si-OH$ and $(\equiv Si-O)_4Si-OH$ moieties of the silica surface. No peak is observed at – 45 ppm, which is the ^{29}Si chemical shift of liquid APS. For the bulk hydrolyzed γ-APS, the line widths are narrow, and the resonance at – 58 ppm appears. For the 100% APS-treated silica, the resonances are much broader, and the band at – 59 ppm appears as a shoulder. The results indicate that CP – MAS ^{29}Si NMR spectroscopy of the solid state is a very powerful tool for studies of silane coupling agents.

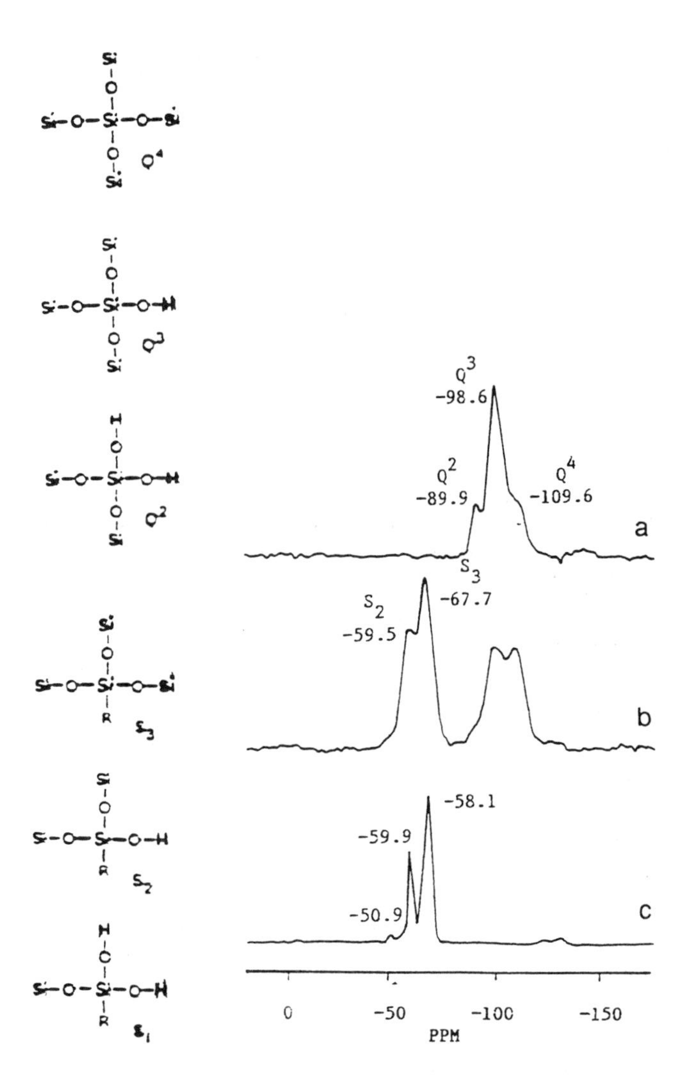

Figure 9.28. The CP–MAS ^{29}Si NMR spectra at 59.6 MHz of pure glass microspheres (a), glass microspheres pretreated with 8.4% γ-APS (b), and poly(γ-APS) (c). (Reproduced from reference 59. Copyright 1989 American Chemical Society.)

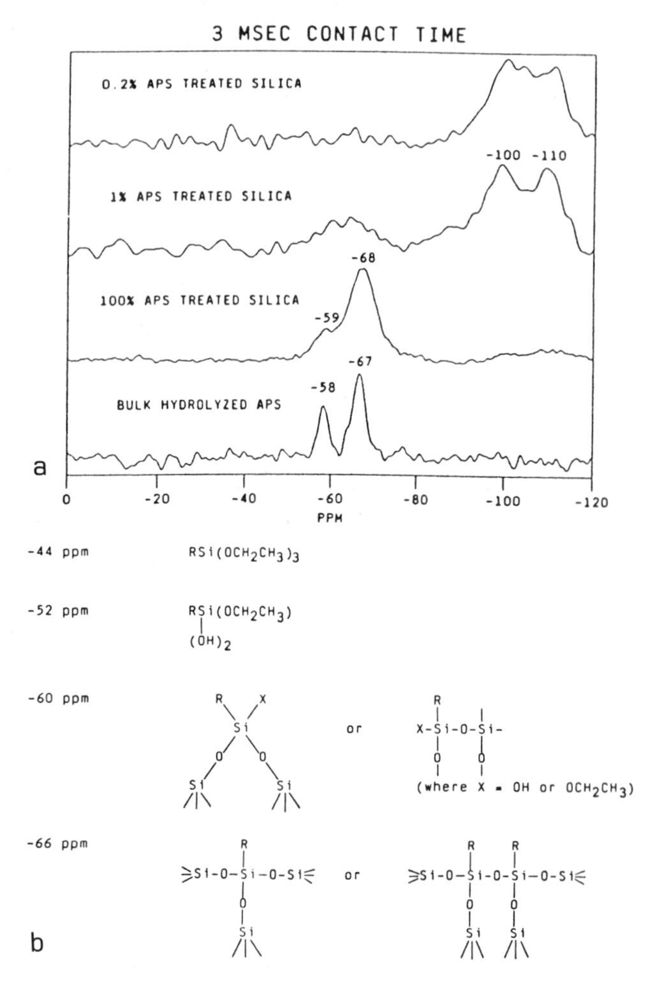

Figure 9.29. (a) The CP–MAS ^{29}Si NMR spectra at 59.6 MHz of bulk hydrolyzed APS, silica treated with 100% APS, silica treated with a 1% solution of APS, and silica treated with a 0.2% solution of APS. (b) Structures assigned to the CP–MAS ^{29}Si NMR chemical shifts of γ-APS. (Reproduced with permission from reference 60. Copyright 1990 Society of Plastic Engineers.)

References

1. Dybowski, C. *Chem. Tech.* **1986,** *15,* 186.
2. Jelinski, J. W. *Chem. Tech.* **1986,** 312.
3. Belfiore, L. A.; Schilling, F. C.; Tonelli, A. E.; Lovinger, A. J.; Bovey, F. A. *Macromolecules* **1984,** *17,* 2561.
4. Earl, W. L.; VanderHart, D. L. *Macromolecules* **1979,** *12,* 762.
5. Tonelli, A. E. *Macromolecules* **1979,** *12,* 255.
6. Grant, D. M.; Cheney, B. V. *J. Am. Chem. Soc.* **1967,** *89,* 5315.
7. Garroway, A. N.; Ritchey, W. M.; Moniz, W. B. *Macromolecules* **1982,** *15,* 1051.
8. Patterson, D. W.; Koenig, J. L. *Polymer* **1987,** *200,* 268.
9. Schaefer, J.; Stejskal, E.; Buchdahl, A. *Macromolecules* **1977,** *10,* 384.
10. Ando, I.; Yamanobe, T.; Sorita, T.; Komoto, T.;

Sato, H.; Deguchi, K.; Imanari, M. *Macromolecules* **1984,** *17,* 1955.

11. Drotloff, H.; Emeis, D.; Waldron, R. F.; Moller, M. *Polymer* **1987,** *28,* 1200.

12. Yamanobe, T.; Sorita, T.; Ando, I. *Makromol. Chem.* **1985,** *186,* 2071.

13. Yamanobe, T.; Sorita, T.; Komoto, T.; Ando, I.; Sato, H. *J. Mol. Struct.* **1985,** *131,* 267.

14. Bunn, A.; Cudby, M. E. A.; Harris, R. K.; Packer, K. J.; Say, B. J., *Polymer* **1982,** *23,* 694.

15. Aujla, R. S.; Harris, R. K.; Packer, K. J.; Parameswaran, M.; Say, B. J. *Polym. Bull.* **1982,** *8,* 253.

16. Gomez, M. A.; Tanaka, H.; Tonelli, A. E. *Polymer* **1987,** *28,* 2227.

17. Bunn, A.; Cudby, M. E. A.; Harris, R. K.; Packer, K. J.; Say, B. J. *J. Chem. Soc., Chem. Commun.* **1981,** 15.

18. Belfiore, L. A.; Schilling, F. C.; Tonelli, A. E.; Lovinger, A. J.; Bovey, F. A. *Macromolecules* **1984,** *17,* 2561.

19. Ferro, D. R.; Ragazzi, M. *Macromolecules* **1984,** *17,* 485.

20. Kurosu, H.; Komoto, T.; Ando, I. *J. Mol. Struct.* **1988,** *176,* 279.

21. Cholli, A. L.; Ritchey, W. M.; Koenig, J. L. *Spectrosc. Lett.* **1983,** *16(1),* 21.

22. Earl, W. L.; VanderHart, D. L. *J. Magn. Reson.* **1980,** *48,* 35.

23. Perry, B. C.; Koenig, J. L.; Lando, J. B. *Macromolecules* **1987,** *20,* 422.

24. Gomez, M. A.; Cozine, M. H.; Tonelli, A. E. *Macromolecules* **1988,** *21,* 388.

25. Taki, T.; Yamashita, S.; Satoh, M.; Shibata, A.; Yamashita, T.; Tabeta, R.; Saito, H. *Chem. Lett. Chem. Soc. Jpn.* **1981,** 1803.

26. Garroway, A. N.; Moniz, W. B.; Resing, H. A. *Org. Coat. Plast. Chem.* **1976,** *36,* 133.

27. Garroway, A. N.; Moniz, W. B.; Resing, H. A. In *Carbon-13 NMR in Polymer Science*; Wallace, M.P., Ed.; ACS Symposium Series 103; American Chemical Society: Washington, DC, 1979; p 67.

28. Sefcik, M. D.; Stejskal, E. O.; McKay, R. A.; Schaefer, J. *Macromolecules* **1979,** *12,* 423.

29. Patterson, D. W.; Shields, C. M.; Cholli, A.; Koenig, J. L. *Polym. Prepr.* **1984,** *25,* 358.

30. Brown, C. E.; Khoury, I.; Bezoari, LM. D.; Kovacic, P. *J. Polym. Sci., Part A: Polym. Chem.* **1982,** *20,* 1697.

31. Jelinski, L. W.; Dumais, J. J.; Watnick, P. I.; Bass, S. V.;Shepherd, L. *J. Polym. Sci., Part A: Polym. Chem.* **1982,** *20,* 3285.

32. Maciel, G. E.; Chuang, I-S.; Myers, G. E. *Macromolecules* **1982,** *15,* 121.

33. Bryson, R. L.; Hatfield, G. R.; Early, T. A.; Palmer, A. R.; Maciel, G. E. *Macromolecules* **1983,** *16,* 1669.

34. Patterson, D. J.; Koenig, J. L.; Shelton, J. R. *Rubber Chem. Technol.* **1983,** *56,* 971.

35. Haw, J. F.; Maciel, G. E. *Anal. Chem.* **1983,** *55,* 1262.

36. Bauer, D. R.; Dickie, R. A.; Koenig, J. L. *J. Polym. Sci., Part B: Polym. Phys.* **1984,** *22,* 2009.

37. Bauer, D. R.; Dickie, R. A.; Koenig, J. L. *Ind. Eng. Chem. Prod. Res. Dev.* **1984,** *24,* 121.

38. Maciel, G. E.; Chuang, I-S.; Gollob, L. *Macromolecules* **1984,** *17,* 1081.

39. Kaplan, S.; Dilks, A. *J. Appl. Polym. Sci.: Appl. Polym. Symp.* **1984,** *38,* 105.

40. Andreis, M.; Koenig, J. L. *Adv. Polym. Sci.* **1989,** *89,* 71.

41. Zaper, A. M.; Koenig, J. L. *Rubber Chem. Technol.* **1987,** *60,* 278.

42. Zaper, A. M.; Koenig, J. L. *Rubber Chem. Technol.* **1987,** *60,* 252.

43. Zaper, A. M.; Koenig, J. L. *Makromol. Chem.* **1988,** *189,* 1239.

44. Andreis, M.; Liu, J.; Koenig, J. L. *J. Polym. Sci.; Part B: Polym. Phys.* **1989,** *27,* 1389.

45. Andreis, M.; Koenig, J. L. *Rubber Chem. Technol.* **1989,** *62,* 82.

46. Clough, R. S.; Koenig, J. L. *Rubber Chem. Technol.* **1989,** *62,* 908.

47. Krejsa, M.; Koenig, J. L. *Rubber Chem. Technol.* **1991,** *62,* 908.

48. Smith, S.; Koenig, J. L. *Rubber Chem. Technol.* (In press)

49. Koenig, J. L.; Patterson, D. J. *Elastomers Rubber Technol.* **1987,** *32,* 31.

50. Komoroski, R. A.; Schockcor, J. P.; Gregg, E. C.; Savoca, J.L. *Rubber Chem. Technol.* **1986,** *59,* 328.

51. Patterson, D. J.; Koenig, J. L. In *Characterization of Highly Cross-linked Polymers;* Labana, S. S.; Dickie, R. A., Eds.; ACS Symposium Series 243; American Chemical Society: Seattle, Washington, 1984; pp 205-232.

52. Barron, P. F.; O'Donnell, J. H.; Whittaker, A. K. *Polym. Bull.* **1985,** *14,* 339.

53. Curran, S. A.; Padwa, A. R. *Macromolecules* **1987,** *20,* 625.

54. Cholli, A. L.; Ritchey, W. M.; Koenig, J. L. *Appl. Spectrosc.* **1987,** *41,* 1418.

55. Horii, F.; Zhu, Q.; Kitamaru, R.; Yamaoka, H. *Macromolecules* **1990,** *23,* 977.

56. VanderHart, D. L.; Earl, W. L.; Garroway, A. N. *J. Magn. Reson.* **1981,** *44,* 361.

57. Mertzel, E.; Perchak, D.; Ritchey, W.; Koenig, J. L. *Ind. Eng. Chem. Res.* **1988,** *27,* 580.

58. Zaper, A. M.; Koenig, J. L. *Adv. Colloid Interface Sci.* **1985,** *22,* 113.

59. Weeding, T. L.; Veeman, W. S.; Jenneskens, L. W.; Gaur, H. A.; Schuurs, H. E. C.; Huysmans, W. G. B. *Macromolecules* **1989,** *22,* 706.

60. Hoh, Ka-Pi; Ishida, H.; Koenig, J. L. *Polym. Compos.* **1990,** *11,* 121.

Suggested Reading

Axelson, D. E. *Solid State Nuclear Magnetic, Resonance of Fossil Fuels: An Experimental Approach;* Multiscience, 1983.

Fyfe, C. A. *Solid State NMR for Chemists;* C. F. C. Press: Guelph, 1984.

High Resolution NMR of Synthetic Polymers in Bulk; Komoroski, R. A., Ed.; VCH Publishers: Deerfield Beach, 1986.

Solid State NMR of Polymers; Mathias, L., Ed.; Plenum: New York, 1991.

10

NMR Relaxation Spectroscopy of Polymers

Relaxation Processes in NMR Spectroscopy

> *Molecular motions of the polymer modulate the dipolar interactions on which all NMR relaxation parameters depend.*
>
> – J. Schaefer et al. (*1*)

Relaxation techniques in NMR spectroscopy yield insight into both structure and molecular dynamics. Each nucleus experiences the magnetic field of its neighbors, and the extent of the interaction depends on the internuclear separations. Molecular motion modulates these magnetic interactions and causes relaxation. Details of the interacting environments of the nuclear spins can be obtained from measurements of the nuclear-relaxation parameters. Of particular interest is the identification of the molecular processes involved in mechanical transitions and energy-dissipation processes of bulk polymers using NMR relaxation measurements.

The relaxation parameters that can be measured include:

- the ^{13}C and ^{1}H spin – lattice relaxation times ($^{C}T_1$ and $^{H}T_1$)
- the spin – spin relaxation time (T_2)
- nuclear Overhauser enhancement (NOE)
- the proton and carbon rotating-frame relaxation times ($^{C}T_{1\rho}$ and $^{H}T_{1\rho}$)
- the C – H cross-relaxation time (T_{CH})
- the proton relaxation time in the dipolar state (T_{1D}).

The ^{13}C NMR relaxation technique allows each carbon of the chain to be observed separately and provides information on the motion of the side groups as well as the motion of the main-chain functional groups of the polymer. The relaxation processes of the nuclear spins act as discriminators, coupling only those transitions whose energies can be matched by the lattice or the spin environment. Fortunately, these relaxation discriminators can be tuned to energies in the range of $10^{-25} - 10^{-15}$ ergs, which corresponds to a frequency range of $10^{2} - 10^{11}$ Hz. The time-scale range for NMR relaxation measurements is from picoseconds to days. The range of correlation times, τ_c, covered by the various NMR relaxation experiments is shown in Figure 10.1 in a schematic representation of the types of NMR experiments and their respective ranges of τ_c (*2*). Spin – lattice relaxation and nuclear Overhauser en-

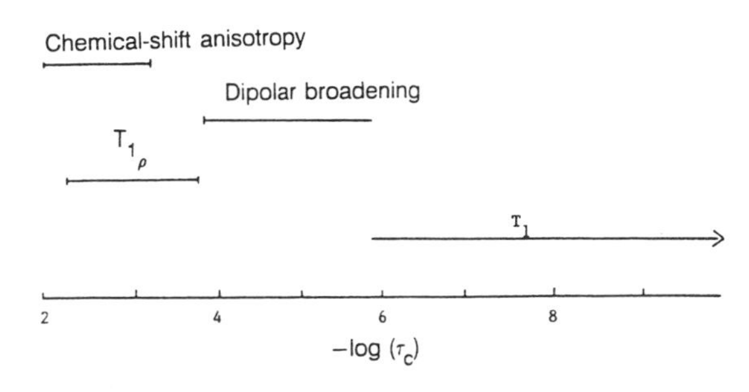

Figure 10.1. A schematic representation of some types of solid-state ^{13}C NMR experiments and the approximate motional frequencies that they probe. (Reproduced from reference 2. Copyright 1985 American Chemical Society.)

1904—4/92/0257$08.00/1

hancement (NOE) measurements are sensitive to spectral densities in the megahertz frequency range (at 500 MHz, $\omega - 1 = 3 \times 10^{-10}$ s). The dipolar-broadening phenomena give information about motions having correlation times of approximately $10^{-5} - 10^{-6}$ s, and the chemical-shift anisotropy and spin–lattice relaxation in the rotating frame can be used to obtain information in the mid-kilohertz range. Motions in the low-kilohertz range can be measured from the broadenings of the resonances that occur during the revolutions of the magic-angle-spinning rotor.

The quantities characterizing the molecular motion, including rate and activation energy, are determined from NMR relaxation experiments. ^{13}C NMR studies also allow the selective observation of a chosen resonance line so that intra- or intermolecular couplings can be observed. Certain NMR pulse sequences isolate the appropriate relaxation parameters and yield specific information about the molecular dynamics in polymers. When variable-temperature measurements are made of these various relaxation parameters, the power of the NMR technique is considerably improved.

Relaxation Processes and Their Origins

The various NMR relaxation parameters will be discussed in terms of their origins, their experimental measurement, and the nature of the results obtained.

The Spin–Lattice Relaxation Process

Spin–lattice relaxation involves a loss of potential energy by the excited nucleus to the surrounding environment or *lattice*. Each nucleus is stimulated to lose its energy in a single quantum jump. Therefore, the potential energy of the population of nuclei will decrease exponentially with time as the number of nuclei remaining in the higher energy level decreases. The rate at which this relaxation process occurs is measured by the spin–lattice relaxation time ($1/T_1$ = rate). The value for T_1 represents the time necessary for the magnetization to return to within $1 - 1/e$, or 63%, of its magnitude at equilibrium.

The important feature of the spin–lattice relaxation process is its dependence on the nature of the molecular motion. The couplings depend on the fluctuating local magnetic fields produced by the motion at the nuclei. The frequency range of motion available for the study by T_1 mechanisms is governed by the precessional or Larmor frequencies. The closer the motional frequencies of the lattice are to the Larmor frequencies of the nuclei, the more efficient the transfer of the magnetization (that is, the shorter the T_1). Thus, T_1 measurements are limited to molecular motions in the 10^7-Hz frequency range. The use of T_1 applies to only a few types of motions, primarily group molecular motions at high frequencies.

For proton NMR spectroscopy, the nuclei are highly coupled, and spin diffusion plays an important role in the relaxation process and can override the effects of molecular motions (*see* the section titled "Interpretation of Polymer Relaxations" in this chapter). For ^{13}C NMR spectroscopy, the results are quite different. Spin diffusion is slow for ^{13}C because of the isolation of the nuclei from each other as a result of the low natural abundance of ^{13}C.

> *The laboratory-frame carbon-13 spin–lattice relaxation times,* T$_1$, *can be directly related with spin–lattice processes (i.e., molecular motions) since spin diffusion and spin–spin interactions are not expected to contribute to the carbon-13 relaxation.*
>
> –E. M. Menger et al. (*3*)

For high-resolution solid-state ^{13}C NMR spectroscopy, the $^{C}T_1$ for the different carbons can be determined, and the molecular motions of the individual carbon atoms in the polymer can be compared.

To measure the spin–lattice relaxation times, the sample magnetization is first prepared in a specified manner, and is then allowed to relax for some time period that is incremented during the course of the experiment.

The inversion-recovery sequence consists of a 180° preparation pulse (to invert the magnetization), followed by a variable time interval (for recovery to occur by relaxation), and then a 90° read pulse (*4*). This pulse sequence is written $-(180°_z - t_1 - 90°_x - \text{PD})$, where t_1 is the variable time and PD is the pulse-delay time between repetitious pulse sequences. The $180°_z$ pulse inverts the magnetization along the z axis so that the magnetization is initially negative. With an increase in time, the magnetization in the negative z axis decreases and passes through zero, and then grows toward its equilibrium value by the spin–lattice relaxation process. The read pulse

($90°_x$ pulse) rotates the magnetization into the y axis for detection. At the end of the inversion-recovery sequence, a resonance may be positive or negative, depending on the value of its T_1. The delay between pulses should be sufficiently long for the spin system to relax between adjacent sequences of pulses and should be 3 – 5 times greater than the longest T_1 in order to observe all of the resonances (*5*).

The magnitude of the magnetization along the z axis at a time t_1 following the 180° pulse is given by

$$M(t_1) = M_o(1 - 2e^{-t_1/T_1}) \qquad (10.1)$$

where M_o is the initial magnetization. The individual T_1s for each spectroscopically resolved carbon are measured by using a semilog plot of the magnetization vs. the time interval. A three-parameter curve-fitting method is used to calculate T_1 from the intensities of the peak maxima and minima (*6*).

In the saturation-recovery method, a magnetic saturation is generated and then allowed to evolve (*7*). In progressive saturation, the evolution period contains a train of 90° pulses separated by a delay time. Once a steady-state magnetization is established, the pulse train is terminated, and the transverse magnetization is detected (*8, 9*).

In order to make ${}^{C}T_1$ measurements for solids, artifacts generated by the inversion-recovery method must be suppressed. T_1 measurements for solids are made by using a pulse sequence that was developed by Torchia (*10*), and that allows cross-polarization (CP) enhancement of the signals. The pulse sequence developed by Torchia begins with a pulse that rotates the proton polarization into the xy plane. The H_1 field is then phase shifted to spin lock the protons in the rotating frame. The ^{13}C nuclei are brought into contact by applying an rf field that satisfies the Hartmann – Hahn condition. After the desired ^{13}C polarization is established, contact is broken by turning off the proton rf field. The carbon field is then phase shifted by 90°, which rotates the carbon magnetization from the x axis to the z axis, at which time the carbon rf field is turned off. The proton-enhanced magnetization changes exponentially from its initial value to its equilibrium value with a time constant equal to ${}^{C}T_1$. Transient NOE effects are suppressed by saturating the proton magnetization during the entire delay time.

Spin – Spin Relaxation Times

The spin – spin relaxation time, T_2, involves the decay of the transverse magnetization to zero by dephasing. This spin – spin relaxation process occurs because of the fluctuating local dipolar magnetic fields that arise from neighboring spins. The magnitudes of these dipolar fields depend on the relative orientations and positions of the neighboring spins. These variations in the local field cause the transverse components of magnetization, which are initially in phase or coherent following the 90° pulse, to become dephased or incoherent in a time scale of exp $(-t/T_2)$. In practice, the rate of decay is faster than that indicated by the intrinsic T_2 because of the inhomogeneity in the static magnetic field. This inhomogeneity causes some nuclei to precess more rapidly than the average precession rate and causes other nuclei to precess more slowly. This dephasing due to the magnetic-field inhomogeneity can be nullified by a spin-echo experiment.

The T_2 is related to the inverse of the resonance line width of an NMR signal. The Fourier transform of the decay in the time domain is the frequency spectrum, and a faster decay in the time domain results in broader resonance line widths.

The T_2 is measured by using a Carr – Purcell pulse sequence. A 180° pulse is applied at a time τ_2 after the 90° pulse to reverse the decay, and this pulse results in a refocused magnetization and an echo maximum at time $2\tau_2$. For solids, this Carr – Purcell pulse sequence results in a slight cross-polarization with each 180° pulse, which makes the T_2 measurements invalid. To correct this problem, a new variable time period, which is equal to an even number of rotor periods, is added after the cross-polarization period. The protons are decoupled during this variable time period. A 180° pulse is applied to the carbons exactly halfway through the interval, and the free induction decay (FID) is recorded with decoupling so that the observed signal is the second half of a rotation-synchronized echo (*11*).

In the absence of molecular motion, T_2 is inversely related to the width of the frequency, $\Delta\nu$.

$$\frac{1}{T_2} = \Delta\nu = (\Delta M_2)^{\frac{1}{2}} \qquad (10.2)$$

where M_2 is the second moment of the absorption spectrum. The term M_2 provides a measure of the strength of the static dipolar interactions and is given by

$$M_2 = (3/5)\gamma^4 h^2 I (I + 1) \Sigma_{ij} r_{ij}^{-6} \quad (10.3)$$

where I is the nuclear spin quantum number, h is Planck's constant, r is bond length, and i and j are nuclei i and j, respectively.

Molecular motion causes the dipolar interactions to be time-dependent, which gives rise to the familiar motional narrowing of resonance lines. The faster the motion, the narrower the line widths.

Spin – Lattice Relaxation in the Rotating Frame

The spin – lattice relaxation time in the rotating frame, $T_{1\rho}$, is similar to T_1 in the static magnetic field. The spin – lattice relaxation time is the rate of relaxation due to spin – lattice interactions in the rotating frame. Because the spin states in the rotating frame are separated by energies in the kilohertz range, fluctuating local fields with components in this frequency region induce the necessary transitions for relaxation in the rotating frame.

As a result of fluctuations in the component of the local field along the H_o direction, transitions that cause relaxation are induced if the fluctuations occur at the frequency γH_1. Figure 10.2 shows the dependence of $T_{1\rho}$, T_1, and T_2 on the molecular correlation times for relaxation as determined by only dipole – dipole interactions (*12*).

The measurement of $^{C}T_{1\rho}$ can be used to characterize motions in solid polymers and to evaluate the chain dynamics in the 10 – 100-kHz frequency range. $^{C}T_{1\rho}$ relaxations are not influenced by spin diffusion because of the small dipolar coupling that arises from the low natural abundance and large separation of the nuclei. Under these conditions, differences in chain motions can be analyzed. This analysis is done by measuring the dispersion in $^{C}T_{1\rho}$.

A simplistic graphical interpretation of the motional information that is available from $^{C}T_{1\rho}$ is presented in Figure 10.3 (*13*). The spectral density of molecular motion is plotted as a function of frequency for a hypothetical polymer in two different states. At the frequency of the carbon rf field, the motion in the two states can be monitored by its effect on the $^{C}T_{1\rho}$ relaxation.

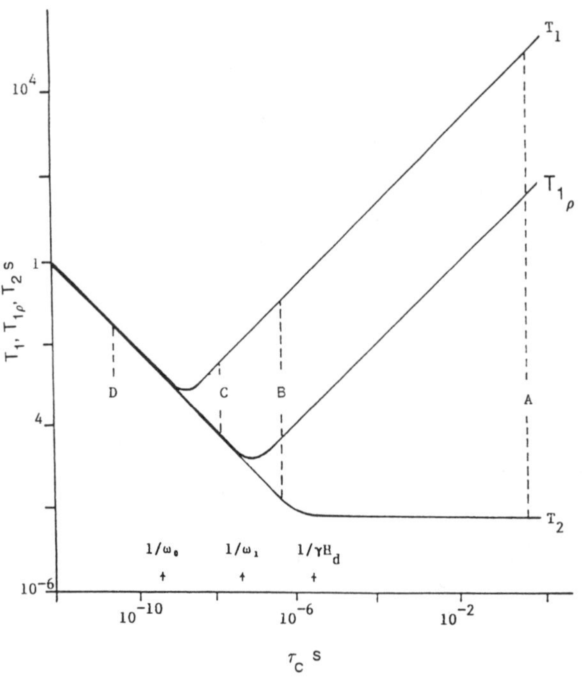

Figure 10.2. Schematic representation of the dependence of relaxation times on the molecular correlation time, τ_c, for relaxation determined by dipole – dipole interactions. Ordinate and abscissa values are approximate. $\omega = \gamma H_o$, $\omega_I = \gamma H_I$, and H_d is the dipolar magnetic field. (a) Rigid lattice, $T_I >> T_{I\rho} >> T_2$; (b) nonrigid solid, $T_I >> T_{I\rho} >> T_2$; (c) viscous liquid, $T_I > T_{I\rho} = T_2$; (d) nonviscous liquid, $T_I = T_{I\rho} = T_2$. (Reproduced with permission from reference 12. Copyright 1971 Academic Press, Inc.)

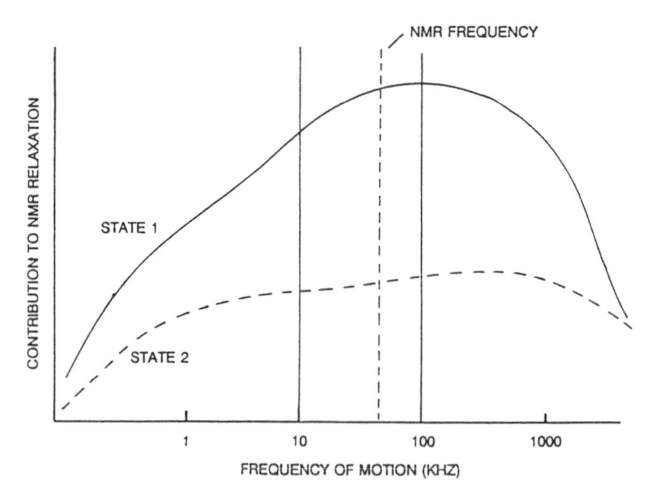

Figure 10.3. The graph of the spectral density of molecular motion as a function of frequency for a hypothetical polymer in two different states. (Reproduced with permission from reference 13. Copyright 1983 Society for Applied Spectroscopy.)

The interpretation of the $^{C}T_{1\rho}$ data is influenced by the fact that the static $^{1}H-^{1}H$ spin–spin processes can contribute to the relaxation and shorten $^{C}T_{1\rho}$, which complicates the information on molecular motion (*14*). Proton–proton spin flips (spin diffusion), which are independent of molecular motion, can cause a decay in the magnetization, especially when the carbon rf field is less than an order of magnitude greater than the local field. For carbons in semicrystalline polymers such as polyethylene, poly(oxymethylene), and poly(ethylene terephthalate), the mutual spin flips dominate $^{C}T_{1\rho}$. For carbons in glassy polymers, in which strong $^{1}H-^{1}H$ static dipolar interactions are absent, the dominant mechanism for $^{C}T_{1\rho}$ is spin–lattice (motional) in character (*15*). The relative contribution to $^{C}T_{1\rho}$ by these two different processes can be measured experimentally. One method is to determine the rf-field dependence of the observed $^{C}T_{1\rho}$ (*14*). An exponential static-field dependence indicates that the spin–spin mechanism dominates, and a weaker dependence suggests that molecular motion is important.

The second method, magic-angle spinning (MAS), can perturb the values of $^{C}T_{1\rho}$ in a number of ways. Spinning places the sample under dilational stresses that can alter molecular motions. MAS modulates the spin–spin interactions and adds mechanical motion to the spin–lattice interactions (*15*).

The $^{C}T_{1\rho}$ is measured from the decay of the rotating-frame ^{13}C magnetization in a resonant rf field. The normal cross-polarization has a time delay during which the proton rf field is turned off, and this delay is introduced after the carbons are pulsed into the rotating frame. The time interval allows the carbon magnetization to decay by dipolar coupling. The rate of decay of the spin-locked carbon magnetization during this period is characterized by the $^{C}T_{1\rho}$ as:

$$M(t) = M_o \exp\left(\frac{-\tau}{^{C}T_{1\rho}}\right) \qquad (10.4)$$

where $M(t)$ is the resonance at time t, and M_o is the resonance intensity at time $t = 0$.

Plots of the $^{C}T_{1\rho}$ values are often nonlinear. The nonlinearity can usually be interpreted as arising from a multiplicity of relaxation times. For glassy polymers, the inhomogeneity in the molecular packing can be the source of the variations because for semicrystalline systems the nonlinearity effects are usually negligible. Consequently, the measurement of $^{C}T_{1\rho}$s for glassy systems is accomplished by estimating the average $^{C}T_{1\rho}$ from the value of the initial slope (*16*).

In general, $^{C}T_{1}$ for glassy amorphous polymers is primarily determined by molecular motion, but for crystalline polymers, the strong proton–proton dipolar interactions and the restrictions on chain movement make the $^{C}T_{1\rho}$ more indicative of spin–spin flips (diffusion).

Motional Basis of the Relaxation Processes

The relaxation rates are governed by two important factors: the strength of local magnetic interactions between nuclei and the molecular motion. For polymers, the predominant relaxation mode occurs through segmental motion and results in intramolecular reorientations, which create a randomly varying field at the site of the observed nucleus.

Near-neighbor contributions give rise to variable *local fields* at each nucleus. Contributions to the local fields can arise from nuclear and electron dipoles, electric quadrupoles, and anisotropies in the chemical-shielding tensors. In synthetic polymers, the nuclear dipole–dipole interactions, which depend on the motions of magnetic dipoles, are the primary sources of relaxation.

Consider the case of an isolated C–H bond. The instantaneous magnetic field, h_{loc}, produced at the ^{13}C by the proton nucleus is given by

$$h_{loc} = \pm \mu_H \left(\frac{3\cos^2 \phi_{CH} - 1}{r_{CH}^{3}}\right) \qquad (10.5)$$

where μ_H is the magnetic dipole moment of the proton, r_{CH} is the nuclear separation, and ϕ_{CH} is the angle of the vector $\mathbf{r}_{CH}$ relative to the external field. The ($\pm$ sign indicates that h_{loc} can add to or subtract from H_o. Thus the local field experienced by the ^{13}C nucleus for a static, isolated C–H bond with this dipolar interaction is $H_o \pm h_{loc}$.

In a solid, a spatial distribution of the C–H bonds produces a range of local fields. This distribution yields the broad lines normally associated with the spectra of solids. In addition, intramolecular reorientations through molecular motion impart a time dependence on h_{loc} through variations in ϕ_{CH}, and intermolecular diffusion and rotation give a time dependence to both ϕ_{CH} and r_{CH}. For motion that is rapid on the time scale of the dipolar interaction (>50 kHz), h_{loc} reduces to its average value of zero and produces narrow lines. The sharp NMR lines observed experimentally for polymer solutions reflect this result. For solids, the local fields have a broad distribution, and motion only partially averages the field, so broad resonance lines are observed.

Nuclei in solids are subjected to four major magnetic interactions that are inherently anisotropic in nature: dipolar coupling, chemical-shielding anisotropy, quadrupole coupling, and scalar coupling. Molecular motion influences each of these interactions differently to produce specific relaxation effects and correlation times.

The dominant mechanism for relaxation in polymers is the dipole – dipole interaction. The relaxation of each carbon is determined primarily by the fluctuations of the C – H vectors with respect to the external magnetic field. The observables are determined by the spectral density (the power spectrum of the local magnetic field) at the relevant frequencies. The effects of molecular motion are generally expressed in terms of the correlation time, τ_c.

> *One way of thinking of the correlation time for motion is as a coherence or memory time. This parameter measures the time required for the loss of memory of a particular spatial location as a result of motion.*
>
> – J. Schaefer et al. (*1*)

The τ_c changes or modulates the dipolar coupling and is determined by the fastest of the following processes:

- rotation of dipoles relative to each other (molecular rotation)
- a reversal of sign of one of the dipoles (a spin-relaxation process)
- a distance change (dissociation)

If the motion involved is rotational, τ_c is the time during which the average angular displacement is 1 rad. For a spherical molecule in solution in an environment of viscosity η, the rotational correlation time, τ_c, can be estimated on the basis of the Einstein – Stokes equation:

$$\tau_c = \frac{\eta \pi D_m{}^3}{6kT} = \frac{V_m \eta}{kT} \quad (10.6)$$

where V_m is the molecular volume, D_m is the molecular diameter, k is the Boltzmann constant, and T is the temperature.

Molecular motion can be represented by an autocorrelation function, $G(t)$, which is expressed in terms of a scalar product of the local field, $h_{loc}(t)$, and the same local field at an earlier time, $h_{loc}(0)$

$$G(t) = h_{loc}(t) \cdot h_{loc}(0) \quad (10.7)$$

The autocorrelation function describes the persistence of the local field relative to the *averaging* process induced by the molecular motion. This process can be described by recognizing that

$$G(t, \tau_c) = \langle h_{loc}(t) \cdot h_{loc}(t + \tau_c) \rangle \quad (10.8)$$

where the time dependence is produced by the molecular motion, and the average is taken over the nuclear-spin system at time t. The autocorrelation function describes the time scale for the decay of inherent motional order in the system.

> *The* G(τ_c) *provides information on the average manner in which the collection of spins (or the collection of vectors between interacting nuclear dipoles) moves about.*
>
> – J. Lyerla (*17*)

The autocorrelation function is generally assumed to be exponential and independent of the time origin. Thus,

$$\begin{aligned} G(t, \tau_c) &= \langle h_{loc}(t) \cdot h_{loc}(t + \tau_c) \rangle_t \\ &= \exp(-t/\tau_c) \end{aligned} \quad (10.9)$$

The spectral density, $J(\omega)$, is the spectrum in the frequency domain corresponding to the autocorrelation function, $G(t)$, in the time domain. So $J(\omega)$ is the Fourier transform of $G(t)$. Therefore,

$$J(\omega) = \int G(\tau_c) \exp(-i\omega\tau_c)\, d\tau_c \quad (10.10)$$

Integration of this equation requires a description of $G(\tau_c)$. If the motional-decay process is exponential, $G(\tau_c)$ becomes

$$G(\tau_c) = \langle h_{loc}^2(0)\rangle e^{(-t/\tau_c)} \quad (10.11)$$

so

$$J(\omega) = \int \langle h_{loc}^2(0)\rangle e^{(-t/\tau_c)}\, e^{(-i\omega\tau)}\, d\tau \quad (10.12)$$

Evaluation of this integral yields

$$J(\omega) = \frac{2\langle h_{loc}^2(0)\rangle \tau_c}{(1 + \omega^2\tau_c^2)} \quad (10.13)$$

$J(\omega)$ is a maximum at $\omega = 0$ and is approximately constant over the range of ω such that $\omega^2\tau_c^2 << 1$ and begins to fall off as ω goes to $1/\tau_c$. The area under the $J(\omega)$ vs. ω curve is constant and is independent of τ_c. Thus variation in τ_c reflects the distribution of spectral densities over the frequency spectrum. In Figure 10.4, $J(\omega)$ is plotted vs. ω for several values of τ_c. The spectral density is a maximum at $\tau_c = 1/\omega_o$ and falls off at both long (slow motion) and short (fast motion) correlation times. Thus, the spin–lattice relaxation time (the inverse of the relaxation rate) will show a minimum when plotted against τ_c. The flat portion of the spectral density curves is termed the region of motional narrowing because

$$J(\omega) = 2\langle h_{loc}(0)\rangle \tau_c \quad (10.14)$$

when $\omega_o << 1/\tau_c$.

The T_1 relaxation of the two-level model is given as follows:

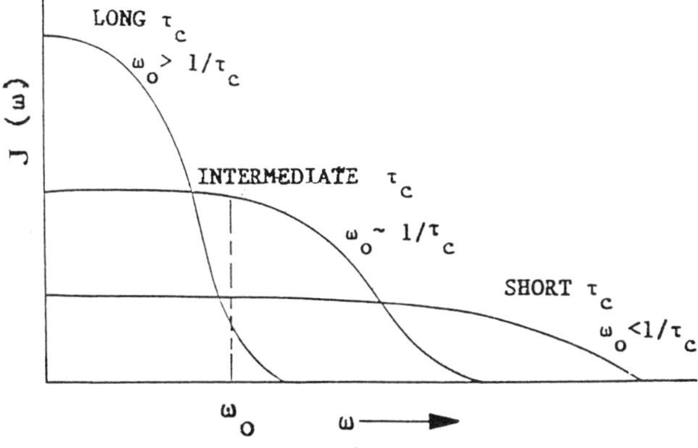

Figure 10.4. Spectral density of J(ω) as a function of frequency for various values of the correlation time, τ_c (ω_o corresponds to the nuclear Larmor frequency). (Reproduced from reference 2. Copyright 1985 American Chemical Society.)

$$\frac{1}{T_1} = \left(\frac{1}{3}\right) \gamma^2\langle h_{loc}^2\rangle \left(\frac{2\tau_c}{1 + \omega^2\tau_c^2}\right) \quad (10.15)$$

For isotropic fluctuations, the T_1 relaxation depends primarily on the transition frequency. Because the Larmor condition requires that $\omega = \gamma H_o$, the measurement of the static-field dependence of T_1 allows the separation of the static term $\langle h_{loc}^2\rangle$ from the dynamic term $\tau_c / (1 + \omega^2\tau_c^2)$. The increase in T_1 with increasing field strength yields τ_c. With this τ_c value, the relaxation rate in any arbitrary field can be used to determine the variance $\langle h_{loc}^2\rangle$ of the local field.

A second method of determining τ_c is to vary the temperature and determine the minimum of T_1 (τ at $\omega\tau = 1$ with constant ω). However, this method determines τ_c at only the temperature of the minimum. The temperature dependence of τ_c must be known to determine the form of the motion.

The T_2 relaxation formula of the two-level model with isotropic fluctuations can be derived in a similar manner:

$$\frac{1}{T_2} = \left(\frac{1}{3}\right) \gamma^2\langle h_{loc}^2\rangle \times \left[\tau_c + \left(\frac{\tau_c}{1 + \omega^2\tau_c^2}\right)\right] \quad (10.16)$$

The first term in this equation for T_2 is independent of frequency. This zero-frequency term gives the T_2 process a slightly different dependence on τ_c than the T_1 process. Experimentally, T_2 and T_1 can be different from each other probably because of the nature of the low-frequency contributions.

Spin–Lattice Relaxation Times for Carbon–Hydrogen Systems

The motional reorientation of a C–H vector provides the relaxation mechanism for the T_1 process in the 15–500-MHz region. Because of the dependence of the dipolar interactions on the inverse sixth power of the separation, intermolecular effects are usually not large enough to be considered. In solution, the carbon T_1 process is dominated by short-range local motions of the chain but not by macro-Brownian motion of the entire chain.

Assuming isotropic random reorientation of C–H

vectors, the spin – lattice relaxation time is given by (*3*):

$$\frac{1}{T_1} = \left(\frac{1}{10}\right) \gamma_H^2 \gamma_C^2 h^2 \Sigma r_i^{-6} \times \left[\left(\frac{\tau_c}{1 + (\omega_H - \omega_C)^2 \tau_c^2}\right) + \left(\frac{3\tau_c}{1 + \omega_C^2 \tau_c^2}\right) + \left(\frac{6\tau_c}{1 + (\omega_H + \omega_C)^2 \tau_c^2}\right)\right] \quad (10.17)$$

A plot of T_1 vs. τ_c is shown in Figure 10.5 for various values of H_o. As H_o increases, the T_1 minimum increases and moves to smaller values of τ_c. If motional narrowing holds (the fast exchange domain), eq 10.17 simplifies to

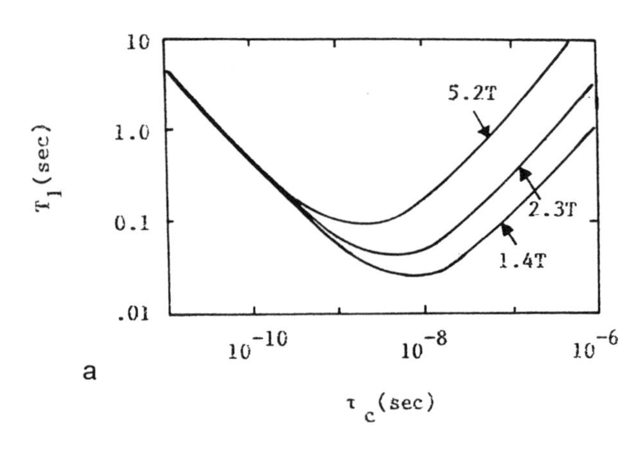

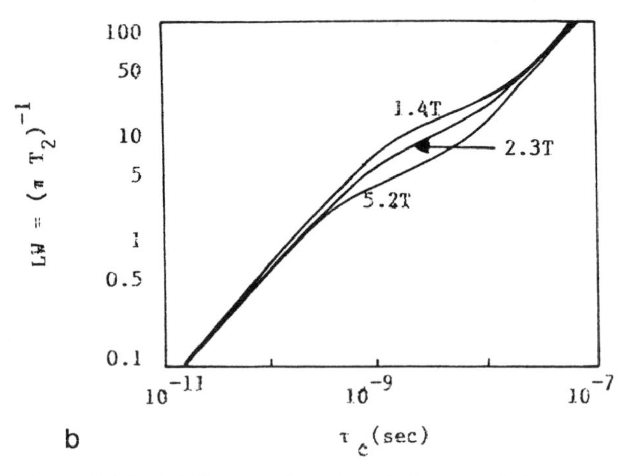

Figure 10.5. (a) T_1 *as a function of the correlation time,* τ_c, *for various magnetic-field strengths. (b) Line width* $(\pi T_2)^{-1}$ *as a function of the correlation time for various magnetic-field strengths. (Reproduced with permission from reference 17. Copyright 1980 Academic Press, Inc.)*

$$\frac{1}{T_1} = \left[\frac{\gamma_H^2 \gamma_C^2 h^2}{r_{CH}^6}\right] \tau_c \quad (10.18)$$

For this motional narrowing condition, T_1 is independent of H_o.

The influence of more than one attached proton on T_1 can be included by multiplying T_1 by the number of protons attached to the carbon (*18, 19*). Because of the dependence of the interaction on the sixth power of the separation distance, nonbonded-proton interactions can generally be neglected for carbon relaxations.

An estimation of the spin – lattice relaxation time can be made for a methine carbon, where r_{CH} = 0.109 nm. The result is

$$\frac{1}{T_1} = 2.03 \times 10^{10} \tau_c \quad (10.19)$$

and, inversely for the correlation time,

$$\tau_c = \frac{4.92 \times 10^{-11}}{T_1} \quad (10.20)$$

where τ_c is in seconds.

For homonuclear dipolar relaxation, the relationship between the T_1 and the spin density is given by

$$\frac{1}{T_1} = \left(\frac{3}{20}\right)\left(\frac{\gamma^4 h^2}{r^6}\right)[J_1(\omega) + 4J_2(2\omega)] \quad (10.21)$$

In the case of motional narrowing, this reduces to

$$\frac{1}{T_1} = \left(\frac{3}{20}\right)\left(\frac{\gamma^4 h^2}{r^6}\right)\tau_c \quad (10.22)$$

Because the protons are on the periphery of molecules, intermolecular interactions can play a substantial role in proton relaxation. Because of the low natural abundance of ^{13}C nuclei and the low probability of $^{13}C - ^{13}C$ bonds, the contributions of $^{13}C - ^{13}C$ dipolar interactions are generally negligible in the relaxation process.

Spin – Spin Relaxation Times for Carbon – Hydrogen Systems

A number of different relaxation effects can influence T_2 (*20*):

There are relaxation contributions to T_{2C} from molecular motions at the frequency ν_{1H}, from low-frequency motions which modulate the resonance frequency via anisotropic chemical shifts, from cross-polarization to the spin-locked protons under conditions of large radiofrequency mismatch, from off-resonance proton irradiation, from weak $^{13}C-{}^{13}C$ dipolar interactions between pairs of nuclei whose chemical-shift principal axes are not colinear, and from molecular motion at the same frequency contributing via $^{13}C-{}^{13}C$ dipolar interactions.

— W. L. Earl and D. L. VanderHart (*20*)

Spin–spin relaxation times in polymers range from 10^{-5} s for the rigid lattice (glassy polymers) to a value greater than 10^{-3} s for rubbers. Below the glass transition, T_2 is temperature independent and insensitive to the motional processes. The temperature dependence of T_2 above T_g is substantial. The values of T_2 decrease with increasing molecular weight because of the corresponding decrease in the mobility of the chains. Long-range motions strongly affect T_2 but not T_1. Consequently, crystallinity retards complete segmental narrowing.

When the motional reorientation rate approaches the frequency of the coherent motion induced by the proton decoupling field, line broadening occurs. An example of line broadening due to the correspondence of the frequency of the motion with the decoupling energy is the temperature-dependent studies of isotactic polypropylene (iPP) (*21*). As the temperature is lowered, the methyl resonance broadens until finally, at 130 °C, the line can no longer be observed (Figure 10.6).

The T_2 for a ^{13}C nucleus relaxed by a single proton can be calculated in the same manner as the longitudinal relaxation time:

$$\frac{1}{T_1} = \left(\frac{1}{20}\right)\left(\frac{\gamma_H^2\gamma_C^2h^2}{r_{CH}^6}\right] \times \Big[2J(0) + J_0(\omega_H - \omega_C) + 3J_1(\omega_C) + 3J_1(\omega_H) + 3J_2(\omega_H + \omega_C)\Big] \quad (10.23)$$

The three different Js correspond to cases involving no net spin, one spin flip, and two flips, respectively. This expression is plotted in Figure 10.5b for various values of H_o. In the motionally narrowed regime, this expression reduces to:

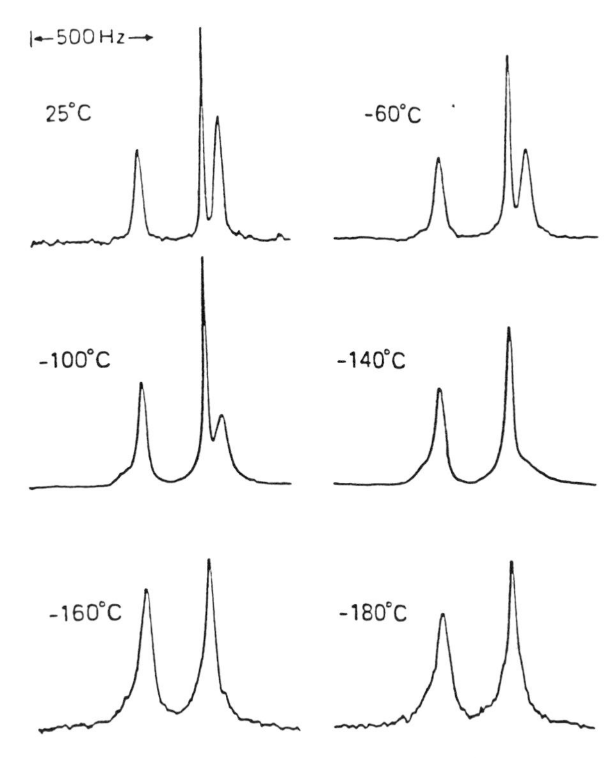

Figure 10.6. Proton decoupled CP–MAS ^{13}C NMR spectra of iPP as a function of temperature from 25 to –180 °C. (Reproduced from reference 11. Copyright 1980 American Chemical Society.)

$$\frac{1}{T_2} = \left(\frac{\gamma_H^2\gamma_C^2h^2}{r_{CH}^6}\right]\tau_c \quad (10.24)$$

which demonstrates that in this regime,

$$T_1 = T_2 \quad (10.25)$$

The frequency independence of $J(0)$ causes T_1 and T_2 to exhibit different behavior in the non-motionally narrowed region.

For H–H dipolar processes, the expression for $^HT_2$ is

$$\frac{1}{^HT_2} = \left(\frac{3}{40}\right)\left(\frac{\gamma^4h^2}{r^6}\right) \Big[3J(0) + 10J_1(\omega_H) + 4J_2(2\omega)\Big] \quad (10.26)$$

An example is a nuclear dipole–dipole interaction. If two identical spins are a distance, r, apart on the same molecule, which is rotating randomly, the spectral densities that arise from the rotating motion are (3)

$$J_0(\omega) = \left(\frac{24}{15r^6}\right)\left[\frac{\tau_c}{(1 + \omega^2\tau^2)}\right] \quad (10.27)$$

$$J_1(\omega) = \left(\frac{1}{6}\right) J_0(\omega) \quad (10.28)$$

$$J_2(\omega) = \left(\frac{2}{3}\right) J_0(\omega) \quad (10.29)$$

The exact shape of the correlation function contains information about the motions involved.

Interpretation of Polymer Relaxations

A single correlation time is insufficient to relate nuclear relaxation to internal motion in polymers. One test of the single correlation time is the requirement that the dynamic parameters (T_1, T_2, and NOE) yield the same correlation time. For polymers, this is clearly not the case.

For most polymers, several correlation times should be used to model the relaxation measurements. The motion of polymer chains is anisotropic as a result of restrictions on the motions by the chain ends. The different orientations of the carbon with respect to the field direction have different relaxation times in a manner analogous to the chemical-shift anisotropy. Because the effect of this orientation on the relaxation of polymers is not resolved adequately, average values of the relaxation parameters are generally reported.

For polymers, the internal motions are often correlated. Consequently, the interpretation of NMR results requires models with multiple correlation times or a distribution of correlation times.

> *A distribution of NMR relaxation times . . . suggests a multiplicity of noninterconverting dynamic environments for chemically, or structurally, nearly equivalent carbons.*
>
> —J. Schaefer et al. (22)

Schaefer (23, 24) has formulated a description of chain motion based on a log χ^2 distribution of isotropic correlation times. The distribution is characterized by a width parameter, p, and an average correlation time as illustrated in Figure 10.7. The logarithmic base accommodates the wide range of correlation times involved in polymer motions. In a statistical sense, p is a measure of the degrees of freedom for the system; that is, a large p suggests independent units and a single correlation time, and a small p is characteristic of the degree of long-range cooperative motion and a broad distribution.

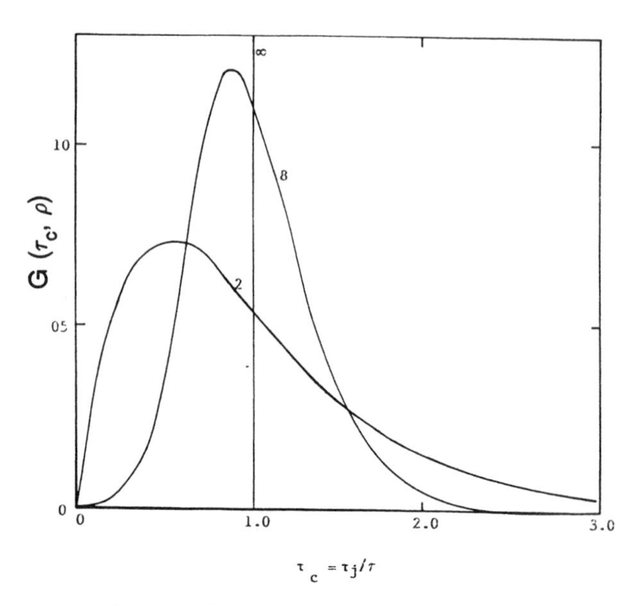

Figure 10.7. A χ^2 distribution of correlation times for three values of the width parameter, p. *(Reproduced with permission from reference 17. Copyright 1980 Academic Press, Inc.)*

Experimental Aspects of Relaxation Measurements for Polymers

The quality of the relaxation data obtained depends on the length, intensity, and shape of the pulses used. For the inversion-recovery technique, the condition for obtaining complete inversion of a spectrum is a short, intense 180° pulse; that is, $t^{\pi} << 1/2\ \Delta$, where t^{π} is the duration of the π pulse, and Δ is the spectral width in hertz. With current probe designs, it is difficult to deliver pulses of sufficient intensity to rigorously satisfy this condition for high-field instruments. Typically $t^{\pi} << 25$ ms (or at least 50 ms if quadrature detection is required) because at 90 MHz, Δ is approximately 20 kHz for polymers. Even when this condition is marginally satisfied, only small errors are introduced into T_1 or T_2 for low-molecular-weight substances when overlap of the lines is minimal and the relaxation times are long. For polymers, the errors accumulate because of the larger spectral widths, extensive overlap of lines, and significant variability of the relative intensities and widths of the spectral lines. This condition is hard to satisfy for polymers because some intensity and phase distortion results, particularly in the wings where the difference between the pulse frequency and the nuclear precession frequency is large. A reduction in the measured T_1 values by a factor of 2 or 3 for some lines can easily result (*25*).

An additional experimental problem seldom encountered with small molecules is the range of values for T_1 or T_2 for the different resonances within the same sample. The values of T_1 can vary in the same polymeric spectrum. The choice of the appropriate interrupt time, τ, is therefore more difficult. A sufficiently large set of τs to span the shortest as well as the longest relaxation times in the spectrum is essential. In addition, care must be exercised when selecting the delay time between pulse sequences. If the delay time is too short, the relaxations of resonances will be distorted to different extents, and significant errors will result. Ribeiro et al. (*25*) suggested that the delay time between pulse sequences should be greater than $7T_{1_{max}}$, where $T_{1_{max}}$ is the longest T_1 in the spectrum.

A fundamental problem with polymers is that the carbon atoms giving rise to a particular resonance are structurally heterogeneous. This heterogeneity can have several different sources, including stereochemical sequences, local geometric conformations, or disorder in chain packing. The measured intensity of a line is the weighted average of all the different carbon nuclei giving rise to the NMR signal at that frequency. Thus those carbon nuclei having the most narrow line widths contribute most to the intensity at the maximum, and those having the greatest line widths contribute the least. All polymer T_1s measured by peak intensities necessarily weigh most heavily the carbon atoms that relax most quickly. Integrated signal areas can be used to calculate T_1 or T_2 values that are more representative of the average carbon nuclei in heterogeneous resonances.

An error that will remain difficult to eliminate is the overlap of the lines with widely divergent T_1 values. The base lines in such cases are variable, and iterative fitting of the line shapes may be necessary.

As a final note of caution, you must always consider whether the measured ^{13}C relaxations are dipolar in origin or are caused by paramagnetic impurities. Commercial polymers contain paramagnetic impurities, such as oxygen, and the question must therefore be raised as to whether such impurities account for the observed ^{13}C relaxation behavior. Because of limited spin diffusion in ^{13}C NMR spectroscopy, these paramagnetic centers, if they exist, should influence the relaxation of only a few nearby ^{13}C spins, and therefore should not make an observable contribution to the relaxation times (*3*).

Spin Diffusion as an Alternative to Relaxation

Spin diffusion is a mechanism for transporting magnetization from one nucleus to another. Spin diffusion is induced by the dipolar interactions of nuclear spins, which produce spin flips between nuclei. This spin flipping leads to an adiabatic transfer of magnetization between neighboring spins. For highly coupled nuclei such as protons, this spin diffusion can be the dominant process and can override motional effects.

When $T_1 >> T_2$, the coupling between the proton spins is much stronger than the coupling with the environment. Consequently, the proton spin system comes to internal equilibrium by spin diffusion more rapidly than it reaches equilibrium with the lattice by relaxation. Under these circumstances, when the proton spins are perturbed by an rf pulse, the excess energy put into the spin system can remain for a long time relative to the time required for the spin system to establish internal equilibrium.

In some circumstances, spin diffusion provides an alternative relaxation mechanism via efficient relaxation centers such as paramagnetic impurities and rotating methyl groups. Thus a rapidly reorienting methyl group can cause all of the protons in a solid to have the same T_1.

For protons, the spin-diffusion effect may be semiquantitatively calculated from the following equation.

$$D = W\langle\omega_o\rangle^2 r^2 \qquad (10.30)$$

where D is the diffusion coefficient, W is the transition probability per unit time that spins exchange energy, and r is the distance between the spins. For protons in a rigid lattice, $W = (M_2^{HH})^{1/2}/30$, and M_2

$= (\ln 2/2)(\Delta\omega)^2$, where $\Delta\omega$ is the full width at half-height for the resonance (*26*). Furthermore,

$$L^2 = 6Dt = 6W <\omega_o>^2 t \quad (10.31)$$

where L is the maximum diffusive path length and t is the diffusion time. If the value of M_2 is $20G^2$ (32 kHz full width at half-height), then $W = 4.0 \times 10^3$ rad/s. For an average internuclear distance of 1.78 Å (e.g., a long-chain methylene sequence such as that found in polyethylene), $L = 275t^{1/2}$ Å. Assuming a relaxation time of 5 s for the crystalline component of polyethylene, the maximum diffusive path length is 600 Å.

The equation describing the distance of the transfer of magnetization is

$$x^2(t) = 4Dt \quad (10.32)$$

Using the value of D for alkanes, $x(t)$ becomes 1.6, 7.0, and 50 nm, corresponding to expected times of 1 ms, 20 ms, and 1 s, respectively (*26*). Typical time constants for proton spin flips are 100 μs to 10 ms. For protons, the spin-exchange times are on the order of T_2. For alkanes, the diffusion constant, D, is 6.2×10^{-12} cm^2 s^{-1} in the absence of the proton rf field (*27*).

Because of a larger gyromagnetic ratio and a higher natural abundance, proton spin diffusion is faster than ^{13}C spin diffusion by several orders of magnitude. For ^{13}C, the diffusion effects are quite different from those of protons. For polyethylene at natural abundance, the most probable ^{13}C – ^{13}C nearest-neighbor distance is 0.7 nm. At this distance, the average lifetime between spin flips is about 45 s (*28*). However, there is a distribution of ^{13}C – ^{13}C internuclear distances, and the lifetimes vary over a wide range. ^{13}C spin diffusion may occur when the spin – lattice times are extremely long, as may be the case for polymeric solids.

At high spinning seeds, magic-angle spinning (MAS) can interfere with the spin-diffusion process and slow it down (*29*). However, the spinning frequency must approach the line width; otherwise, the effect of MAS will be small or even negligible. The presence of spinning sidebands in proton spectra implies that MAS averages the dipolar interactions and thereby suppresses spin diffusion (*30*).

Polymers exhibiting NMR spectra with motionally narrowed sharp resonances sitting on a broad background will have domains whose ^{1}H spin-diffusion coefficients are substantially different. The measurement of spin diffusion requires the development of a gradient in the magnetization. Differences in T_2s or $^{H}T_{1\rho}$s can be used to generate this gradient, and domains in the sample will have different fractions of ^{1}H spins aligned with the magnetic field. The distribution of spins depends on the relaxation characteristics of each domain. The spin-diffusion rate between domains can be measured by using the Goldman-Shen experiment, which uses differences in T_2s to generate the magnetization gradient between domains (*31*). A mixing time (shorter than the spin-lattice relaxation time) occurs, during which the protons interact with each other, and the spin diffusion flows in the direction of equilibration. The magnetization is then observed and resolved into contributions from the different domains on the basis of estimates of the diffusion coefficients. The size and structure of the domain are determined from calculations of the rate of equilibration by mathematically modeling the observed behavior.

Both the magnitude and the intensity of the components of T_1 and $T_{1\rho}$ are affected by spin diffusion. The longer the time scale of the relaxation, the more likely it is that the relaxation will be affected by spin diffusion. Thus T_1s are more likely to be affected than $T_{1\rho}$s. For ^{13}C, spin diffusion must proceed for 10 – 100 s before observable effects are manifested (*32*).

Because of the $1/r^6$ distance dependence of the spin-diffusion rate, magnetization transfer is restricted to the immediate neighborhood and can deliver information on the spatial proximity of different molecules in a solid. Spin diffusion can therefore provide information on the heterogeneity of polymers. Spin diffusion between components in different domains is strongly dependent on domain size, and therefore the measurement of cross-relaxation between different domains provides insight into the domain structure of polymers.

A 2-D experiment can be used to determine the presence of spin diffusion. In the 2-D spectrum, a series of peaks represents the normal spectrum along the diagonal in the absence of spin diffusion. When spin diffusion occurs, crosspeaks are created, and these crosspeaks indicate the connectivity between different resonances. For linear PE, two resonances, corresponding to the crystalline and amorphous phases, are observed. These resonances indicate that spin diffusion occurs, but it is not clear whether the process is sufficient to influence the relaxation processes (*33*). This 2-D spin-diffusion experiment was used to characterize polymer blends for homogeneity (*34*).

Determination of Polymer Dynamics Using Relaxation Methods

Spin – Lattice Relaxation Measurements for Specific Polymers

In ^{13}C NMR spectroscopy, site-specific relaxation measurements can be made for the chemically inequivalent carbon atoms. Measurements of ${}^{C}T_1$ are sensitive to motions with correlation times of approximately 10^{-8} s, which correspond to relatively high-frequency molecular motions. Chemical changes in the polymer chain can affect the ${}^{C}T_1$ values to yield a probe of the chemical environment of the carbons.

The decay of the magnetization for the T_1 measurements is expected to be a single exponential, but deviations from the simple exponential are often observed. Nonexponential decays in polymers arise from several sources: (1) separate contributions from crystalline and amorphous phases, (2) spin diffusion in multiphase polymeric systems, and (3) cross-relaxation under certain circumstances (*35*). Cross-relaxation effects have been observed in the ${}^{H}T_1$ and ${}^{F}T_1$ measurements in poly(vinylidene fluoride) (*35*).

Polyethylene. Carbon spin – lattice relaxation measurements were made separately both on the amorphous phase of polyethylene (PE) using scalar decoupling (*36*) and on the crystalline phase using CP – DD techniques (*20*). The amorphous-phase ^{13}C T_1 values are not affected by changes in molecular weight, branching number, or morphology (*36*). Figure 10.8 shows a plot of NT_1 (N is the number of bonded protons, which is two for methylene) as a function of temperature for the backbone amorphous methylene carbon of branched and linear PEs (*37*). If the samples are in the extreme-narrowing domain, an analysis of the correlation times can be made by using a log χ^2 distribution function. The results are

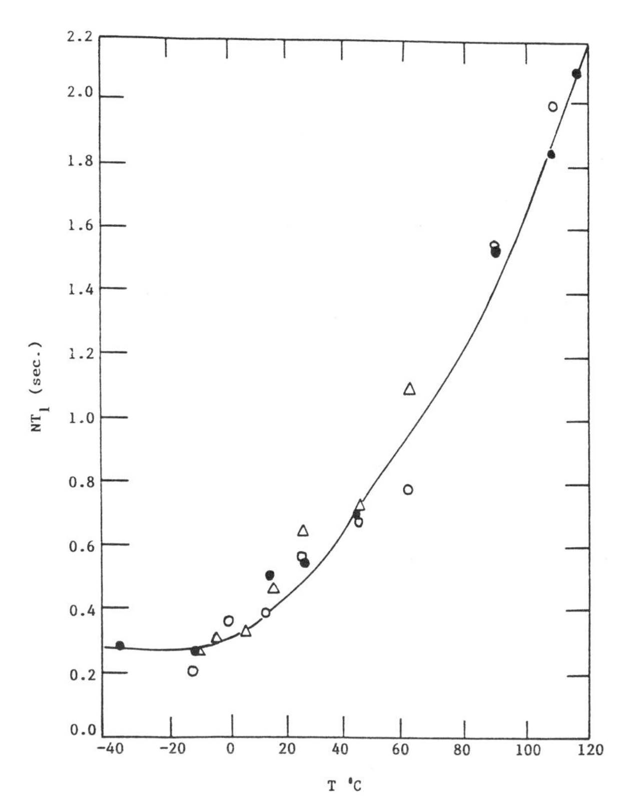

Figure 10.8. Plot of NT_1 *as a function of temperature from the scalar-decoupled* ^{13}C *NMR spectra of the backbone methylene carbons of branched PE (Δ, ○) and linear PE (●). (Reproduced with permission from reference 37. Copyright 1982 John Wiley & Sons, Inc.)*

shown in Figure 10.9 (*37*). The motional correlation times for the linear and branched samples are the same for the amorphous component.

For the crystalline phase of PE, ${}^{C}T_1$ shows a range of values depending on the branch content, the crystalline level, the morphology, and the thermal history of the sample. The inversion-recovery measurement of ${}^{C}T_1$ was made under cross-polarization conditions, and the results are shown in Figure 10.10 for a high-molecular-weight, highly crystalline PE sample (*20*). The crystalline ${}^{C}T_1$ is 1000 s, and the amorphous ${}^{C}T_1$ is 175 ± 5 ms. The short ${}^{C}T_1$ for the amorphous component is probably a result of a 180° jump rotation or flip – flop screw-jump motion about the molecular chain axis. The introduction of branches decreases the crystallinity substantially, and the crystalline ${}^{C}T_1$ decreases to 45 s for a branch content of 12.6 branches per 1000 carbons (*38*). For a PE with a branch content of 5.3 branches per 1000

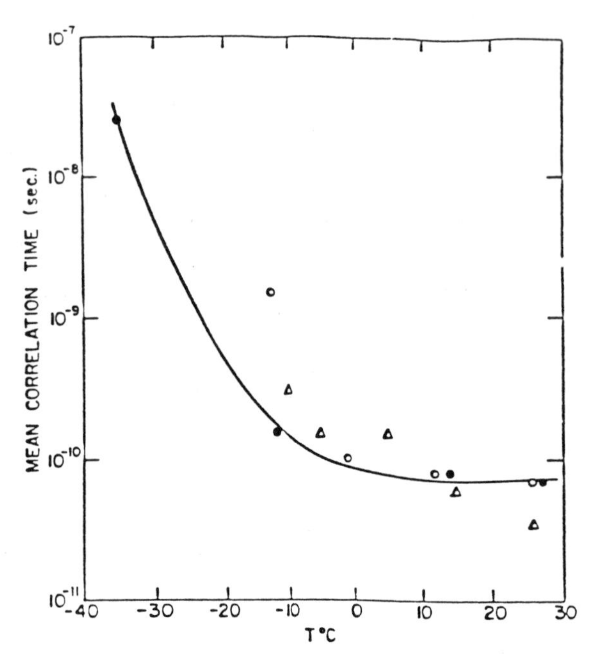

Figure 10.9. Plot of the mean correlation times as a function of temperature from the scalar-decoupled ^{13}C NMR spectra of the backbone methylene carbons of branched PE (Δ , ○) and linear PE (•). (Reproduced with permission from reference 37. Copyright 1982 John Wiley & Sons, Inc.)

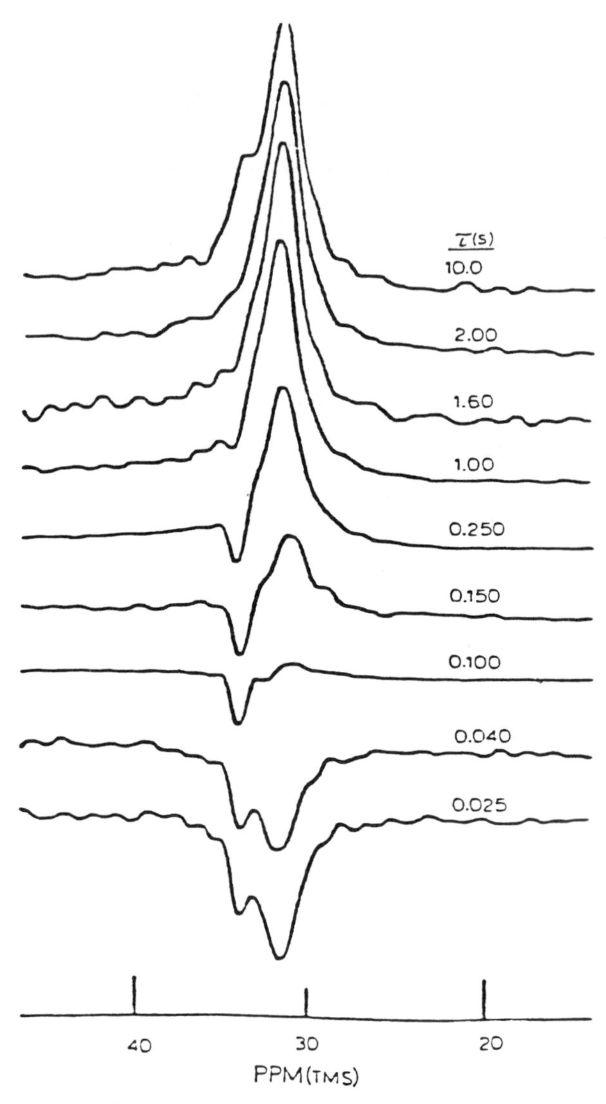

Figure 10.10. The inversion-recovery ^{13}C NMR experiment at 15.08 MHz of high-molecular-weight, highly crystalline PE. The spectra obtained at different times show the large difference in amorphous and crystalline $^{C}T_1$. (Reproduced from reference 20. Copyright 1979 American Chemical Society.)

carbons, a slowly cooled sample has a crystalline $^{C}T_1$ of 239 s, but quenching the sample reduced the crystalline $^{C}T_1$ to 79 s. This decrease in the $^{C}T_1$ values is thought to arise from contributions due to spin diffusion.

> *The greater the disruption of the crystalline order, the greater the "thermal" contact between the amorphous and crystalline phases . . . . On this basis, the* T_1 *value of the crystalline phase within a certain distance from this interface would be reduced in magnitude by the spin-diffusion process.*
> – D. E. Axelson (*38*)

Poly(oxymethylene). The $^{C}T_1$ results are shown in Figure 10.11 of the 45-MHz CP–MAS ^{13}C NMR spectra of poly(oxymethylene) (POM) as a function of the recovery time. The observed resonance has two components: one associated with the crystalline phase ($^{C}T_1$ = 15 s) and the other an overlapping component due to the amorphous regions ($^{C}T_1$ = 75 ms). By using a value of 75 ms for the $^{C}T_1$ of the amorphous POM and by assuming the two protons are r = 1.09 Å apart, a correlation time of 1.3 × 10^{-8} s was calculated. Proton T_1 measurements yield a correlation time of 2 × 10^{-8} s, and this result is in good agreement with the calculated value (*3*).

Polypropylene. Isotactic polypropylene (iPP) was studied, with revealing results (*11*). The CP–MAS spectra of iPP as a function of temperature are shown in Figure 10.6. As the temperature is lowered, all of the resonances broaden, and the the methyl resonance changes most quickly to become

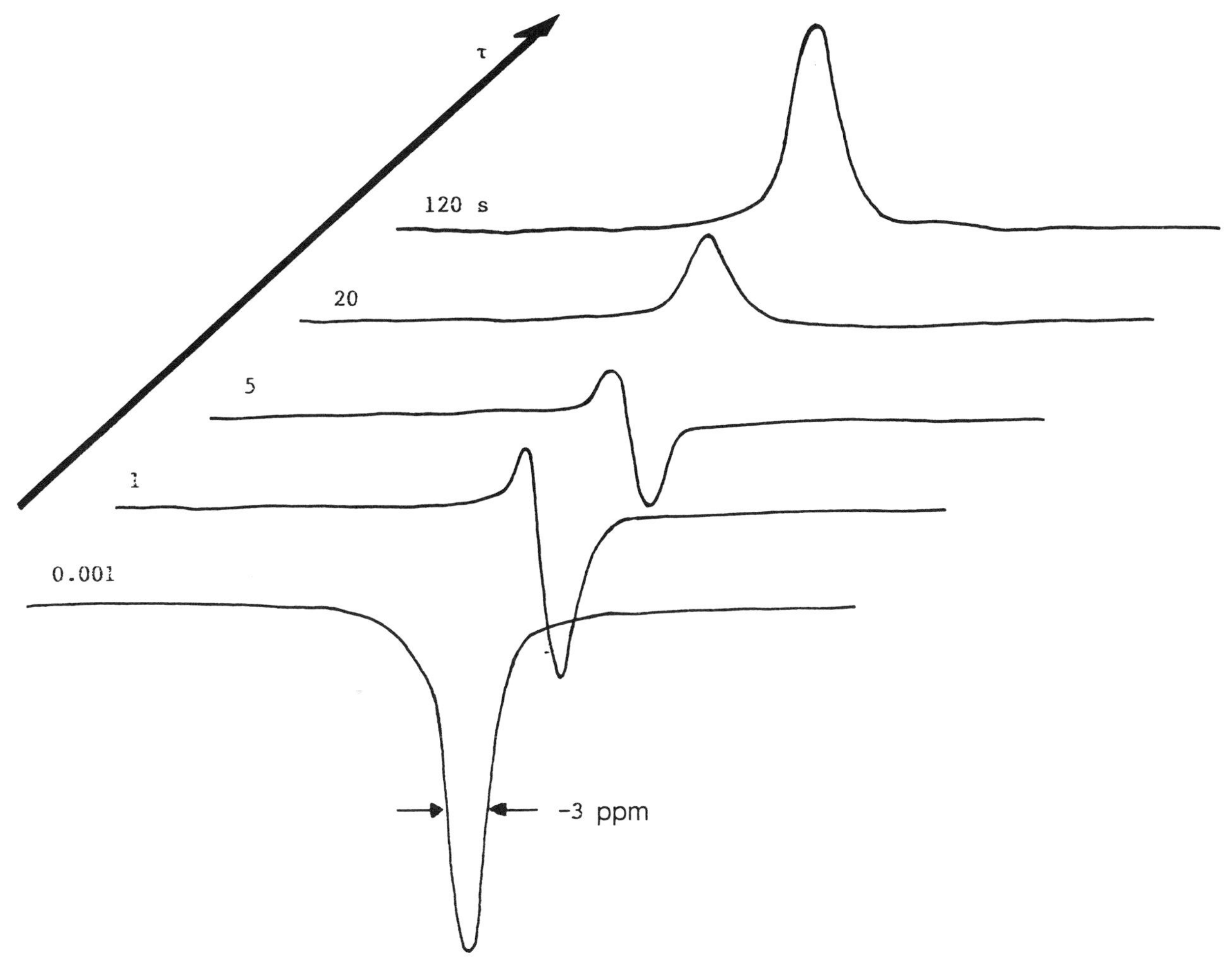

Figure 10.11. The CP–MAS ^{13}C NMR spectra at 45 MHz of POM. The spin–lattice relaxation was recorded as a function of time. (Reproduced from reference 3. Copyright 1982 American Chemical Society.)

completely lost at about −143 °C. The $^{C}T_1$ values can be measured for each of the carbons in the repeat unit because the resonances are experimentally resolvable. The results of these measurements are shown in Figure 10.12. The methyl T_1s are smaller than the methylene and methine carbon resonances by 1 or 2 orders of magnitude, and the methine carbon has a shorter T_1 than the methylene carbon. The r^{-6} distance dependence of dipolar relaxation accounts for the long T_1 values of the methine and methylene carbons as compared to the methyl carbon as well as for the shorter T_1 for the methine carbon relative to the methylene carbon. However, because there are two direct C–H interactions for the methylene carbon, the shorter T_1 for the methine carbon is explained by the modulation of the dipolar-relaxation process by methyl-group rotation. The methine and methylene carbons have $^{C}T_1$ minima at about −113 °C, which is nearly the same temperature as that for the proton T_1 minimum in iPP (*39*).

Poly(tetrafluoroethylene). The ^{13}C NMR T_1 relaxation data of poly(tetrafluoroethylene) (PTFE) at −2 °C is shown on a semilog plot in Figure 10.13 (*11*). The decay is nonexponential, which indicates multiple-relaxation behavior. The long-time relaxation is ascribed to the crystalline regions of the polymer. The faster relaxing component of the noncrystalline regions has a distribution of relaxation times. The $^{C}T_1$ data for the crystalline component of PTFE are plotted vs. the reciprocal of the temperature in the region from −43 to 46 °C in Figure 10.14

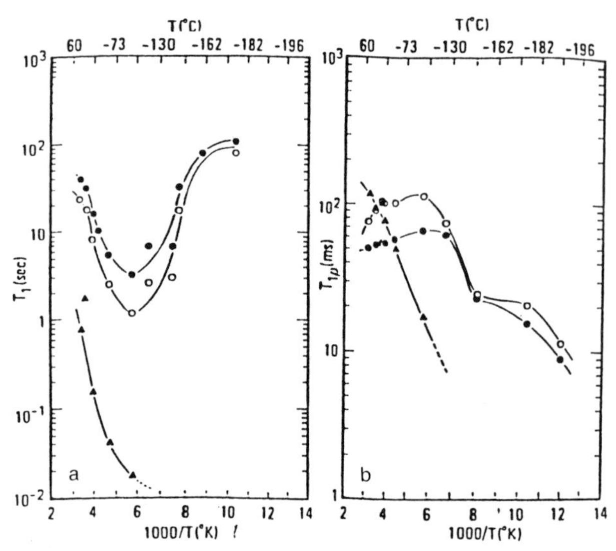

Figure 10.12. The ^{13}C *NMR* T_1 *and* $^{C}T_{1\rho}$ *relaxation data for the methyl (▲), methylene (○), and methine (•) carbons of iPP as a function of temperature at 15.08 MHz. (Reproduced from reference 21. Copyright 1980 American Chemical Society.)*

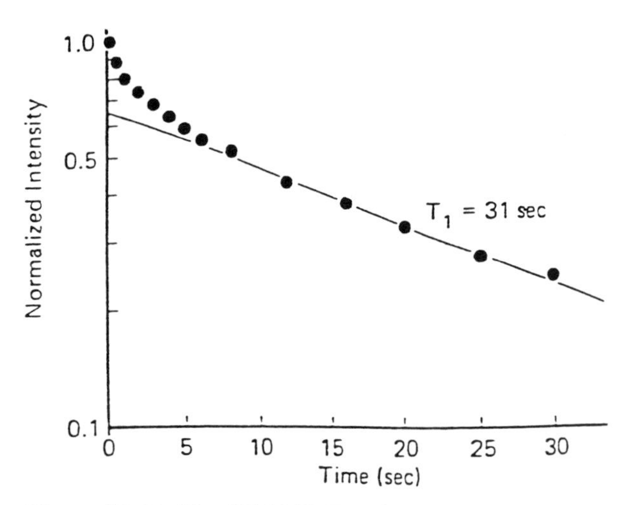

Figure 10.13. The ^{13}C *NMR* T_1 *relaxation data for PTFE at a temperature of −2 °C with 1 ms CP contact time at 15.08 MHz. (Reproduced from reference 21. Copyright 1980 American Chemical Society.)*

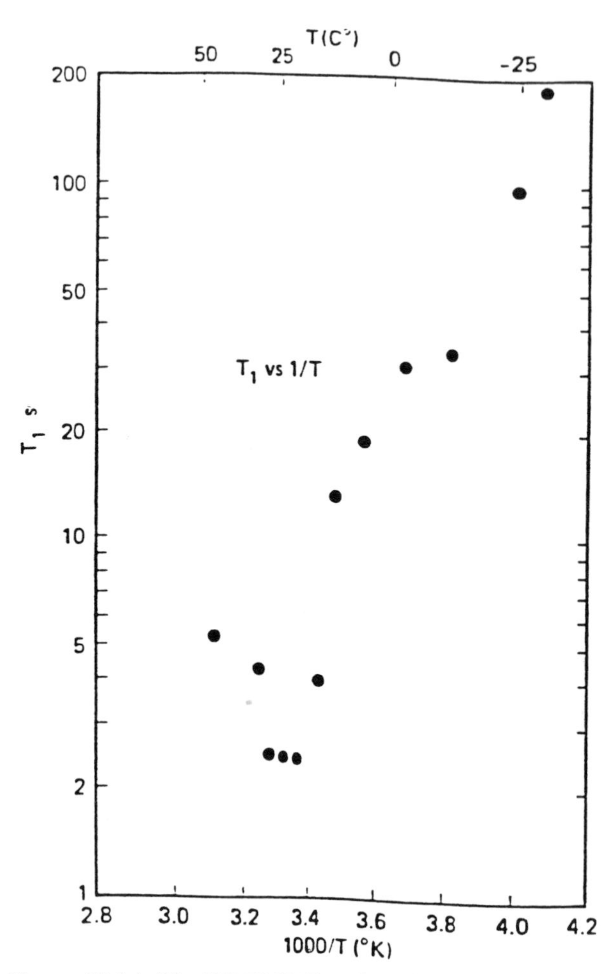

Figure 10.14. The ^{13}C *NMR* T_1 *relaxation times for PTFE as a function of temperature. (Reproduced from reference 21. Copyright 1980 American Chemical Society.)*

(*11*). A shortening of $^{C}T_1$ by 2 orders of magnitude occurs in this temperature region. The sharp decrease in $^{C}T_1$ between 0 and 20 °C is attributed to the crystal-phase transition at 19 °C. Correlation times in the range of $10^{-5} - 10^{-8}$ s are observed (*11*).

Polystyrene. The $^{C}T_1$ plots for polystyrene are highly curved, as shown in Figure 10.15 (*16*), and this result reveals a multicomponent structural basis. The short $^{C}T_1$ component of these plots is approximately 50 ms, which is as short as the $^{C}T_1$s for polymers in solution. In contrast, the longer component is 20 − 30 s, and this results in in a dispersion of 3 orders of magnitude. The motion involved with the short T_1 component is a 180° ring flip that occurs in the 10-MHz range. Apparently, the weak dynamic relaxation γ transition (1 Hz at −120 °C) of the mechanical-loss spectrum of polystyrene arises from these cooperative ring flips. The fraction of carbons contributing to the short T_1 component can be determined by extrapolating to zero time by using the slopes determined at longer times. The fractions of rings undergoing the ring flipping probably arise from differences in the microscopic local free-vol-

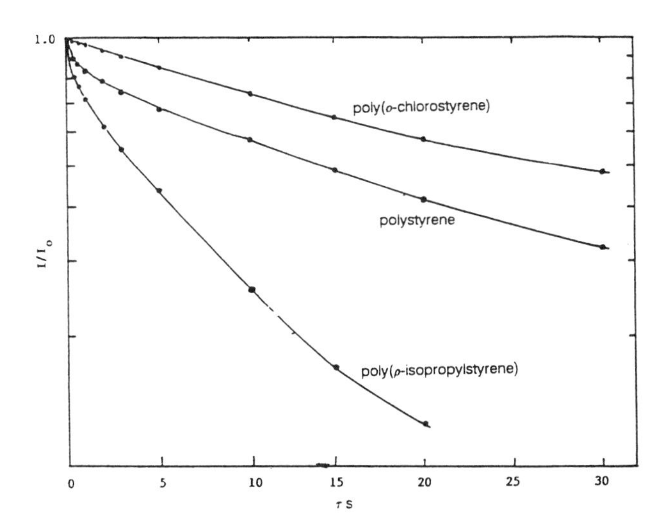

Figure 10.15. 15.1-MHz $^{C}T_1$ *plots of the protonated aromatic carbons of three polystyrene samples. (Reproduced from reference 16. Copyright 1984 American Chemical Society.)*

ume effects due to variable packing of the chains in the glassy state. The fraction of phenyl rings contributing to the short T_1 component varies from 0% for poly(*o*-chlorostyrene) to 11% for poly(*p-tert*-butylstyrene). Apparently, bulky *para* substituents expand the local free volume and promote ring flipping, and main-chain substitution has the opposite effect. For the glassy polystyrene itself, the fraction of phenyls contributing to the short T_1 component ranges from 7 to 10% and is independent of tacticity and thermal history and only weakly dependent on molecular weight (*16*).

Poly(methyl methacrylate). The ^{13}C NMR T_1 relaxation data were obtained for poly(methyl methacrylate) (PMMA) as a function of temperature and are shown in Figure 10.16. The decay curves are not exponential, a result indicating a multicomponent decay. The rapid relaxation of the α-methyl group by rotation about its C-3 axis influences all of the other carbons by spin diffusion as indicated by the similarity of the curves. Figure 10.17 shows the $^{C}T_1$s measured at 50.2 MHz plotted as a function of $1/T$ for two samples of PMMA with different levels of syndiotacticity (the sample designated s-PMMA has 49.5% syndio triads, and the sample designated is-PMMA has 72.0% syndio triads). The values of the $^{C}T_1$ minima are 200 ms for s-PMMA and 300 ms for is-PMMA and are associated with the onset of α-

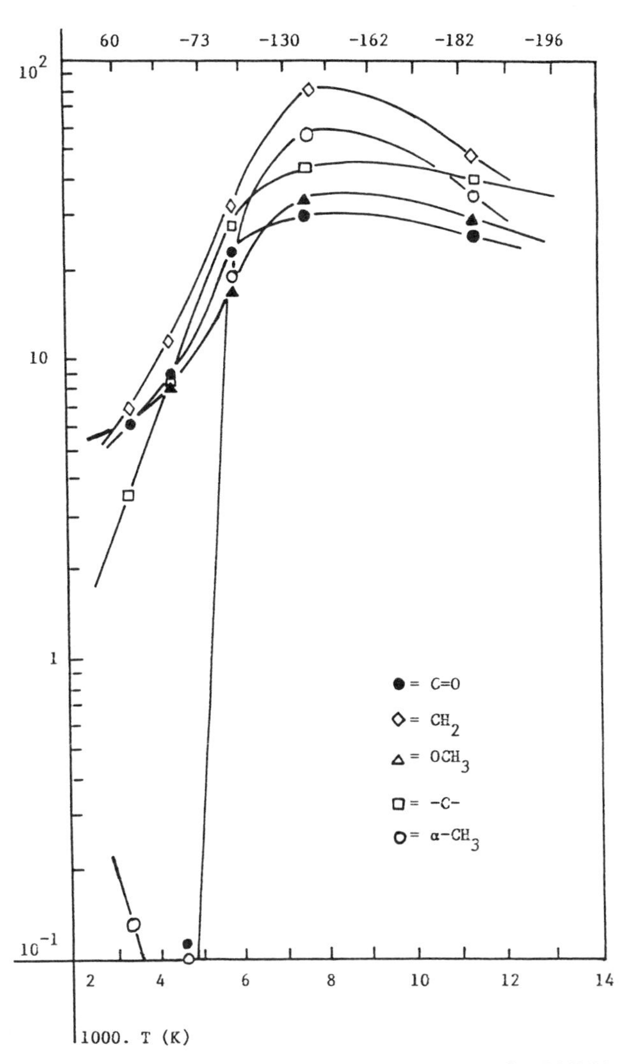

Figure 10.16. The ^{13}C *NMR* T_1 *relaxation time for PMMA as a function of temperature at 1.4* T. *The symbols corresponding to the various functional groups of PMMA are indicated on the figure. (Reproduced from reference 21. Copyright 1980 American Chemical Society.)*

methyl-group rotation with a correlation time of approximately 10^{-9} s (*40*).

Carbon Spin – Lattice Relaxation in the Rotating Frame

> *The* ^{13}C *rotating-frame spin – lattice relaxation time has been shown qualitatively to characterize the room-temperature mid-kilohertz main-chain motions of glassy polymers and has been empirically linked to the mechanical properties of polymers, which presumably depend on this motion.*
>
> – J. Schaefer (*24*)

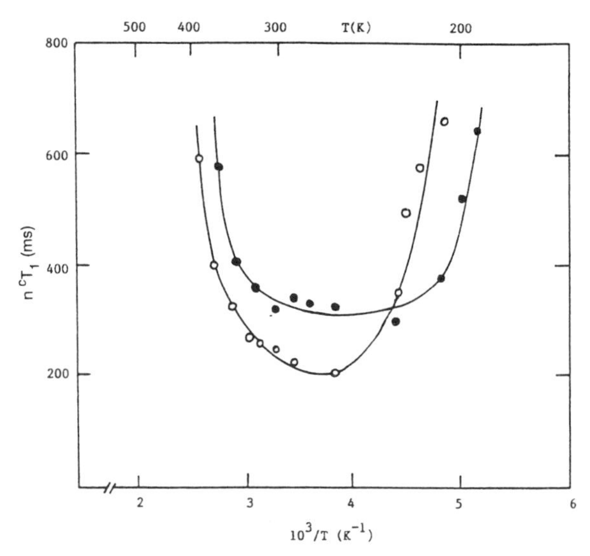

Figure 10.17. The $^{C}T_1$s measured at 50.2 MHz and plotted as a function of 1/T for two samples of PMMA with different levels of syndiotacticity (s-PMMA, ○; and is-PMMA, ●). (Reproduced from reference 40. Copyright 1987 American Chemical Society.)

The ^{13}C spin–lattice relaxation time in the rotating frame, $^{C}T_{1\rho}$, can be used as a motional parameter in the range of frequencies between 20 and 60 kHz. The motion probed is the random orientational motion of dipolar-coupled C–H vectors that occurs at the rotating-frame Larmor frequency. Although $^{C}T_{1\rho}$ is not influenced by spin diffusion, its interpretation is complicated by molecular motions arising from the presence of strong, static dipolar $^1H-^1H$ effects (*see* the section titled "Spin–Lattice Relaxation in the Rotating Frame"). In general, the relaxation is dominated by the strong dipolar $^1H-^1H$ effects for ordered crystalline polymers in which there are short $^1H-^1H$ distances. This situation contrasts with glassy and amorphous polymers in which strong static dipolar $^1H-^1H$ effects are absent because of molecular motion or large interproton distances.

The study of restricted rotational reorientation of carbons in a polymer chain can be accomplished by using measurements of $^{C}T_{1\rho}$.

$$\frac{1}{^{C}T_{1\rho}} = K^2 \sin^2 \theta \, J(\omega) \qquad (10.33)$$

where θ measures the average dipolar fluctuation amplitude due to molecular motion orthogonal to the applied rf field, and K is a constant that includes powder averaging (Chapter 8, ref. 15). For a 100-kHz free rotor in a polymeric solid, a 40-kHz $^{C}T_{1\rho}$ of 1 ms is expected (Chapter 8, ref. 15). From $^{C}T_{1\rho}$ studies of polystyrenes, the ring rotations generate total angular displacements varying from 40° for substituents in the *ortho* position to 70° for bulky nonpolar substituents in the *para* position.

The $^{C}T_{1\rho}$ decay is often nonexponential, and a $<^{C}T_{1\rho}>$, which is a weighted average of all of the $^{C}T_{1\rho}$s present, is normally reported. This effect is shown in Figure 10.18 for poly(butylene terephthalate) (PBT), which illustrates the nonlinearity at long delay times (*41*).

Dispersions in this type of relaxation are generally observed for glassy polymers, and the dispersions have been attributed to dynamic heterogeneity in the glassy state. The dynamic heterogeneity of the solid

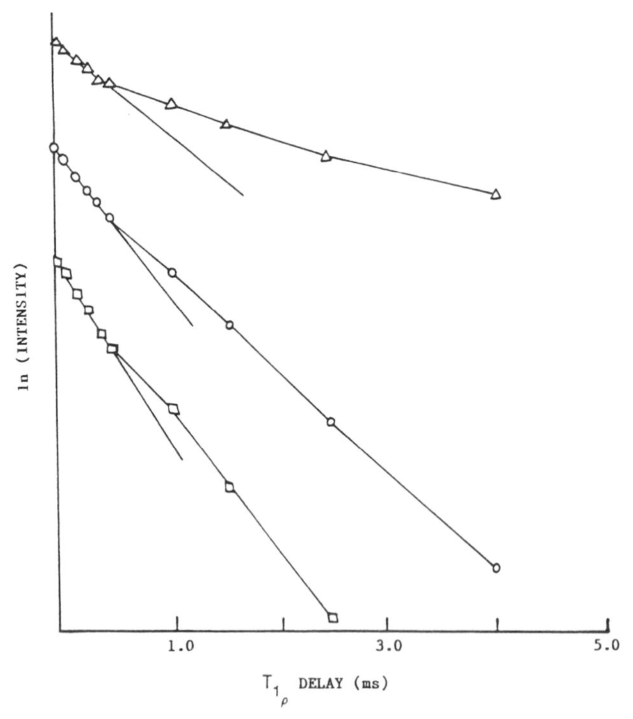

Figure 10.18. Typical $^{C}T_{1\rho}$ values for PBT, illustrating the nonlinearity at long delay times. The average $<^{C}T_{1\rho}>$ is taken from the initial part of the curves. The triangles represent a composite curve for the protonated and non-protonated aromatics, the circles represent $-OCH_2-$, and the squares represent $-CH_2-$. The curves have been displaced vertically for visual clarity. (Reproduced from reference 41. Copyright 1983 American Chemical Society.)

state can be varied by annealing, and the result is changes in the relaxations. This effect is shown in Figure 10.19 for poly(ethylene terephthalate) (PET) (*42*).

The roles of plasticizers and temperature were studied by using ${}^{C}T_{1\rho}$ measurements (*43*). For polycarbonates, the ${}^{C}T_{1\rho}$ values are not sensitive to temperatures between −90 and 20 °C. On the other hand, the ${}^{C}T_{1\rho}$s for the phenyl carbons in the polymer increase as the temperature is increased when the polycarbonate is blended with dibutyl succinate (a relatively low-T_G diluent). The ${}^{C}T_{1\rho}$s at 50 kHz for the five resonances resolved in the solid-state spectrum of bisphenol-A-polycarbonate (BPAPC) are shown in Figure 10.20. The ${}^{C}T_{1\rho}$s of all of the carbons of polycarbonate are increased by the presence of dibutyl succinate, and this result suggests that no specific interactions occur between the BPAPC and dibutyl succinate. Rather, dibutyl succinate increases the spectral density of the 50-kHz motions of the entire repeat unit.

The dispersions in ${}^{C}T_{1\rho}$s are different for various polymers and even for inequivalent carbons in the same polymer. For example, in polycarbonate the ${}^{C}T_{1\rho}$ relaxations of lines arising from the nonproton-

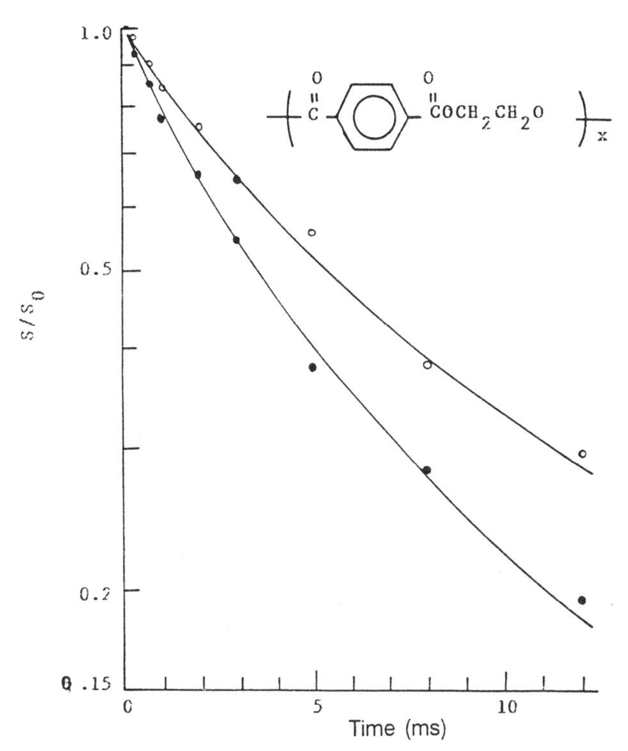

Figure 10.19. The ${}^{13}C$ 37-kHz ${}^{C}T_{1\rho}$ relaxation data for the methylene carbons of PET showing the dispersion that results from the distribution of relaxation times. The average relaxation time, represented by the initial slope of the relaxation curve, is 3.5 ms for the quenched polymer (•) and 6.0 ms for the annealed polymer (○). (Reproduced with permission from reference 42. Copyright 1981 The Royal Society.)

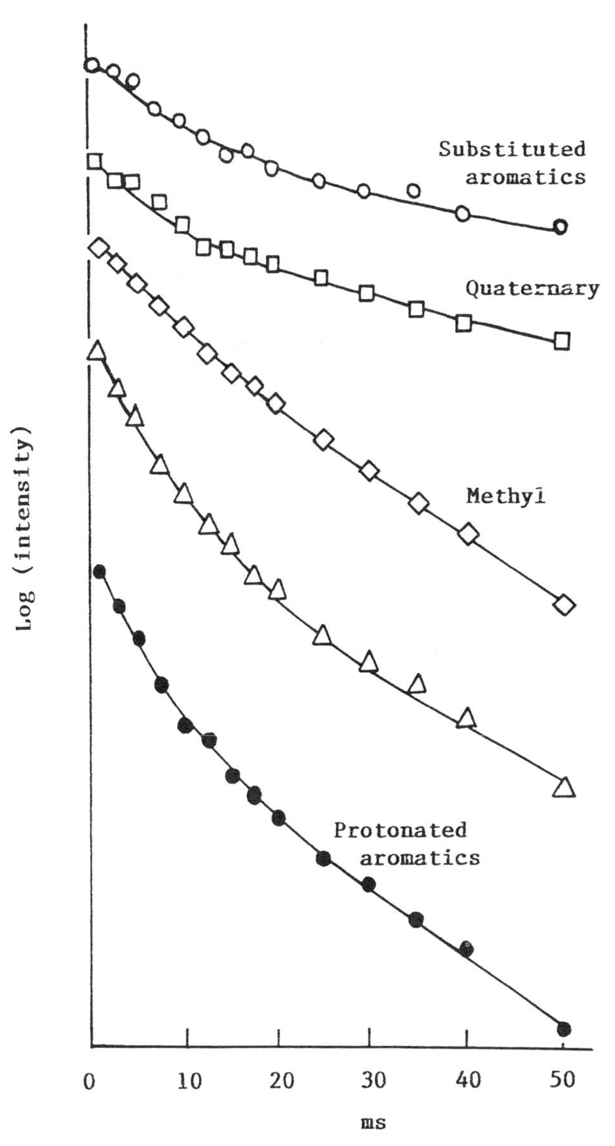

Figure 10.20. The ${}^{C}T_{l}$ relaxation curves at 50 kHz of the five resonances that are resolved in the solid-state spectrum of BPAPC. The logarithm of the peak intensity is plotted as a function of the time, t, *that the carbon magnetization was held in the rotating frame without CP contact with the proton-spin system. (Reproduced from reference 43. Copyright 1983 American Chemical Society.)*

ated carbons are slower than those of the protonated aromatic carbons, even though both types of carbons necessarily have the same motion. This difference results from the dependence of the relaxation on the inverse sixth power of the internuclear separation. The motions involved are cooperative torsional oscillations of phenyl groups.

The ${}^{C}T_{1\rho}$ values were measured for substituted polystyrenes (*16*). The protonated-aromatic-carbon ${}^{C}T_{1\rho}$ changes by a factor of 6 with chemical substitution, but the main-chain carbons are insensitive to substitution.

${}^{C}T_{1\rho}$ measurements were made on the different phases of poly(oxymethylene) (POM) (*3*). The single resonance line has two different ${}^{C}T_{1\rho}$s that are associated with the amorphous and crystalline phases. The amorphous line has a long ${}^{C}T_{1\rho}$ (17.5 ms), and the crystalline line has a ${}^{C}T_{1\rho}$ that ranges from 0.5 to 3 ms, depending on the spinning frequency. This range is not consistent with restricted molecular motions in the crystalline phase having correlation times comparable to the more isotropic motions occurring in the amorphous phase. Instead, a longer ${}^{C}T_{1\rho}$ is expected. This dilemma is resolved by assuming that the ${}^{C}T_{1\rho}$ of the crystalline phase is dominated by spin–spin interactions as revealed by the sensitivity to spinning frequency. The ${}^{C}T_{1\rho}$ of the amorphous phase is dominated by molecular motion that is not perturbed by the slow-spinning frequency. The motions in the amorphous region have large amplitudes and approach isotropic reorientation with a correlation time of 1.3×10^{-8} s.

Proton Spin–Lattice Relaxation Times in the Rotating Frame

> *The high-resolution magic-angle techniques used to obtain* ^{13}*C NMR spectra of solids can be employed to measure individual* $T_{1\rho}$*s for protons attached to different kinds of carbons.*
>
> –J. Schaefer et al. (*42*)

Proton $T_{1\rho}$ relaxation in the rotating frame, ${}^{H}T_{1\rho}$, is the T_1 process in the rotating frame that occurs for the proton nuclei. When proton $T_{1\rho}$ relaxations are measured, the observed relaxation times are averaged over all of the coupled protons because of spin diffusion. However, individual proton $T_{1\rho}$s can be characterized by variations in the normal ^{13}C CP experiments (*44*). Measurements of ${}^{H}T_{1\rho}$s have the advantage that they probe the molecular motion in the kilohertz range, whereas the T_1s reflect motion in the megahertz range. On the basis of differences in the spectral density in the kilohertz region, the ${}^{H}T_{1\rho}$s may have a weak dependence on the rf field (*45*).

The ${}^{H}T_{1\rho}$s are obtained from the final slopes of the plots of observed carbon magnetization generated by long, matched spin-lock cross-polarization transfers from protons to carbons. After the initial increase in the carbon polarization, the carbon signal decreases via the ${}^{H}T_{1\rho}$ relaxation process. Under high-resolution conditions, the individual carbon resonances follow the protons to which they are most closely coupled. Thus, it is possible to resolve ${}^{H}T_{1\rho}$ differences between protons on different types of carbons. The measurement of individual ${}^{H}T_{1\rho}$s involves the introduction of a variable contact time of cross-polarization. A semi-log plot of the carbon magnetization vs. the time of spin-lock cross-polarization transfer yields ${}^{H}T_{1\rho}$.

For PE, the ${}^{H}T_{1\rho}$s vary with the phase and orientation of the sample. For undrawn samples, the ${}^{H}T_{1\rho}$s were 8 ms for the disordered phase and 170 ms for the crystalline phase. For the drawn sample, the corresponding values were 13 and 65 ms, respectively (*26*).

Heterogeneities in blends of polymers can be detected because the ${}^{H}T_{1\rho}$s are sensitive to spin diffusion in polymer systems on the order of an internuclear distance of 1 nm (*46*). For single-phase systems of multicomponent systems, the rate of spin diffusion depends on the nature of the spatial mixing of the polymer chains. When the chains are intimately and homogeneously mixed, only one ${}^{H}T_{1\rho}$ may be observed. If the mixing is nonuniform, several different ${}^{H}T_{1\rho}$s may be detected depending on the system.

The homogeneity of blends of poly(phenylene oxide) and polystyrene were studied by using the measurements of ${}^{H}T_{1\rho}$ (*46*). The ${}^{H}T_{1\rho}$ experiments are sensitive enough to indicate extensive intermixing of poly(phenylene oxide) and polystyrene chains, but there are small domains in the blends where the polystyrene is not uniformly dispersed.

Polybutadiene–polystyrene block copolymers have been studied by using the ${}^{H}T_{1\rho}$ technique with deuterated polystyrene. Magic-angle spinning alone (no dipolar decoupling) is sufficient to produce high-resolution cross-polarization carbon signals from deuterated chains. Also, without dipolar decoupling, no signal is obtained during cross-polarization for protonated chains. Thus, when a mixture of the deuterated and nondeuterated polystyrene is studied, the signal is a direct measure of the *intermolecular* distance between chains. Spectra were measured with both full decoupling and no decoupling. Carbon magnetization in the difference spectra was used to determine ${}^{H}T_{1\rho}$ as a function of contact time. An increase in ${}^{H}T_{1\rho}$ was observed with increasing dilution of the deuterated polystyrene, a result that indicates that interchain motions play a role in the relaxation process. The total evidence suggests a "random or near random" mixing model for the blends (*47*).

The miscibility and subsequent separation of a range of compositions of solution-cast blends of poly(vinylidene fluoride) (PVF_2) and poly(methyl methacrylate) (PMMA) were studied by using the ${}^{H}T_{1\rho}$ technique (*48*). One amorphous phase was found, and the intimate mixing of the polymer chains in this phase existed for all compositions of the blends, even after aging for 2 months at room temperature. The ${}^{H}T_{1\rho}$ technique can be used to detect the presence of phases or domains in the amorphous component of the blends if the domains are larger than ~20 Å. An increase in the measured ${}^{H}T_{1\rho}$ values indicates the occurrence of a subtle separation between unlike chains in the amorphous phase even though a single amorphous phase is present.

Line-Shape Measurements for the Determination of Dynamic Structure

The effect of chemical-exchange processes on NMR line shapes in solution was discovered in 1953 shortly after the discovery of the chemical-shift and spin–spin splitting phenomena (*49*). Line-shape changes have been extensively used to study dynamic molecular processes such as isomerization, ring inversion, and hindered rotation. The basic concept of the line-shape analysis method is that dynamic processes modulate the transition frequencies and affect the line shapes.

Proton Line-Shape Analysis

The proton NMR absorption signal of a solid homogeneous polymer usually consists of a single broad peak. This broad signal is a consequence of the large number of interactions between the various nuclear magnetic moments. Because each proton exists in a slightly different magnetic environment, the resonance envelope consists of a superposition of numerous individual resonances that, when combined, generate a single broad absorption line. The second moment of the NMR line is frequently used, and for a spin system consisting of N spins of spin I, the second moment is related to the internuclear distance r_{ij} (*50*) by

$$\Delta M_2 = \left(\frac{3}{4}\right) \gamma^4 h^2 I\,(I + 1)\left(\frac{1}{N}\right) \times \sum_{ij} \frac{(1 - \cos^2 \theta_{ij})^2}{r_{ij}^{\,6}} \tag{10.34}$$

where θ_{ij} is the angle between r_{ij} and the direction of applied magnetic field. Another important parameter in broad-line NMR is the line width (expressed in terms of frequency), $\Delta\nu$. Variations in broad-line NMR are primarily sensitive to motion. Consequently, changes in $\Delta\nu$ and ΔM_2 are indications of motional processes in the system. If, as a first approximation, the relaxation process can be expressed by a single correlation time, τ_c, the following relation between NMR parameters and τ_c can be established:

$$\tau_c = \frac{\tan\left[\dfrac{\pi\,(\Delta\nu)^2}{2(\Delta\nu_{RL})^2}\right]}{\alpha\,\Delta\nu} \tag{10.35}$$

where $\Delta\nu_{RL}$ is the line width of the rigid lattice (absence of motion on the NMR time scale), and α is a line-shape parameter.

NMR line shapes for polymers vary from the Gaussian type at lower temperatures to the Lorentzian type at higher temperatures. The appearance of two-component spectra is caused by differences in segmental motions in the polymers. The more mobile motions result in the narrow component being observed.

Generally, spin–spin relaxation times are very sensitive to slower relative motions of polymer chains and can provide information on intramolecular couplings such as chemical cross-links and chain entanglements (*51*). For cross-linked polymeric materials, line-width changes are observed with cure. An increase in the line width is observed with an increase in the degree of cure in vulcanized elastomers. The effect of vulcanization on line width is interpreted in terms of the formation of cooperative domains. Both an increase in $\Delta\nu$ and a shift of the transition range to higher temperatures because of the restriction of molecular mobility induced by cross-links are generally observed in cured elastomers. Line width vs. temperature curves for vulcanized natural rubber exhibit two different transition regions. The low-temperature $\Delta\nu$ change (shifted to higher temperatures with increasing cure times) is attributed to methyl-group rotation, and the higher, more abrupt temperature change is attributed to segmental motion. Greater degrees of curing shift the transition region to higher temperatures. Rubbers cross-linked with irradiation have more hindered motion than rubbers cross-linked with sulfur. This difference is explained by the higher potential barrier of rotation about C–C bonds as compared to C–S bonds.

Line widths in ^{1}H NMR spectra of swollen gels are in the range between standard broad-line and high-resolution widths, depending on the degree of restricted molecular motions, that is, the degree of cross-linking (*51*). Usually, the narrow lines are superimposed on a very broad background.

^{13}C Line-Shape Analysis

Line broadenings in ^{13}C NMR spectra can be influenced by relaxation processes such as motional modulation of the chemical-shift anisotropy and motional modulation of the dipolar carbon–proton coupling. The transverse relaxation is

$$(T_2)^{-1} = A_{\mathrm{CH}} J(\omega_{1\mathrm{H}}) \qquad (10.36)$$

where A_{CH} is the square of the carbon–proton dipolar-coupling strength (*52*). For a proton at a distance r from the carbon of interest

$$A_{\mathrm{CH}} = \frac{\gamma_{\mathrm{H}}^2 \gamma_{\mathrm{C}}^2 h^2}{r^6} \qquad (10.37)$$

Because of chain entanglements, not all of the spatial orientations of the polymer chain are readily accessible, and the residual ^{1}H dipolar field remaining from the incomplete averaging broadens the signals.

If the chain entanglements are restricted by cross-linking, the effect on T_2 is even greater. This effect allows the use of T_2 measurements to quantitatively analyze the extent of cross-linking.

Deuterium Line-Shape Analysis

> *It is only fair to state that ^{2}H NMR at present represents one of the most powerful tools for studying polymer dynamics.*
>
> —H. Speiss (*53*)

> *In order to avoid misunderstanding, we stress that, independent of the specific model employed, the ^{2}H line shape provides a spectroscopic measure of the extent to which the number of conformations accessible for a given segment increases as the free volume increases with temperature.*
>
> —D. Hentschel et al. (*54*)

Deuterium NMR spectroscopy in the solid state is well suited for determining the details of the local molecular motions in polymers (*55*). The interaction of the deuteron quadrupole moment with the electric-field-gradient tensor removes the degeneracy of the Zeeman energy levels to give rise to a quadrupolar splitting of the ^{2}H NMR signal. The NMR frequency of the deuteron is given by

$$\omega = \omega_o \pm \delta(3\cos^2\Theta - 1 - \eta\sin^2\Theta\cos 2\phi) \qquad (10.38)$$

where δ is the quadrupole coupling constant (equal to $3e^2Qq/4\eta$), η is the asymmetry parameter ($0 < \eta < 1$), and the orientation of the magnetic field in the principal-axes system of the electric-field-gradient tensor is specified by polar angles Θ and ϕ. In rigid solids, $\delta/2\pi = 62.5$ kHz (*53*).

However, originally there were difficulties in measuring deuteron spectra in FT NMR spectroscopy because of the rapid decay of magnetization following the application of an rf pulse. This rapid decay precludes the use of standard FT methods because a significant part of the signal is lost in the dead time of the receiver. The rapid decay on a time scale on the order of δ^{-1} (i.e., a few microseconds) results from destructive interference of the different spectral components. By applying a second pulse in quadrature with the first one, the magnetization can be refocused to form a solid echo.

Recently, the required pulse techniques have been developed, and currently ^{2}H spectra can now be reliably and easily obtained (*54, 56, 57*). The quadrupolar-echo sequence consists of two 90° pulses that are separated by a time, τ_1, and a 90° relative phase shift:

$$(\pi / 2)_x - \tau_1 - (\pi / 2)_y$$

A refocusing of the transverse magnetization occurs at a time τ_1 later. Fourier transformation of the echo signal from $t = 2\tau_1$ gives a quadrupolar-echo spectrum. Quadrupolar-echo sequences offer a means of studying molecular dynamics in the range 10^{-8} s $< \tau_c < 10^{-4}$ s.

The solid-echo technique just described is limited by the transverse-relaxation time $T_2{}^*$ being a few hundred microseconds at most. A technique called *spin alignment* overcomes this limitation and allows the study of slow motions. The spin-alignment sequence is (*58*)

$$(\pi / 2)_x - \tau_1 - (\pi / 4)_y - \tau_2 - (\pi / 4)_y$$

The first pulse generates transverse magnetization. During the evolution period, τ, the spin system evolves in the presence of the quadrupolar interactions. Spin alignment is created with the second pulse. During the long waiting period, τ_2, spin alignment occurs. The third pulse transforms the spin alignment back into transverse coherence. This coherence evolves again in the presence of the quadrupolar interactions and leads to an alignment echo in phase with the reading pulse after a refocusing time, τ_1. Fourier transformation of the echo yields a spin-alignment spectrum.

In the absence of motion, δ for deuterons in C–D bonds shows only minor variations, and the asymmetry parameter, η, is approximately equal to zero. For this case, Θ is the angle between the respective C–D bond direction and H_o. Thus, for each orientation of the C–D bonds in the sample, a symmetric splitting is observed. The amount of splitting is directly related to the molecular orientation. In an isotropic sample, the resulting *powder-pattern* line shape is the famous *Pake spectrum* of the $I = 1$ spin system. The deuterium NMR spectrum of a static C–D bond (i.e., one that is not undergoing motion on the deuterium-NMR time scale and has a correlation time greater than 10^{-3} s) consists of a Pake doublet that has a quadrupole splitting, $\Delta\nu_q$, of 128 kHz ($\Delta\nu_q$ is 3/4 of the quadrupole coupling constant, or $3e^2Qq/4\eta$). The value of δ is 164 kHz, and η is 0.014 for polyethylene (*59*).

The electric-field-gradient tensor is axially symmetric, which means that there is a one-to-one correspondence between the orientation of the C–D bond with respect to the magnetic field and the frequency at which this orientation resonates. Furthermore, the symmetry axis of the field-gradient tensor is along the C–D bond axis. Because the field-gradient tensor is axially symmetric and because the molecular orientation of the principal axes of the field-gradient tensor is known, the interpretation of deuterium line shapes in the presence of motion becomes straightforward.

In the presence of molecular motion, the deuteron NMR line shape will change. A particularly simple situation arises if the motion is rapid on a time scale defined by the inverse width of the spectrum in absence of motion δ^{-1}. In this rapid-exchange limit, which is reached for correlation times of less than 10^{-7} s, the motion leads to a partially averaged quadupole coupling, and valuable information about the type of motion can be obtained directly from analysis of the resulting line shapes. The motionally averaged NMR frequencies are then

$$\omega = \omega_o \pm \bar{\delta}(3\cos^2\Theta - 1 - \bar{\eta}\sin^2\Theta\cos 2\phi) \tag{10.39}$$

$$\omega = \omega_o \pm \bar{\omega}_Q \tag{10.40}$$

where the average values reflect the averaging of the field-gradient tensor as a result of the motion.

In Figure 10.21 (*53*), three types of motional mechanisms are considered: the kink-3-bond motion, the crankshaft-5-bond motion, and the 180° jump of a phenyl ring. These motions involve interchanges between 2, 3, and 2 C–H bond directions, respectively, leading to values of $\eta = 1$, 0, and 0.6, respectively. The resulting line shapes are shown in the figure.

In addition to depending on the jump angle of the motion, deuterium NMR line shapes are also sensitive to the correlation times for the motion when the correlation times are in the intermediate exchange regime (10^{-8}–10^{-4} s). Therefore, temperature-dependent solid-state deuterium NMR spectra can be used to extract information concerning both the rate

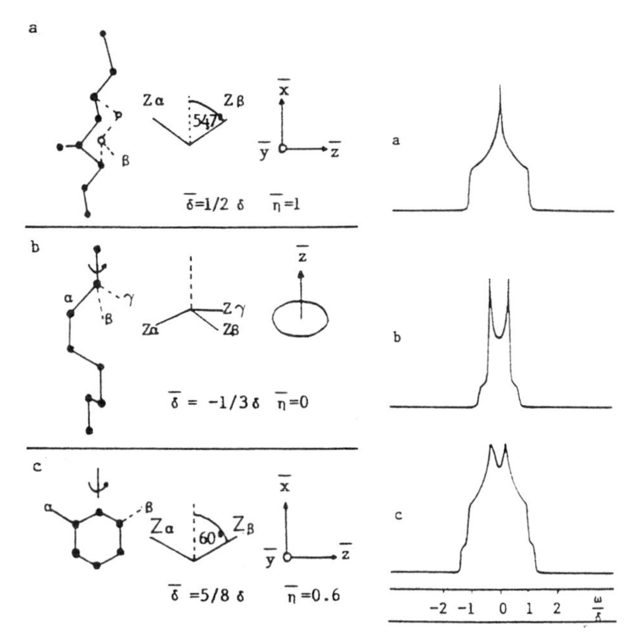

Figure 10.21. Three types of motional mechanisms and their corresponding deuterium line shapes: the kink-3-bond motion (a), the crankshaft-5-bond motion (b), and the 180° jump of a phenyl ring (c). (Reproduced with permission from reference 53. Copyright 1983 Steinkopff Verlag Darmstadt.)

and angular range of local polymer motions. Finally, measurements of spin – lattice relaxation times can be used to verify the correlation times determined from the line-shape analysis and can also be used to determine correlation times when the line shape is no longer sensitive to increased motional rates ($\tau_c < 10^{-8}$ s).

To determine correlation times of various motions, consider as an example a jump motion that will cause the C – D bond direction to fluctuate between two different orientations. In Figure 10.22 (*53*), typical line shapes are displayed. Three important parameters are the splitting in the absence of motion (2Δ), the exchange rate (Ω), and the effective transverse-relaxation time (T_2*). Four different regimes can be observed.

In the fast-exchange limit, $\Omega^2 >> \Delta^2$, and a Lorentzian line shape at the center, ω, is observed with width at half-height of $2/T_2^* + \Delta^2/\Omega$. In the intermediate-exchange region, where Ω^2 ($\simeq \Delta^2$, the signal is spread over the whole frequency range $\omega \pm \Delta$. In the slow-exchange limit, where $\Omega^2 << \Delta^2$, two symmetric Lorentzians centered at $\omega + \Delta$ and $\omega -$

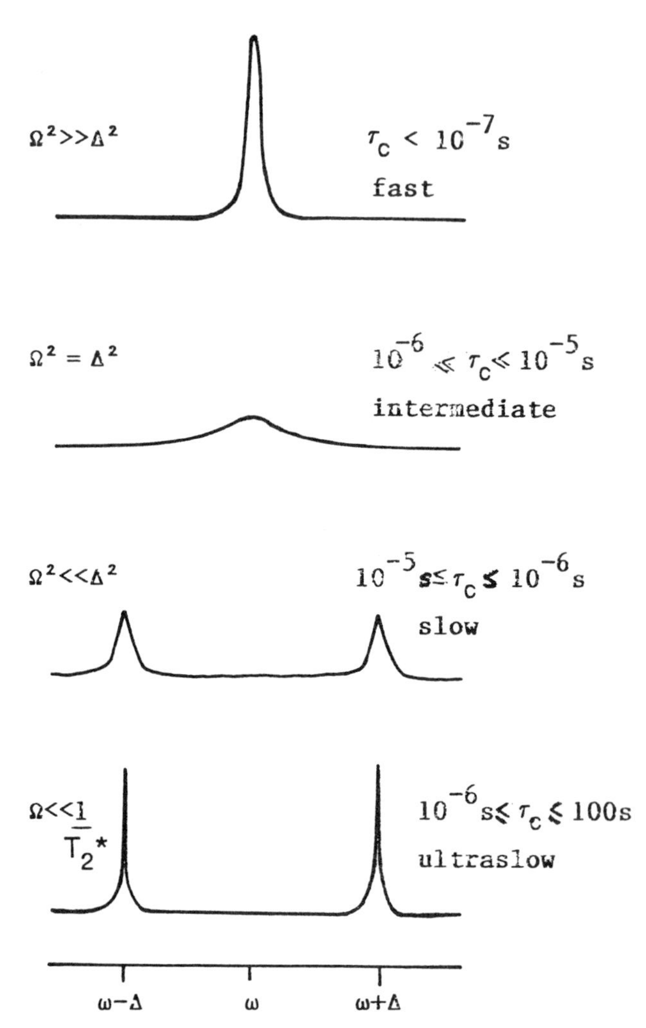

Figure 10.22. Theoretical line shapes resulting from an interchange between two NMR frequencies in the different motional regimes. (Reproduced with permission from reference 53. Copyright 1983 Steinkopff Verlag Darmstadt.)

Δ, with line widths of $2/T_2^* + 2\Omega$, are observed. The exchange rate causes an extra broadening of the lines observed in the absence of motion. In the ultraslow exchange limit $\Omega << 1/T_2^*$, and the line shape becomes insensitive to the motion because the extra broadening is much smaller than the natural line width.

The line-shape analysis of deuteron spectra should provide a means to determine accurate values for correlation times in a range over 3 orders of magnitude. The limits are given by the transverse-relaxation time $T_2^* \sim 5 \times 10^{-4}$ s, which is determined by the dipolar interaction of the deuterons among themselves and with other spins in the system.

The primary limitation of line-shape analysis of deuteron spectra is the need for selectively deuterated sites in the polymer molecule. The chemistry involved in such selective deuteration is often tedious, and encouraging even a capable chemist to make the selective deuteration may often be the most difficult part of the experiment.

Typical ^{2}H spectra for polyethylene (PE) in the temperature range from room temperature to the melting point are shown in Figure 10.23 (*53*). The spectra show the presence of two regions of different mobility. The deuterons in the rigid crystalline regions give rise to a Pake spectrum spanning the full width of 250 kHz. The central portions of the spectra narrow considerably as the temperature is increased and can therefore be attributed to the deuterons in the mobile amorphous regions of the sample. The ^{2}H data can be quantitatively analyzed in "terms of localized motions of flexible units involving not more than 3 – 5 bonds" (*53*). The fraction of flexible units in the amorphous regions as a function of temperature can be calculated. At low temperatures, the kink-3-bond motion is the only active motion near the γ transition. Below room temperature, the average flexible unit involves only three to five bonds, but at higher temperatures seven-bond motions occur.

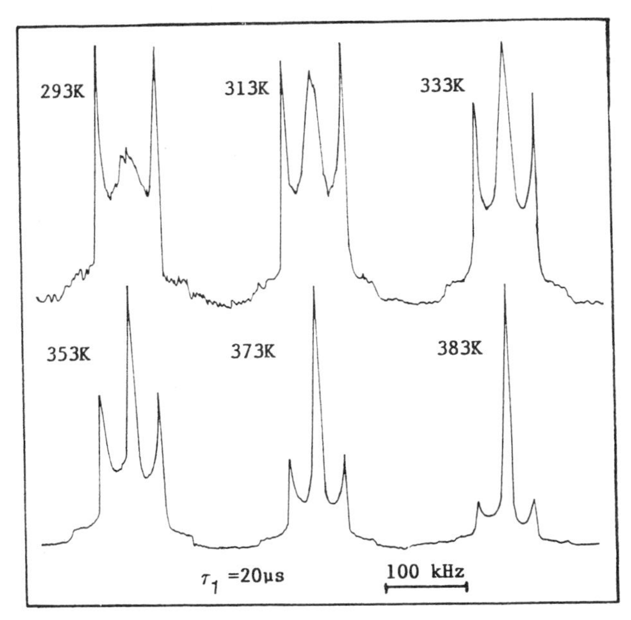

Figure 10.23. The ^{2}H NMR spectra of a linear PE that was isothermally crystallized from the melt at 396 K. The spectra were obtained at 55 MHz for various temperatures up to melting. (Reproduced with permission from reference 53. Copyright 1983 Steinkopff Verlag Darmstadt.)

Applications of the Nuclear Overhauser Enhancement (NOE) Effect for Dynamic Structure Determination

For a two-spin system in which dipolar interactions are the only source of relaxation, the NOE is

$$\mathrm{NOE} = 1 + \left(\frac{\gamma_H^2 \gamma_C^2 h^2}{20 r^6} \right) \times \left[J(\omega_H + \omega_C) - J(\omega_H - \omega_C) \right] {}^{H}T_1 \quad (10.41)$$

This expression allows the C – H distance, r, to be determined readily from the ^{13}C NOE and $^{H}T_1$ measurements in the case of isotropic motion where the spectral densities are simple. This structural case of a single dipole – dipole interaction being the only source of relaxation is extremely rare. If the carbons are relaxed entirely by dipole – dipole interactions, a NOE sensitivity enhancement of 2.9 will be observed. If this maximum is not observed, other relaxations are occurring. The NOE can be zero if dipolar relaxation is totally inefficient.

NOE experiments provide data on internuclear distances resulting from the modulation of the dipole – dipole coupling between different nuclear spins by the Brownian motion of the molecules in solution (*60*). The NOE intensity is related to the distance, r, between the preirradiated and observed spins by an equation of the general form:

$$\mathrm{NOE} \equiv \frac{f(\tau_c)}{\langle r^6 \rangle} \quad (10.42)$$

The function of the correlation time, τ_c, accounts for the influence of the motional-averaging process on the observed NOE. NOE difference measurements can be used to estimate distances between protons. This estimation can be done by following the buildup of NOE until it reaches a steady state (usually over 5 s). The rate at which the NOE grows is proportional to r^{-6}. This method works for rigid systems in which the distances to be estimated are less than 3 Å.

The NOE effect for the CP – MAS spectra of polycarbonate were measured (*22*). The NOE of the carboxyl – carbon resonance was greater than 2, the NOE of the aromatic-carbon resonance was approximately 1.2, and the resonance line due primarily to methyl carbons had a large NOE approaching 2.5. These results were interpreted in terms of different motional characteristics of the carbons.

Relationship Between Molecular Motion as Determined by NMR Spectroscopy and Mechanical Properties of Polymers

A connection between the molecular dynamics as determined by NMR spectroscopy and the mechanical properties of polymers can be expected. However, no connection has been found between dynamic mechanical spectroscopy of polycarbonate and broad-line proton NMR spectroscopy of the same material. Schaefer et al. (*1*) interpreted this lack of correlation on the basis of the suggestion that the motions measured by proton NMR "reflect the long-range dynamic cooperativity of the glassy solid state even though the dipolar interactions themselves are short range. . . . Such motions may have little to do with the main-chain motion responsible for a mechanical loss."

However, CP–MAS ^{13}C NMR experiments can be used to measure the relaxation properties of chemically different types of carbons. Thus, the important main-chain motions can be separated from the unimportant side-chain motions. Additionally, the ^{13}C relaxation parameters are, for the most part, free from the averaging effects of spin-diffusion characteristic of ^{1}H broad-line measurements.

Presently, solid-state NMR spectroscopy and mechanical property measurements can be made on the same sample under end-use conditions. This approach minimizes the possibility of creating artifacts due to different sample preparation procedures.

Dynamic Mechanical Properties

NMR relaxation measurements have been used to identify the two major sub-T_g transitions in the inelastic mechanical-loss response of polystyrene (T_g is the glass transition temperature). The weak tan-delta-gamma transition (1 Hz at −120 °C) of the mechanical-loss spectra of polystyrene has been assigned to cooperative ring flips (10 MHz at 25 °C). The activation energy of the gamma transition of atactic polystyrene is approximately 10 kcal/mol. Thus for a motion with a frequency of 1 Hz at −120 °C, a frequency of 8 MHz is obtained at 25 °C, and this result is determined from the calculation (1 Hz) × exp[−5000(1/298 − 1/153)]. In a similar fashion, the broad intensity tan-delta transition between −60 and 20 °C at 1 Hz is assigned to cooperative ring- and main-chain-restricted oscillations in the range of 10 to several hundred kilohertz (*1*).

Changes in Failure Modes

Quenched PET is a highly ductile material, but annealing at 61 °C for 350 h causes the failure mode to change from ductile to brittle. NMR measurements of $<{}^{C}T_{1\rho}>$ indicate a substantial loss of high-frequency cooperative motions. The ductile behavior appears to arise from cooperative high-frequency components of motion associated with trapped, nonequilibrium ethylene configurations (*42, 61*).

Impact Resistance

A mechanical impact is an energy pulse of a finite duration of approximately 100 ms that generates frequencies on the order of the inverse of this pulse width, that is, frequencies ranging from about 10 to 100 kHz (*1*). High-impact resistance will occur if there is a correspondence between the frequency distribution of the impact and that of the polymer. If the polymer is unable to respond quickly to the impact, local stresses that cause crack formation and failure are generated. If the polymer exhibits sufficient response in the appropriate frequency range to dissipate the energy through internal motions or internal heating, failure can be avoided, and high-impact resistance is observed. The impact requires the utilization of many modes in the polymer in order to dissipate the energy among a large number of segments. To improve impact resistance, the polymer segmental frequencies must be matched with the impact frequencies. NMR spectroscopy allows the characterization of the frequency spectrum of the polymer, particularly in the mid-kHz range, by relaxation experiments.

It was suggested that the ratio, $T_{CH}/T_{1\rho}$, should correlate with the impact strength of glassy polymers (*1*). A short T_{CH} is a measure of those groups suited to coping with the static components of the impact. A short $T_{1\rho}$ measures chain segments inherently capable of dissipating impact energy as heat within the sample. For a homologous series of samples characterized by similar T_{CH} values, $T_{1\rho}$ should correlate with impact resistance. This ratio is a measure of the number of chain segments capable of dissipating the energy of an impact relative to the number of more rigid segments that are unable to respond to the impact. The use of this ratio is an attempt to describe the general shape of the spectral density of the polymer.

References

1. Schaefer, J.; Sefcik, M. D.; Stejskal, E. O.; McKay, R. A.; Dixon, W. T.; Cais, R. E. *Macromolecules* **1984,** *17,* 1107.
2. Bovey, F. A.; Jelinski, L. W. *J. Phys. Chem.* **1985,** *89,* 571.
3. Menger, E. M.; Veeman, W. S.; deBoer, E. *Macromolecules* **1982,** *15,* 1406.
4. Vold, R. L.; Waugh, J. S.; Klein, M. P.; Phelps, D. E. *J. Chem. Phys.* **1968,** *48,* 3831.
5. Freeman, R.; Hill, H. D. W. *J. Chem. Phys.* **1969,** *51,* 3140.
6. Kowalewski, J.; Levy, G. C.; Johnson, L. F.; Palmer, L. *J. Magn. Reson.* **1977,** *6,* 533.
7. Markley, J. L.; Horsley, W. J.; Klein, M. P. *J. Chem. Phys.* **1971,** *55,* 3604.
8. Freeman, R.; Hill, H. D. W. *J. Chem. Phys.* **1971,** *54,* 3367.
9. Freeman, R.; Hill, H. D. W.; Kaptein, R. *J. Magn. Reson.* **1972,** *7,* 82.
10. Torchia, D. A. *J. Magn. Reson.* **1978,** *30,* 613.
11. Fleming, W. W.; Fyfe, C. A.; Kendrick, R. D.; Lyerla, J. R.; Vanni, H.; Yannoni, C. S. In *Polymer Characterization by ESR and NMR*; Woodward, A.E.; Bovey, F.A., Eds.; ACS Symposium Series 142; American Chemical Society: Washington, DC, 1980; 193.
12. Farrar, T. C.; Becker, E. D. *Pulse and Fourier Transform NMR;* Academic Press: New York, 1971; Chapter 6.
13. Havens, J. R.; Koenig, J. L. *Appl. Spectrosc.* **1983,** *37,* 226.
14. Garroway, A. N.; VanderHart, D. L. *J. Chem. Phys.* **1979,** *71,* 2773.
15. Stejskal, E. O.; Schaefer, J.; Steger, T. R. *Faraday Discuss. Chem. Soc.* **1979,** *13,* 56.
16. Schaefer, J.; Sefcik, M. D.; Stejskal, E. O.; McKay, R. A., *Macromolecules* **1984,** *17,* 1118.
17. Lyerla, J. *Methods Exp. Phys. Part A* **1980,** *16,* 241.
18. Abragam, A. *The Principles of Magnetic Resonance* Oxford: New York, 1961; Chapter VIII.
19. Kuhlmann, K. F.; Grant, D. M.; Harris, R. K. *J. Chem. Phys.* **1970,** *52,* 3439.
20. Earl, W. L.; VanderHart, D. L. *Macromolecules* **1979,** *12,* 762.
21. Fleming, W. W.; Fyfe, C. A.; Kendrick, R. D.; Lyerla, J. R.;Vanni, H.; Yannoni, C. S. In *Polymer Characterization by ESR and NMR*; Woodward, A.E.; Bovey, F.A., Eds.; ACS Symposium Series 142; American Chemical Society: Washington, DC, 1980; 193.
22. Schaefer, J.; E. Stejskal; Buchdahl, E. *Macromolecules* **1977,** *10,* 385.
23. Schaefer, J. In *Topics in Carbon-13 NMR Spectroscopy;* Levy, G. C., Ed.; Wiley-Interscience: New York, 1974; Vol. I, Chapter 4.
24. Schaefer, J. *Macromolecules* **1973,** *6,* 882.
25. Ribeiro, A.; Wade-Jardetzky, N. G.; King, R.; Jardetzky, O. *Appl. Spectrosc.* **1980,** *34,* 299.
26. VanderHart, D. L.; Khoury, F. *Polymer* **1984,** *25,* 1589.
27. Douglass, D. C.; Jones, G. P. *J. Chem. Phys.* **1966,** *45,* 956.
28. VanderHart, D. L.; Garroway, A. N. *J. Chem. Phys.* **1979,** *71,* 2773.
29. Andrew, E. R. *Prog. Nucl. Magn. Reson. Spectrosc.* **1971,** *8,* 1.
30. Perez, E.; VanderHart, D. L.; Crist, B., Jr.; Howard, P. R. *Macromolecules* **1987,** *20,* 78.
31. Goldman, M.; Shen, L. *Phys. Rev.* **1966,** *144,* 321.
32. Henrichs, P. M.; Linder, M. *J. Magn. Reson.* **1984,** *58,* 458.
33. Axelson, D. E.; Mandelkern, L.; Popli, R.; Mathieu, P. *J. Polym. Sci., Polym. Phys. Ed.* **1983,** *21,* 2319.
34. Caravatti, P.; Neuenschwander, P.; Ernst, R. R. *Macromolecules* **1985,** *18,* 119.
35. McGarvey, B. R.; Schlick, S. *Macromolecules* **1984,** *17,* 2392.

36. Komoroski, R. A.; Maxfield, J.; Sakaguchi, F.; Mandelkern, L. *Macromolecules* **1979,** *10,* 550.

37. Dechter, J. J.; Axelson, D. E.; Dekmezian, A.; Glotin, M.; Mandelkern, L. *J. Polym. Sci., Polym. Phys. Ed.* **1982,** *20,* 641.

38. Axelson, D. E. *J. Polym. Sci., Polym. Phys. Ed.* **1982,** *20,* 1427.

39. Fleming, W. W.; Lyerla, J. R.; Yannoni, C. S. In *NMR and Macromolecules Sequence, Dynamic and Domain Structure*, Randall, J. C., Jr., Ed.; ACS Symposium Series 247; American Chemical Society: Washington, DC, 1984; 83.

40. Gabrys, B.; Horii, F.; Kitamaru, R. *Macromolecules* **1987,** *20,* 175.

41. Jelinski, L. W.; Dumais, J. J.; Watnick, P. I.; Engel, A. K.; Sefcik, M. D., *Macromolecules* **1983,** *16,* 409.

42. Schaefer, J.; Stejskal, E. O.; Sefcik, M. D.; McKay, R. A. *Philos. Trans. R. Soc. London, A:* **1981,** *299,* 593.

43. Belfiore, L. A.; Henrichs, P. M.; Massa, D. J.; Zumbulyadia, N.; Rotwell, W. P.; Cooper, S. L. *Macromolecules* **1983,** *16,* 1744.

44. Pines, A.; Gibby, M. G.; Waugh, J. S. *J. Chem. Phys.* **1973,** *59,* 569.

45. Schaefer, J.; Stejskal, E. O.; Steger, T. R.. Sefcik, M. D.; McKay, R. A., *Macromolecules* **1980,** *13,* 1121.

46. Stejskal, E. O.; Schaefer, J.; Sefcik, M. D.; McKay, R. A., *Macromolecules* **1981,** *14,* 275.

47. Schaefer, J.; Sefcik, M. D.; Stejskal, E. O.; McKay, R. A., *Macromolecules* **1981,** *14,* 188.

48. Grinsted, R. A.; Koenig, J. L. *J. Polym. Sci., Polym. Phys. Ed.* **1990,** *28,* 177.

49. Gutowsky, H. S.; McCall, D. W.; Slichter, C. P. *J. Chem. Phys.* **1953,** *21,* 279.

50. Slichter, C. P. *Principles of Magnetic Resonance;* Harper & Row: New York, 1980.

51. Andreis, M.; Koenig, J. L. *Adv. Polym. Sci.* **1989,** *89,* 71.

52. Laupretre, F.; Monnerie, L.; Virlet, J. *Macromolecules* **1984,** *17,* 1397.

53. Spiess, H. W. *Colloid Polym. Sci.* **1983,** *261,* 193.

54. Hentschel, D.; Sillescu, H.; Spiess, H. W. *Macromolecules* **1981,** *14,* 1605.

55. Spiess, H. W. *NMR-Basic Principles and Progress;* Springer-Verlag: New York, 1978; 15, pp 55-214.

56. Hentschell, D.; Spiess, H. W. *J. Magn. Reson.* **1979,** *35,* 157.

57. Spiess, H. W.; Sillescu, H. *J. Magn. Reson.* **1980,** *42,* 381.

58. Spiess, H. W. *J. Chem. Phys.* **1980,** *72,* 6755.

59. Hentschel, D.; Sillescu, H.; Spiess, H. W.; Voelkel, R.; Willenberg, B. *Magn. Reson. Relat. Phenom., Proc. Congr. Ampere, 19th* **1976,** *381.*

60. Wuthrich, K. *NMR of Proteins and Nucleic Acids;* Wiley: New York, 1986.

61. Sefcik, M. D.; Schaefer, J.; Stejskal, E. O.; McKay, R. A., *Macromolecules* **1980,** *13,* 1132.

Suggested Reading

Fedotov, V. D.; Schneider, H. *Structure and Dynamics of Bulk Polymers by NMR Methods;* Springer-Verlag: New York, 1989.

Weingaertner, H. *Nucl. Magn. Reson.,* **1989,** *18,* 134.

11

NMR Imaging of Polymeric Materials

Conventional NMR spectroscopy is used to determine the chemical structure of a sample, but it cannot be used to locate the position of the stimulated nuclei in the sample. NMR imaging is a method in which the stimulating signal is encoded so that an image can be reconstructed to show the spatial distribution of the nuclei in the sample. NMR imaging is performed the same way as standard NMR spectroscopy except for spatially encoding the signal.

The key step in NMR imaging is the application of a linear magnetic gradient to the sample in the static magnetic field. In this way, spatial encoding of the stimulated signal is accomplished, and the frequencies of the nuclei reflect their position in the sample. To use an analogy, this process is similar to having a pianist play a chord on a piano. By using frequency analysis of the resulting sound wave, the positions of the piano keys that were struck can be determined.

The NMR imaging technique relies on the detection of the nuclei in a small and controllable region of the sample where the nuclear-resonance frequency is matched to the rf signal. This selected region is systematically moved over the entire sample, and an image is obtained.

The NMR signals depend on the nuclear-relaxation time constants, which respond to the structural environments of the emitting nuclei. NMR imaging can be used to measure the spatial distribution of the NMR parameters, including spin density and the spin – lattice and spin – spin relaxation times. Thus NMR imaging can provide a variety of structural factors measured in situ. NMR imaging may be considered a type of chemical microscope that can obtain images of heterogeneous samples.

Medical Magnetic Resonance Imaging (MRI)

The past decade has witnessed a tremendous growth in the medical applications of NMR imaging for the purpose of clinical diagnosis. Medical professionals usually refer to magnetic resonance imaging as MRI in order to prevent negative thinking on the part of the patient.

> Here I will continue to use the term nuclear magnetic resonance imaging or NMRI in order to differentiate it from medical imaging or MRI. It is a little late to try to reduce the "negative thinking" on the part of the reader relative to NMR.

Medical applications of NMR imaging rely on tissue contrasts due to differences in relaxation times. The hypothesis is that the relaxation times of malignant tumors are longer than those of the healthy tissue. This property is apparently not unique to cancerous tissue, but rather is indicative of the change in the molecular-level structure and content of water associated with certain disease states. In general, the T_1 is longer in malignant tissue by a factor of 1.5 – 2.0. Most lesions have long T_1s and T_2s as compared to adjacent tissue. Tumors, infarctions, hemorrhages, and demyelinating diseases give good contrast in images because of differences in T_2s. At present, the most effective procedure for detecting soft-tissue tumors is the use of MRI. NMR images of cancer patients display good discrimination of bone, brain, and liver tumors. MRI is particularly useful for discovering growths that are hidden by bone. Other pathologies detected by MRI imaging include hydro-

1904—4/92/0285$09.25/1

cephalus, carotid artery aneurysm, edema associated with kidney transplants, liver cirrhosis, and multiple sclerosis plaques.

Image contrast due to differences in relaxation times can be artificially enhanced by the addition of very low concentrations of paramagnetic elements. The paramagnetic agents affect only the relaxation times of accessible segments, and this effect highlights regions deprived of flow or suffering ischemia (ischemia is a condition of biological tissue whereby the supply of oxygen is inadequate to maintain normal tissue function). By monitoring the change in relaxation rate as a function of time, the rate of clearance of the paramagnetic substance from the blood, muscle, abdominal organs, or myocardium can be measured. Subsequently, the liver and gall bladder functions can be assessed. With real-time imaging, the rate of physiological processes in tissues can be recorded.

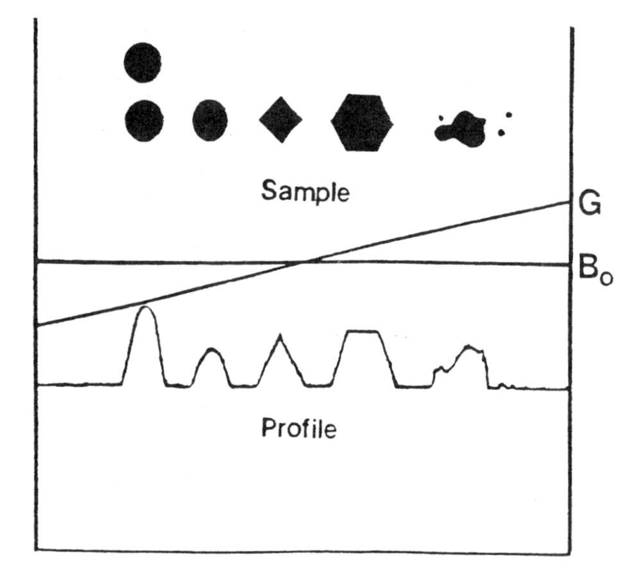

Figure 11.1. The NMR imaging experiment in one dimension with a linear magnetic gradient. The experiment generates a profile of the spin density of the sample.

Basis of NMR Imaging

In conventional NMR spectroscopy, the sample is placed in a homogeneous magnetic field, H_o, so that all chemically equivalent spins experience the same field and hence precess at the same Larmor angular frequency, ω_o, given by

$$\omega_o = \gamma H_o \quad (11.1)$$

where γ is the magnetogyric ratio. Under these circumstances, the observed proton NMR spectrum of water consists of a single sharp line with an intrinsic width of about 0.1 Hz.

In contrast with conventional NMR spectroscopy, NMR imaging involves placing a sample in a deliberately nonuniform magnetic field (*1*), which is achieved by modifying the homogeneous magnetic field with a system of gradient magnetic coils to generate linear gradients on the order of a few gauss per centimeter (1 G/cm = 10^{-2} T/m). The purpose of the nonuniform field is to label, or *encode*, different regions of the sample linearly with different NMR frequencies. Because the magnetic field is varied in a known manner at specific positions within the sample, the frequency of the NMR signal indicates the spatial position of the resonating nuclei (Figure 11.1). In one dimension (1-D), the position of the sample is related to a frequency by the relationship

$$\Delta\omega_z = \omega_z - \omega_o = \gamma G_z z \quad (11.2)$$

where the magnetic-field gradient $G_z = \partial B_z/\partial z$. A tailored rf pulse with a narrow frequency range is used to excite only those nuclei at corresponding positions in the z dimension. The amplitude of the NMR signal received from the z-axis line is a measure of the number of resonant nuclei on that line, and the NMR spectrum represents a graph of spin density vs. distance (neglecting relaxation effects). The field gradient is described by a tensor with nine components, but for a large H_o you need only be concerned with the three components $G_\alpha = \partial B_\alpha / \partial\alpha$, where $\alpha = x, y, z$.

Three-dimensional (3-D) NMR imaging requires a 3-D gradient field. The frequency spectrum (still obtained by Fourier transforming the FID) gives the number of resonating spins along a specific direction of the field gradient. In fact, each plane perpendicular to the direction of the field gradient has a different resonance frequency, and the signal intensity at that frequency will be proportional to the number of nuclei contained in that plane. In other words, the frequency spectrum is just a projection of the spin density (neglecting relaxation effects for the moment) along the field direction.

Imaging Methods

In the 2-D Fourier transform imaging method (2-D FT), a slice is defined in the xy plane by selective excitation (*2*). A field gradient, G_y, is then applied along the y direction for a time t_y, after which a gradient, G_x is applied along the x direction for a time t_x. The NMR signal is read as a function of t_x to n data points. The process is repeated for n times t_y to give an array of n^2 data points. If proton density at any point (x, y) in the defined slice is $\rho(x, y)$, and if relaxation effects are neglected, the NMR signal from an element of area $dx\,dy$ at (x, y) in the slice will be proportional to

$$\rho(x, y) \exp (i\gamma Bt)\, dx\, dy \qquad (11.3)$$

During the time t_x,

$$B = B_o + xG_x \qquad (11.4)$$

During the subsequent time t_y,

$$B = B_o + yG_y \qquad (11.5)$$

After eqs 11.4 and 11.5 are substituted in eq 11.3 and the signal is demodulated at the angular frequency $\omega_o = \gamma B_o$, the NMR signal from the element is proportional to

$$\rho(x, y) \exp [i\gamma(xG_xt_x + yG_yt_y)]\, dx\, dy \qquad (11.6)$$

For the whole slice, the NMR signal is

$$S(t_x, t_y) = \int\int \rho(x, y) \exp [i\gamma(xG_xt_x + yG_yt_y)]\, dx\, dy \qquad (11.7)$$

on which a 2-D FT can be performed to give an image of $(n \times n)$ pixels with values of $\rho(x, y)$.

The spin-warp variation of 2-D Fourier imaging successively increments G_y through n values while holding t_y constant, rather than incrementing t_y through n successive values (*3*).

2-D Fourier imaging can be extended to 3-D imaging. A gradient, G_z, is applied along the z axis for a time t_z, then G_y is applied along the y axis for a time t_y, and finally, G_x is applied along the x axis for a time t_x. The NMR FID is recorded to n data points during t_y. The procedure is repeated for n values of t_y, and for each of these, for n values of t_z. The NMR signal for the whole object is

$$S(t_x, t_y, t_z) = \int\int\int \rho(x, y, z) \exp [i\gamma(xG_xt_x + yG_yt_y + zG_zt_z)]\, dx\, dy\, dz \qquad (11.8)$$

The 3-D Fourier transform yields the 3-D image, $\rho(x, y, z)$.

Volume Selection in NMR Imaging Methods

From the preceding discussion, the methods of volume selection in NMR imaging can be classified in a simple scheme based on how the volume elements are selected in each experiment (*1*). In this classification scheme, the sample volume is divided into $n_x\, n_y\, n_z$ elements, with the n_x, n_y, and n_z volume elements, called *voxels*, along the three Cartesian axes.

The *sequential point method* is conceptually the simplest imaging experiment. Each individual voxel is sampled by a separate experiment. The sampled voxel is moved electronically throughout the sample volume. Hence, the number of experiments, N, required for an image is equal to $n_x\, n_y\, n_z$. Although this is time consuming, the sequential point method has the advantage of simple image reconstruction.

In the *sequential line method*, volume elements along an entire selected line are observed simultaneously. This method reduces the number of experiments required for the complete image to meet the requirement that $N = n_xn_z$ (for the n_y line). To isolate a specific line in a sample, a second rf pulse and gradient coil are activated after the first gradient field and rf pulse. Because the two coils are perpendicular, the planes selected are also perpendicular, and the planes intersect in the line that is sampled. Fourier transformation of the FID gives the spin density in that line.

Simultaneous observation of an entire imaging plane is termed the *sequential plane method*, for which $N = n_z$. By applying a gradient during the application of a narrow rf pulse, the nuclei are excited in a plane whose position and thickness depend on the rf-pulse shape. Once the plane has been selected by the z gradient, the x and y gradients are employed to determine the location within the selected xy plane of the spin-density and relaxation parameters.

Finally, the optimum experiment is one in which all voxels are sampled simultaneously, leading to a 3-D NMR imaging method with $N = 1$.

Use of Gradient Magnetic Fields in Imaging

The spatial distribution of spins in one dimension can be easily determined by acquiring the NMR signal in the presence of a constant magnetic-field gradient. The gradient will cause a linear dependence of the resonant frequency as a function of position along the gradient axis. Consequently, spins at one end of the axis will be in a high magnetic field and will have a higher resonance frequency, and spins at the other end will be in lower fields and will have lower resonant frequencies. However, the presence of the gradient dephases the spin magnetization and prevents the detection of any signal. Fortunately, the spin magnetization can be made in phase by using a *gradient echo*.

To initiate a gradient echo, a negative gradient is applied in the readout direction to produce dephasing of the affected volume of spins. This is followed immediately by a positive gradient to produce the rephasing phenomenon. All of the spins are back in phase at the center of the readout gradient.

Magnetic fields are generated by means of quadrupolar coils that are designed to introduce linear gradients along the static magnetic field and along the perpendicular directions. By adjusting the currents, stepped gradients of different magnitudes can be produced. The direction of the gradient field depends on the direction of the current flowing in the wire, and the magnitude of the gradient field is inversely proportional to the distance from the coil. When an electric current flows through a wire coil, a magnetic field encircles the coil to produce a net magnetic vector in the center of the coil. If two coils (termed a *Maxwell pair*) are used, and the current flows clockwise in one and counterclockwise in the other, the opposing magnetic fields sum to zero in a plane halfway between the two coils. By superimposing two opposing-coil fields on the uniform static field, the net magnetic field can be made to vary in a particular direction.

The three gradient coils fit coaxially inside the bore of the magnet. The z gradient is generated by two opposing coils, and the x and y gradient fields require a four-coil arrangement. As with the z gradient, the small field variations caused by the gradient coils are superimposed on the strong static field that is oriented along the z axis. The variation in the z-oriented field along the y axis produces the y gradient. The geometry of the x-gradient coil is identical to that of the y-gradient coil, except that the coils are rotated 90° so that the variation is along the x axis.

Selective-Excitation Techniques

In ordinary FT NMR spectroscopy, an rf pulse rotates the magnetization of all the voxels in the sample. For NMR imaging, selective excitation requires an rf pulse that rotates the magnetization of only selected voxels within a slice of the object from along the z axis to the y axis.

Selective-excitation techniques restrict a specific volume by the following method (*4*). A gradient coil that permits the sample volume to be separated into a series of slices or planes is activated. Each slice has a characteristic resonant frequency that, except for the central plane, varies slightly from the resonant frequency of the static magnetic field because of the superimposed gradient field (Figure 11.2). If a tailored rf pulse (in this case a sine-function-modulated rf pulse) of exactly one frequency is applied, only protons in one very thin slice perpendicular to the applied gradient will resonate with a frequency corresponding to $\omega = \gamma(B_o + G_z z)$ (Figure 11.2). Spins on either side of this slice will be progressively less affected the further they lie from the isochromatic slice.

In practice, a range of frequencies is contained in the rf pulse. For a particular gradient, the mean frequency of the pulse determines the position of the slice, and the range of frequencies in the pulse determines the thickness of the slice (Figure 11.3).

The transverse magnetization under these gradient conditions does not end up on the y axis but rather is at an angle $2\pi\nu'\Delta t$ to produce a dephasing of the signal. This dephasing can be reversed by applying a gradient of the same amplitude but opposite in sign for a time Δt. This is known as a *rephasing gradient* (Figure 11.3).

Imaging by the Slice

Signals can be obtained simultaneously from all points in a plane by using projection-reconstruction

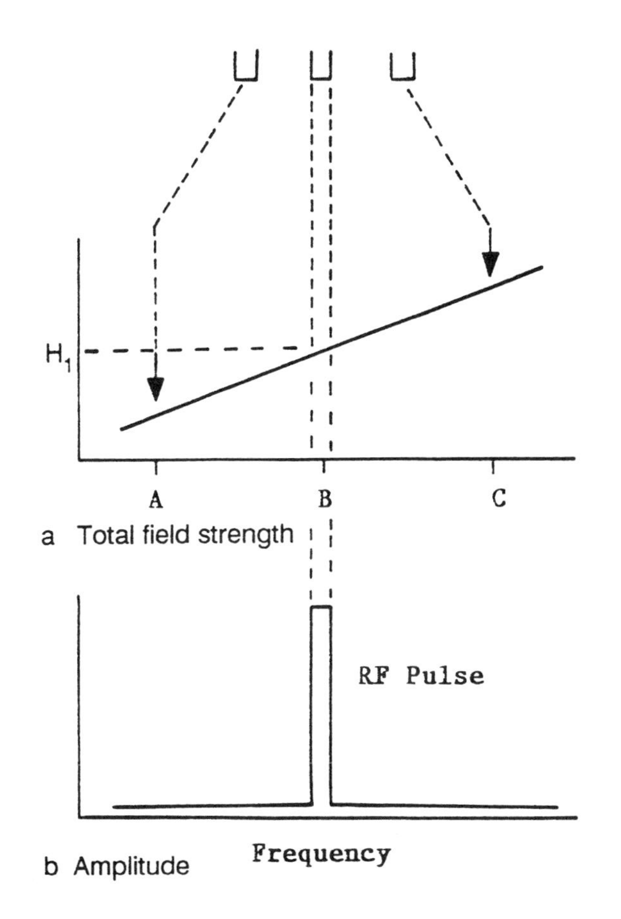

Figure 11.2. (a) Ideal selective 90° rotation H_1 *and gradient profiles. (b) The slice profile corresponding to the* H_1 *profile in a. (Reproduced with permission from reference 9. Copyright 1984 Taylor & Francis, Ltd.)*

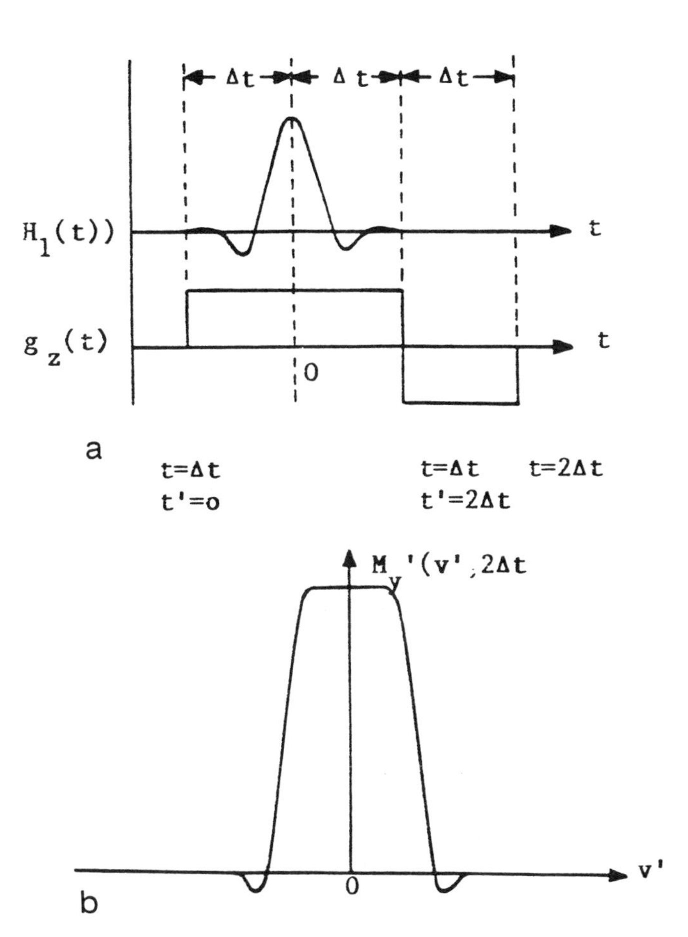

Figure 11.3. (a) Selective 90° rotation H_1 *and gradient profiles. (b) The slice profile corresponding to the* H_1 *profile in a. (Reproduced with permission from reference 9. Copyright 1984 Taylor & Francis, Ltd.)*

(*5*) or 2-D Fourier transform (2-D FT) techniques (*3*).

Projection reconstruction produces a 2-D image from back projections of a plane onto a line. Each projection is obtained by exciting all the nuclei in an *xy* plane and then observing the resulting signal in the presence of a linear gradient (Figure 11.4) (*6*). Fourier analysis of the NMR signal separates the frequencies along the gradient to give a set of points that is the projection of the plane onto a line. This process is repeated many times with different gradient directions. When sufficient projections have been accumulated, a 2-D image of the slice can be reconstructed (Figure 11.4).

To obtain images with the 2-D FT technique, the volume is first divided into a series of *xy* planes by application of a *z* gradient coupled with selective irradiation. During the phase-encoding period, Δt_y, nuclei in a column of voxels experience different magnetic fields. At the end of the phase-encoding period, the nuclei in different voxels have built up a phase angle difference, ϕ_y, from voxels with a lower G_y gradient. After the variable *y* gradient has been turned off, the spins again precess at the same frequency. However, they "remember" the previous phase-encoding event by retaining their characteristic *y*-coordinate-dependent phase angles. As the *y* gradient is increased on subsequent repetitions, the phase relationships along the *y* axis increase linearly.

Following the G_y gradient, the readout gradient, G_x, is activated to create a distribution of frequencies along *x*. The presence of the G_x gradient causes increasing magnetic field strength in the *x* direction. The precessional frequencies of the nuclei in the row of voxels in the *x* direction vary depending on their

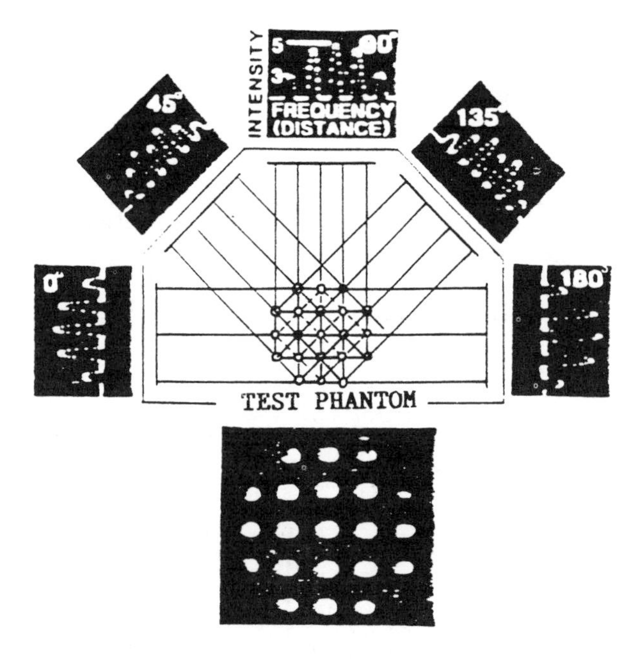

Figure 11.4. The projection-reconstruction method. A linear magnetic-field gradient is applied to the sample so that the resonance frequencies of the nuclear spins take on a spatial dependence. The resulting NMR spectrum represents a projection of nuclear-spin densities perpendicular to the direction of the applied gradient. The figure shows experimentally obtained projections taken at 0°, 45°, 90°, 135°, and 180° from a test phantom. A 2-D image is produced from a series of projections distributed over a 180° arc in analogy to X-ray computerized tomography. (Reproduced with permission from reference 6. Copyright 1985 Optical Society of America.)

location along the x axis. The frequencies caused by the G_x gradient combine to form the FID signal that is sampled. Although this signal contains information from all voxels in the imaging slice, the information gathered from one such cycle is not sufficient to determine the signal amplitude that is required in each voxel to reconstruct the image. Consequently, the cycle must be repeated with a different setting of the phase-encoding gradient, G_y. By using Fourier analysis, any shape can be approximated by a sum of sine waves of a given amplitude and frequency. Similarly, the projection of spin density along a column can be derived from the superposition of these different phase-shifted signals.

Image-Reconstruction Methods

The domain of signal acquisition in NMR imaging comprises a product of the magnetic-field gradient and a time, and this product is termed k space. An appropriate sampling of k space followed by Fourier transformation yields the desired spin density in the spatial domain. Two image-reconstruction techniques in common use are *Fourier imaging* (FI) and *filtered back projection* (FBP), which employ Cartesian and radial sampling rasters, respectively. FI requires the sequential application of orthogonal gradients. In contrast, FBP involves the simultaneous application of gradients.

For FI, the magnetic-field gradient in the longitudinal field has transverse G_x and G_y gradients. These components are applied for durations of t_x and t_y, respectively. The signal, S_o, arising from the transverse magnetization persisting at the time of origin of sampling ($k_x = 0$) is a function of the apparent nuclear-spin density $\rho(x, y)$. The k space is defined by the components

$$k_x = (2\pi)^{-1}\gamma G_x t_x \qquad (11.9)$$

$$k_y = (2\pi)^{-1}\gamma G_y t_y \qquad (11.10)$$

The signal obtained with FI is

$$S(k_x, k_y) = \int \int_{\infty}^{-\infty} \rho(x, y) \times \exp\,[2\pi i(k_x x + k_y y)]\, dx\, dy \qquad (11.11)$$

and Fourier transformations lead to

$$\rho(x, y) = \int \int_{\infty}^{-\infty} S\,(k_x, k_y) \times \exp\,[-2\pi i(k_x x + k_y y)]\, dx\, dy \qquad (11.12)$$

Similar results are obtained for FBP when the coordinates are expressed in polar coordinates k and ϕ. The polar Fourier transformation can be written as

$$\rho(x, y) = \int \int_{\infty}^{-\infty} S\,(k, \phi) \times \exp\,[-2\pi ikR]\,|k|\, dk\, d\phi \qquad (11.13)$$

where $R = x \cos \phi + y \sin \phi$ and corresponds to the component of the vector (x, y) along the line inclined at the angle ϕ to the x axis. All of the elements lying along a normal to this line at a distance R from the origin will possess a common contribution from the integral over k space, which is an effect known in the image-reconstruction process as *back projection* (*7*).

Relaxation Parameters in NMR Imaging

The spin densities and the molecular environments of the nuclei are reflected in the time variation of the amplitude of the measured rf signal and hence are reflected in the intensity of each voxel in the image. When the T_1s and T_2s are different in the voxels of a heterogeneous sample, these differences can be exploited to develop contrast in the NMR images. The pulse sequence that is usually used to measure the T_2-relaxation phenomena in images is called *multiple spin-echo*. At a given repetition time, T_R, the NMR signal is measured at several different echo times, T_E. These echoes provide a measure of the T_2 relaxation. By repeating the process at different T_R values, the T_1 relaxation can also be measured.

Because differences in relaxation times and spin densities determine image contrast, data on relaxation times are important for the selection of the optimal rf pulse sequence for imaging a selected sample. Relaxation times can be measured at any point on an image. The ability to accurately quantify relaxation rates is important in understanding and optimizing image contrast. Spin density, T_1, and T_2 images can be computed from measurements by using pulse sequences with predetermined variations (*8*). These fundamental images represent the inherent data in the system and can be recombined to reconstitute computed images for a given pulse sequence.

Pulse Sequences for Generating Contrast in Imaging

Contrast in NMRI depends on both material-specific and operator-selected parameters. The material-specific parameters include the spin density and the relaxation times T_1 and T_2. The operator-selected parameters include the pulse sequence (inversion-recovery or spin-echo) and the pulse delay and repetition times (timing parameters). For a given imaging system and pulse sequence, the delay and repetition times, in conjunction with the intrinsic material parameters, dictate the appearance of the final image. If the correct pulse sequence is employed and if the relaxation times of the two materials are known, the delay and repetition times that will produce the maximum difference in signal intensity between those materials can be calculated.

Partial Saturation Images

Partial saturation images are produced by repeated 90° rf pulses:

$$(90° - T_R)_n$$

where T_R is the repetition time. In general, *partial saturation* refers to any pulsing sequence that is recycled prior to complete longitudinal recovery of the magnetization. Subsequent sequences do not begin with equilibrium magnetization throughout the sample. The partial-saturation sequence is generally the method of choice in terms of efficiency, which is defined by signal-to-noise ratio per unit time. Repetition of the pulse sequences increases the signal-to-noise ratio and improves the contrast in the resulting images.

When the first 90° pulse is followed by a second 90° pulse, the value of the second FID is determined by the amount of longitudinal relaxation that has occurred in the interval, T_R, between the pulses:

$$\mathrm{FID}_o = M_o\left(1 - e^{\frac{-T_R}{T_1}}\right) \qquad (11.14)$$

The amount of recovery allows differentiation of samples with different T_1 values. A volume element with a short T_1 will recover more fully than a volume element with a long T_1. When partial saturation is used, a voxel with a short T_1 returns a stronger FID signal than a voxel with a long T_1.

Partial-saturation images depend on spin density and T_1 but are independent of T_2 (*9*). The signal intensity, I, in a partial-saturation sequence can be expressed as

$$I \propto N(\mathrm{H})\left(1 - e^{\frac{-T_R}{T_1}}\right) \qquad (11.15)$$

where N(H) represents the proton density. If the

pulse-repetition time is much longer than the T_1s of all the protons being sampled ($T_R >> T_1$), then e^{-T_R/T_1} is approximately equal to 0, and the signal is no longer dependent on the T_1 relaxation times. Assuming a homogeneous proton density, image contrast is caused by differential saturation arising from differences in T_1s.

Inversion-Recovery Images

Inversion-recovery images are generally produced by repetitions of a 180° rf pulse followed by a 90° rf pulse to generate an FID:

$$(180° - T_I - 90° - T')_n$$

where T' is the recovery-time interval between the 90° pulse and the subsequent 180° pulse, and T_I is the interpulse interval. The repetition time, T_R, is defined as the time between successive cycles ($T_R = T' + T_I$).

The dependence of the NMR image intensity on both the T_1 relaxation time and the time interval between pulses in the inversion-recovery technique is

$$I \propto N(\mathrm{H})\left(1 - 2e^{\frac{-T_I}{T_1}} e^{\frac{-T_R}{T_1}}\right) \qquad (11.16)$$

If the T_I between the 180° pulse and the subsequent 90° detection pulse is much shorter than the sample T_1 relaxation times, the signal will have a negative value (inverted magnetization). Phase-sensitive images provide greater contrast between fast- and slow-recovering species than magnitude images. Phase-sensitive images also provide the ability of isolating one component of a two-component system by nulling the appropriate signal through an appropriate choice of T_I.

Spin-Echo Images

The spin-echo (SE) technique is the most common pulse sequence applied in NMRI today. Images are constructed by acquiring a multitude of projections (typically 256 per image), each with an identical setting of a readout gradient during which the sequence is sampled. Each projection is differentiated from the others by a phase difference that is produced by advancing the phase-encoding gradient.

As shown in Figure 11.5, the method consists of a series of rf pulses that are repeated many times in order to achieve a sufficient signal-to-noise ratio. Each projection is produced by a 90° pulse, followed by a 180° pulse for induction of the spin echo. The 90° rf pulse tips the magnetization into the xy plane where the magnetization begins dephasing. The 180° rf pulse is applied after a time, τ, and forces the magnetization to refocus at a time 2τ (also known as the echo time, T_E) after the 90° rf pulse, at which time the data are collected. The frequency-encoding gradient, G_X, causes the spins to precess at different frequencies depending on their position in the static magnetic field. The phase-encoding gradient, G_Y, is orthogonal to G_X. Varying the intensity of G_Y causes the spins to dephase at different rates, and this result provides the second dimension of a 2-D image. The slice-selection gradient, G_Z, and the Gaussian-shaped 90° rf pulse determine the position and thickness of the region of interest. The data are Fourier transformed in two dimensions to produce the image of the selected slice. The time delay between the observation pulse and the observation is called the *echo time* (T_E). The time between two consecutive pulse sequences, the repetition time (T_R), usually ranges from 250 ms to 10 s.

Spin-echo techniques have a unique position in NMR spectroscopic applications. The main problem with NMR imaging is the long data-collection time, which is mainly due to the spin–lattice relaxation

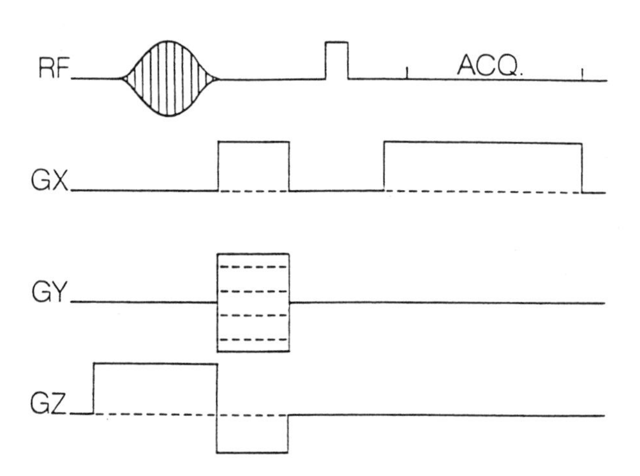

Figure 11.5. The pulse sequence for the spin-echo method.

time, T_1. Each measurement necessitates a time period on the order of T_1 (which is approximately 0.5 s for aqueous systems) for the system to return to equilibrium magnetization. By using spin-echo repetition, a large number of spin echoes can be repeated within a T_1 or T_2 decay period.

The spin-echo experiment starts with a 90° rf pulse (Figure 11.5). A magnetic-field gradient is switched on in the x direction. After some incremented time, $T_{I/2}$, a 180° pulse is applied. The phase of the different isochromats is then inverted. The gradient is switched on with an opposite sign, then is switched off after another $T_{I/2}$ period. At the end of the evolution period, the different components of the magnetization are dispersed due to only their location. The dispersions due to either field inhomogeneity or chemical shift are refocused.

The spin-echo sequence can be expressed as

$$(90° - T_I - 180° - T')_n$$

The amplitude of the spin echo, which determines the pixel value, is given by

$$I \propto N(\mathrm{H}) \left[1 - 2e^{\frac{-(T_R - T_I)}{T_1}} + e^{\frac{-T_R}{T}} \right] e^{\frac{-T_E}{T_2}} \quad (11.17)$$

where T_E is called the *echo delay* ($T_E = 2T_I$). The signal intensity is related to both T_1 and T_2. This equation can be simplified under the condition that T_I is short compared to the repetition time, T_R (i.e., $T_I << T_R$), then $(T_R - T_I) \sim T_R$.

$$I \propto N(\mathrm{H}) \left(1 - e^{\frac{-T_R}{T_1}} \right) e^{\frac{-T_E}{T_2}} \quad (11.18)$$

Hence, the spin-echo signal intensity increases if T_2 increases, T_1 decreases, or T_E is shortened. The intensity increases if the T_R between successive 90° pulses increases. These results indicate that the pixel intensities depend on T_1 and T_2 as well as on the operator-variable parameters T_R and T_I.

If T_R is long compared to the longest T_1 of the voxels, the exponential e^{-T_R/T_1} goes to 0, and the signal intensity becomes independent of T_1 and is modulated by only T_2 relaxation processes. If the T_E is shortened relative to T_2, then e^{-T_E/T_2} approaches 1, and the signal is related only to T_1.

Multiecho Method

A variation of the spin-echo technique is the use of a series of echos. A nonselective 180° rf pulse is applied at a time, τ, after the acquisition. A second echo is produced at time 2τ after the second 180° rf pulse. The total echo time for the second echo is $2T_E$. A second incremented gradient in the y direction after acquisition is used to rephase the spins before the next 180° rf pulse. The cycle is repeated until the desired number of echoes is acquired. This allows the total time of the experiment to be reduced by the total number of different echoes acquired because all echoes are collected within one delay time (rather than one echo per delay).

The later echoes produce progressively darker images because the intensity of the echoes gradually decreases from the T_2 decay. The intensity of all voxels decreases in each successive echo, and the signals from voxels with long T_2 values decay more slowly than those with short T_2 values. Thus voxels with long T_2s have higher absolute intensities in the later echoes. The contrast at a given echo delay is more prominent for longer pulse-repetition times because T_1 recovery at the time of the initial 90° pulse is more complete.

A T_2 image can be calculated from a single spin-echo scan without substantially increasing the scan time if at least two echoes are sampled. The signals from the first and second echoes represent two points on the T_2 decay curve. The resulting signals I_1 and I_2 are given by

$$I_1 \propto N(\mathrm{H}) \left(1 - e^{\frac{-T_R}{T_1}} \right) e^{\frac{-T_E}{T_2}} \quad (11.19)$$

$$I_2 \propto N(\mathrm{H}) \left(1 - e^{\frac{-T_R}{T_1}} \right) e^{\frac{-2T_E}{T_2}} \quad (11.20)$$

Combining equations leads to the equation for T_2:

$$T_2 = \frac{T_E}{\ln\left(\frac{I_1}{I_2}\right)} \quad (11.21)$$

T_1 images are more difficult to generate. If the image is obtained by using the spin-echo sequence, a minimum of two images, which are obtained with different repetition times (while keeping the same value for the echo delay), are required.

Spatial Resolution in NMR Imaging

An NMR image is characterized by three factors: its spatial resolution, its object-contrast level, and its signal-to-noise ratio. Resolution of an image is determined by the minimum volume that contains a sufficient number of nuclei to give a detectable signal. In practice, this signal is determined by the sample band width, the slice thickness, the number of encoding steps, the number of data points required, the sensitivity of the spectrometer, and the strength of the magnetic-field gradients. For any given resolution and contrast, the signal-to-noise ratio needs to reach only a level that is sufficient for detection. The signal from each voxel decreases as the resolution is enhanced, and the decreased signals lead to an imaging time that is dependent on the sixth power of resolution.

To obtain a spatially resolved image, the gradient-imposed spread in NMR frequencies, $(\Delta x)(\gamma G)$, must be greater than the natural spread in frequencies involved in the normal (absence of field gradient) line width $(\Delta \nu_{1/2})$, that is, $(\Delta x)(\gamma G) > (\Delta \nu_{1/2}) > 1 / T_2^*$, where (Δx) is the spatial resolution, and $(\Delta \nu_{1/2})$ is the experimental line width. Current imaging hardware is based on biological tissues, in which the T_2s are greater than 30 ms, so adequate resolution can be achieved by using moderate gradient strengths (i.e., a γG of 0.1 kHz/mm). The highest resolution that has been reported for a single cell (ova from *Xenopus*) was a 10- $\times$ 13-μm pixel that was 250 μm thick (*10*).

The current commercial hardware places a severe limitation on the types of samples that can be imaged. In particular, solids cannot currently be imaged because the line widths are broad due to very short T_2s. The broad line widths require large linear gradients that cannot be generated with current technology. In a strongly protonated solid in which molecular motion is restricted, a typical value of the local dipolar field might be 5 G, and a gradient greater than 50 G/cm (0.5 T/m) is needed to achieve a resolution of 1 mm. An alternate approach is to reduce the effective local field by using line-narrowing techniques of the type used in high-resolution solid-state NMR spectroscopy.

However, the interactions of mobile liquids in solid polymers can be investigated, and spatial information about the structure of the polymer can be derived. Technological advances have allowed a spatial resolution of 20 μm, and further improvements are expected to increase the signal-to-noise ratio. The signal-to-noise ratio improves with a higher magnetic field (according to a 7/4 power) and a larger spin magnetization, which is inversely proportional to the sample temperature.

Imaging Nuclei Other Than Hydrogen

In addition to protons, several other important nuclei have been used to generate images, including ^{13}C, ^{19}F, ^{23}Na, ^{31}P, ^{15}O, and ^{39}K. The problem with using these nuclei rather than protons for imaging applications is that, except for protons, these nuclei have much lower inherent NMR sensitivity. In biological studies, probably the most interesting nucleus is ^{31}P because it is contained in the high-energy phosphates adenosine triphosphate (ATP) and phosphocreatine, which are necessary for sustaining the energy-transfer process within the cell. These phosphorus-containing metabolites can be observed in vivo in normal and abnormal tissues (*11*). The problem with phosphorus imaging is the low physiological concentration of phosphorus, which is only about 0.4% of that of protons. This low level reduces the overall sensitivity by a factor of 3×10^{-4}.

The NMR sensitivity of fluorine is almost as high as that of protons, making it an attractive system. Fluorine images have been obtained (*12*).

NMRI experiments on ^{13}C nuclei are obviously desirable for polymers because of the wide chemical-shift range and the potential to image a number of molecules in a single heterogeneous system through their specific chemical shifts. However, natural-abundance ^{13}C imaging is limited by the low sensitivity, low natural abundance, and long spin–lattice relaxation times in addition to spin–spin couplings to protons and large chemical shifts. The first ^{13}C images were obtained with 99% ^{13}C-enriched methanol and with a sample-scanning period of 2 h. The images of the spherical phantoms were separated into four shadows with a 1:3:3:1 intensity ratio as expected (*13*). In the same study, natural-abundance uncoupled ^{13}C images were obtained for ethylene glycol. Quite recently, multiecho acquisition (10 echoes) was used with the standard 2-D FT imaging sequence to produce natural-abundance ^{13}C images for ethylene glycol in 9 min (*14*).

Computer Processing of Images

Computer processing of images involves three phases: digitization of the image, mathematical processing of the digitized data, and display of the processed image.

Digitization and Storage of NMR Images

In digital imaging, the sampling theorem states that at least two numerical data points must be recorded (sampled) per line pair (or cycle). The pixel is the smallest unit of information in the NMR image. Typically, a 256 × 256 image will result in a total of 1.6 Mbytes of data. Thus, there exists the potential problem of transfer and storage of the images as well as of processing time per image due to the large amount of data. When multiecho and multislice images are recorded, the amount of data is correspondingly larger.

The end result of NMR imaging is a collection of 2-D slice images of a predetermined thickness. Typically, 10 – 60 slices are required for 3-D studies.

Mathematical Processing of NMR Images

Most images are interpreted on the basis of the fine art of visual science, namely examining pictures. Historically, images were correlated by visual side-by-side comparison. Initially, the image is examined, and regions of interest are located, and then these regions are further examined at various levels of higher magnification. For routine work, visual comparison extends only as far as coarse recognition during analysis of the sample under different imaging conditions. For materials imaging, this visual method is expensive, slow, and often unreliable because of limited statistical sampling and subjective assessment by the analyst. The lack of the required precision and consistency that occurs with visual examination suggests the need for digital analysis of the images and automation of the image-analysis process (*15*).

Humans concentrate most of their attention on the borders between more or less homogeneous regions. Consequently, a natural first step in image analysis is to convert the image into an outline drawing. Outlines are edges, and edges by definition are transitions between two markedly different intensities. An edge is a region of the xy plane where $f(x, y)$ has a large gradient. This gradient can be estimated by the directional derivatives of the function along any two orthogonal directions. The process of obtaining a gradient picture is usually known as *spatial differentiation*, *edge enhancement*, or *sharpening*. Such an image is shown in Figure 11.6 for a water-filled polyurethane foam (*16*). The image on the left is the original image, and the image on the right was obtained by allowing only two signals, 0 and 1 (above and below a threshold) for the pixels, and by using an edge-location algorithm on the resulting image.

One of the most obvious uses of image analysis is *feature extraction*, where some specific feature of the image is isolated and interpreted. For object identification, a variety of algorithms can be used, including edge finding, line finding, noise removal, deblurring, classification, image matching, and shape identification. The simplest way to select objects is to use a threshold or simple boundary-detection technique that will yield the outline of the object in an NMR image. The boundary-detection algorithm traces the perimeter of the structure. The resulting boundary is always a closed loop composed of (x, y) coordinate pairs on the surface of the object. Boundary extraction is performed automatically on each slice until the entire sample is reduced to a collection of object loops with known spacing (the slice thickness). Once the boundary information is collected, the loops on adjacent levels must be connected to form an actual surface.

Display of NMR Images

The 2-D images can be displayed by plotting the values of the gray scale in two dimensions or in color codes when proper equipment is available. The color presentation is more striking and, at first glance, easier to interpret. However, color can be misleading because the eye has different sensitivities to the various colors, and different individuals perceive color differently.

Three-dimensional display requires a complete 3-D presentation of the structure. In NMR imaging, thin cross-sectional images of the sample (i.e., a collection of slice images) are produced. The process involves filtering and recombining the data to form a realistic representation of the structure of the sample.

Applications of NMR Imaging to Polymers

Contrast in NMR images of polymers depends on differences in the environments of the nuclei. These environments are influenced by a wide variety of chemical and physical effects. If the relationships between the images and the differences in chemical and physical effects can be extracted from the NMR images, the potential applications of the NMR imaging technique to materials are numerous. Examples of such applications are contained in the following preliminary list:

- adsorption and diffusion processes of fluids in materials
- detection of internal defects and voids
- characterization of heterogeneous mixtures of different materials
- study of molecular interactions between materials
- detection of internal gradients in materials
- determination of spatially distributed structural changes
- composition profiling from surfaces and other sources
- study of variations in molecular mobility throughout a sample
- evaluation of homogeneity of mixing processes
- determination of internal flow processes

In the following sections some of these applications will be demonstrated, and others will be discussed in terms of the potential impact of NMR imaging.

Measurements of Internal Defects in Polymeric Systems

Foams. Because NMR imaging allows the image of a slice of a polymeric sample to be obtained, internal defects can be measured if they are larger than the resolution of the technique (currently >20 μm). An example is shown in Figure 11.6a for a polyurethane foam filled with water. The light portions of the image represent pores filled with water, and the dark areas arise from either solid foam or an absence of material (air). With an edge-detection algorithm, the outline of the pores can be observed as shown in Figure 11.6b. The distribution of the pore sizes can be observed. A histogram of the pore sizes vs. the number of pores of that size can be constructed. Such a histogram can be correlated with the

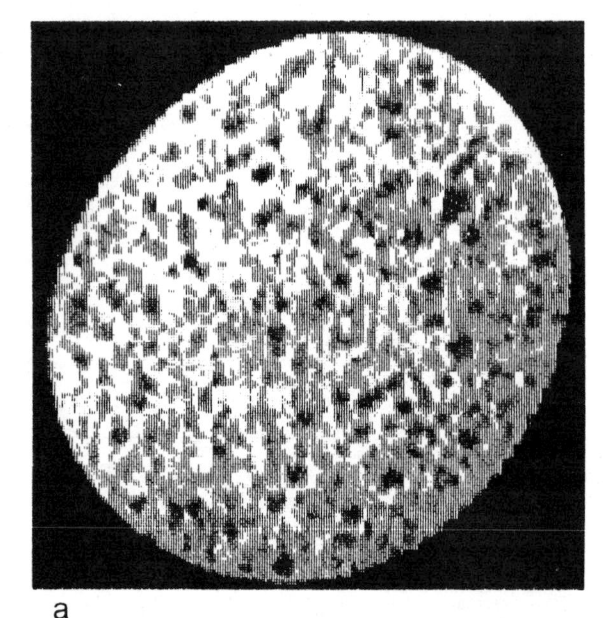

a

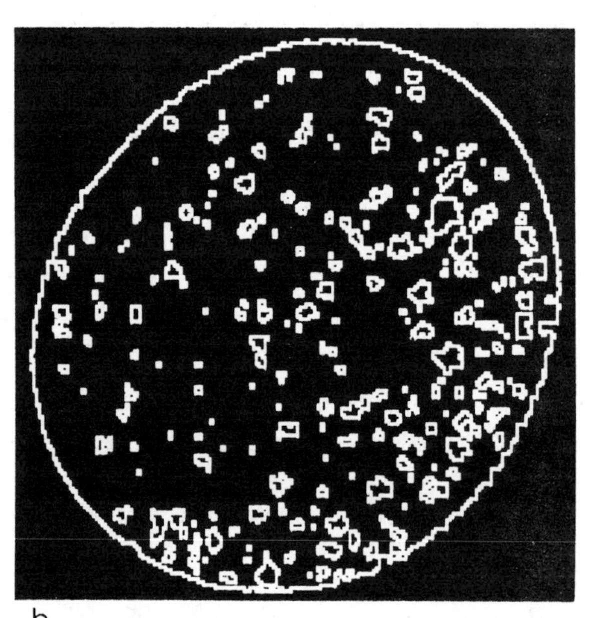

b

Figure 11.6. (a) The NMR image of a polyurethane foam filled with water. (b) The image that was obtained after using an edge-detection algorithm on the image in a. (Reproduced with permission from reference 16 Copyright 1989 John Wiley & Sons, Inc.)

chemical-foaming process variables, and this correlation could lead to an understanding of the mechanism. Such knowledge could promote better process control and an improved foamed product.

Composites. NMR imaging is a means of detecting internal material imperfections in fabricated articles. The applications of NMR imaging in the field of processing and fabrication of polymeric materials are diverse. These applications include the detection of subsurface defects, including interfacial flaws and microcracks, as well the detection and characterization of areas that are modified through the introduction of mobile foreign substances such as additives, degradation products, and contaminants.

Because of the sophistication of the structure of engineering articles of polymers and the complexity of the process and fabrication procedures, noninvasive tests are necessary to ensure the quality and integrity of the manufactured articles. Defective or damaged areas of polymeric materials can be made to appear in the NMR image by sorption of a mobile liquid such as water. The uniformity of the polymeric materials can be evaluated because improperly manufactured engineering articles have different NMR images. A materials-acceptance criterion could therefore be written as a function of tolerances in the NMR images. Such an image-based materials-acceptance protocol would ensure proper manufacture and performance of polymeric engineering components.

In Figure 11.7, the two images shown were taken 0.5 cm apart through the same pultruded rod that was made with glass fibers and a nylon matrix (*17*). The rod was soaked in water, and the light areas in the image represent void areas filled with water. Comparison of the sizes of the voids in the pultruded rod

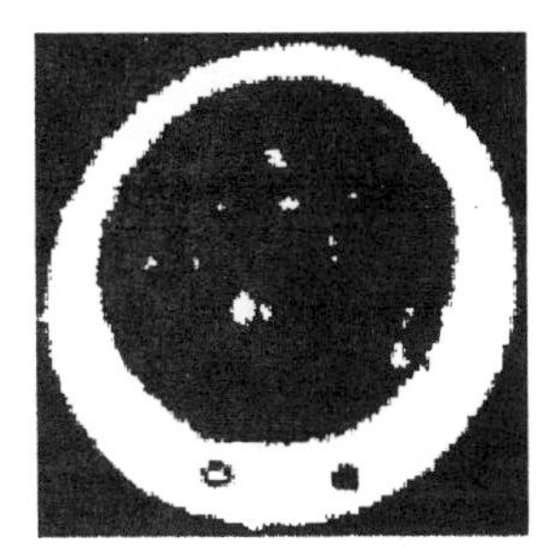
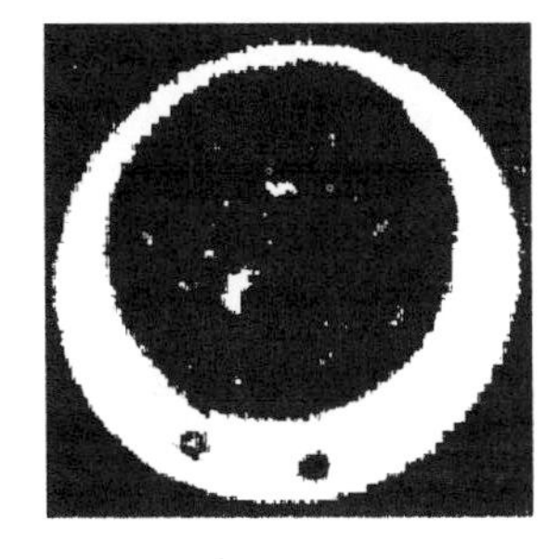

Figure 11.7. Two images taken 0.5 cm apart through the same rod. (Reproduced with permission from reference 17. Copyright 1989 Gordon and Breach Scientific Publishers.)

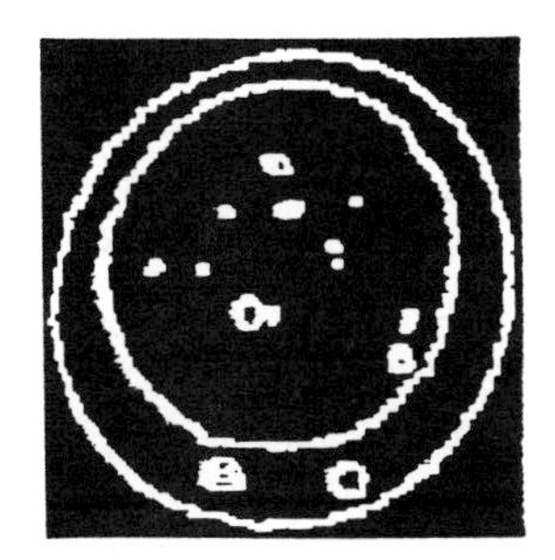
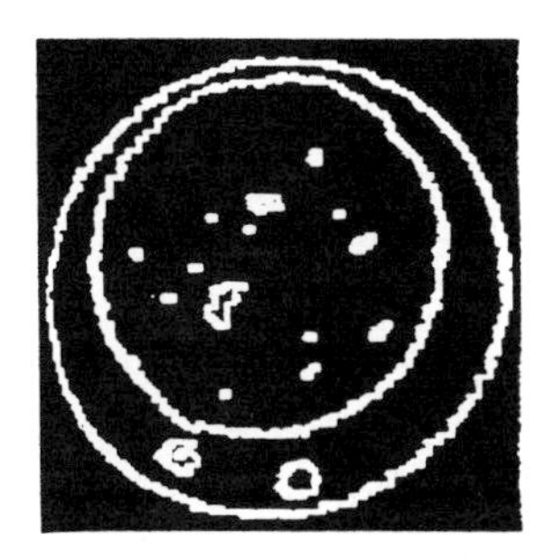

Figure 11.8. Profiles of a nylon RIM-pultruded composite rod showing the distribution of the defects for images taken 0.5 cm apart. (Reproduced with permission from reference 17. Copyright 1989 Gordon and Breach Scientific Publishers.)

indicates that some of the voids approach 1 mm in magnitude. A comparison of the corresponding edge-enhanced images (*see* Figure 11.8) shows that some of the voids in the images occur in the same location, a result that indicates that the voids are connected, or tubular in shape. Thus, a channellike void region over a length of 0.5 cm is suggested. By using a computer, a representation of the defects in the image can be made (Figure 11.9). From the computer comparison of the two images taken 0.5 cm apart, a tubular shaped void running from one image to the other within the nylon rod can be identified. Such a void could be obtained if an air bubble was trapped in the matrix during the pultrusion process.

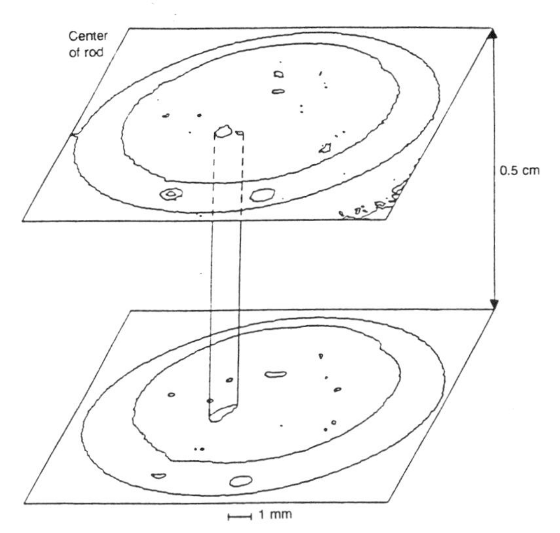

Figure 11.9. An edge-enhanced contour plot of the images in Figure 11.8. A correlation of these images shows that some of the voids are tubular shaped, that is, a channellike void region exists over the dimension of 0.5 cm. (Reproduced with permission from reference 17. Copyright 1989 Gordon and Breach Scientific Publishers.)

Industrial Inspection. Industrial inspection requires the establishment of dimensional or structural tolerances for manufactured objects. The critical features are those that make the objects unacceptable and lead to their rejection. Unfortunately, many tolerances or other inspection criteria are beyond the ability of most inspection systems to visually measure. NMR imaging probes the internal homogeneity of the object and allows the geometric location of the inhomogeneities to be determined. This information might be used to selectively reject portions of a continuous manufacturing process, such as an extruded polymer structure. In this manner, NMR imaging can be used to provide internal quality-control measures for polymer-manufacturing processes.

Standardized techniques to compare the image from the object being inspected with the stored ideal model and a plan of action based on the comparison must be developed. The confidence limits could be based on previous manufacturing defects recorded in a product database or determined from initial estimates based on experience with similar objects. Initially, this information can be estimated, and then the information can be refined as the manufacturing process is better understood. As soon as an unacceptable feature is detected, the part can be rejected. All specified features of a good object can be inspected by image analysis. This inspection must be able to detect a bad part as quickly and as efficiently as possible.

Nondestructive Testing. The aim of nondestructive testing (NDT) is to make a direct correlation between an observed structural measurement and failure properties. The mere detection of defects is not good enough to use as a method for NDT. The engineer must ask: "How critical is the flaw? What does this flaw mean to the life of the component?" To test the detection capability of an NDT method, the nature of the defect most commonly present must be known. Probability of detection (POD) tables or graphs must be developed. POD is a statistical concept in which a relatively small number of defects of a given type are selected and inspected, and then the results are treated by using the theorems of sampling statistics as a way to derive an estimate of the POD for the entire "population" of such defect types. The measurement of NMR images is potentially a fast and convenient method to detect and measure the size and distribution of defects. Although NMR imaging has not been used for this purpose to date, the potential for this application exists.

Adsorption of Liquids in Polymers

Liquid adsorption in solids can be followed by using NMR imaging. Differences in the images reflect differences in the internal structure of the solid, as well as molecular interactions between the solute and the polymer. An example is the adsorption of water into glass-fiber epoxy resins (*18*). In Figure 11.10, images are shown of the water-in-epoxy systems whose polymerization was initiated with either an anhydride or an amine.

The distribution of the adsorbed moisture in the composites is substantially different. The structural differences within the epoxy networks are apparently due to differences in the cross-link density or *mesh* of the epoxy system because of the initiator chemistry.

Crazing of many glassy thermoplastics is readily observed when these materials are under stress. Considerable efforts have been made to understand crazing behavior and the necessary requirements to produce crazes. However, to date, a guide to the craze resistance of materials that are subjected to concurrent environmental factors has not been developed. NMR imaging can contribute to a better understanding of environmental solvent crazing.

The solvent crazing of polystyrene by toluene was studied by using imaging techniques (*19*). A polystyrene rod with a length of 5 cm and a diameter of 1.9 cm was submerged in neat toluene for various times (Figure 11.11). The resulting image cross-sections clearly demonstrate the time-dependent ingress of toluene into the polymer matrix and the development of environmental stress cracks.

NMR images obtained from the earliest echoes (Figure 11.12) in the spin-echo sequence are primarily governed by spin density and are insensitive to T_2. However, T_2 effects dominate images arising from later echoes, rendering the long-T_2 species preferentially displayed in the image. The aforementioned approach is demonstrated for toluene-exposed polystyrene in Figure 11.12. Comparison of the strongly T_2-weighted image set (echo 4) with the spin-density images (echo 1) indicates the presence of two types of absorbed toluene: (1) a more mobile phase (long T_2) confined to a narrow ($1-2$ mm) annular region

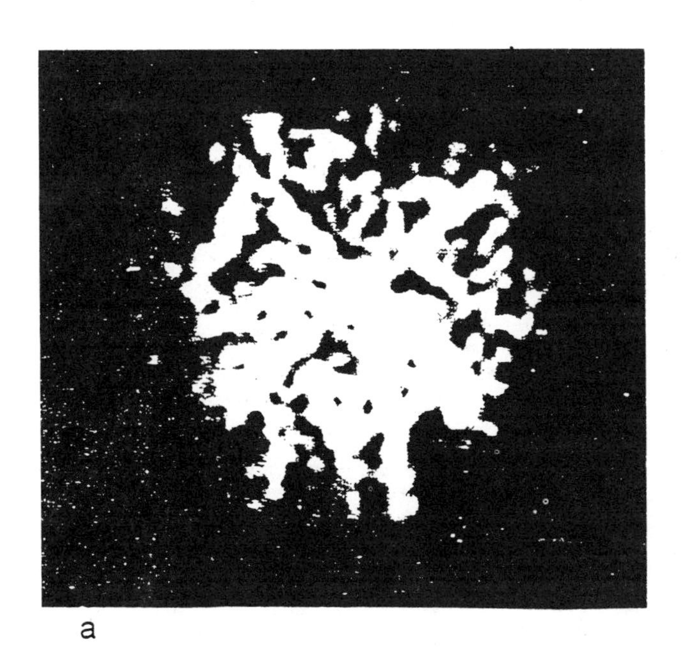

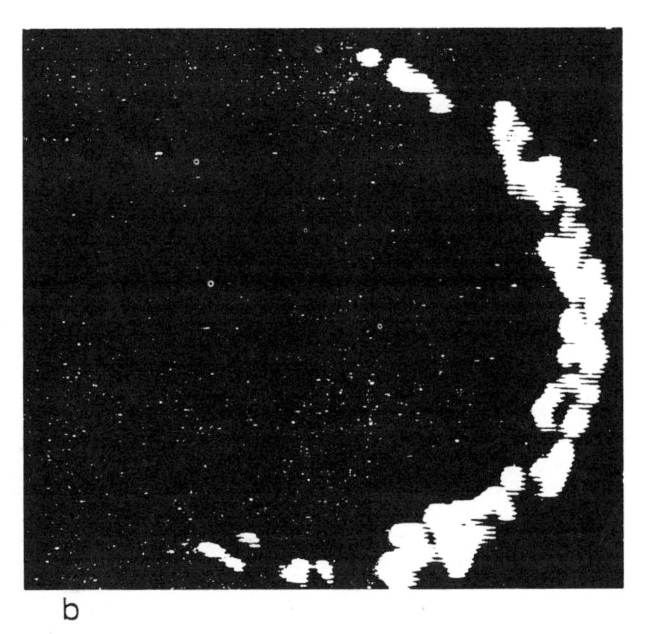

Figure 11.10. The water image of the epoxy systems that were initiated with an anhydride (a) and an amine (b). (Reproduced with permission from reference 18. Copyright 1984 John Wiley & Sons, Inc.)

on the outside of the sample, and (2) the absorbed phase of toluene that has penetrated deeper into the sample matrix and is much less mobile (short T_2). These images suggest that the toluene associated with the crazed outer regions of the polystyrene is an interstitial fluid primarily localized in microscopic voids in the polymer. By contrast, the short T_2 phase represents toluene that is more intimately associated with the polymer matrix. It was suggested (*15*) that the toluene associated with the polymer matrix induces crazing because the macroscopic environmental stress cracks provide stress relief.

Diffusion of Liquids in Polymers

The diffusion of small molecules is important to the material polymer properties ranging from processing and production to end-use applications and shelf life. Polymer-penetrant systems exhibit a wide range of diffusion phenomena, from common Fickian diffusion to case II diffusion. These diffusion processes represent the extreme responses of polymers to a solvent, and the different processes have substantially different characteristics.

Fickian diffusion is characterized by an exponential increase in concentration in going from the glassy polymer core to the fully swollen regions of the polymer. Fickian diffusion is also characterized by a weight gain that increases with the square root of time in a polymer sheet. The diffusion constant can be a function of concentration, C, in which case a series of relative concentrations vs. distance, x, yields a master curve when C/C_o is plotted vs. $x/t_{1/2}$, where C_o is the initial concentration (*20*).

Case II behavior has the following four characteristics: (1) the concentration increases rapidly from 0 in the glassy polymer core to an equilibrium value in the swollen regions of the sample; (2) the concentration throughout the swollen regions is constant; (3) the front advances through the glass at a constant

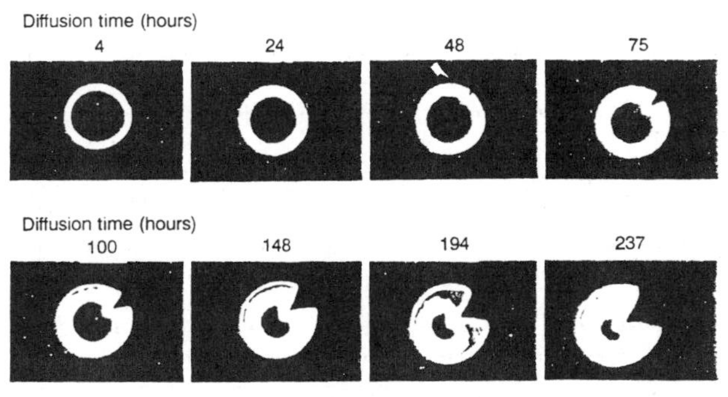

Figure 11.11. Cross-sectional images of a toluene-immersed polystyrene. The images were derived from the first echoes of the pulse train. These images display spin-density information. The environmental stress cracks arise from solvent-induced swelling. (Reproduced with permission from reference 19. Copyright 1985 Bruker Reports.)

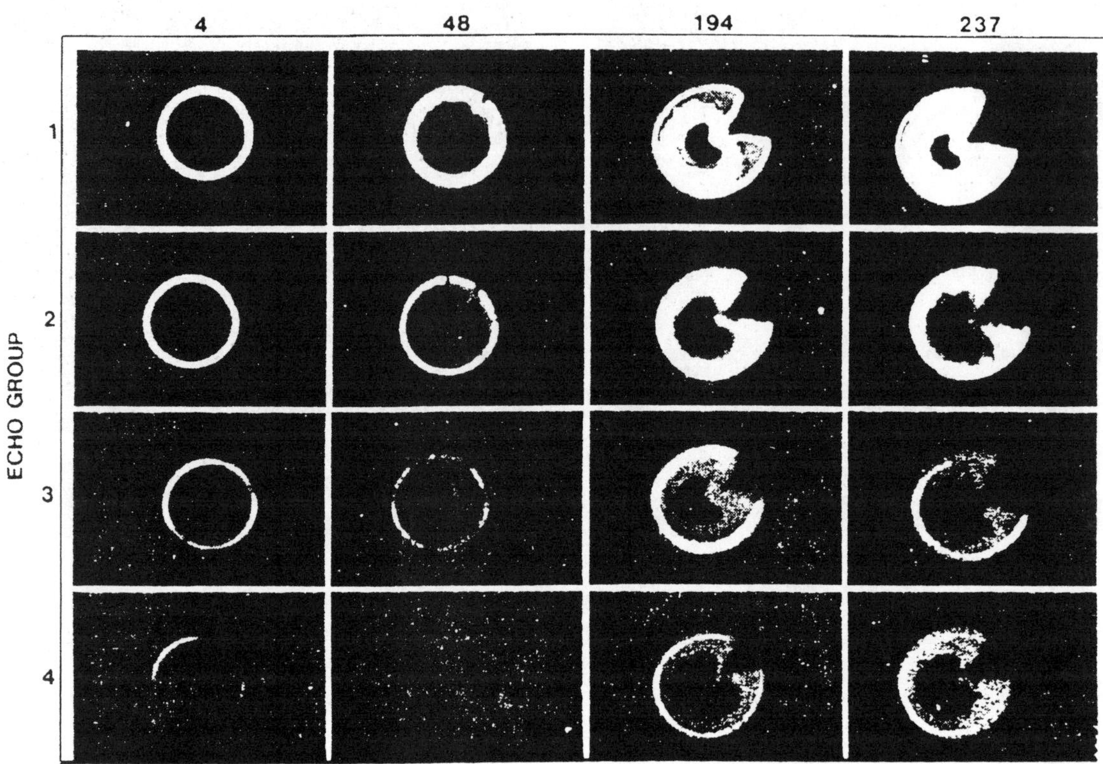

Figure 11.12. Selected T_2*-weighted NMR images of toluene diffusion in polystyrene. Four echo groups, each consisting of three coadded echoes, were acquired. Images generated from echo group 1 are primarily defined by spin density, but* T_2 *effects dominate the later echoes, producing images which preferentially map the longer* T_2 *species (interstitial toluene). (Reproduced with permission from reference 19. Copyright 1985 Bruker Reports.)*

velocity; and (4) a Fickian concentration profile precedes the sharp front as it advances through the glass.

NMR imaging for the study of diffusion has two advantages. First, NMR imaging provides a visual representation of the spatial distribution of diluent by acquiring signals directly from the protons of the solvent without interrupting the diffusion process or destroying the sample. Second, regions of inhomogeneous diffusion can be detected and analyzed. With other techniques, these discontinuities are averaged in the data and add to the error. This advantage extends to sample geometries as well. NMR imaging is not restricted in geometry except for size. A 3-D image can be used to monitor a diffusion process for all regions of a sample.

The geometry best suited for NMR imaging is a cylinder. A cylinder fills the sample chamber so that the best resolution per unit volume is obtainable. Although a sphere might be expected to be the best geometry because the sphere has the most isotropic magnetic susceptibility, spheres present a problem in the slice-selection process because signal contributions occur from diffusion that is not parallel to the slice plane when the slice thickness is small in comparison with the sphere diameter (*21*).

To study diffusion in polymers with NMR imaging, an image that accurately reflects the quantity of solvent per unit volume must be acquired. This is known as a *spin-density image*. However, in order to acquire a spin-density image, several conditions must be satisfied. First, the spins must return to equilibrium between subsequent repetitions of the pulse sequence in order to prevent saturation effects. To recover 99% of the original magnetization, the repetition time, T_R, must be equal to $5 \times T_1$. The second condition is that T_E must be short in comparison to the spin – spin relaxation time, T_2, so that little signal loss occurs because of spin – spin interactions. To maintain 99% of the magnetization, T_E must be less than 1% of T_2. Using computer-simulation studies of case II diffusion (*21*), it was determined that the image must be acquired in less time than it takes the case II front to traverse about one-tenth of an image pixel. This requirement severely limits the velocity range in which NMR imaging is effective. The maximum front velocity is 2.2 Å s^{-1} using current hardware and the spin-echo method. Fortunately, other imaging experiments are available to reduce the imaging time. One such experiment is termed *FLASH* (fast low-angle shot), and this method reduces the imaging time by a factor of 10 by the use of a small tip angle and a gradient echo (*22*).

Self-diffusion coefficients provide important information concerning the extent of the translation of solvent molecules in their environment. This information can be interpreted in terms of molecular organization and phase structure. The self-diffusion coefficients are sensitive to structural changes as well as to binding and association phenomena. Measurements of the self-diffusion coefficients can be made by using the conventional Stejskal–Tanner pulsed-gradient spin-echo (PGSE) pulse sequence, which is a spin-echo experiment (*23*). The first gradient pulse in this sequence causes the spins to dephase at a spatially dependent rate. After the 180° rf pulse, the second gradient causes the spins to refocus at the same spatially dependent rate. Under static conditions, all of the spins refocus at the correct time and generate a spin echo whose intensity is attenuated by only spin–spin effects. However, all molecules are subject to Brownian motions and move into regions of different effective fields. The result is that the spins do not refocus correctly, and the echo is attenuated according to

$$M = M_o \exp\left(\frac{-T_E}{T_2}\right) \exp\left(-\gamma^2 G^2 \beta D\right) \quad (11.22)$$

The first exponential term is the attenuation due to T_2 occurring during T_E. The second exponential term accounts for the attenuation due to diffusion in the presence of a gradient pulse, G, which is on for a duration, δ. The variable β is the combination of the duration time of the gradient pulses and the interval between them, Δ, as shown in eq 11.23.

$$\beta = \delta^2\left(\Delta - \frac{\delta}{3}\right) \quad (11.23)$$

For spectra collected at the same T_E, the first term cancels. A plot of the natural log of the normalized signal vs. $\gamma^2 G^2 \beta$ is linear, with the slope equal to the self-diffusion coefficient, D.

Normal image-contrast measurements were made to visualize the effects of diffusion on NMR images (*20*). The imaging sequence is easily adapted to image-contrast by adding two additional gradients in the x direction. The equations are the same as the spectroscopic version, and the read gradients have little effect if they are much smaller than the motion-probing gradients.

The diffusion coefficients are quantitatively evaluated from a series of images recorded with different gradient-field strengths. Analysis involves the simulation of the effects of diffusion using the dynamic-magnetization equations to calculate for each pixel the magnetization that ultimately yields an image whose intensities represent the spatially resolved diffusion coefficients. Finally, a true diffusion-constant image can be obtained, and the calculated diffusion coefficients are encoded into an intensity scale on the diffusion-constant image (*24*). On this scale, high intensities correspond to fast diffusion. In this manner, the spatial diffusion of a liquid into a solid material can be characterized in a quantitative fashion.

Self-diffusion constants for water and polyethylene oxide (PEO) were measured by using imaging techniques (*25*). The calculated diffusion coefficients were

2.3×10^{-5} cm^2 s^{-1} for H_2O and
0.5×10^{-5} cm^2 s^{-1} for PEO at 293 K.

NMR imaging was used to study the diffusion of water into solid blocks of nylon 6,6 at 100 °C at a spatial resolution of 240 μm (*20*). Penetrant profiles of relative concentration C/C_o at 100 °C were measured as a function of depth for different times. These profile curves collapse onto a single master curve of C/C_o vs. $x/t_{1/2}$, a result that establishes the diffusion of water in nylon 6,6 as Fickian.

Case II diffusion was studied by using NMR imaging for methanol diffusing into poly(methyl methacrylate) (PMMA) (*24*). Figure 11.13 shows the image from the PMMA in methanol after 48 h. The diffusion coefficient can be calculated by measuring the thickness of the sorbed layer as a function of time. With data-processing techniques, the measure-

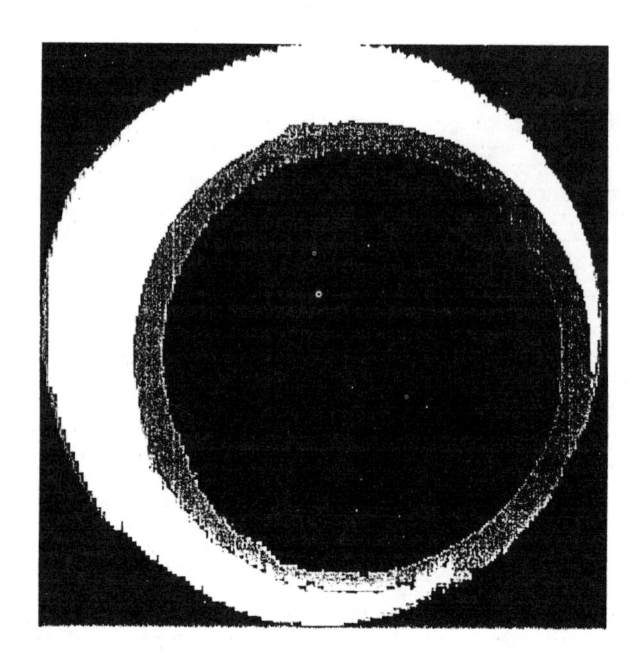

Figure 11.13. The image from PMMA in methanol after 48 h. The constant level of methanol in the penetrant front is indicative of case II diffusion. (Reproduced with permission from reference 22. Copyright 1989 John Wiley & Sons, Inc.)

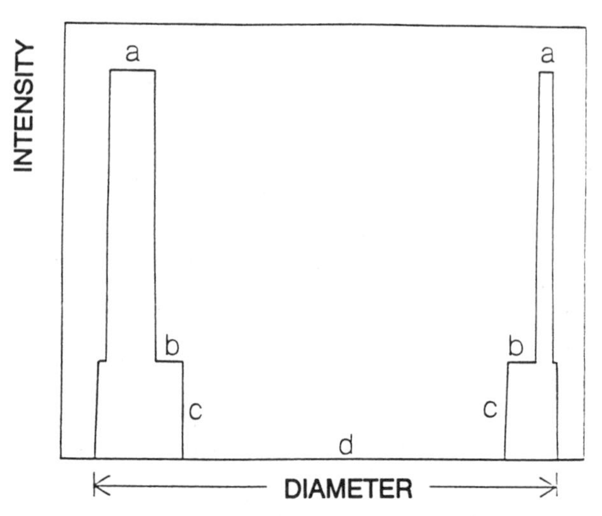

Figure 11.14. A three-level gray scale of the image in Figure 11.13 obtained by drawing a profile across the sample. In this figure, a represents the solvent, b is the imbibed region, c gives the concentration of methanol in the imbibed region, and d is the core region of the PMMA rod. (Reproduced with permission from reference 22. Copyright 1989 John Wiley & Sons, Inc.)

ments can be simplified by giving the images a three-level gray scale and then drawing a profile across the sample as shown in Figure 11.14. The results are shown in Figure 11.15, in which the thickness of the layer is plotted vs. time. The linearity of this plot with time confirms that case II diffusion is occurring. The nonzero intercept at time zero is indicative of an initial Fickian-diffusion process followed by case II diffusion. The constant level of methanol in the penetrant front is also indicative of case II diffusion.

NMR relaxation parameters are useful probes of molecular motions in polymers. Each correlation time represents the average value of the system with some distribution around that average. NMR imaging permits the determination of the spatial distribution of NMR relaxation times. This distribution provides information concerning the local motions of the system. In this case, the polymer is partially swollen with solvent, and the spatial distributions of the relaxation times reveal the interactions between the solvent and the polymer in the diffusion process.

Figure 11.16 (p 305) shows a series of images of methanol in PMMA at different echo times from a multi-echo experiment. The intensities are represented by colors, red being the highest intensity and blue being the lowest. The T_2 attenuation of the intensity is more apparent for the longer echo times. Figure 11.17 (p 306) is the T_2 image calculated from the images in Figure 11.16. In this case, blue represents the longer T_2s, and red represents the shorter T_2s. Figure 11.18 (p 306) is a profile of the image of Figure 11.17 showing an almost linear decrease of T_2 to the center core. Because the volume available to the methanol molecules is the same from the surface

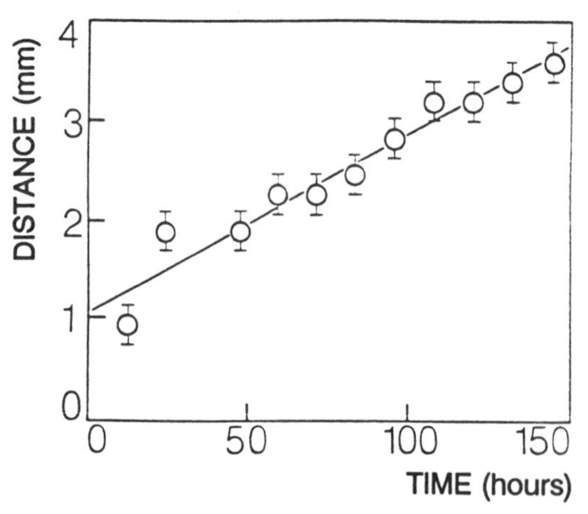

Figure 11.15. A plot of the thickness of the layer of methanol in PMMA vs. time. (Reproduced with permission from reference 22. Copyright 1989 John Wiley & Sons, Inc.)

to the core, the relaxation time should be constant throughout the region if no other interactions exist. The change in T_2 from the surface to the core indicates that some physical effect other than the presence of free volume is affecting the mobility of the solvent. The polymer core effectively fixes the position of the chains at the interface between the swollen and glassy regions of the polymer. The polymer-chain dynamics change from the usual anisotropic motions, with a variety of frequencies and amplitudes, to anisotropic motions that depend on position along the polymer chain. Frequencies and amplitudes decrease toward the fixed point of the chain, that is, the glassy core. This change in polymer motion influences the motion of the solvent in PMMA. The result is a change in the T_2s with distance from the glassy core.

The self-diffusion-coefficient images are generated by using the magnitude images. The images are fit to the following equation:

$$I = I_o \exp\left(\frac{d}{B}\right) \quad (11.24)$$

as a function of the gradient attenuation factor, d, where $d = -\gamma^2 G^2 \beta$. This equation generates an image that is based on a dummy variable, B, which is equal to $1/D$. The inverse of the image is taken to generate the self-diffusion-coefficient image based on D. Figure 11.19 (p 306) is a series of images of methanol in PMMA taken at different gradient strengths. Each successive image is the result of a more intense motion-probing gradient. The first image has a more intense signal on the outer regions of the rod with the intensity decreasing towards the core. Figure 11.20 (p 307) is the calculated self-diffusion image that was generated from the images in Figure 11.19. The region within 100 μm of the glassy core has a self-diffusion coefficient of 3.2 ($\pm$ 0.9 $\times 10^{-7}$ cm^2 s^{-1}. The outer region of the swollen polymer has a self-diffusion coefficient of 9.2 $\pm$ 0.9 $\times 10^{-7}$ cm^2 s^{-1}. Acetone swells PMMA to a greater extent than methanol, and the self-diffusion coefficients of the acetone–PMMA system are approximately 2 orders of magnitude greater than those of the methanol–PMMA system. This result is apparently due to the increased volume available to the acetone molecules. The self-diffusion coefficients decrease by 35% from equilibrium in the outer regions to the region near the glassy core. The decreasing motions of the polymer chains as the core is approached reduce the solvent mobility as reflected in the self-diffusion coefficients.

NMR imaging of the diffusion process should be useful for multicomponent systems based on differences in relaxation times or chemical shifts. A most interesting example of multicomponent images was generated by using images of different nuclei (^{23}Na, ^{27}Al, and ^{2}H) in aqueous silica dispersions (pastes) (*26*). Time-dependent images that were obtained indicate that the ions diffusing across ion-rich and ion-free paste boundaries obey Fickian laws of diffusion.

The major limitation of NMR imaging for diffusion studies is the long measurement time, which restricts the measurements to relatively slow diffusing systems. When the diffusion is too fast, there is a misregistration of the spins in the diffusion gradients, and this misregistration produces artifacts. One method that has been used for rapid NMR imaging of molecular self-diffusion coefficients allows the measurement of a 256 $\times$ 256 image in 15 s (*27*).

Desorption of Diluents from Polymers

Desorption is one diffusion process that has been given little attention primarily because of the lack of adequate analytical techniques. Desorption measurements above the glass-transition temperature of an unswollen polymer are expected to follow Fickian characteristics. Likewise, a polymer swollen so that the T_g is below the experimental temperature initially exhibits Fickian desorption. The solvent is thought to desorb rapidly from the surface of the polymer and thereby raise the T_g of the surface layer. When the surface T_g is above the experimental temperature, the desorption process slows, and the process is controlled by the diffusion through the glassy surface layer. NMR imaging provides the spatial distribution of solvent in the polymer as well as the spatial distribution of the rate of desorption.

The desorption process can be related to T_d, which is the inverse of the rate of net solvent loss for a given pixel, through the following equation:

$$M = M_o \exp\left(\frac{-T_{exp}}{T_d}\right) \quad (11.25)$$

A nonlinear least-squares fit of the experimental data

is used to calculate a T_d image on a pixel-by-pixel basis.

Images of the desorption of methanol from swollen rods of PMMA were obtained (*28*). The methanol volume fraction was 0.26. After immersion in methanol, the rods were placed in fully deuterated cyclohexane. Image acquisition began 6 min after submersion. Images were collected in 1-h increments over a 104-h period. A part of the series of images acquired after submersion in cyclohexane is shown in Figure 11.21 (p 307). In these images, red represents the highest intensity, blue represents the lowest intensity, and the various shades of color between red and blue represent the intermediate-intensity levels. The signal intensity decreases with time, as can be seen from the comparison of the two images obtained 100 h apart. The maximum intensity of the later image is only 50% that of the initial image. The diameter of the rod decreases by 816 ± 68 μm over the 100-h period as measured from the images.

A T_d image that was calculated from 20 images taken at 5-h intervals over 100 h is shown in Figure 11.22 (p 308). In this image, red represents the largest T_d, and blue represents the smallest T_d, and these extremes correspond to the slowest and fastest decreases, respectively. This image shows that the faster intensity decreases are near the surface of the rod and the slower intensity decreases are near the glassy PMMA core. The T_d at the surface is 58 h, and the T_d at the glassy core is 450 h as determined from this image. This result agrees with the Fickian characteristics and indicates that the imbibed solvent near the surface desorbs quickly and the desorption rate decreases toward the sample core. However, for this system there is no evidence of the development of a glassy skin on the polymer surface (*28*).

Swelling Behavior of Polymer Systems

Solvent absorption and swelling behavior have been used for the purpose of determining the cross-link density in elastomeric systems. The method is based on the fact that the higher the cross-link density, the less solvent is imbibed in the system and the lower the degree of swelling. NMR imaging allows this idea to be pursued further by examining the homogeneity of the swelling process. The intensities of the mobile protons of the swelling agent probe the homogeneity and spatial distribution of the cross-links of the network system (*29*). Figure 11.23 (p 308) shows the benzene – proton image obtained by using a spin-echo pulse sequence on a highly cross-linked sulfur-vulcanized rubber sample that was swollen in benzene for 2 days. The black spot in the image is an air-bubble artifact. Figure 11.24 (p 308) shows a three-level magnified contour plot of a portion of the the swollen-rubber image. This contour plot indicates that there is a benzene background; an intermediate level of benzene, which indicates a moderate level of cross-linking; and regions of little benzene, which are indicative of a high level of cross-linking. Considerable inhomogeneity in the cross-linking occurs in this rubber sample. Such inhomogeneities could arise from improper mixing, thermal gradients, or variations in the vulcanization chemistry.

Samples of cured butyl rubber swollen in cyclohexane have been obtained, and the NMR images demonstrate poor mixing of the formulation (Figure 11.25, p 309). The image shows regions of high cross-link density and entrapped air bubbles (*30*).

Because of the high mobility and narrow proton line width of swollen elastomer samples, images can be obtained directly from the rubber portion of the sample (*31*). Single- and multiecho (T_2 resolved) images were obtained for samples of high *cis*-1,4-polybutadiene that was vulcanized with thiram (TMTD) (*32*). In preparation for imaging, all samples were extracted to remove nonnetwork material, and the samples were then swollen with deuterated cyclohexane. Consequently, the NMR images of the swollen materials were obtained through spatial variations in the proton-signal intensities of the cross-linked rubber. The images were obtained by using standard Carr – Purcell spin-echo (selective 90° – nonselective 180°) pulse sequences. These T_2-weighted images are shown in Figure 11.26 (p 310) for a polybutadiene sample containing 2 phr of TMTD that was cured for the times indicated on the figure. The signal intensity of each pixel serves to characterize the degree of segmental motion of its contents. For example, the decline in the signal-to-noise ratio of the images with increasing cure demonstrates that as the average length of chain segments between effective cross-link sites decreases, so does the extent of their relative motion. Histograms that yield the number of pixels with a particular value of T_2 can be obtained as shown in Figure 11.27 (p 311). A significant feature of these histograms is the skewed nature of the distributions at all cure times favoring the long-T_2 range. The histo-

Figure 11.16. A series of images of methanol in PMMA at different echo times from the multiecho experiment. T_E *(reading from top left to lower right) is equal to 5.25, 10.50, 15.75, 21.00, 26.25, 31.50, 36.75, and 42.00 ms. (Reproduced from reference 24. Copyright 1990 American Chemical Society.)*

grams sharpen with cure time to reflect the loss of uncured chains. These histograms can be interpreted in terms of the distribution of cross-links along the chain (*32*).

Adhesive Systems

Adhesive-bonded structures are important in almost all engineering fields. All adhesives are in a liquid or liquidlike state before the final cure process. The liquid state, a requirement for good flow, leveling, and wetting, is typically achieved by dissolving or dispersing the actual binding substances in water or solvents. In solventless adhesives, the liquid state is obtained by using heat or by using low-molecular-weight or low-viscosity resins that cure with a curing agent, heat, or irradiation. This liquid state is ideal for NMR imaging, and abundant information about the adhesive system can be obtained.

In the formation of adhesive bonds, the adhesive goes from a viscous liquid layer in the glue line to a bonded solid layer by drying or cross-linking. Both of these processes result in a decrease in the volume of the adhesive. In water-based systems such as poly(vinyl acetate) (PVAc) (as used in Elmer's glue),

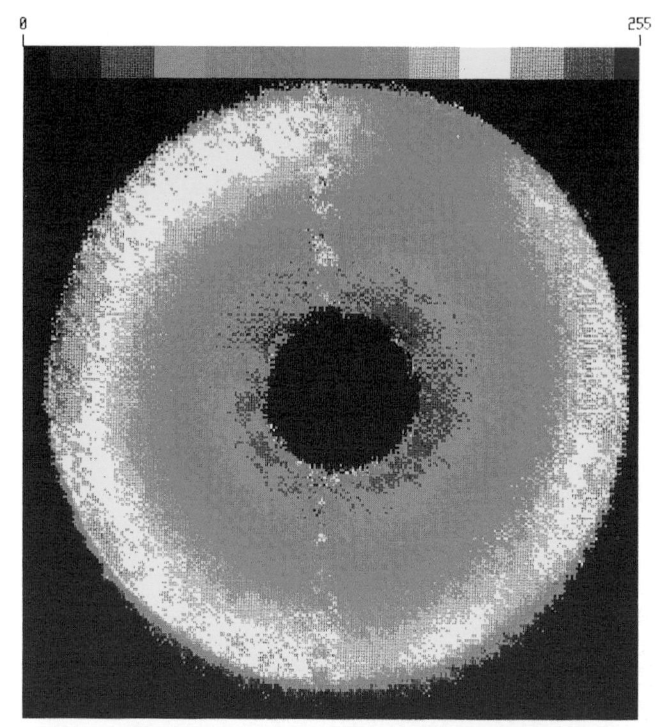

Figure 11.17. T_2 image calculated from the images in Figure 11.16. Color scale represents T_2s ranging from 7 ms (red) to 24 ms (dark blue). (Reproduced from reference 24. Copyright 1990 American Chemical Society.)

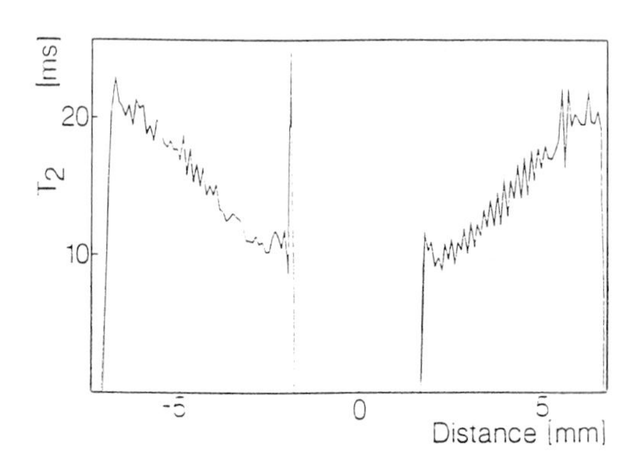

Figure 11.18. Profile of Figure 11.17 shows a nearly linear decrease of T_2s to the center core. (Reproduced from reference 24. Copyright 1990 American Chemical Society.)

Figure 11.19. A series of images of methanol in PMMA taken at different gradient strengths of (reading from upper left to lower right) 0, 13.9, 27.6, 41.4, 55.1, and 68.8 G/cm. (Reproduced from reference 24. Copyright 1990 American Chemical Society.)

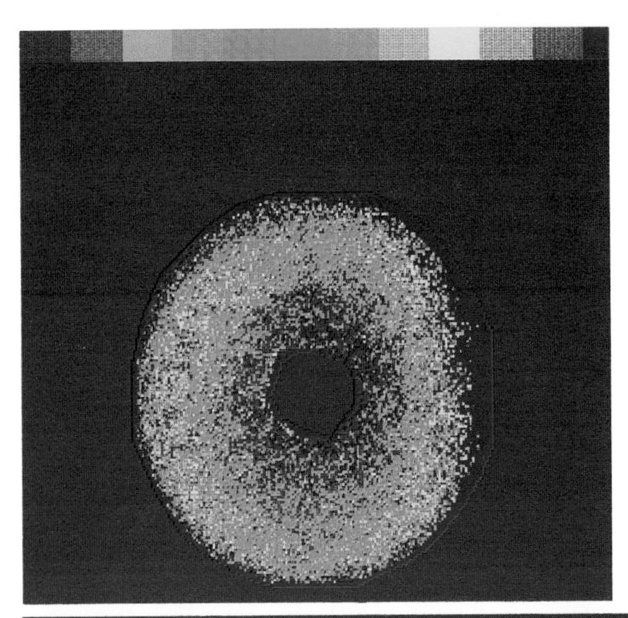

Figure 11.20. The calculated self-diffusion image generated from the images of methanol in PMMA in Figure 11.19. (Reproduced from reference 24. Copyright 1990 American Chemical Society.)

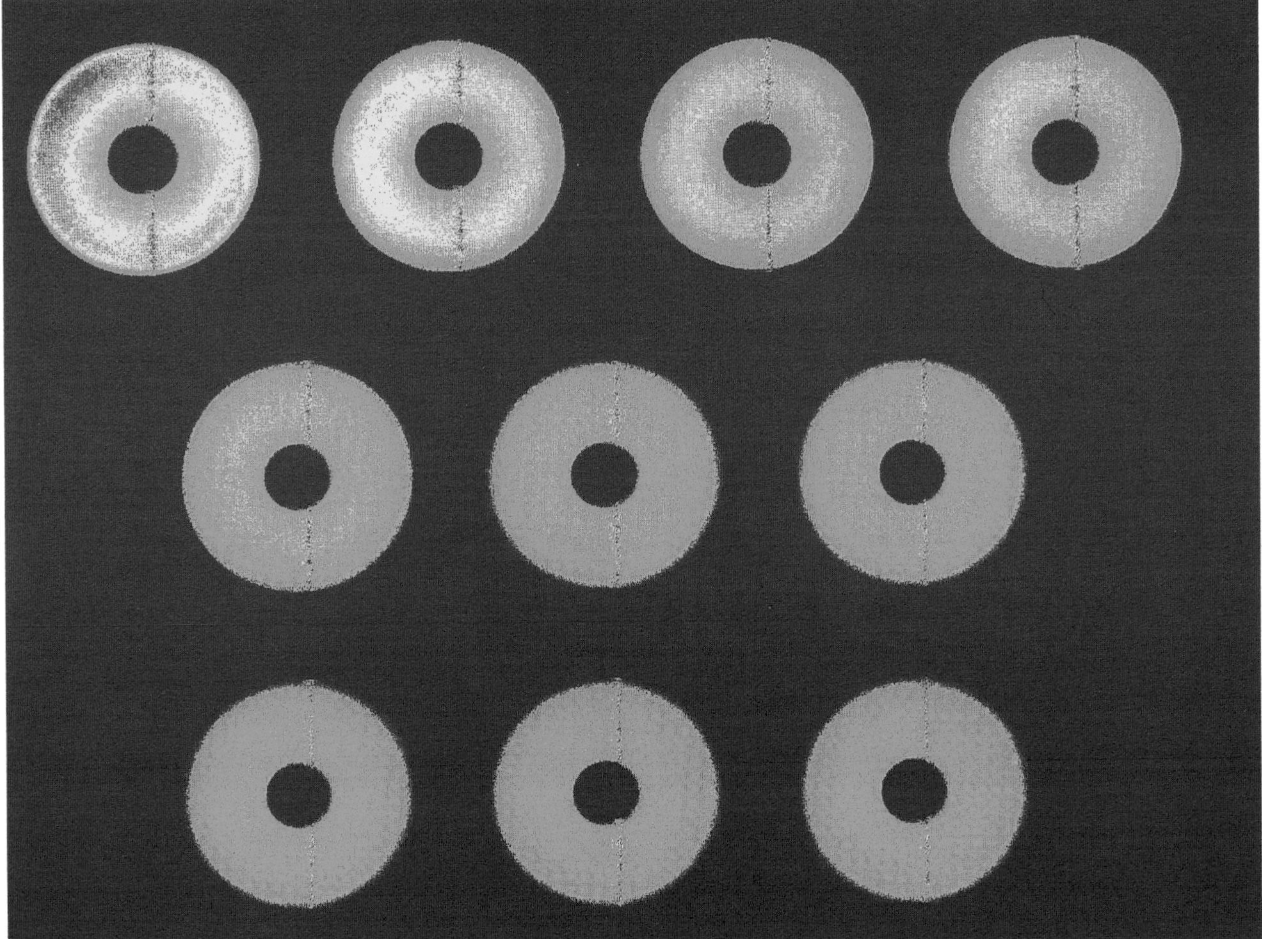

Figure 11.21. A representative set of spin-echo images acquired after different exposure times to cyclohexane (reading from upper left to lower right): 6 min, and 10, 20, 30, 40, 50, 60, 70, 80, and 90 h. (Reproduced from reference 28. Copyright 1990 American Chemical Society.)

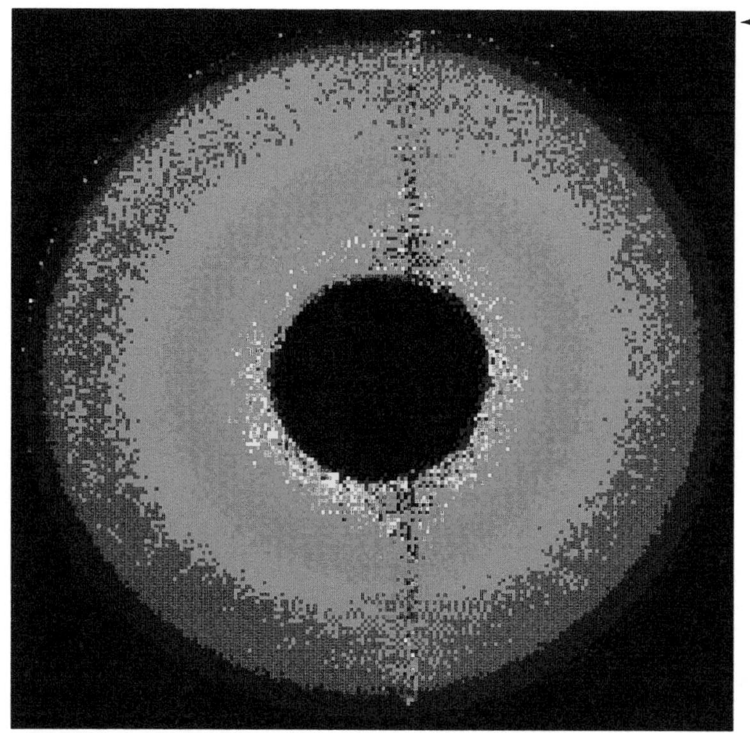

Figure 11.22. A T_d image calculated from 20 images taken at 5-h intervals over a period of 100 h. This image shows that the faster intensity decreases are near the surface of the rod, and the slower intensity decreases are near the glassy PMMA core. (Reproduced from reference 28. Copyright 1990 American Chemical Society.)

the solids content of the adhesive is approximately 50 – 60%, and at set the adhesive occupies only 50% of the available volume in the glue line. For adhesive-bond integrity, the distribution of the nonload-bearing volume or voids with respect to the adhesive joint must be known. Control of the adhesive-setting process is required so that the excluded volume of the adhesive does not disrupt the bond integrity of the adhesive joint.

The mechanical properties of adhesive-bonded systems depend primarily on the nature, number, and distribution of defects such as voids, resin-rich areas, poor surface-bonding sites, and debonding sites. Until now, the relative role of these defects in the performance of the adhesives has not been determined because no techniques have been available to monitor the nature, number, and distribution of the defects and also allow mechanical testing of the samples that have been evaluated.

NMRI has the capability of measuring imperfec-

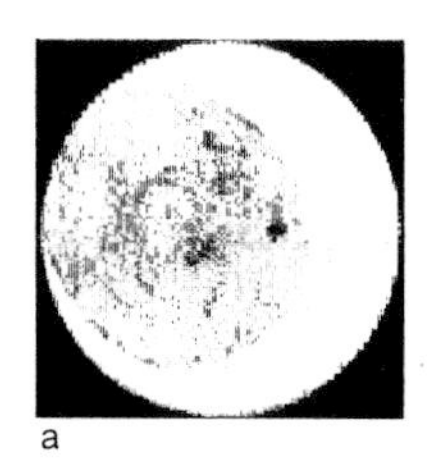
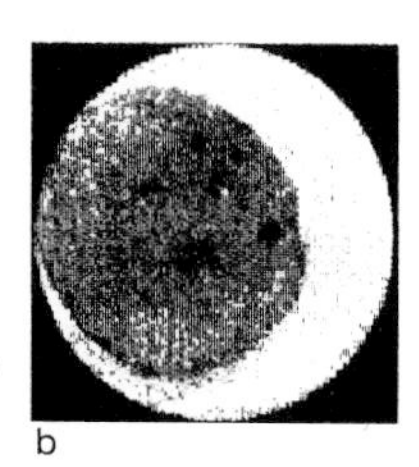
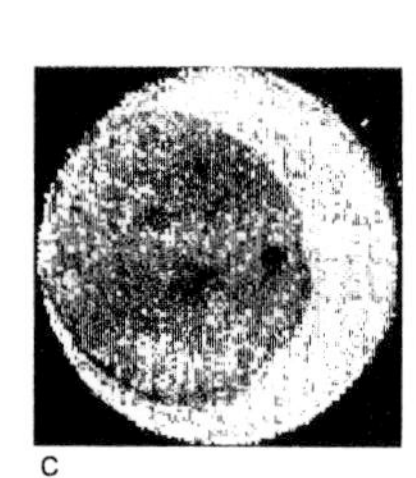

a b c

Figure 11.23. Proton NMR images of 1,4-dioxane swollen in polybutadiene rubber. The rubber was fully swollen. The inside diameter of the vial was approximately 18.5 mm, and the slice thickness was approximately 2 mm. (a) T_E = 30 ms, (b) T_E = 70 ms, and (c) the computer-generated T_2 image obtained from the images in a and b. (Reproduced with permission from reference 29. Copyright 1989 John Wiley & Sons, Inc.)

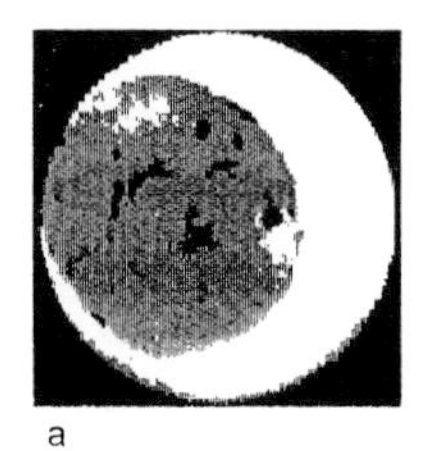
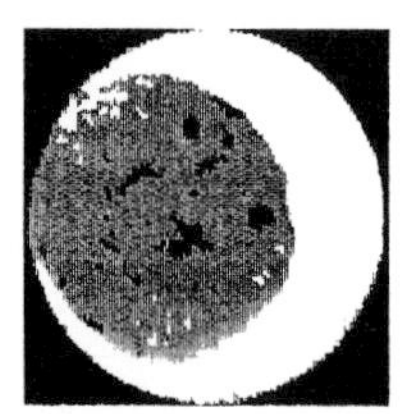
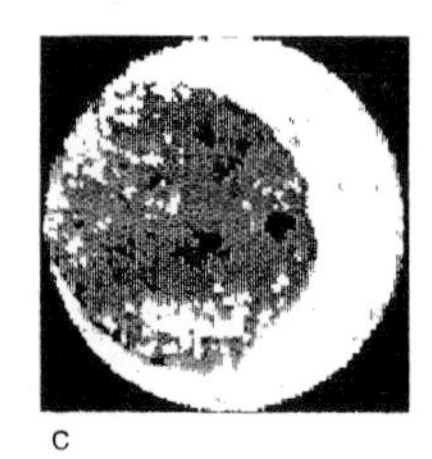

a b c

Figure 11.24. Contour plots of the images in Figure 11.23 were obtained after the images had undergone an imaging process. Signal intensities and T_2 relaxation times are scaled from 0 (lowest value) to 256 (highest value). The contour plots of Figure 11.23 have the following values: (a) dark, 0 – 160; medium, 161 – 197; and light, 198 – 256; (b) dark, 0 – 105; medium, 106 – 177; and light, 178 – 256; (c) dark, 0 – 95; medium, 96 – 165; and light, 166 – 256. (Reproduced with permission from reference 29. Copyright 1989 John Wiley & Sons, Inc.)

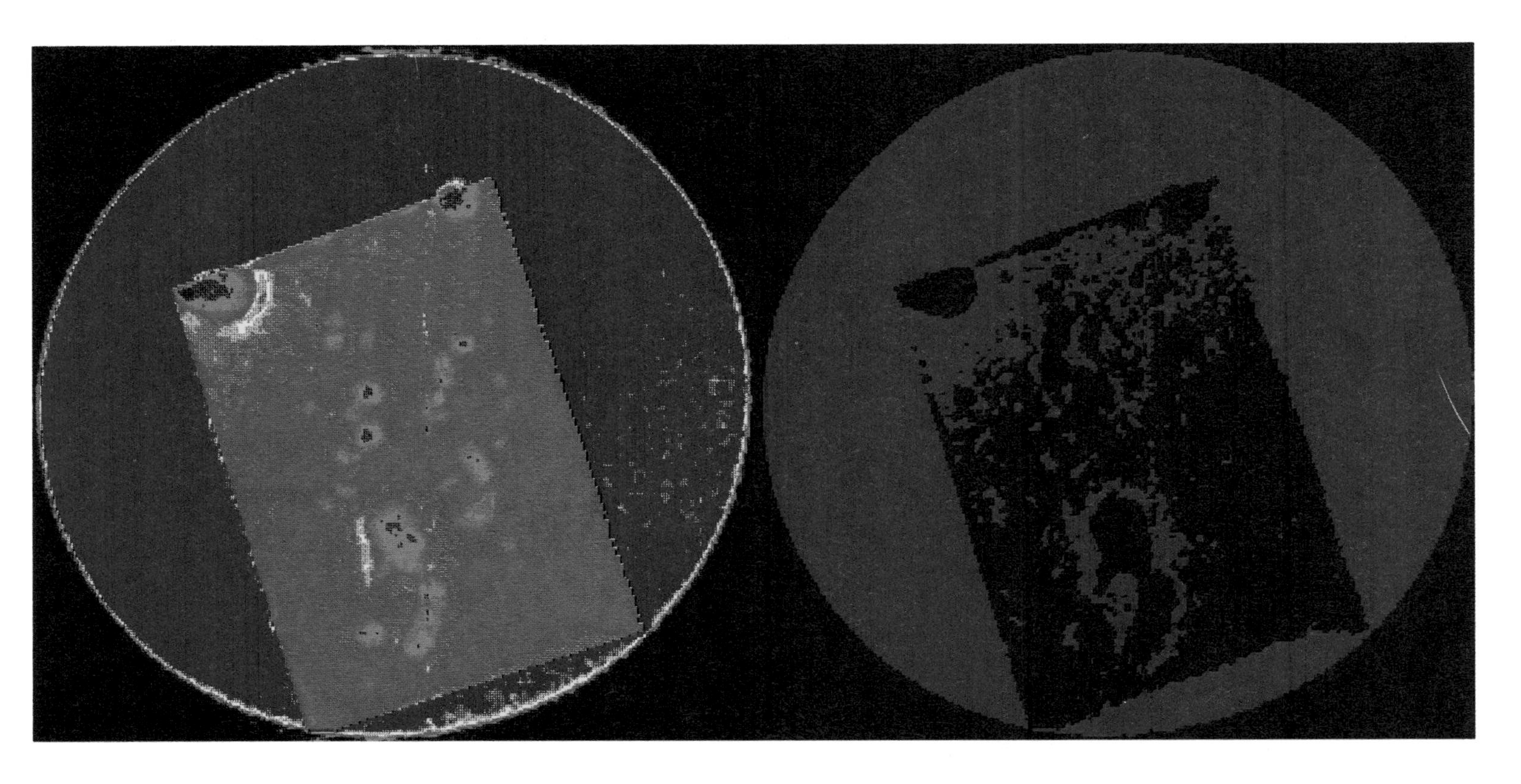

Figure 11.25. The proton image of vulcanized butyl rubber sample that was cured for 25 min and swollen in cyclohexane. The image on the right is a two-color plot of the image on the left cut at 100 (on a 0 – 256 scale). The sample was swollen in a 20-mm diameter tube, and the resolution is 78 μm/pixel. (Reproduced with permission from reference 30. Copyright 1991 Rubber Chemistry and Technology.)

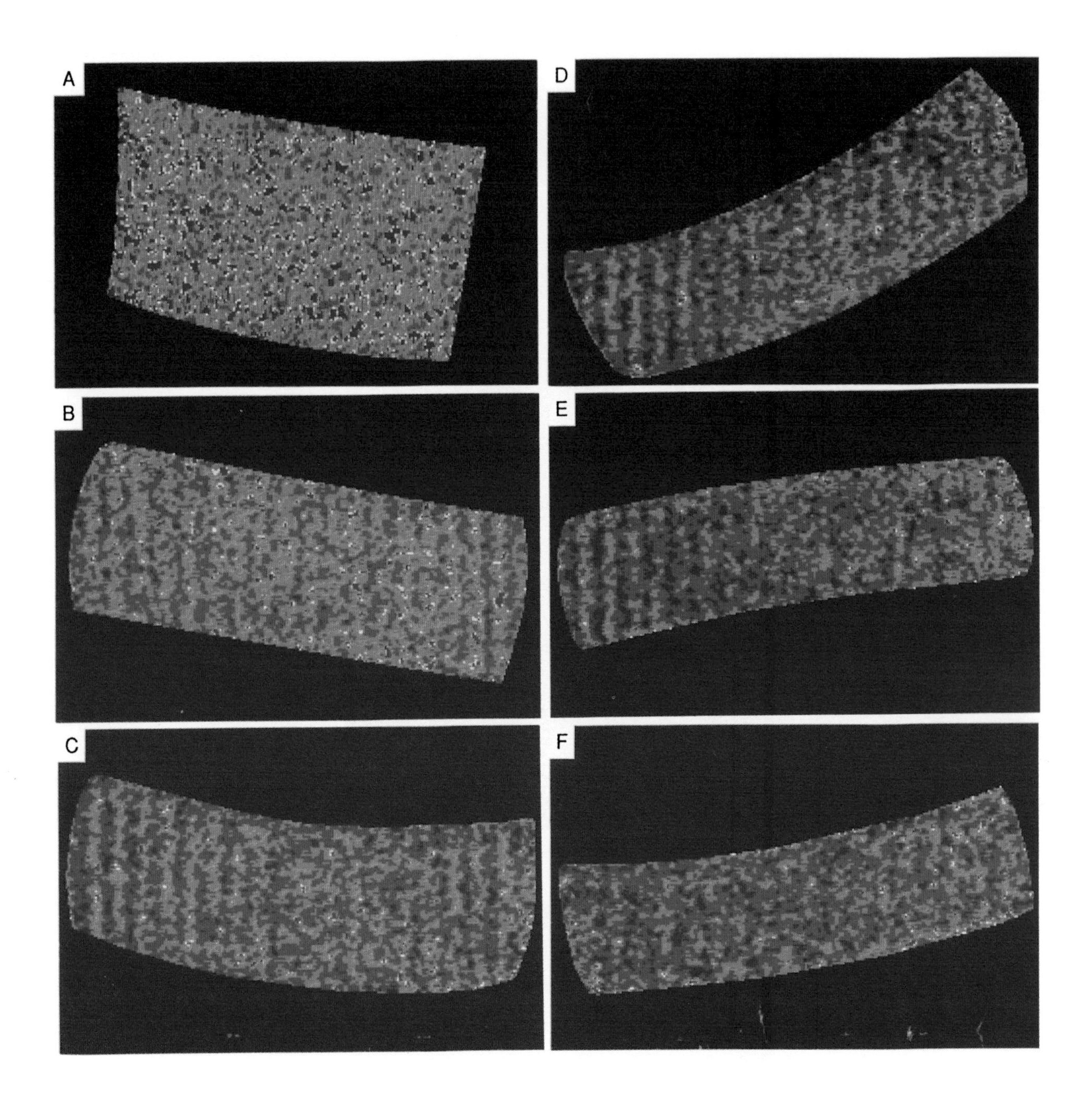

Figure 11.26. T_2-weighted images for a PB sample containing 2 phr of TMTD. The sample cure times are (A) 2 min, (B) 5 min, (C) 10 min, (D) 30 min, (E) 1h, and (F) 2h. (Reproduced from reference 32. Copyright 1991 American Chemical Society.)

tions in adhesive systems and is a noninvasive and nondestructive method. Adhesives that have a solids content of less than 100% contain volatile substances, such as water, that leave voids in the adhesive if there is no external pressure applied to force the adherends together. The voids in the adhesive decrease the strength of the structure. Defects, voids, or damaged areas of the adhesive materials will be shown clearly in the NMR image.

NMRI measurements were made by using the water-based PVAc adhesive system to join wood blocks (*33*). The changes that occur in the adhesive as the bond is being formed can be observed as shown in Figure 11.28. The image in Figure 11.28a

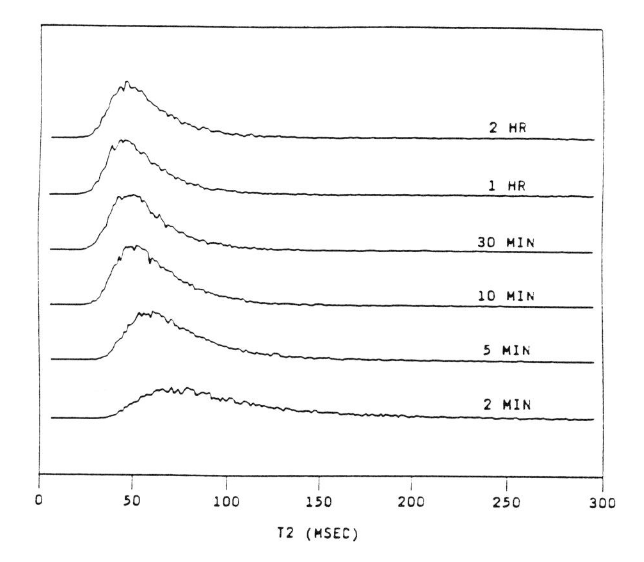

Figure 11.27. Histogram of T_2 *values of TMTD-cured polybutadiene for the six sample images presented in Figure 11.26. (Reproduced from reference 32. Copyright 1991 American Chemical Society.)*

was taken 10 min after application of the adhesive. This image shows a very thin glue line between the wood pieces and a 3-mm deep, 11-mm wide trough in one piece of wood. After drying for 4 h (Figure 11.28b), no wet adhesive remains in the thin glue line, although the adhesive in the trough is still wet and mobile. Even after 16 h (Figure 11.28c), the PVAc emulsion in the trough has not dried totally, although most of the water has penetrated into the wood. The image in Figure 11.28c shows an empty space or void in the adhesive.

NMRI is a sensitive method for finding imperfections because the technique allows detection of even minor differences in physical and chemical properties of adhesives (*34*–36). Figure 11.29a shows a homogeneous glue line of PVAc emulsion 1 (solids content 50%, water content 50%, and viscosity 35 Pa) between a wood block and a ceramic tile (*34*). The intensity of the signal from the adhesive is uniform throughout the glue line. Small empty spaces within the adhesive layer are due to the bottom pattern of the ceramic tile. Figure 11.29b represents a glue line made of emulsions 1 and 2 (solids content 60%, water content 40%, and viscosity 35 Pa). These emulsions have the same viscosity but a different solids content. Because this image was acquired with a relatively long repetition time and a short echo time, the observed contrast is mainly due to differences in proton spin densities, that is, water concentration. The higher signal intensity corresponds to emulsion 1, and the weaker intensity corresponds to emulsion 2. Figure 11.29c shows a comparison of images obtained from emulsions 1 and 3 (solids content 50%, water content 50%, and viscosity 2 Pa), which differ in their viscosities. In this image, the stronger signals are due to emulsion 3.

Voids that result from absorption or evaporation of water from water-based adhesives are abundant in thick joints. The location of the voids can be critical to the final joint strength. NMRI was used to study structural variations found in single lap joints made between ceramic tiles with water-based styrene–acrylate adhesives (*36*). The loss of water in these adhesives creates voids that represent approximately 20% of the total joint volume. Bonded samples were immersed in water at 30 °C, and the images were collected. Several samples were both imaged and tested mechanically. Each image represents a 1-mm slice of the sample in the joint area, and successive images were taken 3 mm apart. Typical images after the adhesive had set are shown in Figure 11.30. The images show that the drying of the adhesive caused

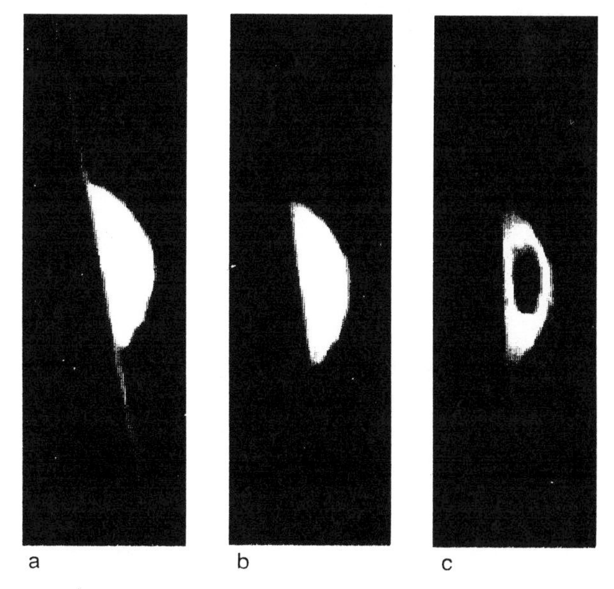

Figure 11.28. Plasticized PVAc emulsion dried between two wood blocks for (a) 10 min, (b) 4 h, and (c) 16h. The adhesive in the joint of <*100 μm thickness is visible in a. The black space in the middle of the trough in c is caused by water absorbed from the adhesive into the wood. (Reproduced with permission from reference 33. Copyright 1988 Journal of Adhesion Science and Technology.)*

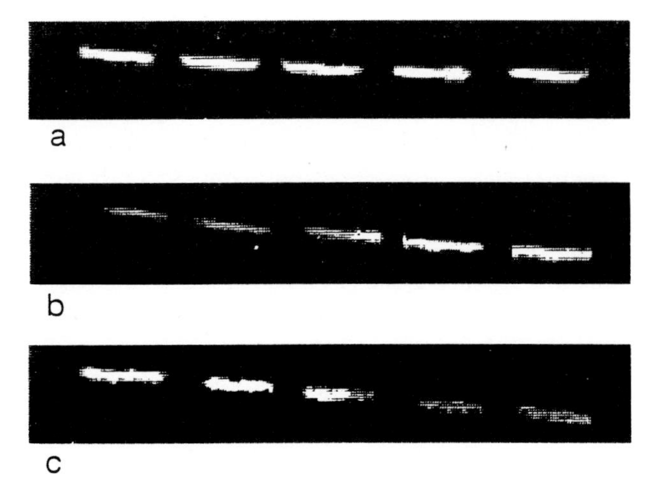

Figure 11.29. (a) A proton image of a homogeneous glue line of PVAc emulsion 1. (b) A proton image of a glue line made of PVAc emulsions 1 and 2. (c) A proton image obtained from PVAc emulsions 1 and 3. (Reproduced with permission from reference 34. Copyright 1989 Journal of Adhesion Science and Technology.)

Figure 11.30. Three images of a typical joint geometry after setting of the adhesive. The three slices were collected at different positions in the sample. (Reproduced with permission from reference 36. Copyright 1990 New Polymeric Materials.)

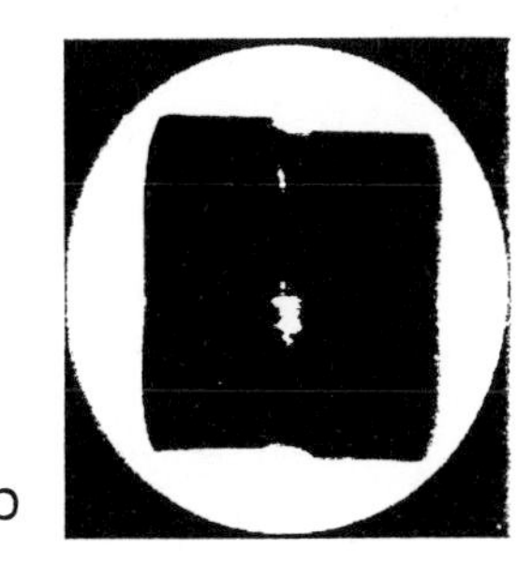

←*Figure 11.31. Two images of samples with lower compressive strengths. (a) The image shows a large void and a crack. (b) The image was obtained at a position where there was a visible hole in the sample. (Reproduced with permission from reference 36. Copyright 1990 New Polymeric Materials.)*

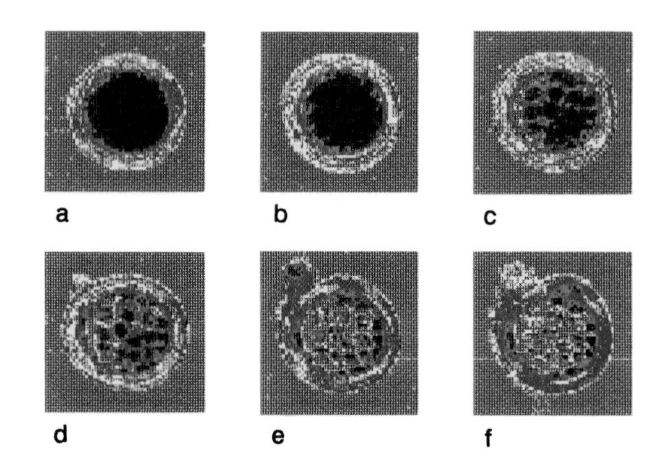

Figure 11.32. Proton images of water penetration through a 300-μm film of PVAc into wood. The NMR images were acquired after the following times: (a) 1, (b) 2, (c) 4, (d) 6, (e) 10, and (f) 24 h. Light color represents the highest signal intensity. (Reproduced with permission from reference 39. Copyright 1990 Journal of Adhesion.)

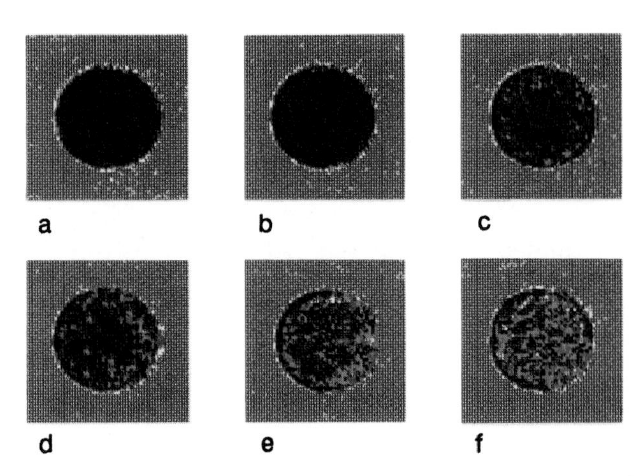

Figure 11.33. Proton images of water penetration through a 300-μm film of a non-cross-linked polyurethane dispersion coating into wood. The NMR images were acquired after the following times: (a) 1, (b) 2, (c) 4, (d) 6, (e) 10, and (f) 24 h. Light color represents the highest signal intensity. (Reproduced with permission from reference 39. Copyright 1990 Journal of Adhesion.)

the edges to shrink approximately 500 μm. This volume corresponds to about 5% of the total joint volume and indicates that most of the voids are located in the middle of the joint. However, because water cannot penetrate these adhesives, the voids in the central area of the joint could not be detected. The images in Figure 11.31 are of two samples with different compressive strengths. The image in Figure 11.31a shows a large void and a crack. Image b was obtained at a position where a hole approximately 300 μm in diameter was visible. These and other NMR images were correlated with the compressive-strength measurements to show that the void content and location are critical to the strength. The major mechanism for water intrusion into joints is penetration through voids and cracks. The cracks that reach the surface are the most damaging to the compressive strength.

NMRI images of glue lines of epoxy adhesive joints of aluminum – epoxy – aluminum were observed (*37*), and voids were also detected.

In the absence of moisture, interfacial bonds are generally stable and do not fail under normal stresses on the bonded components. However, water has the ability to compete with the chemical functional groups on the adhesive molecules for interface-bonding sites on the substrate surfaces. This reaction with water weakens the adhesive bonds. In the worst case, water can completely displace the adhesive from the substrate surface. From a practical point of view, the stability of the interfacial adhesive bond is directly related to the adhesive's resistance to the deteriorating effects of water. The successful use of long-lived adhesive systems requires matching the adhesive to the environment. More information is needed about the chemistry of the interfacial bonds formed, the importance of moisture adsorption, and the possible effects of field aging on equilibria involving competitive absorption and bond stability as a function of time.

NMRI can be used to evaluate water resistance of adhesive joints and has been specifically used to evaluate water penetration at 25 °C for polyurethane-dispersion coatings compared to a plasticized PVAc emulsion and a 5-min cure epoxy adhesive (*38, 39*). The water-penetration period was followed by the acquisition of cross-sectional images of the samples. Wooden rods with coatings of 300 (± 50 μm were immersed in water, and images were obtained at different times after immersion. The NMR images of PVAc and polyurethane-dispersion coated rods are shown in Figures 11.32 and 11.33. The images of the epoxy-coated rods did not show water penetration through the coating in 24 h. Water rapidly penetrated the PVAc coating (Figure 11.32a). The PVAc coating swelled, and separate layers of the coating partially delaminated, a result that indicates poor interfacial adhesion of the layers (Figure 11.32b – d). After immersion for 6 h, the coating shows a deformation (Figure 11.32d) that continues to grow. While the coating further swells and degrades, the wood reaches saturation after about 10 h of immersion. The non-cross-linked polyurethane dispersion has a much better water resistance than the PVAC emulsion (Figure 11.33). The mechanism of the water penetration is also different for the polyurethane. The polyurethane coating did not swell appreciably during the 24-h period but rather allowed a slow diffusion of water. The differences in water-penetration mechanisms arise from two factors. First, the PVAc contains poly(vinyl alcohol), which is water soluble and which allows the rapid intrusion of water. Second, the average particle size of the colloidal polyurethane dispersion is much smaller, a condition that leads to a tighter packing of the dried film. The addition of the polyaziridine cross-linker decreases the water absorption of the polyurethane coating by a factor of approximately 14.

Chemical-Shift Imaging

NMRI usually assumes that the spins (usually that of the protons of water) precess at the same frequency, but due to chemical-shift differences arising from different chemical types of protons in substances, some of the spins experience slightly different local fields and hence precess at different frequencies. The local-field change is written as δH_o, where H_o is the static field, and δ is the chemical shift in parts per million (ppm). In imaging, the presence of two different types of resonanting nuclei can lead to overlapping images and artifacts. Each individual image is centered at its resonant frequency in the absence of a magnetic-field gradient, and therefore the resulting image is smeared. Because the read or frequency-encoding gradient spreads resonance frequencies out according to positions along the gradient direction, the observed image actually consists of two or more partially overlapping sets of data (one corresponding to each type of nucleus). If the resonances are due to different species, two or more different images will be obtained. If all resonances arise from the same molecule, the resonances will have identical spatial distributions and images.

The usual imaging schemes apply a linear gradient, G_r, to frequency encode the data. Applying an inverse Fourier transform maps the spin density as a function of frequency linearly to spatial location. The linear relation between frequency, ω, and position, x, is

$$\omega = \gamma G_r x \tag{11.26}$$

where γ is the gyromagnetic ratio for hydrogen. In a gradient-free environment, the precessional frequency of the proton of molecule A decreases by

$$\Delta\omega_A = \gamma\delta_A H_o \tag{11.27}$$

This leads to a shift in the image position of the molecule A with respect to that of the protons of water by

$$\Delta x_A = \frac{\Delta\omega}{\gamma G_r} \tag{11.28}$$

Consequently, the image of molecule A could overlap that of water in the region of interest and cause an artifact in the image, and this artifact might be incorrectly interpreted as an actual spatial feature. By increasing G_r, the pixel shift due to chemical differences is reduced. However, much valuable information is contained in the image if the chemical shifts can be sorted out correctly. An image can be formed from only a selected portion of the total NMR spectrum. This process is called *chemical-shift imaging*.

A particular resonance peak can be selectively excited by rf irradiation to the exclusion of others in the chemical-shift spectrum. A long, low-power amplitude-shaped rf pulse can be used to excite a narrow range of resonant frequencies distributed about a particular frequency. Such a *soft* pulse is more frequency sensitive than a short, square *hard* pulse.

High-resolution NMR spectra displaying chemically shifted resonances provide information on the chemical species present in the system and their relative concentrations. The magnetic-resonance response can be simultaneously obtained from all regions of a heterogeneous sample by using a four-dimensional Fourier-transform technique in which the high-resolution spectrum obtained during the data acquisition defines one dimension, and the other three dimensions form a Cartesian coordinate system.

The application of various spatially resolved NMRI techniques for the observation of high-resolution spectra has had limited success. This is largely due to the mutually exclusive requirements of both the highly homogeneous magnetic field, which is necessary for the observation of chemical-shift information, and the inhomogeneous field, which is applied as a linear magnetic-field gradient and is necessary to obtain spatially resolved data. Chemical-shift-imaging techniques use pulsed magnetic-field gradients, which, in the standard configuration of superconducting magnets, generate sufficiently large eddy currents upon gradient removal to temporarily degrade the field homogeneity. This is one of the reasons that the implementation of high-resolution spatial spectroscopy is difficult.

Currently, there are several approaches to the problem of the chemical-shift effects in NMRI. First, an image that corresponds to a preselected chemical shift of a sample may be constructed either locally or globally. When different chemical shifts originate from different chemical species, an image taken at a specific chemical shift will provide information on the spatial distribution of the corresponding species while excluding the interference of other species in the image. A local method assumes knowledge of the chemical shift and usually produces an image of the chemical species under consideration. The in-phase and out-of-phase experiments can be used for this purpose (*40*). In addition, chemical-shift-selective suppression of an unwanted species or selective excitation of the species to be imaged (*41, 42*) as well as a method based on chemical-shift specific-slice selection (*43*) have been proposed as local methods. A global method essentially produces a chemical-shift spectrum for each localized region or volume element and thus creates a stack of chemical-shift images. A global deconvolution-calculation technique that uses a combination of a Wiener filter and an apodization function has been proposed (*44*). A method of convolution in which the image is deconvoluted by the NMR spectrum of the sample has also been proposed (*45*). This latter method shifts the image of each individual resonance such that it is centered about the carrier frequency. No totally adequate method of suppressing the chemical-shift artifacts has yet been developed, but all of the aforementioned methods improve the quality of the images when multiple chemical shifts are present.

On the other hand, chemical-shift imaging is highly desirable. Selective-excitation chemical-shift imaging is possible only if the spectrum of the sample is resolvable for the entire imaging volume. It has been suggested (*46*) that chemical-shift-sensitive NMR images can be obtained with spectral simplification by tailoring the excitation pulses. Chemical-shift images were reported for two rubbery polymers, polybutadiene and poly(dimethylsiloxane) (*47*) as well as for polyether polyol with an isocyanate curing agent (*38*).

NMR Imaging of Solids

NMRI of solids is a potentially valuable measurement technique that can be used to monitor inhomogeneities of chemical species and to identify defects in solids. The problem with imaging solids is mainly due to the much larger NMR spectral line widths of solids (10 – 100 kHz) as compared to liquids (<10 Hz). The imaging of solids is desirable, but in a strongly protonated solid where molecular motion is restricted, a typical value of the local dipolar field might be 5 G, and this dipolar field requires a gradient that is greater than 50 G/cm (0.5 T/m) to achieve a resolution of 1 mm. The fundamental problem is the need to narrow broad spectral lines in order to separately resolve them in a magnetic-field gradient. Several strategies could be used to image materials with T_2 values shorter than 10 ms, including shortening the gradient-switching times or raising the temperature of the sample in order to increase the T_2 values. Many solid polymers have components of their proton magnetization with a T_2 of approximately 0.5 ms. To obtain an image by using spin-echo techniques, a T_E of less than a millisecond is required. This in turn requires gradient-switching times no greater than 100 μs. Gradient-switching times of 100 μs have been obtained by using special gradient-coil windings. Images of polypropylene and poly(vinyl chloride) were obtained (*48*) with a resolution of 2 mm.

The most important factors that contribute to proton line widths are spin – spin dipolar coupling, chemical-shift anisotropy (CSA), and the bulk magnetic susceptibility (BMS). A few studies attempting to reduce the line-width problem have been performed. One approach is to reduce the effective local field by using multiple-pulse line-narrowing sequences (*49*). Typically, proton line widths of solids under multiple-pulse sequences are >0.25 ppm. High-power decoupling was used to obtain ^{13}C images of solid compounds with small CSAs, and this method has the potential for imaging nuclei in solids with dilute spins (*50*). A method that phase encodes the FID while incrementing the applied gradient during a fixed evolution time has been developed (*51*). Garroway et al. have increased the effective gradient strengths with multiple-quantum NMR spectroscopy (*52*).

Two-dimensional images of solids have been obtained by using a combination of 2-D Fourier imaging (2-D FI) and multiple-pulse methods. This experiment results in a major reduction in line widths due to dipolar broadening (*51*). However, the multiple-pulse techniques do not eliminate broadening due to CSA or BMS. Cory et al. (*53*) recognized that these broadening effects could be essentially eliminated with magic-angle spinning (MAS), and that in some cases, high-speed MAS could also remove dipolar broadening. Cory et al. have obtained ^1H NMR images of polybutadiene – polystyrene blends as shown in Figure 11.34 (*54*). Images of solids can be obtained by using large gradients (up to 100 G/cm) that are not switched but rather are varied in steps while the decoupling field is kept at moderate levels (*55*). A phantom image that consisted of three sheets of PMMA and PE was obtained.

Another approach is to use the solid-echo method. During pulse sequences leading to the observation of solid echoes, the effects of local dipolar fields are reversed for half the time. Thus, the use of echoes increases the time after which coherent magnetization can be seen, and the result is effective line narrowing (*56*).

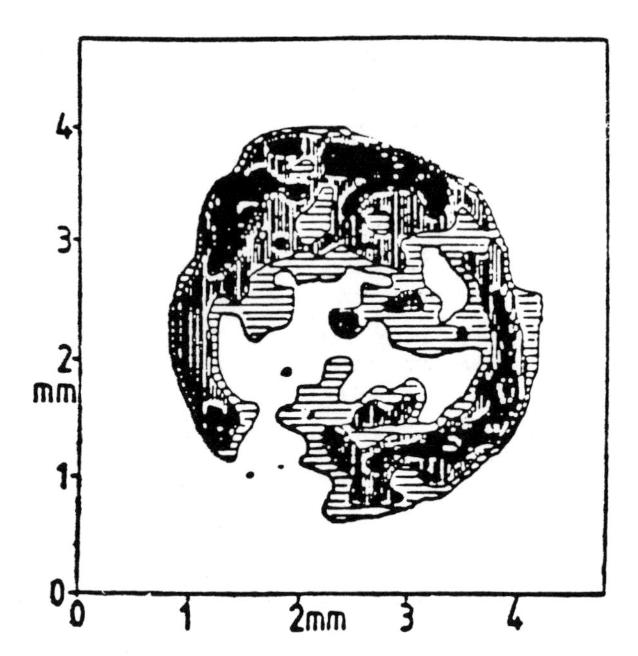

Figure 11.34. ^{1}H MAS image of a polybutadiene–polystyrene blend (Reproduced from reference 54. Copyright 1989 American Chemical Society.)

Chemical-shift imaging of solids is a useful extension of NMRI. Chemical-shift imaging via 2-D FI techniques can be used for liquids, but this method is not easily adapted to solids. MAS averages the chemical-shift anisotropies, and these averaged chemical shifts can be used to generate chemical-shift-selective images (*53*).

Back projection (BP) has been suggested (*57*) as a method for chemical-shift-resolved extensions because BP does not utilize switched gradients. Chemical-shift and spatial information are encoded separately by incrementing the magnetic-field-gradient strength over a series of experiments, and then the data are Fourier transformed. Studies in this difficult area are continuing, and positive results can be expected.

Chemical-Shift ^{13}C NMR Imaging in the Solid State

The ultimate NMRI experiment for heterogeneous polymer systems is chemical-shift ^{13}C NMR imaging in the solid state. It has been done! The ^{13}C chemical-shift image of organic matrix – carbon fiber composites was reported (*58*). The image obtained (*see* Figure 11.35) required 26.7 h and had a resolution of approximately 0.5 mm. Thus, the potential is there, but much work remains in order to produce such images routinely.

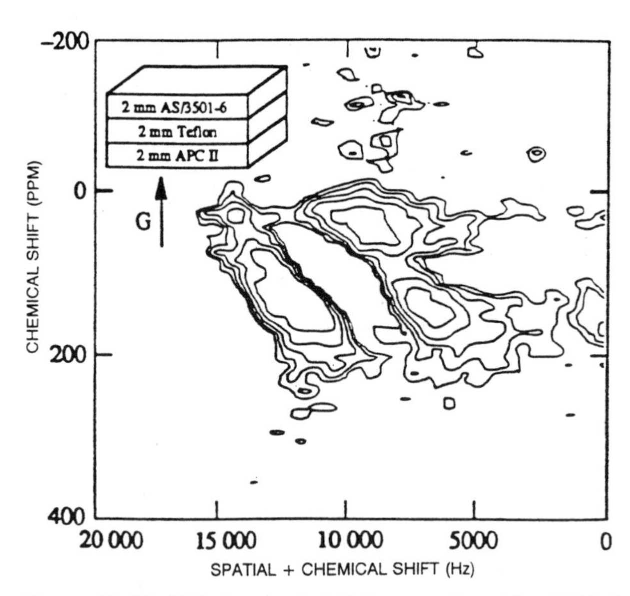

Figure 11.35. ^{13}C chemical-shift image of an AS – 3501-6 epoxy laminate – Teflon – APC-II PEEK laminate "sandwich". (Reproduced with permission from reference 58. Copyright 1989 Academic Press.)

References

1. Brunner, P.; Ernst, R. R. *J. Magn. Reson.* **1979,** *33,* 83.
2. Kumar, A.; Welti, D.; Ernst, R. R. *J. Magn. Reson.* **1975,** *18,* 69.
3. Edelstein, W.; Hutchison, J. M. S.; Johnson, G.; Redpath, T. *Phys. Med. Biol.* **1980,** *25,* 751.
4. Garroway, A. N.; Grannell, P. K.; Mansfield, P. *J. Phys. C: Solid State Phys.* **1974,** *7,* L 457.
5. Lauterbur, P. C. *Nature* **1973,** *242,* 190.
6. Rothwell, W. P. *Appl. Opt.* **1985,** *24,* 3958.
7. Lauterbur, P. C.; Lai, C. -M. *IEEE Trans.* **1980,** *27,* 1227.

8. Liu, J.; Nieminen, A. O. K.; Koenig, J. L. *J. Magn. Reson.* **1989,** *85,* 95.

9. Bailes, D. R.; Bryant, D. J. *Contemp. Phys.* **1984,** *25,* 441.

10. Aguayo, J. B.; Blackhand, S. J.; Schoeninger, J.; Mattingly, M. A.; Hintermann, M. *Nature* **1986,** *322,* 190.

11. Ackerman, J. J. H.; Grove, T. H.; Won, G. G.; Gadian, D. G.; Radda, G. K. *Nature* **1980,** *283,* 167.

12. Holland, G. N.; Bottomley, P. A.; Hinshaw, W. S. *J. Magn. Reson.* **1977,** *28,* 133.

13. Kormos, D. W.; Yeung, H. N.; Gauss, R. C. *J. Magn. Reson.* **1987,** *71,* 159.

14. Hall, L. D.; Webb, A. G.; Williams, S. C. R. *J. Magn. Reson.* **1989,** *81,* 565.

15. Tanaka, H.; Hayashi, T.; Nishi, T. *J. Appl. Phys.* **1986,** *59,* 3627.

16. Perry, B. C.; Koenig, J. L. *J. Polym. Sci., Part A: Polym. Chem.* **1989,** *27,* 3429.

17. Hoh, Ka-Pi; Perry, B.; Rotter, G.; Ishida, H.; Koenig, J. L. *J. Adhes.* **1989,** *27,* 245.

18. Rothwell, W. P.; Holecek, D. R.; Kershaw, J. A. *J. Polym. Sci., Polym. Lett. Ed.* **1984,** *22,* 241.

19. Rothwell, W. P.; Gentempo, P. P. *Bruker Reports* **1985,** *1,* 46.

20. Blackband, S.; Mansfield, P. *Solid State Phys.* **1986,** *19,* L49.

21. Weisenberger, L. A.; Koenig, J. L. *Appl. Spectrosc.* **1989,** *43,* 98.

22. Weisenberger, L. A.; Koenig, J. L. *J. Polym. Sci., Part C* **1989,** *27,* 55.

23. Stejskal, E. O.; Tanner, J. E. *J. Chem. Phys.* **1965,** *42,* 288.

24. Weisenberger, L. A.; Koenig, J. L. *Macromolecules* **1990,** *23,* 2445.

25. Merboldt, K.; Hanicke, W.; Frahm, J. *J. Magn. Reson.* **1985,** *64,* 479.

26. Smith, E. G.; Rockliffe, J. W.; Riley, P. I. *J. Colloid Interface Sci.* **1989,** *131,* 29.

27. Merboldt, K.; Hanicke, W.; Gyngell, M. L.; Frahm, J.; Bruhn, H. *J. Magn. Reson.* **1989,** 115.

28. Weisenberger, L. A.; Koenig, J. L. *Macromolecules* **1990,** *23,* 2454.

29. Clough, R. S.; Koenig, J. L. *J. Polym. Sci., Lett. Ed.* **1989,** *27,* 451.

30. Krejsa, M.; Koenig, J. L. *Rubber Chem. Technol.* **1991,** *64,* 635.

31. Chang, C.; Komoroski, R. A. **1989,** *Macromolecules, 22,* 600.

32. Smith, S.; Koenig, J. L. *Macromolecules* **1991,** *24,* 3496.

33. Nieminen, A. O. K.; Koenig, J. L. *J. Adhes. Sci. Technol.* **1988,** *2,* 407.

34. Nieminen, A. O. K.; Liu, J.; Koenig, J. L. *J. Adhes. Sci. Technol.* **1989,** *3,* 445.

35. Nieminen, A. O. K.; Koenig, J. L. *Adhes. Age* **1989,** *32,* 17.

36. Nieminen, A. O. K.; Evans, M.; Koenig, J. L. *New Polym.* **1990,** *2,* 197.

37. Nieminen, A. O. K.; Koenig, J. L. *J. Adhes.* **1989,** *30,* 47.

38. Nieminen, A. O. K.; Koenig, J. L. *Appl. Spectrosc.* **1989,** *43,* 153.

39. Nieminen, A. O. K.; Koenig, J. L. *J. Adhes.* **1990,** *32,* 105.

40. Dixon, W. T. *Radiology* **1984,** *153,* 189.

41. Hall, L. D.; Sukumar, S.; Talagala, S. L. *J. Magn. Reson.* **1984,** *56,* 275.

42. Hall, L. D.; Rajanayagam, V. *J. Magn. Reson.* **1987,** *74,* 139.

43. Volk, A.; Tiffon, B.; Mispelter, J.; Lhoste, J. M. *J. Magn. Reson.* **1987,** *71,* 168.

44. Liu, J.; Nieminen, A. O. K.; Koenig, J. L. *Spectrosc.* **1989,** *43,* 1260.

45. Cory, D. G.; Reichwein, A. M.; Veeman, W. S. *J. Magn. Reson.* **1988,** *80,* 259.

46. Bornert, P.; Dreher, W.; Gossler, A.; Klee, G.; Peter, R.; Schneider, W. *J. Magn. Reson.* **1989,** *81,* 167.

47. Garrido, L.; Mark, J. E. *Polym. Reprints* **1989,** *30,* 217.

48. Carpenter, T. A.; Hall, L. D.; Jezzard, P. *J. Magn. Reson.* **1989,** *84,* 383.

49. Wind, R. A.; Yannoni, C. S. *J. Magn. Reson.* **1979,** *36,* 269.

50. Szeverenyi, N. M.; Maciel, G. *J. Magn. Reson.* **1984,** *60,* 460.

51. Chingas, G. C.; Miller, J. B.; Garroway, A. N. *J. Magn. Reson.* **1986,** *66,* 530.

52. Garroway, A. N.; Baum, J.; Munowitz, M. G.; Pines, A. *J. Magn. Reson.* **1984,** *60,* 337.

53. Cory, D. G.; van Os, J. W. M.; Veeman, W. S. *J. Magn. Reson.* **1988,** *76,* 543.

54. Cory, D. G.; de Boer, J. C.; Veeman, W. S. *Macromolecules* **1989,** *22,* 1618.

55. Corti, M.; Borsa, F.; Rigamonti, A. *J. Magn. Reson.* **1988,** *79,* 21.

56. McDonald, P. J.; Attard, J. J.; Taylor, D. G. *J. Magn. Reson.* **1987,** *72,* 224-229.

57. Cory, D. G.; Miller, J. B.; Garroway, A. N. *J. Magn. Reson.* **1989,** *85,* 219.

58. Fry, C. G.; Lind, A. C.; Davis, M. F.; Duff, D. W.; Maciel, G. E. *J. Magn. Reson.* **1989,** *83,* 656.

Suggested Reading

Andrew, E. R. *Acc. Chem. Res.* **1983,** *16,* 114.

Foster, M. A.; Huthchinson, J. M. S. *J. Biomed. Eng.* **1985,** *7,* 171.

King, K. F.; Moron, P. R. *Med. Phys.* **1984,** *11,* 1.

Manified, P.; Morris, P. G. *NMR Imaging in Biomedicine;* Academic Press: New York, 1982.

Pykett, L. *Semin. Nucl. Med.* **1983,** *13,* 319.

Smith, S. L. *Anal. Chem.* **1985,** *57,* 595A.

Twieg, D. B. *Med. Phys.* **1983,** *10,* 610.

Index

D

E

F

K

L

M

N

O

P

Q

R